Materials for Devices

From everyday applications to the rise of automation, devices have become ubiquitous. Specific materials are employed in specific devices because of their particular properties, including electrical, thermal, magnetic, mechanical, ferroelectric, and piezoelectric. *Materials for Devices* discusses materials selection for optimal application and highlights current materials developments in gas sensors, optical devices, mechanoelectrical devices, and medical and biological devices.

- Explains how to select the right material for the right device
- Includes 2D materials, thin films, smart piezoelectric films, and more
- Presents details on organic solar cells
- Describes thin films in sensors, actuators, and LEDs
- Covers thin films and elastic polymers in biomedical devices
- Discusses growth and characterization of intrinsic magnetic topological insulators

This work is aimed at researchers, technologists, and advanced students in materials and electrical engineering and related fields who are interested in developing sensors or devices.

Advances in Materials Science and Engineering

Series Editor
Sam Zhang

Thin Films and Coatings: Toughening and Toughness Characterization
Sam Zhang

Semiconductor Nanocrystals and Metal Nanoparticles: Physical Properties and Device Applications
Tupei Chen and Yang Liu

Advances in Magnetic Materials: Processing, Properties, and Performance
Sam Zhang and Dongliang Zhao

Micro- and Macromechanical Properties of Materials
Yichun Zhou, Li Yang, and Yongli Huang

Nanobiomaterials: Development and Applications
Dong Kee Yi and Georgia C. Papaefthymiou

Biological and Biomedical Coatings Handbook: Applications
Sam Zhang

Hierarchical Micro/Nanostructured Materials: Fabrication, Properties, and Applications
Weiping Cai, Guotao Duan, and Yue Li

Biological and Biomedical Coatings Handbook, Two-Volume Set
Sam Zhang

Nanostructured and Advanced Materials for Fuel Cells
San Ping Jiang and Pei Kang Shen

Hydroxyapatite Coatings for Biomedical Applications
Sam Zhang

Carbon Nanomaterials: Modeling, Design, and Applications
Kun Zhou

Materials for Energy
Sam Zhang

Protective Thin Coatings and Functional Thin Films Technology, Two-Volume Set
Sam Zhang, Jyh-Ming Ting, Wan-Yu Wu

Fundamentals of Crystallography, Powder X-ray Diffraction, and Transmission Electron Microscopy for Materials Scientists
Dong ZhiLi

Materials for Devices
Sam Zhang

For more information about this series, please visit: https://www.routledge.com/Advances-in-Materials-Science-and-Engineering/book-series/CRCADVMATSCIENG

Materials for Devices

Edited by
Sam Zhang

CRC Press is an imprint of the
Taylor & Francis Group, an **informa** business

First edition published 2023
by CRC Press
6000 Broken Sound Parkway NW, Suite 300, Boca Raton, FL 33487-2742

and by CRC Press
4 Park Square, Milton Park, Abingdon, Oxon, OX14 4RN

CRC Press is an imprint of Taylor & Francis Group, LLC

ISBN: 978-0-367-67930-9 (hbk)
ISBN: 978-0-367-69323-7 (pbk)
ISBN: 978-1-003-14135-8 (ebk)

DOI: 10.1201/9781003141358

Typeset in Times
by KnowledgeWorks Global Ltd.

Contents

Preface

As the society progress to its current stage, automation, artificial intelligence…come to take part and immerge deep into every aspect. To realize those functionalities, devices are everywhere. Realization of the devices needs materials, the right material for the right function, the right material for the right device. This book illustrates just that. All types of devices need a material as a carrier. Be it a bulk material or a coating or a thin film. A certain function needs a certain material in a correct form. This certain device could be some kind of gas sensors for antiterror applications or medical applications, or thermal sensors for high temperature or low temperature applications, or mechanical devices such as electromagnetic or piezoelectric sensors, or optoelectrical sensors. Materials for devices are materials employed in devices because of their particular properties, such as electrical, thermal, magnetic, mechanical, ferroelectric, or piezoelectric properties. Examples of materials for devices are polymers, oxides, semiconductors, and liquid crystals. There are some ten chapters covering the following aspects.

Two-Dimensional Materials in Photoconductive Detectors, UV Stability of Organic Solar Cells, Absorption Enhancement in Photonic-structured Organic Solar Cells, Recent Advances in Gas Sensors for Device Applications, The Sputtered Thin Films as the Sensing Materials for the MEMS Gas Sensors, Electron Retarding Materials Making More Efficient Light-Emitting Devices, Thin-Film Thermoelectrics: Materials, Devices, and Applications, Smart Piezoelectric Films for Sensors and Actuators Applications, Nanomaterials for Lithium(-ion) Batteries, Applications of Thin Films in Metallic Implants, and finally Flexible Sensors for Biomedical Applications Based on Elastic Polymers.

Contact or noncontact sensing of a problem or realization of a function relies on the creation of a right material for the device. This book highlights current materials development in gas sensors, optical devices, mechanoelectrical devices, medical and biological devices. Electronic "nose" could "smell" out minute explosives in antiteorror applications, flexible sensors for biomedical implants, etc. Researchers, technologists, graduates, and senior students who are interested in current day sensors or devices should be interested in this book.

I'd like to take this opportunity to acknowledge Fundamental Research Funds for the Central Universities SWU118105, to thank all the chapter authors for their support and hard work. I'd like to thank Ms Allison Shatkin, Gabrielle Vernachio, and the editorial team of the publishing company for their continued support in getting the book to its current form.

Thank you all!

Dr. Prof. Sam Zhang
Southwest University
Chongqing, China
January 15, 2022
samzhang@swu.edu.cn

About the Author

Professor Sam Zhang Shanyong (張善勇), FRSC, FTFS, FIoMMM, academically known as Sam Zhang, was born and brought up in the famous "City of Mountains" Chongqing, China. He received his Bachelor of Engineering in Materials in 1982 from Northeastern University (Shenyang, China), Master of Engineering in Materials in 1984 from Iron & Steel Research Institute (Beijing, China), and Ph.D. degree in Ceramics in 1991 from The University of Wisconsin-Madison, USA. He was a tenured full professor (since 2006) at the School of Mechanical and Aerospace Engineering, Nanyang Technological University Singapore. Since January 2018, he joined School of Materials and Energy, Southwest University, China. He serves as the founding and current director of the *Centre for Advanced Thin Films and Devices* (http://fmae.swu.edu.cn/s/fmaenew/yjzx/).

Prof. Zhang was elected as Fellow of Royal Society of Chemistry (FRSC) and Fellow of Thin Films Society (FTFS) in 2018, Fellow of Institute of Materials, Minerals and Mining (FIoMMM) in 2007.

Prof. Zhang was a Principal Editor for Journal of Materials Research (USA) for thin films and coating field (since 2003). Prof. Zhang has been serving the world's first "Thin Films Society" (www.thinfilms.sg) as its founding and current president since 2009. Prof. Zhang has authored/edited 13 books, of them 11 are published with CRC Press/Taylor & Francis. The most recent titles are *Materials for Energy* published in October 2020 (https://www.routledge.com/Materials-for-Energy/Zhang/p/book/9780367350215) and in August 2021: *Protective Thin Coatings Technology* (https://www.routledge.com/Protective-Thin-Coatings-Technology/Zhang-Ting-Wu/p/book/9780367542504) and *Functional Thin Films Technology* (https://www.routledge.com/Functional-Thin-Films-Technology/Zhang-Ting-Wu/p/book/9780367541774). Currently, Prof. Zhang is working on 2 new books: "Materials in Advanced Manufacturing," edited by Yingquan Yu and Sam Zhang, to be published by CRC Press, and a 4-volume "Materials for Energy Handbooks" in Chinese to be published by Chongqing University Press. Of these books, *Materials Characterization Techniques* has been adopted as textbook by more than 30 American and European universities alone as of October 2015. That book had also been translated into Chinese and published by China Science Publishing Co in October 2010 and distributed nationwide in China (also available online at Amazon.cn). Prof. Zhang is also the Series Editor for Advances in Materials Science and Engineering Book Series published by CRC Press/Taylor & Francis. A full list of his books are at amazon: https://www.amazon.com/-/e/B001JPBX42

Prof. Zhang's current researches center in the areas of Films and Coatings for solar cells, hard yet tough nanocomposite coatings/high entroy alloy coatings for tribological applications by physical vapor deposition; Over the years, he has published 13 books and authored/co-authored over 360 peer-reviewed international journal papers with 31.7 citations per article and h-index: 56 and total citation of over 11,000 times as at 11 Jan 2022. (https://publons.com/researcher/2817766/sam-zhang/).

Contributors

Jing Chen
Chinese Academy of Sciences
Shenzen, People's Republic of China

Yu Duan
Southwest University
Chongqing, People's Republic of China

Shuanglong Feng
Chinese Academy of Sciences
Chongqing, People's Republic of China

Yong-Qing Fu
Northumbria University
Newcastle upon Tyne,
United Kingdom

Katayoon Kalantari
Northeastern University
Boston, Massachusetts

Weixia Lan
Shanghai University
Shanghai, People's Republic of China

Hui Li
Chinese Academy of Sciences
Shenzhen, People's Republic of China

Lin Li
Chinese Academy of Sciences
Shenzhen, People's Republic of China

Yi Li
Southwest University
Chongqing, People's Republic of China

Hua-Feng Pang
Xi'an University of Science and
Technology
Xi'an, People's Republic of China
and
Northumbria University
Newcastle upon Tyne, United Kingdom

Bahram Saleh
Northeastern University
Boston, Massachusetts

Ashutosh Sharma
Department of Materials Science and
Engineering
Ajou University
Gyeonggi-do, Republic of Korea

Bharat Sharma
Institute of Microstructure Technology
Karlsruhe Institute of Technology
Eggenstein-Leopoldshafen,
Germany

Ady Suwardi
Institute of Materials Research and
Engineering (IMRE)
Singapore, Republic of Singapore

Yankun Tang
Xi'an Jiaotong University
Xi'an, People's Republic of China

Xin Tian
Xi'an Jiaotong University
Xi'an, People's Republic of China

Hairong Wang
Xi'an Jiaotong University
Xi'an, People's Republic of China

Lei Wang
Chinese Academy of Sciences
Shenzhen, People's Republic of China

Xizu Wang
Institute of Materials Research and Engineering (IMRE)
Singapore, Republic of Singapore

Thomas J. Webster
Hebei University of Technology
Tianjin, China
and
UFPI
Teresina, Brazil
and
School of Engineering
Saveetha University
Chennai, India

Dong-Sing Wuu
National Chi Nan University
Nantou, Taiwan, Taipei

Jianwei Xu
Institute of Materials Research and Engineering (IMRE)
Singapore
and
National University of Singapore
Singapore
and
Institution of Sustainability for Chemical, Energy and Environment (ISCE2)
Jurong Island, Singapore

Maowen Xu
Southwest University
Chongqing, People's Republic of China

Tao Xu
Shanghai University
Shanghai, People's Republic of China

Sam Zhang
Southwest University
Chongqing, People's Republic of China

Furong Zhu
Hong Kong Baptist University
Hong Kong, People's Republic of China

Qiang Zhu
Institute of Materials Research and Engineering (IMRE)
Singapore, Republic of Singapore

1 Two-Dimensional Materials in Photoconductive Detectors

Yu Duan[a,b]*, Shuangling Feng*[b]*, and Sam Zhang*[a]
[a]Southwest University, Chongqing,
People's Republic of China
[b]Chinese Academy of Sciences, Chongqing,
People's Republic of China

CONTENTS

DOI: 10.1201/9781003141358-1

1.1 INTRODUCTION TO PHOTODETECTORS AND TWO-DIMENSIONAL MATERIALS

In the development of human science, light has always been a critical source of information. People's understanding of light has gradually expanded from visible light to ultraviolet-infrared spectrum through the advancement of technology. According to the wavelength of the incident light, people divide light into three parts: ultraviolet (10–400 nm), visible light (400–780 nm), and infrared (780 nm–1 mm). The ultraviolet can be further divided into ultraviolet radiation A (UVA, 320–400 nm), ultraviolet radiation B (UVB, 275–320 nm), ultraviolet radiation C (UVC, 200–275 nm), and ultraviolet radiation D (UVD, 100–200 nm); the infrared can be further divided into near-infrared, mid-infrared, and far-infrared (the range varies according to the requirements). Although humans can only see visible light with the naked eye, ultraviolet and infrared light have always been present in our lives.

Photodetectors play an essential role in the photoelectric system. Photodetectors convert light signals into electrical signals, which are processed by electronic circuits to realize light detection function.[1–3] Nowadays, photodetectors have been widely used in photoelectric display, imaging, environmental monitoring, optical communication, military, safety inspection, biomedicine, and many other fields.[4–7] According to the different detection bands, their functions are also different. For instance, deep ultraviolet light detectors are used for ultraviolet lithography and living cell detection. Visible range detectors are used in digital cameras and visual imaging, and infrared detectors are used in night vision, optical communication, and atmospheric quality inspection. Since the discovery of the photoelectric effect in 1887,[8] scientists have devoted themselves to exploring the interaction between light and semiconductor materials, which has laid the foundation for many theories of optoelectronic systems and become the key to modern industrial and scientific applications.

Photodetectors based on photoelectric effect have two categories: the external photoelectric effect devices and the internal photoelectric effect devices according to different working mechanisms. External photoelectric effect refers to the phenomenon that the electrons in the material escape from the surface of the material under light, such as the phototube and photomultiplier tube. Internal photoelectric effect refers to the phenomenon that the carriers (free electrons or holes) generated by light excitation still move in the material, which changes the conductivity of the material or produces photogenerated volts, divided into photovoltaic and photoconductive effect. Photovoltaic devices mainly use the junction area (PN junction, Schottky junction, etc.) to work.[5] When light is irradiated to the junction area, photogenerated carriers are generated. Under the action of the built-in electric field in the junction area, electrons and holes are separated. The photocurrent is formed after the circuit is turned on, and the open circuit voltage is established under the open circuit condition. Photoconductive devices produce photoelectric response mainly because the incident light changes the conductivity of the material. The working mechanism will be introduced in detail later. Due to the existence of potential barrier and depletion region, the dark current and noise of photovoltaic devices are significantly reduced, but the gain and photocurrent of photovoltaic devices are also limited. However, due to the absence of transmission barrier in the channel, photoconductive devices are more likely to form a high photoconductive gain under external bias, which makes the photocurrent and response of the detector relatively higher.

Although traditional 3D thin-film photodetectors, such as silicon photodetectors and InGaAs photodetectors,[9,10] have dominated the commercial photodetector market for many years with their high performance, skilled integration technology and large-scale production, however, there are still many problems, such as difficulty in preparing flexible devices, need to work in ultra-low temperature environment, toxic elements and so on. With the increasing demand of the industry and the changing requirements, people's demand for new materials is becoming more urgent.

Since the successful preparation of graphene by Andre Geim and Konstantin Novoselov of the University of Manchester in 2004, graphene has been highly sought after by researchers due to its remarkable quantum Hall effect, extremely large specific surface area, high Young's modulus, high carrier mobility and high thermal conductivity, and other excellent physical and chemical properties. However, the zero-band gap of graphene seriously affects its application in the field of optoelectronics. Fortunately, scientists have developed more two-dimensional (2D) semiconductor materials, e.g., black phosphorus (BP), 2D transition metal dichalcogenides (TMDs), which have been extensively studied in photonics and optoelectronics.

Compared with the traditional thin-film materials and bulk materials used in optoelectronics, 2D materials have many unique advantages. First of all, the light–matter interactions of 2D materials are strong. Despite the atomic thickness, the quantum constraint in the vertical direction leads to spikes in the conduction band and valence band near the edge. When the energy of the incident photon approaches the bandgap, valence electrons can be successfully excited into the conduction band, and free electron–hole pairs can be generated, thus realizing photoelectric conversion.[11] Secondly, there is no dangling bond on the surface of 2D materials, and layers are bound together by van der Waals force, so there is no need to worry about the impact of lattice mismatch.[12] The solid covalent bond provides in-plane stability of 2D crystals, while weak van der Waals force can ensure that the wafers stack together.[13,14] Unlike the traditional heterostructures, the van der Waals heterojunction has a smooth interface without dangling bonds, which reduces the capture or scattering of charge in the heterojunctions and can realize ultra-fast carrier transmission.[15,16] Thirdly, the 2D material has a wide bandgap width range, and its optical absorption band covers ultraviolet to infrared and even terahertz. In addition, the bandgap width can be adjusted in a certain range utilizing layer number dependence, chemical doping,[17] gate voltage control,[18] and so on, which is conducive to achieving coordinated optoelectronic devices and meeting different application conditions. Last, but not the least, the 2D material has good light transmittance and flexibility. The light transmittance allows it to build various structures by stacking without considering the light-shielding from the upper layers. At the same time, flexibility enables it to be better used in flexible and wearable devices.[19,20]

Hitherto, 2D materials occupy a prominent position in the field of photodetectors because of their exceptional performance. This chapter summarizes the typical applications of 2D materials in photoconductive detector (PCD) from the perspective of the device structure. First, the parameters and basic structure of the PCD are introduced. Then, according to the functions of different structural parts, the application of 2D materials in different parts of the PCD is reviewed. Finally, the development of 2D materials in the field of PCD is summarized.

1.2 MECHANISM, PARAMETERS, AND STRUCTURE OF THE PCD

1.2.1 The Working Mechanism of PCD

Photoconductive effect is one of the crucial characteristics of semiconductor materials, which mainly refers to the change in the conductivity of the material under the action of incident photons (usually manifested as the increase of the conductivity of the material). However, in practical applications, the change of photoconductance is a relatively complicated process, involving several mechanisms running continuously or simultaneously, including absorption of incident photons to generate photogenerated carriers, the transmission of photogenerated carriers, the recombination and capture of carriers in the transmission process, etc. Therefore, many factors are affecting the conductivity of semiconductor materials under the action of incident photons, such as the bandgap width of semiconductor materials, the absorption coefficient of specific band light radiation, the number and saturation degree of carriers generated by the absorbed photons, and the carrier mobility of the material itself. Meanwhile, the response time of photoconductance variation also depends on many factors, such as carrier lifetime, carrier transit time, carrier capture time by trap state, etc. For these reasons, semiconductor materials are extremely sensitive to the effects of light and are suitable as candidate materials for PCD.

The primary working mechanism of the PCD is as follows: when the detector is under the condition of external working bias and no light, as shown in Figure 1.1(a), a small current exists between the two electrodes, which is called dark current (I_{dark}). When light is irradiated on the device, as shown in Figure 1.1(b), photons with energy higher than the semiconductor bandgap ($E_{ph} > E_g$) can excite electrons in the valence band into the conduction band, and holes are generated in the valence band in response, thereby changing the conductivity of the material. The

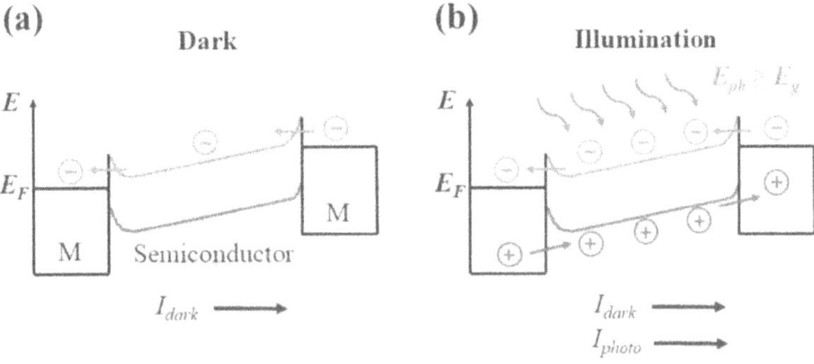

FIGURE 1.1 Diagram of photoconductive effect of semiconductor materials, (a) dark condition, (b) illumination condition. Incident light excites the carriers and increases the conduction current.[21]

generated electron–hole pairs are separated by the working bias, drifted to the electrodes at both ends, and exported to the external circuit for work. This will increase the current between the electrodes, which is called photocurrent (I_{photo}).[21] The external circuit senses the current change of the detector to determine whether there is light. Devices that use this mechanism to detect optical radiation are called PCD, also called photoresistors. This kind of device has a simple structure, stable performance, high light responsivity, and the most comprehensive application.

In the photoconductive effect, there is a particular case called the photogate effect. When the material absorbs incident photons, two kinds of photogenerated carriers are produced. The photogenerated electrons (or photogenerated holes) are trapped by the local state, which causes the local state to be negatively charged (or positively charged). The charged local state will play a role similar to the gate voltage, adjusting the carrier concentration in the material through electrostatic coupling, thereby affecting the conductivity of the material. This localized state of bound carriers usually exists on the surface of semiconductor materials, defects and material interfaces, especially for low-dimensional materials such as quantum dots, nanowires or 2D materials. The characteristics of large specific surface area and atomic layer thickness of low-dimensional materials make them extremely sensitive to surface states and defects. Therefore, this kind of photogating gain is widely present in low-dimensional nanomaterial devices and their heterojunction devices. In general, after the local state captures a carrier, it is slower to escape the bondage, which prevents it from recombining with another kind of carrier, prolongs the life of the carrier, and even being induced to do work by the external circuit for many times. It will provide high photoconductivity gain for the device but also prolong the response time of the device.

1.2.2 Performance Parameters

1.2.2.1 Dark/Light Current and Switching Ratio

Dark current (I_{dark}) refers to the current flowing through the detector without light irradiation. Light current (I_{light}) refers to the current value flowing through the detector under light irradiation. I_{light}–I_{dark} represents the current completely caused by sunlight, which is called photogenerated current. And the switching ratio of photodetector is the ratio of light current to dark current when the device is under the same bias voltage. Photodetectors with excellent performance need to have low dark current, high light current and large switching ratio.

1.2.2.2 Responsivity

Responsivity is a physical quantity that describes the intensity of the output signal of a device under unit radiation. Its calculation formula is as follows:[22]

$$R = \frac{I_{light} - I_{dark}}{P} \tag{1.1}$$

R stands for responsivity, and the unit is A/W, P is the effective incident light power irradiated on the device channel. Responsivity is one of the most important indexes to describe the detector performance. It directly reflects the intensity of the photoelectric response signal when the optical signal is incident on the device, and it is the most basic index to evaluate the detection performance. Although the extracted current signal can be amplified by an external circuit when the responsivity is low, additional noise will also be introduced, so a better responsivity is the premise of a high-performance detector.

1.2.2.3 Response Time

Response time is a physical quantity that describes the response speed of the detector to the light signal. The detector has a stable dark current in dark conditions. When light is incident on the detector, the detector current changes (usually rise) due to the photoconductivity effect and turns into photocurrent. The transition of the current signal requires a certain period called the response time or rising edge time of the detector. In contrast, when the photocurrent reaches a stable value, the incident light signal is removed, and the output signal of the detector is restored from the photocurrent to the dark current level. This period is called the recovery time or falling edge time of the detector.

The usual calculation method is: the response time is the time required for the intensity of the photogenerated current to rise from 10% to 90%; the recovery time is the time required for the intensity of the photogenerated current to fall from 90% to 10%.[23]

1.2.2.4 Cut-Off Wavelength

The cut-off wavelength represents the longest wavelength of the optical signal that the detector can detect. Because the longer the wavelength of light, the smaller the energy of the photon. The photon energy must be greater than the energy of electrons crossing the bandgap to excite electrons from the valence band to the conduction band. Therefore, the cut-off wavelength of the detector is inversely proportional to the bandgap width of the semiconductor material. The formula is:[24]

$$\lambda = \frac{1240}{E_g} \tag{1.2}$$

E_g represents the bandgap of the material in eV; λ represents the maximum response wavelength of the semiconductor material, the cut-off wavelength is in nm.

1.2.2.5 Noise Equivalent Power (NEP)

The equivalent noise power is the minimum optical signal power that the detector can distinguish from the total noise (environmentally induced, internally generated, etc.) at a bandwidth of 1 Hz. The relationship between NEP and responsiveness is[5]

$$NEP = \frac{I_n}{R} \tag{1.3}$$

where I_n is the noise current at 1 Hz bandwidth, and R is the device responsivity. NEP is a reference index to measure the detection ability of the detector. The smaller the NEP is, the higher the sensitivity and the better the performance of the detector.

1.2.2.6 Specific Detectivity (D^*)

The specific detectivity is another critical parameter of the detector, which is used to measure the performance of the detector by normalization. It integrates the detector's responsivity, the effective area of the device, noise, circuit bandwidth, and other important indicators and achieves normalized comparison through formula calculation. Its unit is cm $Hz^{1/2}$ W^{-1} or Jones. The higher the value of D^* is, the more sensitive the detector is. The formula is:[25]

$$D^* = \frac{\sqrt{\Delta f \times S}}{NEP} \tag{1.4}$$

where Δf represents the bandwidth and S represents the channel area of the device.

1.2.3 Structure of PCD

The simple device structure of the 2D material PCD prepared in the laboratory is shown in Figure 1.2(a). The lowest layer is generally SiO_2/Si substrate covered by an inductive light absorption layer of semiconductor materials. The light absorption layer is flanked by metal electrodes connected to external circuits when light is irradiated to the inductively absorbing layer, the conductivity of the semiconductor material changes. The current of the device changes from dark current to photocurrent, which the external circuit can detect. The performance of the device depends entirely on the intrinsic performance and quality of the inductively absorbing layer material, and of course, the material performance can also be regulated through doping, defect engineering and other processes.

However, the generation, separation, and transport of photogenerated carriers need to be carried out using inductive light absorption layer materials. In this series of processes, the photogenerated electron–hole pairs are easy to compound, which reduces the carrier separation efficiency and affects the response performance of the device. Given this phenomenon, the researchers updated the device structure of the PCD, as shown in Figure 1.2(b), adding a layer of carrier transport layer above (or below) the induction absorption layer to connect the two electrodes.

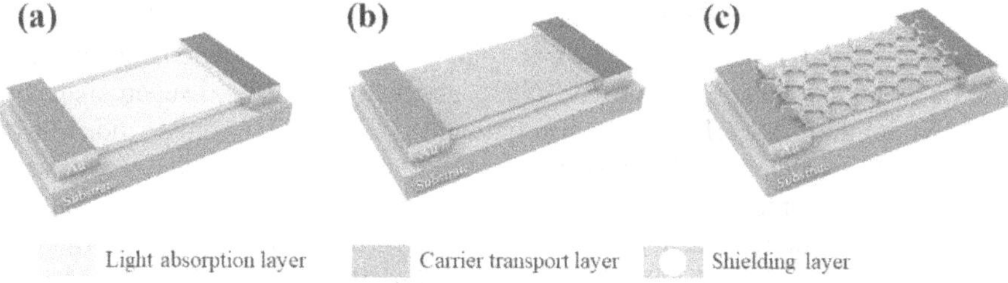

FIGURE 1.2 Structure of photoconductance detector, (a) simple 2D material photoconductance detector structure, (b) structure of 2D material photoconductance detector with heterojunction, (c) structure 2D material photoconductance detector with a heterojunction and a shielding layer.

The built-in electric field is constructed at the contact interface based on the difference of work function between the light absorption layer and the transmission layer. Photogenerated electrons and holes are effectively separated under the built-in electric field, which effectively prevents the recombination of carriers, thus improving the performance of the device.

To improve the anti-jamming capability of the PCD, some researchers will overlay a shielding layer on the outermost layer of the device, as shown in Figure 1.2(c), to shield part of the electromagnetic interference in the environment, thus improving the stability and sensitivity of the PCD. In addition, researchers have developed more characteristic detector structures based on the properties and synergies of different materials.

1.3 LIGHT ABSORPTION LAYER

As the name suggests, the light absorption layer is the part used to absorb incident photons and convert them into photogenerated carriers. Among the components of the PCD, the light absorption layer is an essential part of determining the detection band and the efficiency of producing photogenerated carriers, which can be called the "heart" of the detectors. The most important property of semiconductor material as a light absorption layer is that it has a noticeable bandgap, and the bandgap determines the cut-off wavelength of the PCD. Semiconductors with different bandgaps can detect light in different wavelength ranges. Therefore, the work type of PCD determines the choice of semiconductors for the light absorption layer. It can also be said that the selection of the semiconductor determines the work type of PCD. Compared with traditional semiconductor materials, many 2D materials have not only apparent bandwidth but also show excellent tunability within a certain range, so they are widely used in the research of PCD, mainly including TMDs, transition metal carbides (or nitrides, carbonitrides) (MXenes), BP and other 2D single-element nanomaterials (Xenes) materials.

1.3.1 TMDs

Transition metal chalcogenide is a huge branch of the 2D material family. They are usually represented by the chemical formula MX_2, where M represents the transition metal elements, and X represents the chalcogenide elements (S, Se, and Te),[14] as shown in Figure 1.3(a). This class of compounds includes semiconductors (e.g., MoS_2, WSe_2, and $PdTe_2$), metals (e.g., $NbTe_2$ and $TaTe_2$), as well as superconductors (e.g., $NbSe_2$ and TaS_2). For these materials, different element combinations bring different electronic properties. Structurally, the transition metal M atom layer separates the chalcogenide X atom layers in the upper and lower hexagonal planes, and covalent bonds connect the atoms in the layer. In contrast, the monoatomic layers are connected by weak van der Waals force. Therefore, the superposition of the transition metal atomic layer and chalcogenide atomic layer forms an X–M–X sandwich structure (Figure 1.3(b)).

In recent years, TMDs have attracted much attention due to their excellent physical properties.[27] Because of the layer-dependent property, some semiconductor

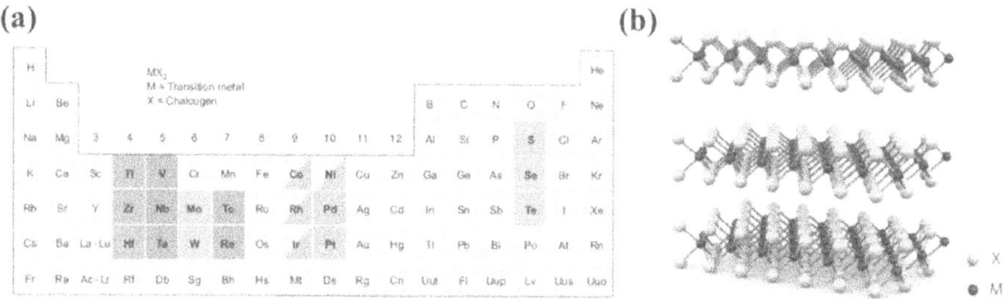

FIGURE 1.3 (a) The elemental composition of TMDs in the periodic table, including several transition metal elements and chalcogenide elements. Among them, Co, Rh, Ir, and Ni are partially colored, indicating that only part of dichalcogenides are 2D-layered structures.[26] (b) 3D structure diagram of TMDs. The transition metal elements are sandwiched between two layers of chalcogenide elements.[14]

TMDs exhibit a tunable bandgap from the indirect bandgap of bulk materials to the direct bandgap of single-layer materials. And, the bandgap width decreases as the number of layers increases. On the whole, the bandgap range of semiconductor TMDs is about 0.5–2 eV,[5] and the bandgap can also be modulated utilizing defect engineering,[28,29] alloy preparation,[30,31] and electric field modulation.[32,33] So their response range to light signal cover from visible light to infrared light. Therefore, 2D layered TMDs are widely used to prepare new generation optoelectronic devices to meet the requirements of high responsivity.

MoS_2, as a representative of TMDs materials, has long become candidate for light absorption layer of PCD. The single-layer MoS_2 has a direct bandgap of ~1.8 eV,[34] the field-effect electron mobility of ~260 $cm^2\ V^{-1}\ s^{-1}$, field-effect hole mobility of 175 $cm^2\ V^{-1}\ s^{-1}$,[35] a large current on-off ratio of 1×10^8,[36] good stability and outstanding flexibility.[37,38] These are all necessary conditions for the steady development of MoS_2 in the field of optoelectronics. Lee et al.[39] reported a few-layer MoS_2 PCD based on a transparent top gate electrode (Figure 1.4(a)). The experimental results show that the triple-layer MoS_2 device has an excellent ability to detect red light. In contrast, the single-layer and double-layer MoS_2 devices are very useful in the detection of green light. The different functions are attributed to the regulation of MoS_2 layers on the bandgap, as shown in Figure 1.4(b). The comparison test results confirmed that the single-layer MoS_2 has a significant bandgap of 1.8 eV, while the band gaps of the double-layer and triple-layer MoS_2 are 1.65 eV and 1.35 eV, respectively. In addition, Jian et al.[40] confirmed the effect of annealing and sulfurization on the photoelectric properties of oxygen-doped MoS_2 by combining experiments with thermodynamic calculation. On one hand, annealing can increase the oxygen doping concentration and narrow the bandgap, thus improving the optical response and specific detectivity. On the other hand, sulfurization treatment can transform oxygen-doped MoS_2 into pure MoS_2, changing the indirect bandgap into a direct bandgap, thus reducing the response time. However, in recent years, relying solely on MoS_2 has been unable to meet the needs for high response rate and fast response time of PCD. Researchers began to consider building heterojunctions to improve further the performance parameters of PCD. For instance, Gwang et al.[41]

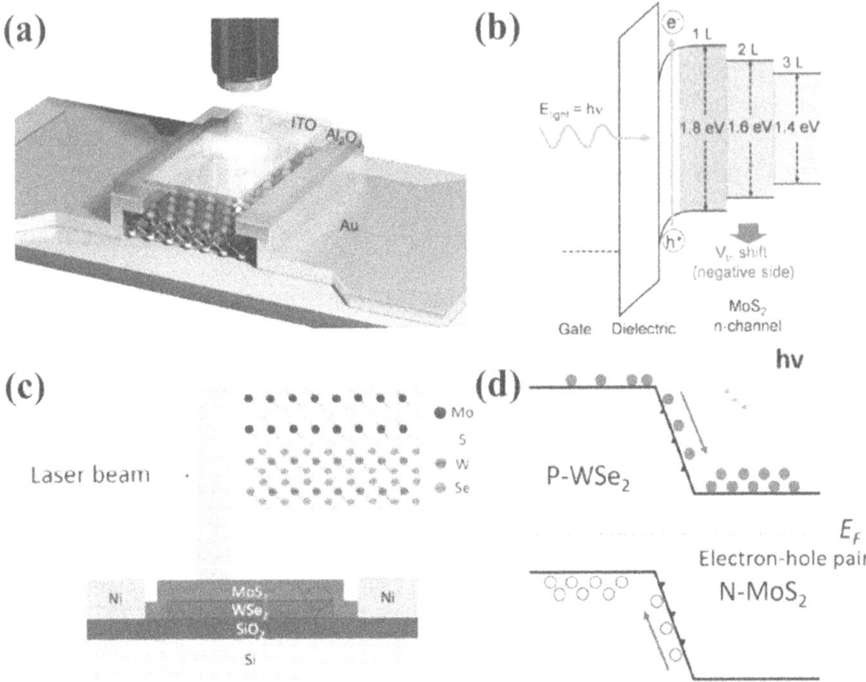

FIGURE 1.4 (a) Schematic 3D view of a single-layer MoS_2 transistor with Al_2O_3 dielectric and ITO top gate. (b) The energy band diagram of ITO (gate)/Al_2O_3 (dielectric)/monolayer, double layers, and triple layers MoS_2 under the action of light (E_{light}) illustrate the photoelectric effect of different MoS_2 layers.[39] (c) Structure of MoS_2/WSe_2 heterojunction PCD under laser irradiation. (d) Schematic diagram of band and carrier flow of WSe_2/MoS_2 van der Waals heterojunction. The trajectories of carrier generated by heteroplasms are shown.[41]

developed high sensitivity PCD based on WSe_2 and MoS_2 van der Waals heterostructure (Figure 1.4(c)). The vertical built-in electric field of the PN junction separates the photogenerated carriers (Figure 1.4(d)), thus generating the photogating effect and obtaining a high photoconductive gain of 10^6. The PCD has excellent performance, a high responsivity of 2700 A/W, a specific detection rate of 5×10^{11} Jones, and a response time of 17 ms. This scheme is combined with the large-area synthesis technology of 2D materials, which is of great significance to the practical application of PCD. Furthermore, there are many other TMDs materials and their heterogeneous junction that have been sought after for photoconduction detectors in recent years, such as HfS_2,[42,43] $MoTe_2$,[44] PtS_2,[45] and so on. But some of them are sensitive to water and oxygen in the air, thus their atmospheric stability needs to be improved.[46]

The unique physical and chemical properties of TMDs materials play an essential role in the field of solar cells, lasers, and other energy storage devices and the area of photodetectors, and are expected to become the main materials for nanoelectronics optoelectronics, and electrochemical applications in the future. However, the lack of methods to achieve large-area, high-quality industrial production of TMDs has severely limited its application. So, scholars will continue the pace of their research.

1.3.2 MXenes

In many 2D materials, MXenes was first discovered and termed by Barsoum et al.[47] in 2011. The typical general formula of MXenes is $M_{n+1}X_n$ or $M_{n+1}X_nT_z$ ($n = 1, 2,$ or 3), where M represents transition metal elements such as Sc, Ti, Ta, Hf, Zr, V, Nb, Cr, Mo, etc., X represents C and/or N, and Tz represents active functional groups such as fluorine ion (F^-), oxygen ion (O^{2-}), or hydroxyl (OH^-) which are connected to the surface of MXenes.[48] The MXenes originally evolved from the layered and hexagonal MAX phases by selectively etching A elements, mainly from III-A and IV-A groups.[49] There are only more than 20 MXenes reported so far, which is far lower than the number of MAX phases reported.[50,51] Therefore, the research on MXenes will continue.

The surface functional groups of MXenes have interesting properties, which can significantly affect the intrinsic physical, chemical, and mechanical properties of materials, including work function, electronic band structure, etc.[49,52,53] The bandgap of MXenes varies from 0.04 eV to 3.23 eV, which can be a direct bandgap or indirect bandgap according to the different surface functional groups.[54–56] A proper bandgap provides a proper light response band of MXenes. According to the research results, MXenes has strong light absorption performance, which can absorb ultraviolet light to infrared light. For example, $Ti_3C_2T_z$ film absorbs light from 300 to 500 nm, from UV to visible spectrum.[57] The transmittance of the film can be changed by changing the thickness of the film and using different intercalating agents such as DMSO and urea. When the thickness of the film is appropriate, it can also have strong spectral absorption in the 700–800 nm light range.[58]

Velusamy et al.[59] prepared five different MXenes (Mo_2CT_x, $Ti_3C_2T_x$, Nb_2CT_x, T_2CT_x, and V_2CT_x,) PCD (Figure 1.5(a)), among which Mo_2CT_x MXene has the best performance (Figure 1.5(b)). The prepared Mo_2CT_x PCD have high responsivity (up to 9 A W^{-1}) and a specific detection rate (~5×10^{11} Jones) in the 400–800 nm range. The spatially resolved electron energy loss spectrum and the ultrafast femtosecond transient absorption spectrum of MXene nanosheets indicate that the optical response of Mo_2CT_x strongly depends on surface plasmon-assisted hot carriers (Figure 1.5(c) and (d)). In addition, the Mo_2CT_x thin-film device exhibits relatively stable performance under environmental conditions, continuous light, and mechanical stress, indicating that it has a durable light detection capability in the visible spectrum.

1.3.3 BP

BP is a single-element 2D layered semiconductor material, which is also called phosphorene when it is a monolayer. Its atomic structure is shown in Figure 1.6.

Each phosphorus atom in the layer covalently bonds with three adjacent phosphorus atoms to form a folded hexagonal crystal structure, and the layers are connected by van der Waals force. This structure brings a modest band gap,[61,62] excellent optical absorption efficiency,[63,64] and high room temperature carrier mobility

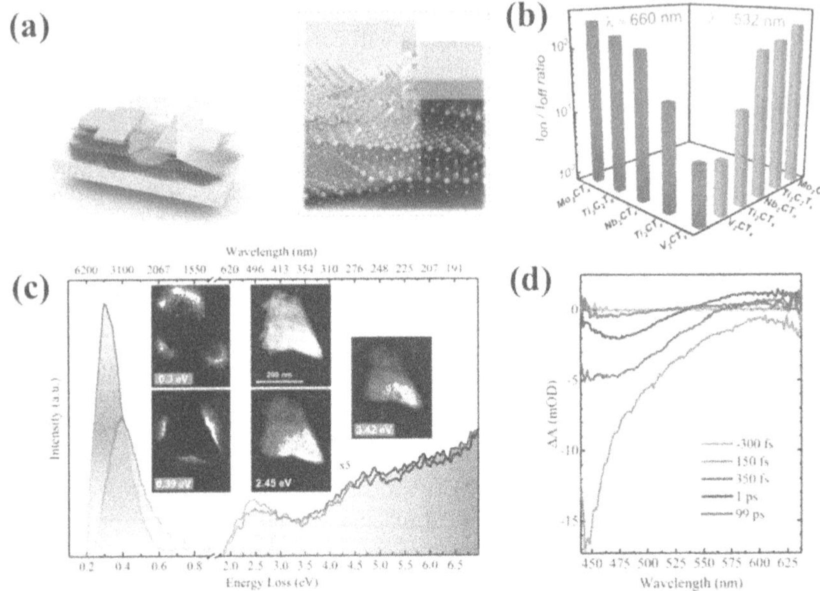

FIGURE 1.5 (a) The schematic diagram of the MXene PCD under illumination shows the process of carrier migration to the biased gold electrode. (b) The ratio of photocurrent to dark current of different MXene PCD, measured under conditions of 660 nm 0.39 W cm^{-2} (red bars) and 532 nm 0.41 W cm^{-2} (green bars). (c) Zero-loss-peak subtracted EELS acquired on a truncated triangular nanosheet of Mo_2CT_x (≈58 nm ± 1 at the center). The spectra are normalized to the transversal SP peak at 2.45 eV (highlighted in red) and magnified by a factor of 5 above 1.7 eV. Insets: ADF-STEM micrograph of the Mo_2CT_x nanosheet on a Si_3N_4 supporting membrane (black area), and the EELS fitted intensity maps of the corresponding longitudinal SP modes; namely, dipole (0.3 eV) and quadrupole (0.39 eV), in addition to the transversal SP mode (2.45 eV) and the IBT (starting at 3.42 eV). (d) Time-resolved visible transient absorption spectra of Mo_2CT_x in different time slots.[59]

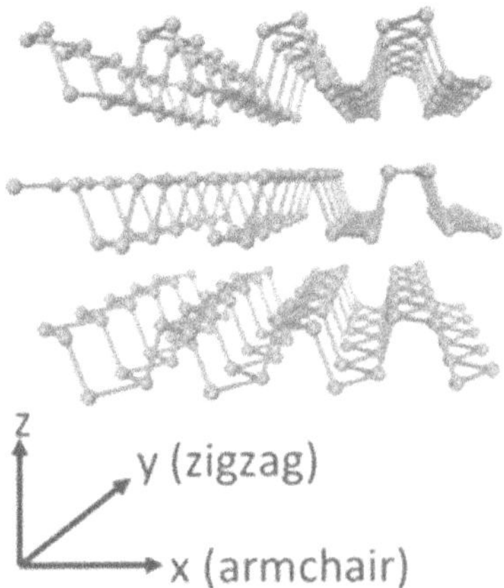

FIGURE 1.6 Structure of BP with the armchair and zigzag crystal directions.[60]

(~1000 $cm^2\ V^{-1}\ s^{-1}$),[65,66] which attracts the attention of researchers. Due to the interaction between layers, the bandgap width of BP is also affected by the number of layers. The bandgap of single-layer BP is ~2.0 eV, and with the increase in layers, the bandgap gradually decreases to ~0.33 eV.[67,68] The wide bandgap range enables its light absorption band to cover from visible light to mid-infrared. Notably, unlike TMDs, BP is a direct bandgap material regardless of the number of layers, making it easier for electrons to be excited from the valence band to the conduction band.[69] In addition, the folded crystal structure of BP reduces its symmetry, resulting in its unique anisotropy. However, due to the low separation efficiency of photogenerated electron–hole pair caused by the high mobility of carriers, the light response rate of BP is still very low compared with traditional near-infrared detectors. Constructing p–n junctions by superimposing BP with other materials has proved to be an effective way to improve the performance of BP-PCD.

Guo et al.[70] produced BP mid-infrared PCD (the structure is shown in Figure 1.7(a)) with high gain and a responsivity of up to 82 A W^{-1} (at 3.39 μm, shown in Figure 1.7(b) at room temperature. The outstanding performance mainly comes from the photogating effect brought by the heterojunction, as shown in Figure 1.7(c) and (d). The built-in electric field separates the photogenerated electron-hole pairs at the BP/SiO_2 interface. Shallow trap state trapping photocarriers induce the photogate effect in the BP channel and SiO_2/BP interface, thus effectively extending the lifetime of photogate holes. Combined with the high carrier mobility of BP, the holes are quickly exported to achieve a very high photoelectric response. In addition, the appropriate channel length and the number of BP layers are also important parameters for high performance.

Although BP has excellent photoelectric properties, stability issues have always hindered its practical application. Under environmental conditions, O_2 and H_2O can irreversibly and quickly react with BP to form phosphorus compounds or phosphoric

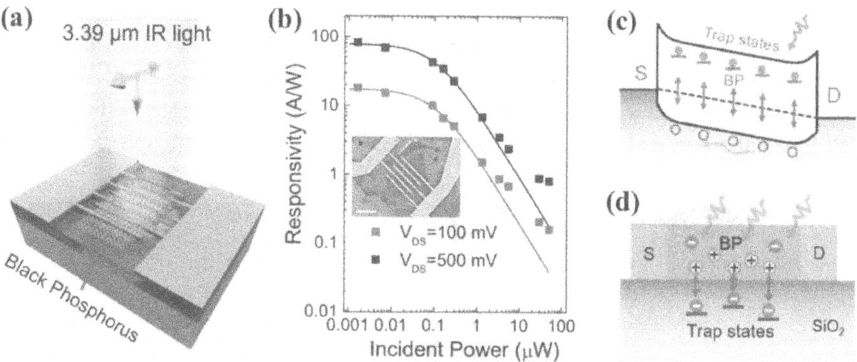

FIGURE 1.7 (a) 3D schematic of BP-PCD under 3.39 μm light illumination. (b) The relation points of BP-PCD between the response rate and the incident light power. Solid line: the fitting result of Hornbeck Haines model. Insert: False-color scanning electron micrograph of BP-PCD. Scale: 5 μm.[70] The principle diagram of electron trap induced light control in (c) BP channel material and (d) SiO_2/BP interface.[71]

acid, which severely degrades the performance of BP devices. To improve the stability of the device, the researchers tried to use a variety of methods to enhance the device (such as element doping,[72] oxide layer protection,[73] hBN encapsulating,[74] etc.[75–77]) and achieved remarkable results. In the future, BP will indeed have outstanding performance in the field of photodetectors.

1.3.4 Other Xenes

In addition to graphene and BP, researchers have also discovered a series of new single-element 2D materials. They are called Xenes, where "X" represents a chemical element and "ene" represents its unique sp^2 hybrid alkene bond structure.[78] So far, about 15 Xenes materials have been synthesized through theoretical calculations and experimental studies, including group IIIA (borophene,[79] aluminene,[80] gallenene,[81] indiene[82]), IVA (graphene, silicone,[83] germanene,[84] stanene,[85] plumbene[86]), VA (phosphorene,[87] arsenene,[88] antimonene,[89] bismuthene[90]), and VIA (selinene,[91] tellurene[92]) elements. The synthetic methods of Xenes are various, including mechanical stripping, physical vapor deposition, pulsed laser deposition, chemical vapor deposition, liquid phase stripping, wet synthesis, and so on.

Xenes in groups III and IV are metallic and semi-metallic materials with low photoelectric response. In contrast, Xenes in groups VA and VIA have excellent semiconductor properties and are suitable for use as light-absorbing materials for PCD. For example, single-layer arsenene and antimonene have indirect band gaps of 2.49 eV and 2.28 eV, respectively, and they can be transformed from indirect bandgap semiconductors to direct bandgap semiconductors under biaxial strain.[93–95] The DFT calculation shows that the single-layer γ-selenene and β-tellurene have indirect band gaps of 1.05 eV[96] and 1.17 eV,[97] respectively, and both have high carrier mobility. The monolayer γ-selenene has an electron mobility of 6.97×10^3 cm^2 V^{-1} s^{-1} and hole mobility of 9.48×10^3 cm^2 V^{-1} s^{-1}.[98] While the carrier mobility of β-tellurene is expected to reach hundreds to thousands.[97]

In 2017, Qin et al.[99] used physical vapor deposition to successfully synthesize high-quality 2D selenene nanosheets with large lateral dimensions, up to 30 inches, and a minimum thickness of 5 nanometers (Figure 1.8(a)). The growth mechanism of 2D selenium nanosheets was studied through crystal structure, and back-gated FETs based on selenene nanosheets, as shown in Figure 1.8(b), were fabricated. Although the response

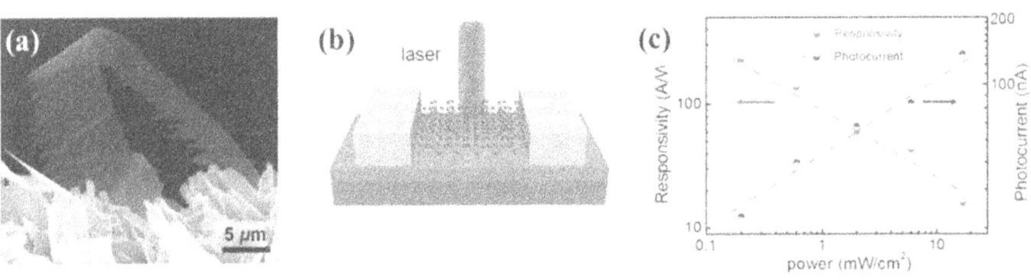

FIGURE 1.8 (a) SEM image of 2D selenene nanosheets with saw-like structure. (b) Back-gate Se nanosheet PCD structure diagram. (c) The relationship between photocurrent and responsivity and laser lighting power at $V_{ds} = 3$ V and $V_g = -80$ V.[99]

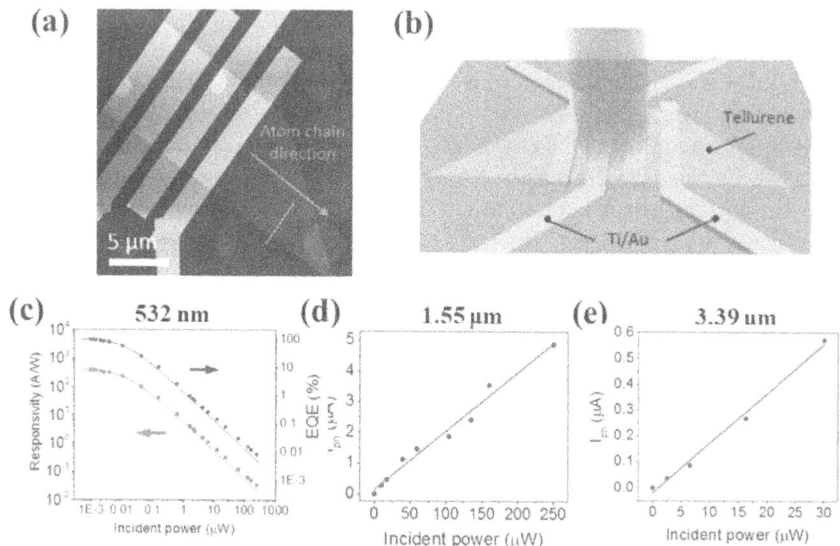

FIGURE 1.9 (a) AFM image of tellurene field-effect transistor. (b) Schematic diagram of tellurene field-effect transistor for photoelectric detection. (c) Responsivity and EQE curve with incident power of 532 nm light ($V_g = 5$ V and $V_{ds} = 1$ V). (d) Curve of photocurrent with incident light power of 1.55 μm light ($V_g = 100$ V and $V_{ds} = 1$ V). (e) Curve of photocurrent with incident light power of 3.39 μm light ($V_g = 5$ V and $V_{ds} = 1$ V).[100]

speed is slow due to its low intrinsic carrier mobility (0.26 $cm^2\ v^{-1}\ s^{-1}$ at 300K), it has a high responsivity of 263 A/W (Figure 1.8(c)). This proves that selenene has great potential in the field of electronics and optoelectronics. Recently, Shen et al.[100] reported high-sensitivity mid-infrared PCD based on 2D tellurene, as shown in Figure 1.9(a) and (b). The responsivity of tellurene PCD is 383A/W, 19 mA/W, and 18 mA/W at wavelengths of 520 nm, 1.55 μm, and 3.39 μm, respectively (Figure 1.9(c)–(e)). Its detection band covers the range from visible light to short-wave infrared and even mid-infrared. At the same time, the photogating effect due to local trapping state was used to obtain a high gain of 1.9×10^3 and 3.15×10^4 at 520 nm and 3.39 μm wavelengths, respectively, so that its unique polarization sensitivity, ultra-fast responsivity, wide detection range, and ability to work in conjunction with other materials make tellurium an exciting 2D semiconductor for mid-infrared optoelectronic detection.

Generally speaking, 2D Xenes materials have great potential in the field of photoelectric detection. With these novel Xenes materials, PCD with different structures can also be constructed, which can improve R, D^*, expand the detection band range, shorten the response time, and achieve high performance.

1.4 CARRIER TRANSPORT LAYER AND TRANSPARENT ELECTRODE

Although the carrier transport layer and the electrode are the two parts of the PCD, their primary function is to export the photogenerated carriers to the external circuit for work before recombining. Therefore, the 2D material used for the carrier transport layer or the transparent electrode must have high carrier mobility, so that the photogenerated carriers can be transported quickly. Besides, the high transmittance of 2D material can be used as a good transparent electrode material. The carrier transport layer has another key function: to form a built-in electric field with the light absorption layer, thereby effectively separating photogenerated electron–hole pairs. Because the 2D materials are combined by van der Waals force and the surface is flat, the carriers can transport each other more efficiently and ignore the influence of lattice mismatch. The main carrier transport layer and transparent electrode materials include graphene and MXene.

1.4.1 Graphene

As a representative of 2D materials, graphene has many excellent physical properties, such as a long mean free path and easy-to-regulate Fermi energy level. Moreover, its special 2D structure of a single atomic layer also determines its high mechanical strength, ultra-thin transparency, and other physical properties. However, due to its high transparency and zero bandgap characteristics, graphene has a low optical absorption rate (only 2.3%) in the ultraviolet to near-infrared region and a short interaction length which is not conducive to optical applications.[101,102] Nevertheless, its high carrier concentration (10^{13} cm^{-2}) and extremely high carrier mobility (about 200,000 cm^2 $V^{-1}S^{-1}$) make it the preferred carrier transport layer and transparent electrode in 2D materials.

So far, many heterogeneous structures using graphene as a carrier transport layer have been developed to achieve high detector responsiveness. In traditional semiconductor PCD, interface/surface states are detrimental to the performance of devices such as detection efficiency and responsitivity, which act as carrier recombination/charge centers and cause many effects such as carrier recombination or surface scattering.[103,104] However, in the hybrid graphene/semiconductor PCD, the surface state at the heterojunction interface may introduce a special carrier multiplication mechanism, resulting in high responsitivity.[105,106] Under the irradiation of light, the interface states become the trapping center for photogenerated carriers. When the electron/hole is captured by the interface states, the Fermi level on the interface will be raised/lowered due to the carrier interface localization, thus changing the Fermi level of graphene correspondingly through the Fermi level alignment effect. Huijun Tian et al.[107] constructed a heterostructure with undoped GaAs and graphene, and used the ionic liquid as the top gate to control the graphene carrier concentration (structure is shown in Figure 1.10(a)), thereby achieving a high response rate of ≈1321 A W^{-1} at room temperature. The results show that the gain mechanism is mainly due to the effective trapping of holes by the graphene/GaAs interface state, which adjusts the Fermi level of GaAs surface and the graphene layer (shown in Figure 1.10(b)). In addition, the back gate voltage can also adjust the photocurrent and response characteristics of the device based on this structure. Wenjun Chen et al.[108] spin-coated PbS quantum dots on graphene to form a graphene/PbS quantum dots (QDs) intercalation structure, shown in Figure 1.10(c) and (d). Based on the embedded graphene transmission layer, a strong and uniform EQE can be achieved in the range of 600–950 nm, which increases the photoelectric conversion efficiency of PbS-QDs by 20%. The

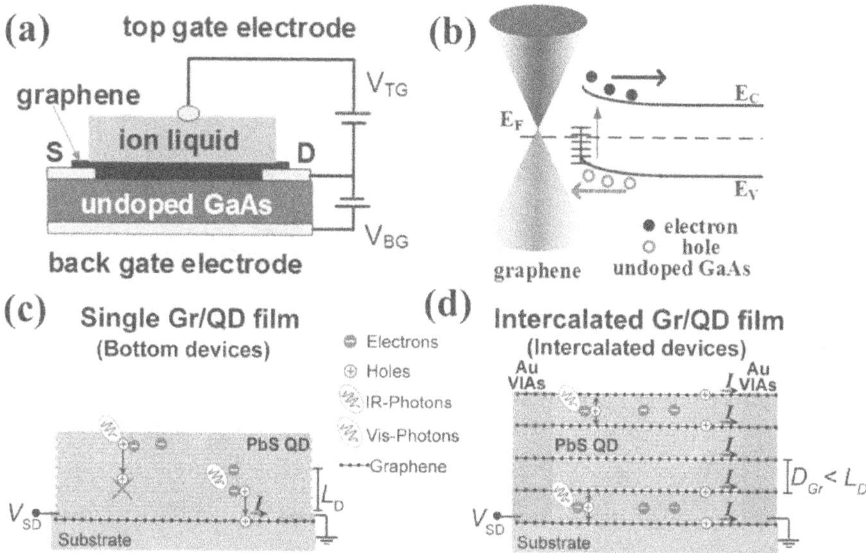

FIGURE 1.10 (a) Structural diagram of graphene/GaAs heterojunction PCD. (b) Energy band diagram of graphene and undoped GaAs contact. Electrons and holes are effectively separated to both ends of GaAs and graphene through the built-in electric field of heterojunction.[107] (c) Schematic diagram of the single-layer graphene/PbS QDs PCD structure. (d) Schematic diagram of the intercalated graphene/PbS QDs PCD structure.[108]

graphene layer interspacing is studied to improve the charge transfer and stability between QDs and graphene, thereby improving the performance of this intercalated structure device. Wei-Long Xu et al.[109] prepared the hybrid perovskite/reduced graphene oxide film by antisolvent treatment. The graphene network interspersed in the perovskite nanoparticles promotes charge transfer and serves as a transmission pathway to improve the performance of the device.

As for the detector transparent electrodes, most of them are made of metals with good conductivity, such as gold and silver. The Schottky barrier at the metal/semiconductor contact interface also affects the detection performance of the device. In addition, the low light transmittance of the upper metal electrode in the Schottky device and the complicated preparation process of the Schottky contact are also problems to be faced for PCD.[110] Graphene and its composites can be used as conductive electrodes to solve the above problems. On the one hand, there are several ways to regulate the graphene Fermi level to improve its contact characteristics with semiconductors.[111–113] On the other hand, the properties of graphene allow it to be used to prepare highly transmittable and flexible electrodes. A variety of graphene preparation and transfer methods have been developed, making the fabrication of graphene transparent electrodes more convenient. Abir Hossain et al.[114] used graphene as the transparent electrode and combined with TMDs to design a 2D heterojunction PCD. Its structural schematic diagram and optical image are shown in Figure 1.11(a) and (b). By creating nearly periodic folds to the 2D material and using van der Waals heterojunction slip to release the bending stress, the

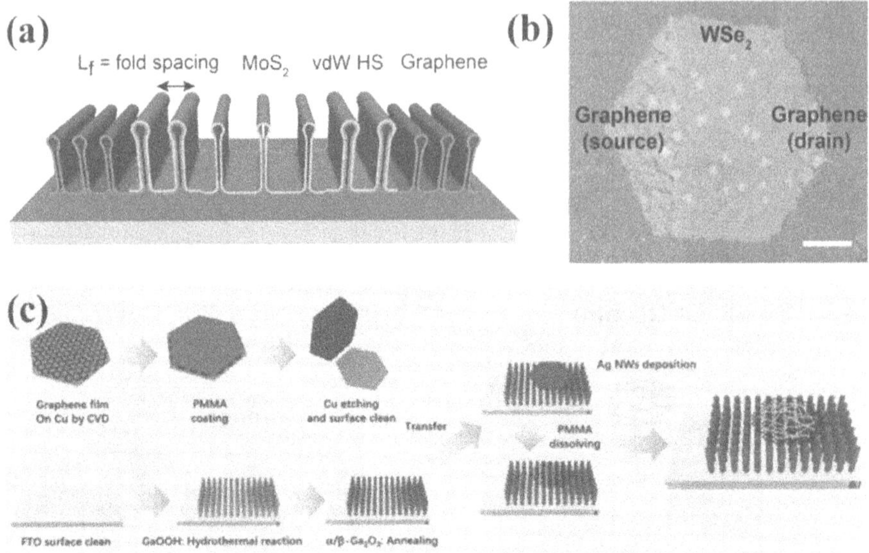

FIGURE 1.11 (a) Schematic diagram of the fold structure of graphene and TMDs heterojunction. The fold spacing of TMDs material is wide, and the spacing of graphene folds is narrow. (b) Optical image of WSe_2 PCD with graphene as source and drain electrodes.[114] (c) Schematic diagram of the preparation process of α/β-Ga_2O_3 nano-array PCD based on the graphene/Ag nanowire hybrid upper transparent electrode.[115]

mechanical flexibility and electronic movement can be improved simultaneously. Photoluminescence spectroscopy shows that the optical band gap of WSe_2 shifts by less than 2 meV between flat and 15% biaxial crumpling, corresponding to a change in strain of less than 0.05%. The photoresponsivity reached 20 A/W under an illumination power density of 4 $\mu W/cm^2$ at 20 V bias, a performance comparable to flat photosensors. Wu et al.[115] prepared a graphene/silver transparent electrode on α/β-Ga_2O_3 composite film. The preparation process is shown in Figure 1.11(c). Compared with the traditional metal electrode, the graphene/Ag nanowire transparent electrode with α/β-Ga_2O_3 has good conductivity and optical transparency, which improves the separation and transfer of photogenerated carriers.

Therefore, combining graphene with other organic/inorganic optical materials to form heterogeneous structures is undoubtedly an effective way to improve the original semiconductor properties and achieve high response PCD.

1.4.2 MXenes

As a new member of the 2D material family, MXenes can be used as a semiconductor material to absorb and convert photons and as a carrier transport layer and transparent electrode material for PCD. According to the first principle, some MXenes with terminal functional groups have semiconductor properties, while all MXenes without terminal functional groups have metallic properties.[116,117] For example,

$Ti_3C_2F_2$ and $Ti_3C_2(OH)_2$ are semiconductors with band gaps of 0.1 eV and 0.05 eV, respectively, while Ti_3C_2 is a metal conductor.[47] At the same time, the specific surface area, conductivity, and carrier mobility of MXenes are also high. Its electrical conductivity is at least one order of magnitude higher than carbon and other conventional electrode materials,[118] and it largely depends on the preparation method and layering method of the material, and exhibits a high degree of anisotropy.[57,119] In addition, MXenes have good light transmittance (the transmittance of single-layer Ti_3C_2Tz nanosheets reaches 97%), which makes MXenes thin films widely used in transparent electrodes and optoelectronic devices.

In 2017, Kang et al.[120] designed a PCD based on $Ti_3C_2T_x$/n-Si vertical heterostructure. The device schematic and energy band structure are shown in Figure 1.12(a) and (b). The test results show that the MXenes crystal represented by $Ti_3C_2T_x$ is a new type of transparent conductive 2D material with a wide range of adjustable work function and can form a good Schottky contact with silicon under the action of van der Waals force. Among them, the thickness of $Ti_3C_2T_x$ has a greater impact on the performance of the heterojunction. As the number of layers increases, the transparency of $Ti_3C_2T_x$ decreases, but the conductivity increases. Under 405 nm incident light with a power of 15.17 mW/cm^2,

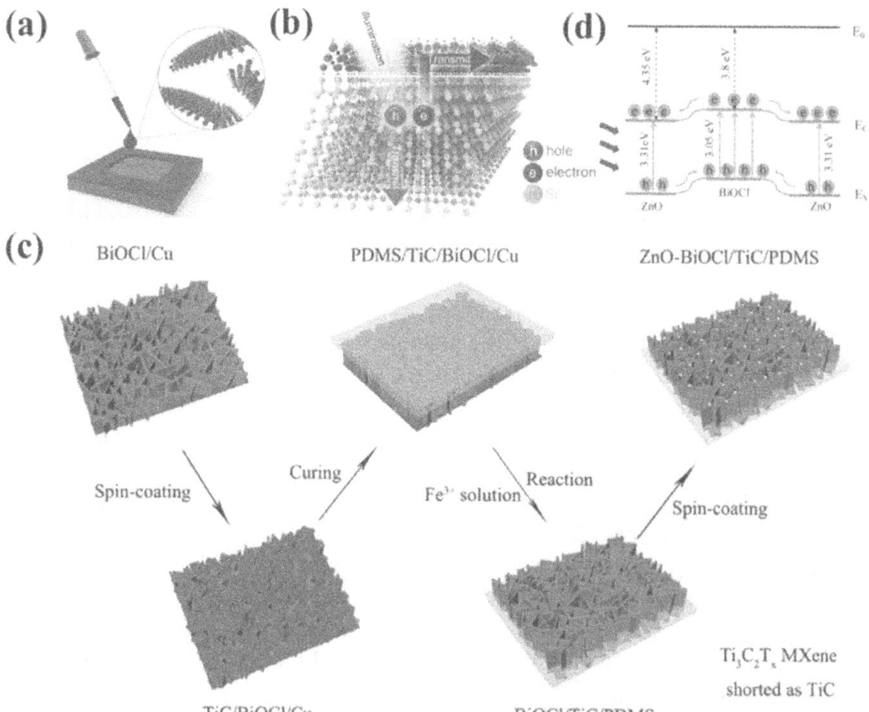

FIGURE 1.12 (a) Fabrication of $Ti_3C_2T_x$/n-Si van der Waals heterojunction photoelectric devices. (b) Schematic diagram of $Ti_3C_2T_X$/N-Si heterojunction structure and carrier transport.[120] (c) Preparation process of BiOCL/TiC/PDMS and ZnO-BiOCL/TiC/PDMS device. (d) Schematic diagram of Zno-BiOCL/TiC/PDMS carrier separation mechanism.[121]

the photoresponse and switching ratio are 26.95 mA/W and 1×10^5, respectively, and have high response and recovery speed. This research proves the application value of MXenes material in the field of PCD. Recently, Ouyang et al.[121] have also introduced MXenes as a transport layer into PCD. First of all, they fabricated a BiOCl/polydimethylsiloxane (PDMS) detector by the transfer method, but only showed a small photocurrent under UV light irradiation. After introducing tic as the transport layer of the detector, the photocurrent of the device increased by 2 to 3 orders of magnitude. However, the dark current increases with the increase in photocurrent. The device has a large dark current (6.7 pA), low on-off ratio (2.4), and long response time (6.87 s) under 5 V and 350 nm light illumination. After that, they modified BiOCl with ZnO QDs. The dark current of the ZnO-BiOCl/TiC/PDMS detector is reduced to 86 fA, the on-off ratio is increased to 7996.5, and the decay time is shortened to 0.93 s. However, the photocurrent of ZnO-BiOCl/PDMS PCD decreased, and the decay time was prolonged after the tic transport layer was removed. The experimental process is shown in Figure 1.12(c). The results fully verify that TiC as a transport layer can increase the carrier transmission efficiency of the device, thus effectively improving the detector performance. Its working mechanism is shown in Figure 1.12(d). ZnO and BiOCl contact produces band bending, and electrons flow easily from BiOCl to ZnO, while holes flow more easily to BiOCl, thus effectively separating electron hole pairs and improving the photoelectric response.

Although nearly 10 years have passed since the discovery of the first MXenes member, the development of MXenes materials is still in its infancy. In the future, more MXenes materials will be developed and utilized in PCD.

1.5 STRUCTURE OPTIMIZATION OF PCD

The development of 2D materials makes it occupy an important position in the light absorption layer and carrier transport layer of PCD. In addition, researchers also use the unique physical properties of 2D materials to optimize the structure of PCD, thereby further improving device performance.

Guo et al.[122] reported a light-driven junction field-effect transistor (LJFET) composed of n-type ZnO band as channel material and p-type WSe_2 nanosheets as photoactive gate material breaks the mutual restriction between quantum efficiency and response time by decoupling the gain and carrier lifetime. The schematic diagram of the device structure and working states are shown in Figure 1.13(a) and (b). Under light conditions, the resistance of WSe_2 decreases due to the photoconductivity effect, which provides a conductive path for the applied gate voltage, thus effectively modulating the depletion layer width in the ZnO channel. Interestingly, when an optical signal is detected, the response current of the device decreases, shown in Figure 1.13(c), which is contrary to the traditional PCD. The gain and response time are determined by the field-effect modulation and switching speed of LJFET, respectively. Therefore, it shows a high response rate of 4.83×10^3 A W^{-1} and a gain of ~ 10^4. At the same time, the fast response time is ~10 μs. The LJFET architecture provides a new strategy for simultaneously achieving high gain and fast response of PCD.

Hexagonal boron nitride (h-BN) is a graphene-like 2D atomic layered material with a series of valuable properties, such as thermal shock resistance, high resistivity (10^{12}–10^{14} Ω·cm), nontoxicity, chemical inertness, and so on.[123–126] And it is

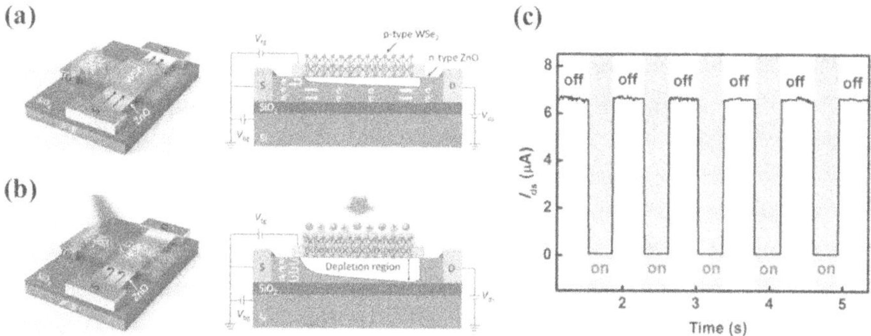

FIGURE 1.13 Diagram of structure and working state WSe_2/ZnO LJFET in (a) dark and (b) light environment, depicts the mechanism of the device in different states. (c) The response curve of LJFET to 637 nm illumination is obtained at $V_{tg} = -1$ V and $V_{bg} = 15$ V. (The power intensity is 367.5 mW cm^{-2}.)[122]

a wide bandgap semiconductor with a bandgap higher than 5.9 eV. Because h-BN nanosheets have an atomically smooth surface and no dangling bonds and charge traps, it becomes ideal packaging material.

In 2017, Chen et al.[127] designed a dual-gate BP-PCD with a dynamically adjustable detection range (Figure 1.14(a)). Because the BP layer is sandwiched between two pieces of h-BN, it has high quality and minimal charge trap density, which enables BP to work in its inherent photoconductive region and have fast response speed. After applying the displacement field through the double gate, the responsivity of PCD at the charge-neutrality point was 518, 30, and 2.2 mA W^{-1}, and the corresponding NEP was 0.03, 35, and 672 pW $Hz^{-1/2}$ at the incident light with the wavelength of 3.4, 5, and 7.7 μm, respectively. Recently, Leonardo et al.[128] obtained room temperature terahertz PCD with high sensitivity (NEP ~ 160 pW $Hz^{-1/2}$). The main structure is that a single layer of graphene is encapsulated in h-BN to form a clean h-BN/SLG/h-BN heterostructure, as shown in Figure 1.14(b), thereby

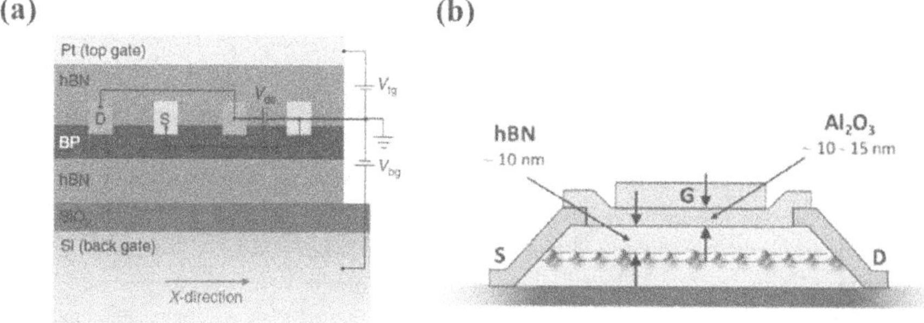

FIGURE 1.14 (a) The structure diagram of the infrared photoelectric detector in adjustable BP based on double-gate transistor structure.[127] (b) Schematic diagram of an h-BN encapsulated, graphene-based, room-temperature terahertz receiver.[128]

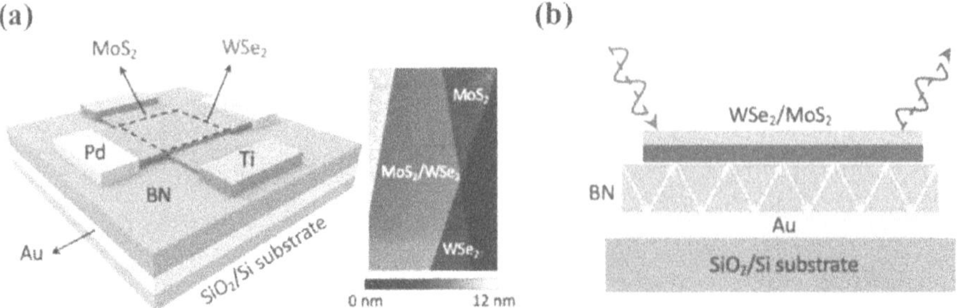

FIGURE 1.15 (a) Left: schematic diagram of vertical van der Waals heterostructure PCD consisting of 5 nm WSe_2/5 nm MoS_2/40 nm h-BN/Au on SiO_2. Right: AFM image of WSe_2/MoS_2 van der Waals heterostructure. (b) Schematic diagram of Fabry-Perot interference mechanism, the reflected light on the metal substrate is again absorbed by WSe_2/MoS_2 and reflected at MoS_2/h-BN interference.[129]

achieving ultra-high charge mobility (>50 000 $cm^2\ V^{-1}\ s^{-1}$). The response time of this detector is only 3 ns, which is an order of magnitude faster than any other low NEP ($<10^{-9}$ W $Hz^{-1/2}$) layered terahertz PCD at room temperature.

Furthermore, the broad bandgap and transmittance of h-BN can be used for another purpose. Huang et al.[129] novelly prepared a PCD based on WSe_2/MoS_2/h-BN/Au/SiO_2 with a Fabry-Perot cavity structure using a layer-by-layer stacking method (Figure 1.15(a)). The Raman and light absorption of WSe_2/MoS_2 heterojunction in the hybrid structure have been greatly enhanced. This phenomenon is attributed to the resonance effect induced by the Fabry Perot cavity constructed by the h-BN layer, which can capture photons-specific wavelength for a long time and increase the path of interaction between WSe_2/MoS_2 absorption layer and light, as shown in Figure 1.15(b). The report showed a promising method to effectively enhance the light absorption of few-layer 2D materials, which can improve the optoelectronic performance of devices. Although the device prepared in this article is a photovoltaic detector, the method is also applicable to PCD.

1.6 SUMMARY

In the past decade, breakthroughs have been made in the research of two-dimentional (2D) materials. The unique physical and chemical properties brought by its special layered atomic structure have great application potential, which can meet the performance requirements that traditional materials cannot reach, and has attracted the attention of researchers. This chapter introduces the main working principle and performance parameters of photoconductive detector (PCD), and summarizes the typical applications of 2D materials in various parts of PCD. Transition metal dichalcogenides, transition metal carbides (or nitrides, carbonitrides) (MXenes), black phosphorus, and other 2D single-element nanomaterials have strong light–matter interaction and appropriate and adjustable band gap. They can efficiently absorb incident photons and convert them into photogenerated carriers, so they are used as ideal materials for light absorption layers. For graphene and some metallic MXenes, although their photoelectric conversion efficiency is affected due to high light transmittance and no obvious bandwidth, their high carrier mobility and no effect of lattice mismatch are important advantages as a carrier transport layer. Furthermore, due to the high transmittance, easy stack, and clean atomic level flat surface, they can also be used as packaging materials or build special structures to improve the performance of devices. These functions can well meet the needs of PCD for wide response band, high response rate, and fast response speed. But, the large-area and high-quality preparation of 2D materials is a huge production problem. Although many preparation methods have been reported, they still cannot meet the actual requirements of large-scale and low-cost production. In addition, due to the unique properties of 2D materials, almost all process flows, such as interconnection contact, doping, gate processing, patterning, etching, will have great challenges, and this process steps integration into an ultra-clean environment process is also necessary to promote large-scale manufacturing applications. Nevertheless, the research enthusiasm for 2D materials continues unabated. With the joint efforts of scientists from all over the world, 2D materials will shine surely in the field of optoelectronics.

REFERENCES

1. Qiao H, Huang ZY, Ren XH, Liu SH, Zhang YP, Qi X, Zhang H. Self-powered photodetectors based on 2D materials. *Adv. Opt. Mater.*, 2020, 8: 1900765.
2. Wang XT, Cui Y, Li T, Lei M, Li JB, Wei ZM. Recent advances in the functional 2D photonic and optoelectronic devices. *Adv. Opt. Mater.*, 2019, 7: 1801274.
3. Wang B, Zhong SP, Zhang ZB, Zheng ZQ, Zhang YP, Zhang H. Broadband photodetectors based on 2D group IVA metal chalcogenides semiconductors. *Appl. Mater. Today*, 2019, 15: 115–138.
4. Dong T, Simoes J, Yang ZC. Flexible photodetector based on 2D materials: processing, architectures, and applications. *Adv. Mater. Interfaces*, 2020, 7: 1901657.
5. Long MS, Wang P, Fang HH, Hu WD. Progress, challenges, and opportunities for 2D material based photodetectors. *Adv. Funct. Mater.*, 2019, 29: 1803807.
6. Zhuo RF, Zuo SY, Quan WW, Yan D, Geng BS, Wang J, Men XH. Large-size and high performance visible-light photodetectors based on two-dimensional hybrid materials SnS/RGO. *RSC Adv.*, 2018, 8: 761–766.
7. Konstantatos G. Current status and technological prospect of photodetectors based on two-dimensional materials. *Nat. Commun.*, 2018, 9: 5266.
8. Brongersma ML, Halas NJ, Nordlander P. Plasmon-induced hot carrier science and technology. *Nat. Nanotechnol.*, 2015, 10: 25–34.
9. Djeffal F, Boubiche N, Ferhati H, Faerber J, Le Normand F, Javahiraly N, Fix T. Highly efficient and low-cost multispectral photodetector based on RF sputtered a-Si/Ti multilayer structure for Si-photonics applications. *J. Alloys Compd.*, 2021, 876: 160176.
10. Zhu H, Chen Y, Zhao Y, Li X, Teng Y, Hao XJ, Liu JF, Zhu H, Wu QH, Huang Y. Growth and characterization of InGaAs/InAsSb superlattices by metal-organic chemical vapor deposition for mid-wavelength infrared photodetectors. *Superlattices Microstruct.*, 2020, 146: 106655.
11. Britnell L, Ribeiro RM, Eckmann A, Jalil R, Belle BD, Mishchenko A, Kim YJ, Gorbachev RV, Georgiou T, Morozov SV, Grigorenko AN, Geim AK, Casiraghi C, Castro Neto AH, Novoselov KS. Strong light-matter interactions in heterostructures of atomically thin films. *Science*, 2013, 340: 1311–1314.

12. Lee CH, Lee GH, van der Zande AM, Chen WC, Li YL, Han MY, Cui X, Arefe G, Nuckolls C, Heinz TF, Guo J, Hone J, Kim P. Atomically thin p-n junctions with van der Waals heterointerfaces. *Nat. Nanotechnol.*, 2014, 9: 676–681.
13. Geng HJ, Yuan D, Yang Z, Tang ZJ, Zhang XW, Yang K, Su Y. Graphene van der Waals heterostructures for high-performance photodetectors. *J. Mater. Chem. C*, 2019, 7: 11056–11067.
14. Wang QH, Kalantar-Zadeh K, Kis A, Coleman JN, Strano MS. Electronics and optoelectronics of two-dimensional transition metal dichalcogenides. *Nat. Nanotechnol.*, 2012, 7: 699–712.
15. Rivera P, Schaibley JR, Jones AM, Ross JS, Wu SF, Aivazian G, Klement P, Seyler K, Clark G, Ghimire NJ, Yan JQ, Mandrus DG, Yao W, Xu XD. Observation of long-lived interlayer excitons in monolayer $MoSe_2$-WSe_2 heterostructures. *Nat. Commun.*, 2015, 6: 6242.
16. Hong XP, Kim J, Shi SF, Zhang Y, Jin CH, Sun YH, Tongay S, Wu JQ, Zhang YF, Wang F. Ultrafast charge transfer in atomically thin MoS_2/WS_2 heterostructures. *Nat. Nanotechnol.*, 2014, 9: 682–686.
17. Han HV, Lu AY, Lu LS, Huang JK, Li HN, Hsu CL, Lin YC, Chiu MH, Suenaga K, Chu CW, Kuo HC, Chang WH, Li LJ, Shi YM. Photoluminescence enhancement and structure repairing of monolayer $MoSe_2$ by hydrohalic acid treatment. *ACS Nano*, 2016, 10: 1454–1461.
18. Yang J, Lu TY, Myint YW, Pei JJ, Macdonald D, Zheng JC, Lu YR. Robust excitons and trions in monolayer $MoTe_2$. *ACS Nano*, 2015, 9: 6603–6609.
19. Zhou YH, An HN, Gao C, Zheng ZQ, Wang B. UV-Vis-NIR photodetector based on monolayer MoS_2. *Mater. Lett.*, 2019, 237: 298–302.
20. Zheng ZQ, Yao JD, Wang B, Yang YB, Yang GW, Li JB. Self-assembly high-performance UV-vis-NIR broadband beta-In2Se_3/Si photodetector array for weak signal detection. *ACS Appl. Mater. Interfaces*, 2017, 9: 43830–43837.
21. Buscema M, Island JO, Groenendijk DJ, Blanter SI, Steele GA, van der Zant HSJ, Castellanos-Gomez A. Photocurrent generation with two-dimensional van der Waals semiconductors. *Chem. Soc. Rev.*, 2015, 44: 3691–3718.
22. Lu YJ, Lin CN, Shan CX. Optoelectronic diamond: growth, properties, and photodetection applications. *Adv. Opt. Mater.*, 2018, 6: 1800359.
23. Kan H, Zheng W, Lin RC, Li M, Fu C, Sun HB, Dong M, Xu CH, Luo JT, Fu YQ, Huang F. Ultrafast photovoltaic-type deep ultraviolet photodetectors using hybrid zero-/two-dimensional heterojunctions. *ACS Appl. Mater. Interfaces*, 2019, 11: 8412–8418.
24. Oshima T, Okuno T, Fujita S. Ga_2O_3 thin film growth on c-plane sapphire substrates by molecular beam epitaxy for deep-ultraviolet photodetectors. *Jpn. J. Appl. Phys.*, 2007, 46: 7217–7220.
25. Zheng W, Lin RC, Zhang ZJ, Liao QX, Liu JJ, Huang F. An ultrafast-temporally-responsive flexible photodetector with high sensitivity based on high-crystallinity organic-inorganic perovskite nanoflake. *Nanoscale*, 2017, 9: 12718–12726.
26. Chhowalla M, Shin HS, Eda G, Li LJ, Loh KP, Zhang H. The chemistry of two-dimensional layered transition metal dichalcogenide nanosheets. *Nat. Chem.*, 2013, 5: 263–275.
27. Shim J, Park HY, Kang DH, Kim JO, Jo SH, Park Y, Park JH. Electronic and optoelectronic devices based on two-dimensional materials: from fabrication to application. *Adv. Electron. Mater.*, 2017, 3: 1600364.
28. Xie Y, Liang F, Wang D, Chi SM, Yu HH, Lin ZS, Zhang HJ, Chen YX, Wang JY, Wu YC. Room-temperature ultrabroadband photodetection with MoS_2 by electronic-structure engineering strategy. *Adv. Mater.*, 2018, 30: 1804858.
29. Xie Y, Zhang B, Wang SX, Wang D, Wang AZ, Wang ZY, Yu HH, Zhang HJ, Chen YX, Zhao MW, Huang BB, Mei LM, Wang JY. Ultrabroadband MoS_2 photodetector with spectral response from 445 to 2717 nm. *Adv. Mater.*, 2017, 29: 1605972.

30. Zhang XM, Xiao SQ, Shi LH, Nan HY, Wan X, Gu XF, Ni ZH, Ostrikov K. Large-size $Mo_{1-x}W_xS_2$ and $W_{1-x}Mo_xS_2$ (x = 0–0.5) monolayers by confined-space chemical vapor deposition. *Appl. Surf. Sci.*, 2018, 457: 591–597.
31. Wang ZQ, Liu P, Ito Y, Ning SC, Tan YW, Fujita T, Hirata A, Chen MW. Chemical vapor deposition of monolayer $Mo_{1-x}W_xS_2$ crystals with tunable band gaps. *Sci. Rep.*, 2016, 6: 21536.
32. Chu T, Ilatikhameneh H, Klimeck G, Rahman R, Chen ZH. Electrically tunable bandgaps in bilayer MoS_2. *Nano Lett.*, 2015, 15: 8000–8007.
33. Ramasubramaniam A, Naveh D, Towe E. Tunable band gaps in bilayer transition-metal dichalcogenides. *Phys. Rev. B*, 2011, 84: 205325.
34. Mak KF, Lee C, Hone J, Shan J, Heinz TF. Atomically thin MoS_2: a new direct-gap semiconductor. *Phys. Rev. Lett.*, 2010, 105: 136805.
35. Liu Y, Guo J, Zhu EB, Liao L, Lee SJ, Ding MN, Shakir I, Gambin V, Huang Y, Duan XF. Approaching the Schottky-Mott limit in van der waals metal-semiconductor junctions. *Nature*, 2018, 557: 696–700.
36. Radisavljevic B, Radenovic A, Brivio J, Giacometti V, Kis A. S Single-layer MoS_2 transistors. *Nat. Nanotechnol.*, 2011, 6: 147–150.
37. Kotekar-Patil D, Deng J, Wong SL, Lau CS, Goh KEJ. Goh, Single layer MoS_2 nanoribbon field effect transistor. *Appl. Phys. Lett.*, 2019, 114: 013508.
38. Mak KF, Shan J. Photonics and optoelectronics of 2D semiconductor transition metal dichalcogenides. *Nat. Photonics*, 2016, 10: 216–226.
39. Lee HS, Min SW, Chang YG, Park MK, Nam T, Kim H, Kim JH, Ryu S, Im S. MoS_2 nanosheet phototransistors with thickness-modulated optical energy gap. *Nano Lett.*, 2012, 12: 3695–3700.
40. Jian JY, Chang HL, Dong PF, Bai ZW, Zuo KN. A mechanism for the variation in the photoelectric performance of a photodetector based on CVD-grown 2D MoS_2. *RSC Adv.*, 2021, 11: 5204–5217.
41. Shin GH, Park C, Lee KJ, Jin HJ, Choi SY. Ultrasensitive phototransistor based on WSe_2-MoS_2 van der waals heterojunction. *Nano Lett.*, 2020, 20: 5741–5748.
42. Wang DG, Zhang XW, Guo GC, Gao SH, Li XX, Meng JH, Yin ZG, Liu H, Gao ML, Cheng LK, You JB, Wang RZ. Large-area synthesis of layered $HfS_{2(1-x)}Se_{2x}$ alloys with fully tunable chemical compositions and bandgaps. *Adv. Mater.*, 2018, 30: 1803285.
43. Yan CY, Gan L, Zhou X, Guo J, Huang WJ, Huang JW, Jin B, Xiong J, Zhai TY, Li YR. Space-confined chemical vapor deposition synthesis of ultrathin HfS_2 flakes for optoelectronic application. *Adv. Funct. Mater.*, 2017, 27: 1702918.
44. You JW, Ye Y, Cai K, Zhou DM, Zhu HM, Wang RY, Zhang QF, Liu HW, Cai YT, Lu D, Kim JK, Gan L, Zhai TY, Luo ZT. Enhancement of $MoTe_2$ near-infrared absorption with gold hollow nanorods for photodetection. *Nano Res.*, 2020, 13: 1636–1643.
45. Li L, Wang WK, Chai Y, Li HQ, Tian ML, Zhai TY. Few-layered PtS2 phototransistor on h-BN with high gain. *Adv. Funct. Mater.*, 2017, 27: 1701011.
46. Mirabelli G, McGeough C, Schmidt M, McCarthy EK, Monaghan S, Povey IM, McCarthy M, Gity F, Nagle R, Hughes G, Cafolla A, Hurley PK, Duffy R. Air sensitivity of MoS_2, $MoSe_2$, $MoTe_2$, HfS_2, and $HfSe_2$. *J. Appl. Phys.*, 2016, 120: 125102.
47. Naguib M, Kurtoglu M, Presser V, Lu J, Niu JJ, Heon M, Hultman L, Gogotsi Y, Barsoum MW. Two-dimensional nanocrystals produced by exfoliation of Ti_3AlC_2. *Adv. Mater.*, 2011, 23: 4248–4253.
48. Li M, Lu J, Luo K, Li YB, Chang KK, Chen K, Zhou J, Rosen J, Hultman L, Eklund P, Persson POA, Du SY, Chai ZF, Huang ZR, Huang Q. Huang, Element replacement approach by reaction with lewis acidic molten salts to synthesize nanolaminated MAX phases and MXenes. *J. Am. Chem. Soc.*, 2019, 141: 4730–4737.

49. Agresti A, Pazniak A, Pescetelli S, Di Vito A, Rossi D, Pecchia A, Maur MAD, Liedl A, Larciprete R, Kuznetsov DV, Saranin D, Di Carlo A. Titanium-carbide MXenes for work function and interface engineering in perovskite solar cells. *Nat. Mater.*, 2019, 18: 1264–1264.
50. Sun WW, Xie Y, Kent PRC. Double transition metal MXenes with wide band gaps and novel magnetic properties. *Nanoscale*, 2018, 10: 11962–11968.
51. Li XQ, Wang CY, Cao Y, Wang GX. Functional MXene materials: progress of their applications. *Chem.-Asian J.*, 2018, 13: 2742–2757.
52. Yu ZM, Feng W, Lu WH, Li BC, Yao HY, Zeng KY, Ouyang JY. MXenes with tunable work functions and their application as electron- and hole-transport materials in non-fullerene organic solar cells. *J. Mater. Chem. A*, 2019, 7: 11160–11169.
53. Dall'Agnese C, Dall'Agnese Y, Anasori B, Sugimoto W, Mori S. Oxidized Ti_3C_2 MXene nanosheets for dye-sensitized solar cells. *New J. Chem.*, 2018, 42: 16446–16450.
54. Liu JH, Kan X, Amin B, Gan LY, Zhao Y. Theoretical exploration of the potential applications of Sc-based MXenes. *Phys. Chem. Chem. Phys.*, 2017, 19: 32253–32261.
55. Khazaei M, Ranjbar A, Arai M, Yunoki S. Topological insulators in the ordered double transition metals M-2 ' M '' C-2 MXenes (M ' = Mo, W; M '' = Ti, Zr, Hf). *Phys. Rev. B*, 2016, 94: 125152.
56. Khazaei M, Arai M, Sasaki T, Estili M, Sakka Y. Two-dimensional molybdenum carbides: potential thermoelectric materials of the MXene family. *Phys. Chem. Chem. Phys.*, 2014, 16: 7841–7849.
57. Zhang CF, Anasori B, Seral-Ascaso A, Park SH, McEvoy N, Shmeliov A, Duesberg GS, Coleman JN, Gogotsi Y, Nicolosi V. Transparent, flexible, and conductive 2D titanium carbide (MXene) films with high volumetric capacitance. *Adv. Mater.*, 2017, 29: 1702678.
58. An HS, Habib T, Shah S, Gao HL, Radovic M, Green MJ, Lutkenhaus JL. Surface-agnostic highly stretchable and bendable conductive MXene multilayers. *Sci. Adv.*, 2018, 4: eaaq0118.
59. Velusamy DB, El-Demellawi JK, El-Zohry AM, Giugni A, Lopatin S, Hedhili MN, Mansour AE, Di Fabrizio E, Mohammed OF, Alshareef HN. MXenes for plasmonic photodetection. *Adv. Mater.*, 2019, 31: 1807658.
60. Abate Y, Akinwande D, Gamage S, Wang H, Snure M, Poudel N, Cronin SB. Recent progress on stability and passivation of black phosphorus. *Adv. Mater.*, 2018, 30: 1704749.
61. Zhou Y, Zhang MX, Guo ZN, Miao LL, Han ST, Wang ZY, Zhang XW, Zhang H, Peng ZC. Recent advances in black phosphorus-based photonics, electronics, sensors and energy devices. *Mater. Horiz.*, 2017, 4: 997–1019.
62. Castellanos-Gomez A. Black phosphorus: narrow gap, wide applications. *J. Phys. Chem. Lett.*, 2015, 6: 4280–4291.
63. Yuan HT, Liu XG, Afshinmanesh F, Li W, Xu G, Sun J, Lian B, Curto AG, Ye GJ, Hikita Y, Shen ZX, Zhang SC, Chen XH, Brongersma M, Hwang HY, Cui Y. Polarization-sensitive broadband photodetector using a black phosphorus vertical p-n junction. *Nat. Nanotechnol.*, 2015, 10: 707–713.
64. Xia FN, Wang H, Jia YC. Rediscovering black phosphorus as an anisotropic layered material for optoelectronics and electronics. *Nat. Commun.*, 2014, 5: 4458.
65. Qiao JS, Kong XH, Hu ZX, Yang F, Ji W. High-mobility transport anisotropy and linear dichroism in few-layer black phosphorus. *Nat. Commun.*, 2014, 5: 4475.
66. Li LK, Yu YJ, Ye GJ, Ge QQ, Ou XD, Wu H, Feng DL, Chen XH, Zhang YB. Black phosphorus field-effect transistors. *Nat. Nanotechnol.*, 2014, 9: 372–377.
67. Tran V, Soklaski R, Liang YF, Yang L. Layer-controlled band gap and anisotropic excitons in few-layer black phosphorus. *Phys. Rev. B*, 2014, 89: 235319.

68. Miao JS, Zhang L, Wang C. Black phosphorus electronic and optoelectronic devices. *2D Mater.*, 2019, 6: 032003.
69. Zhang GW, Huang SY, Chaves A, Song CY, Ozcelik VO, Low T, Yan HG. Infrared fingerprints of few-layer black phosphorus. *Nat. Commun.*, 2017, 8: 14071.
70. Guo QS, Pospischil A, Bhuiyan M, Jiang H, Tian H, Farmer D, Deng BC, Li C, Han SJ, Wang H, Xia QF, Ma TP, Mueller T, Xia FN. Black phosphorus mid-infrared photodetectors with high gain. *Nano Lett.*, 2016, 16: 4648–4655.
71. Zhang L, Wang B, Zhou YH, Wang C, Chen XL, Zhang H. Synthesis techniques, optoelectronic properties, and broadband photodetection of thin-film black phosphorus. *Adv. Opt. Mater.*, 2020, 8: 2000045.
72. Yang BC, Wan BS, Zhou QH, Wang Y, Hu WT, Lv WM, Chen Q, Zeng ZM, Wen FS, Xiang JY, Yuan SJ, Wang JL, Zhang BS, Wang WH, Zhang JY, Xu B, Zhao ZS, Tian YJ, Liu ZY. Te-doped black phosphorus field-effect transistors. *Adv. Mater.*, 2016, 28: 9408–9415.
73. Zhou QH, Chen Q, Tong YL, Wang JL. Light-induced ambient degradation of few-layer black phosphorus: mechanism and protection. *Angew. Chem., Int. Ed.*, 2016, 55: 11437–11441.
74. Avsar A, Vera-Marun IJ, Tan JY, Watanabe K, Taniguchi T, Neto AHC, Ozyilmaz B. Air-Stable transport in graphene-contacted, fully encapsulated ultrathin black phosphorus-based field-effect transistors. *ACS Nano*, 2015, 9: 4138–4145.
75. Zhu WN, Yogeesh MN, Yang SX, Aldave SH, Kim JS, Sonde S, Tao L, Lu NS, Akinwande D. Flexible black phosphorus ambipolar transistors, circuits and AM demodulator. *Nano Lett.*, 2015, 15: 1883–1890.
76. Wood JD, Wells SA, Jariwala D, Chen KS, Cho E, Sangwan VK, Liu XL, Lauhon LJ, Marks TJ, Hersam MC. Effective passivation of exfoliated black phosphorus transistors against ambient degradation. *Nano Lett.*, 2014, 14: 6964–6970.
77. Hsieh YL, Su WH, Huang CC, Su CY. In situ cleaning and fluorination of black phosphorus for enhanced performance of transistors with high stability. *ACS Appl. Mater. Interfaces*, 2020, 12: 37375–37383.
78. Wang B, Zhong SP, Ge YQ, Wang HD, Luo XL, Zhang H. Present advances and perspectives of broadband photo-detectors based on emerging 2D-Xenes beyond graphene. *Nano Res.*, 2020, 13: 891–918.
79. Rubab A, Baig N, Sher M, Sohail M. Advances in ultrathin borophene materials. *Chem. Eng. J.*, 2020, 401: 126109.
80. Yeoh KH, Yoon TL, Rusi, Ong DS, Lim TL. First-principles studies on the superconductivity of aluminene. *Appl. Surf. Sci.*, 2018, 445: 161–166.
81. Steenbergen KG, Gaston N. Thickness dependent thermal stability of 2D gallenene. *Chem. Commun.*, 2019, 55: 8872–8875.
82. Singh D, Gupta SK, Lukacevic I, Muzevic M, Sonvane Y, Ahuja R. Effect of electric field on optoelectronic properties of indiene monolayer for photoelectric nanodevices. *Sci. Rep.*, 2019, 9: 17300.
83. Zhao HX, Sun QQ, Zhou J, Deng X, Cui JX. Switchable cavitation in silicone coatings for energy-saving cooling and heating. *Adv. Mater.*, 2020, 32: 2000870.
84. Suzuki S, Iwasaki T, De Silva KKH, Suehara S, Watanabe K, Taniguchi T, Moriyama S, Yoshimura M, Aizawa T, Nakayama T. Direct growth of germanene at interfaces between van der waals materials and Ag(111). *Adv. Funct. Mater.*, 2020: 2007038.
85. Pang WH, Nishinoa K, Ogikuboa T, Araidai M, Nakatake DM, Le Lay G, Yuhara J. Epitaxial growth of honeycomb-like stanene on Au(111). *Appl. Surf. Sci.*, 2020, 517: 146224.
86. Yuhara J, He BJ, Matsunami N, Nakatake M, Le Lay G. Graphene's latest cousin: plumbene epitaxial growth on a "nano watercube." *Adv. Mater.*, 2019, 31: 1901017.

87. Arcudia J, Kempt R, Cifuentes-Quintal ME, Heine T, Merino G. Blue phosphorene bilayer is a two-dimensional metal and an unambiguous classification scheme for buckled hexagonal bilayers. *Phys. Rev. Lett.*, 2020, 125: 196401.
88. Shah J, Wang W, Sohail HM, Uhrberg RIG. Experimental evidence of monolayer arsenene: an exotic 2D semiconducting material. *2D Mater.*, 2020, 7: 025013.
89. Zhang J, Ye SA, Sun Y, Zhou FF, Song J, Qu JL. Soft-template assisted synthesis of hexagonal antimonene and bismuthene in colloidal solutions. *Nanoscale*, 2020, 12: 20945–20951.
90. Wang YM, Feng W, Chang MQ, Yang JC, Guo YD, Ding L, Yu LD, Huang H, Chen Y, Shi JL. Engineering 2D multifunctional ultrathin bismuthene for multiple photonic nanomedicine. *Adv. Funct. Mater.*, 2020: 2005093.
91. Ding YZ, Huffaker A, Kollner TG, Weckwerth P, Robert CAM, Spencer JL, Lipka AE, Schmelz EA. Selinene volatiles are essential precursors for maize defense promoting fungal pathogen resistance. *Plant Physiol.*, 2017, 175: 1455–1468.
92. Wang DW, Yang AJ, Lan TS, Fan CY, Pan JB, Liu Z, Chu JF, Yuan H, Wang XH, Rong MZ, Koratkar N. Tellurene based chemical sensor. *J. Mater. Chem. A*, 2019, 7: 26326–26333.
93. Singh D, Gupta SK, Sonvane Y, Lukacevic I. Antimonene: a monolayer material for ultraviolet optical nanodevices. *J. Mater. Chem. C*, 2016, 4: 6386–6390.
94. Zhang SL, Yan Z, Li YF, Chen ZF, Zeng HB. Atomically thin arsenene and antimonene: semimetal-semiconductor and indirect-direct band-gap transitions. *Angew. Chem., Int. Ed.*, 2015, 54: 3112–3115.
95. Lee J, Tian WC, Wang WL, Yao DX. Two-dimensional pnictogen honeycomb lattice: structure, on-site spin-orbit coupling and spin polarization. *Sci. Rep.*, 2015, 5: 11512.
96. Liu C, Hu T, Wu YB, Gao H, Yang YL, Ren W. 2D selenium allotropes from first principles and swarm intelligence. *J. Phys.-Condes. Matter*, 2019, 31: 235702.
97. Zhu ZL, Cai XL, Yi SH, Chen JL, Dai YW, Niu CY, Guo ZX, Xie MH, Liu F, Cho JH, Jia Y, Zhang ZY. Multivalency-driven formation of Te-based monolayer materials: a combined first-principles and experimental study. *Phys. Rev. Lett.*, 2017, 119: 106101.
98. Wang D, Tang LM, Jiang XX, Tan JY, He MD, Wang XJ, Chen KQ. High bipolar conductivity and robust in-plane spontaneous electric polarization in selenene. *Adv. Electron. Mater.*, 2019, 5: 1800475.
99. Qin JK, Qiu G, Jian J, Zhou H, Yang LM, Charnas A, Zemlyanov DY, Xu CY, Xu XF, Wu WZ, Wang HY, Ye PDD. Controlled growth of a large-size 2D selenium nanosheet and its electronic and optoelectronic applications. *ACS Nano*, 2017, 11: 10222–10229.
100. Shen CF, Liu YH, Wu JB, Xu C, Cui DZ, Li Z, Liu QZ, Li YR, Wang YX, Cao X, Kumazoe H, Shimojo F, Krishnamoorthy A, Kalia RK, Nakano A, Vashishta PD, Amer MR, Abbas AN, Wang H, Wu WZ, Zhou CW. Tellurene photodetector with high gain and wide bandwidth. *ACS Nano*, 2020, 14: 303–310.
101. Xie C, Mak C, Tao XM, Yan F. Photodetectors based on two-dimensional layered materials beyond graphene. *Adv. Funct. Mater.*, 2017, 27: 1603886.
102. Sun ZH, Chang HX. Graphene and graphene-like two-dimensional materials in photodetection: mechanisms and methodology. *ACS Nano*, 2014, 8: 4133–4156.
103. Ali ST, Ghosh S, Bose DN. Ruthenium and sulphide passivation of GaAs. *Appl. Surf. Sci.*, 1996, 93: 37–43.
104. Yin X, Chen HM, Pollak FH, Chan Y, Montano PA, Kirchner PD, Pettit GD, Woodall JM. Photoreflectance study of surface photovoltage effects at (100)gaas surfaces interfaces. *Appl. Phys. Lett.*, 1991, 58: 260–262.
105. Tian HJ, Liu QL, Zhou CX, Zhan XJ, He XY, Hu AQ, Guo X. Hybrid graphene/unintentionally doped GaN ultraviolet photodetector with high responsivity and speed. *Appl. Phys. Lett.*, 2018, 113: 121109.

106. Guo XT, Wang WH, Nan HY, Yu YF, Jiang J, Zhao WW, Li JH, Zafar Z, Xiang N, Ni ZH, Hu WD, You YM, Ni ZH. High-performance graphene photodetector using interfacial gating. *Optica*, 2016, 3: 1066–1070.
107. Tian HJ, Hu AQ, Liu QL, He XY, Guo X. Interface-induced high responsivity in hybrid graphene/GaAs photodetector. *Adv. Opt. Mater.*, 2020, 8: 1901741.
108. Chen WJ, Ahn S, Balingit M, Wang JY, Lockett M, Vazquez-Mena O. Near full light absorption and full charge collection in 1-micron thick quantum dot photodetector using intercalated graphene monolayer electrodes. *Nanoscale*, 2020, 12: 4909–4915.
109. Xu WL, Ding C, Niu MS, Yang XY, Zheng F, Xiao J, Zheng M, Hao XT. Reduced graphene oxide assisted charge separation and serving as transport pathways in planar perovskite photodetector. *Org. Electron.*, 2020, 81: 105663.
110. Chen X, Liu KW, Zhang ZZ, Wang CR, Li BH, Zhao HF, Zhao DX, Shen DZ. Self-powered solar-blind photodetector with fast response based on Au/beta-Ga_2O_3 nanowires array film Schottky junction. *ACS Appl. Mater. Interfaces*, 2016, 8: 4185–4191.
111. Jessen BS, Gammelgaard L, Thomsen MR, Mackenzie DMA, Thomsen JD, Caridad JM, Duegaard E, Watanabe K, Taniguchi T, Booth TJ, Pedersen TG, Jauho AP, Boggild P. Lithographic band structure engineering of graphene. *Nat. Nanotechnol.*, 2019, 14: 340–346.
112. Sugawara K, Suzuki K, Sato M, Sato T, Takahashi T. Enhancement of band gap and evolution of in-gap states in hydrogen-adsorbed monolayer graphene on SiC(0001). *Carbon*, 2017, 124: 584–587.
113. Yang H, Heo J, Park S, Song HJ, Seo DH, Byun KE, Kim P, Yoo I, Chung HJ, Kim K. Graphene barristor, a triode device with a gate-controlled Schottky barrier. *Science*, 2012, 336: 1140–1143.
114. Hossain MA, Yu J, van der Zande AM. Realizing optoelectronic devices from crumpled two-dimensional material heterostructures. *ACS Appl. Mater. Interfaces*, 2020, 12: 48910–48916.
115. Wu C, He C, Guo D, Zhang F, Li P, Wang S, Liu A, Wu F, Tang W. Vertical alpha/beta-Ga_2O_3 phase junction nanorods array with graphene-silver nanowire hybrid conductive electrode for high-performance self-powered solar-blind photodetectors. *Mater. Today Phys.*, 2020, 12: 100193.
116. Naguib M, Mochalin VN, Barsoum MW, Gogotsi Y. 25th anniversary article: MXenes: a new family of two-dimensional materials. *Adv. Mater.*, 2014, 26: 992–1005.
117. Khazaei M, Arai M, Sasaki T, Chung CY, Venkataramanan NS, Estili M, Sakka Y, Kawazoe Y. Novel electronic and magnetic properties of two-dimensional transition metal carbides and nitrides. *Adv. Funct. Mater.*, 2013, 23: 2185–2192.
118. Pomerantseva E, Bonaccorso F, Feng XL, Cui Y, Gogotsi Y. Energy storage: The future enabled by nanomaterials. *Science*, 2019, 366: eaan8285.
119. Shahzad F, Alhabeb M, Hatter CB, Anasori B, Hong SM, Koo CM, Gogotsi Y. Electromagnetic interference shielding with 2D transition metal carbides (MXenes). *Science*, 2016, 353: 1137–1140.
120. Kang Z, Ma YA, Tan XY, Zhu M, Zheng Z, Liu NS, Li LY, Zou ZG, Jiang XL, Zhai TY, Gao YH. MXene-silicon van der waals heterostructures for high-speed self-driven photodetectors. *Adv. Electron. Mater.*, 2017, 3: 1700165.
121. Ouyang WX, Chen JX, He JH, Fang XS. Improved photoelectric performance of UV photodetector based on ZnO nanoparticle-decorated BiOCl nanosheet arrays onto PDMS substrate: the heterojunction and $Ti_3C_2T_x$ MXene conduction layer. *Adv. Electron. Mater.*, 2020, 6: 2000168.
122. Guo N, Xiao L, Gong F, Luo M, Wang F, Jia Y, Chang HC, Liu JK, Li Q, Wu Y, Wang Y, Shan CX, Xu Y, Zhou P, Hu WD. Light-driven WSe_2-ZnO junction field-effect transistors for high-performance photodetection. *Adv. Sci.*, 2020, 7: 1901637.

123. Perevislov SN. Structure, properties, and applications of graphite-like hexagonal boron nitride. *Refract. Ind. Ceram.*, 2019, 60: 291–295.
124. Cai DL, Yang ZH, Duan XM, He PG, Wang SJ, Yuan JK, Rao JC, Jia DC, Zhou Y. Inhibiting crystallization mechanism of h-BN on alpha-cordierite in BN-MAS composites. *J. Eur. Ceram. Soc.*, 2016, 36: 905–909.
125. Zhang X, Chen JX, Zhang J, Wan DT, Zhou YC. High-temperature mechanical and thermal properties of h-BN/30 vol% Y_2SiO_5 composite. *Ceram. Int.*, 2015, 41: 10891–10896.
126. Zhang X, Chen JX, Li XC, Zhang J, Wan DT, Zhou YC. Microstructure and mechanical properties of h-BN/Y_2SiO_5 composites. *Ceram. Int.*, 2015, 41: 1279–1283.
127. Chen XL, Lu XB, Deng BC, Sinai O, Shao YC, Li C, Yuan SF, Tran V, Watanabe K, Taniguchi T, Naveh D, Yang L, Xia FN. Widely tunable black phosphorus mid-infrared photodetector. *Nat. Commun.*, 2017, 8: 1672.
128. Viti L, Purdie DG, Lombardo A, Ferrari AC, Vitiello MS. HBN-encapsulated, graphene-based, room-temperature terahertz receivers, with high speed and low noise. *Nano Lett.*, 2020, 20: 3169–3177.
129. Huang X, Feng XW, Chen L, Wang L, Tan WC, Huang L, Ang KW. Fabry-Perot cavity enhanced light-matter interactions in two-dimensional van der Waals heterostructure. *Nano Energy*, 2019, 62: 667–673.

2 UV Stability of Organic Solar Cells

Tao Xu[a] *and Furong Zhu*[b]
[a]Shanghai University, Shanghai, People's Republic of China
[b]Hong Kong Baptist University, Hong Kong, People's Republic of China

CONTENTS

2.1 INTRODUCTION

Today, about 90% of the worldwide solar cell production relies on wafer-based crystalline silicon technology. However, the thin-film approach is gaining attention and is likely to represent about 10% of the overall production soon. Currently, there are intense research activities on organic solar cells (OSCs), for the obvious reason of searching for new and efficient renewable energy sources to reduce our dependence on fossil fuel. OSCs can be prepared by solution-fabrication processes, having the advantages of low cost, lightweight, mechanical flexibility, and easy fabrication.[1–3] In recent years, many studies focused on improving the power conversion efficiency (PCE) of OSCs through material innovation, such as developing novel narrow-bandgap polymer donors and nonfullerene acceptors, employing binary and ternary bulk heterojunction active layers, incorporating appropriate carrier-transporting layers and optimizing the fabrication process.[4–8] Solution-processable OSCs with an encouraging PCE of >18% have been reported.[9–12]

DOI: 10.1201/9781003141358-2

Much progress has been made in the enhancement of PCE of OSCs; however, considerable improvement in the long-term stability of OSCs is needed if this technology is to become a viable option for commercialization.[13–15] The understanding of degradation mechanisms is still very nebulous, which in turn limits further improvement of device performance. The stability of OSCs is closely associated with the intrinsic material properties, stratification of donor and acceptor in bulk heterojunction (BHJ), interfacial reaction, photo-degradation in the functional layers, and the built-in potential across the BHJ. It has been reported that ultraviolet (UV) exposure would cause a continuous deterioration in the performance of OSCs due to an inevitable UV-induced degradation process in the functional active layer. Although in real application, encroachment of the moisture and oxygen is one of the reasons causing OSC degradation, the influence can be minimized by incorporating a proper encapsulation. But the UV-induced degradation can only be eliminated by preventing UV light in solar irradiance reaching the active region in OSCs. This involves the use of a UV filter in OSCs, which complicates the manufacturing process and also adds an additional cost. The use of the different nanostructured transparent electrodes, having a poor transparency in UV range, helps in creating a built-in UV filter inside OSCs, and thereby prolonging the lifetime of OSCs.

The high-energy UV photons in the sunlight have less contribution to the energy conversion in OSCs, due to the limited absorption by the organic materials.[16,17] Instead, the exposure to the energetic UV photons is one of the main reasons leading to the photo-degradation in OSCs.[18] Change in the structures of the organic photoactive materials can create sub-bandgap state and disorganize the energy level alignment, causing deterioration in the performance of OSCs. An obvious improvement in the operational stability of OSCs has been observed when the UV portion of the incoming light is filtered out. For instance, a short exposure to UV photons is sufficient to induce a 50 % decrease in device efficiency.[19] UV-induced aggregation of [6,6]-phenyl C71-butyric acid methylester ($PC_{71}BM$), a commonly used acceptor, and deterioration of chemical bonds in the donor polymers were observed,[20] resulting in the degradation of the $PC_{71}BM$-based OSCs.[21] The decomposition of the photoactive layer due to the exposure of the UV irradiation also leads to a dramatic irreversible change in the nanomorphology of the active layer and a catastrophic failure in OSCs.[22,23] The improved understanding and development of effective solutions to enhance the UV stability of OSCs have led to the creation of a wide spectrum of device knowledge and process integration technologies with enormous commercial potential, contributing to the building of a sustainable knowledge-based economy. The ubiquitous adoption of OSC technology as one of the clean energy sources will reduce greenhouse gas emissions, thus contributing to the preservation of our environment.

In this chapter, we discuss effective approaches for enhancing the UV stability of OSCs through incorporating different external and built-in UV filters. The use of the nanostructured transparent electrodes, serving as both the transparent contact and the internal built-in UV filter, for achieving UV-durable OSCs is also presented. It shows that the nanostructured transparent electrodes are suitable for applications in large-area solar cells that can be prepared using the solution-fabrication process at a low cost.

2.2 ORGANIC SOLAR CELLS WITH A UV FILTER

Adding a UV-cut filter (UCF) into OSCs is a common method to protect the cells from the degradation of performance after being exposed to UV. UCF refers to a filter that absorbs UV photons below a fixed wavelength, also known as a defogging filter, which are mostly used in photography. Jaehoon et al.[24] reported extremely high stability for inverted-type fullerene solar cells, which have BHJ layers consisting of poly[4,8-bis(5-(2-ethylhexyl)thiophen-2-yl)benzo[1,2-b:4,5-b′]dithiophenealt-3-fluorothieno[3,4-b]thiophene-2-carboxylate] (PTB7-Th) and $PC_{71}BM$, by employing UCF that is mounted on the front of glass substrates, as it is shown in Figure 2.1(a). The UCF can block most

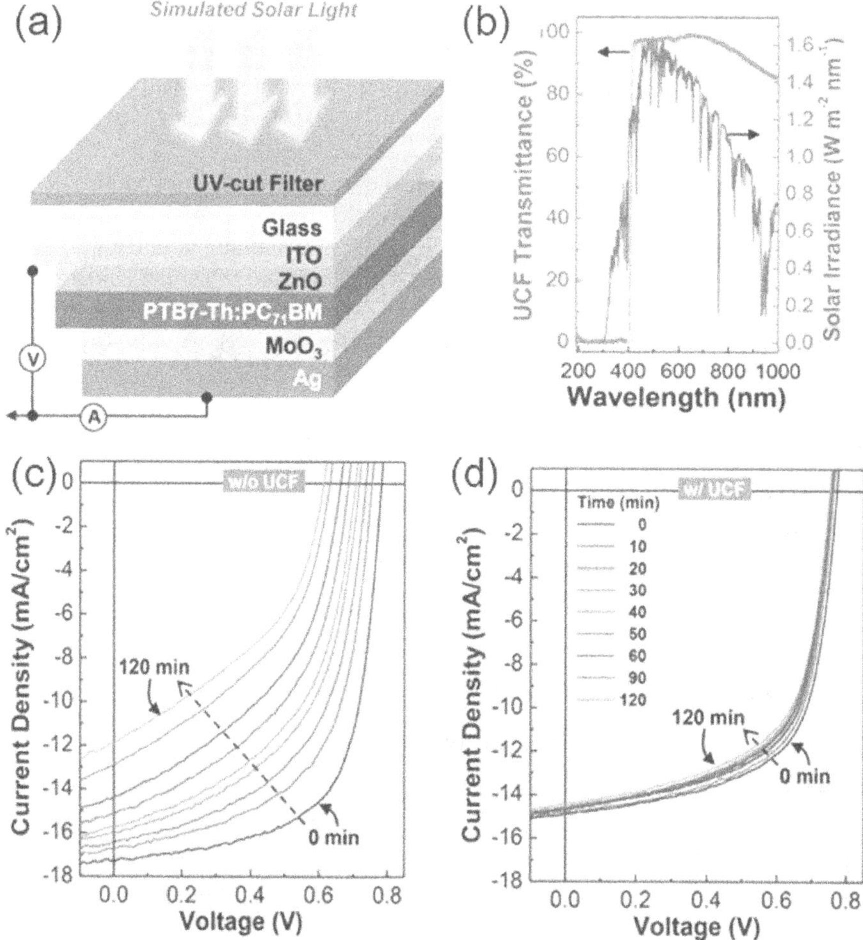

FIGURE 2.1 (a) A UCF-based OSC comprising a layer configuration of glass/ITO/ZnO/PTB7-Th:$PC_{71}BM$/MoO_3/Ag. (b) Optical transmittance of a UCF (left) and the spectrum of solar irradiance (right). Current density–voltage (*J–V*) characteristics of OSCs (c) without a UCF and (d) with a UCF, measured under one Sun illumination (AM1.5G, 100 mW cm^{-2}) for 120 min. (Reproduced with permission.[24] Copyright 2016, WILEY-VCH.)

of UV photons below 403 nm at the expense of ~20% reduction in the total intensity of solar light (Figure 2.1(b)). In terms of PTB7 derivatives, the conjugated heterocyclic groups including bithienyl-benzodithiophene (BDT) and thienothiophene (TTP) units in the polymer main chains are considered to be quite vulnerable to photodegradation under illumination with solar radiation. As most conjugated organic compounds have double bonds (pi-orbitals), their photodegradation undergoes by cleavage of double bonds under UV lights. The BDT and TTP units have pi-orbitals in a fused heterocyclic structure so that their double bonds can be strongly influenced by the UV light that is about 5% of total solar flux.[25] In addition, the morphology of polymer:fullerene BHJ layers can be affected by the degradation of the BDT and TTP units under sunlight because the initial conformation of polymer chains can be deformed by the disruption of conjugated bond structures in the polymer main chains. The results show that OSCs with UCF exhibit an extremely slow decay in PCE compared with that without UCF, as shown in Figure 2.1(c) and (d). The poor device stability without UCF is ascribed to the oxidative degradation of constituent materials in the BHJ layers, due to the formation of $PC_{71}BM$ aggregates.

In another work, Weng et al.[26] also studied the effect of the UCF on the UV stability of OSCs. Acrylic is a material with very high transmittance for the wavelength longer than 300 nm. Since 95% of the UV radiation in sunlight is in the ultraviolet A (UVA) wavelength range (between 315 and 400 nm), the UVA can penetrate the acrylic roof and shine directly on OSCs. The acrylic roof is therefore not expected to be a good UVA shelter. The influence of UVA in sunlight on the stability of OSCs with and without an optical UCF was analyzed. PBDTTT-EFT:$PC_{71}BM$-based OSCs retained 95% of the initial PCE, aged in air over a period of 13 days, when a UCF was integrated with the cells. The results indicate that the use of a UCF is an effective approach to improve the UV stability of OSCs. However, the loss of light intensity is still inevitable due to the addition of a UCF in OSCs, leading to an inferior initial PCE as compared to OSCs without a UCF. In addition, the incorporation of a UCF also increases manufacturing cost.

2.3 METAL OXIDE ELECTRODES

2.3.1 An Overview of Metal Oxide Electrodes

Transparent conducting electrodes (TCEs), which transmit light and conduct electrical current simultaneously, mostly in the visible spectral range, are of increasing importance for flat panel displays and solar cells.[27] Such materials have been known since the end of the nineteenth century, the first example being thin metal films prepared by evaporation and sputtering. The investigation of transparent conducting oxides (TCOs), including CdO, Cu_2O, and PbO, was first reported in 1907.[28] The broad industrial application of TCO materials began when scientists used infrared light filters composed of tin or In_2O_3 on low-pressure sodium discharge lamps to increase the lamp efficiency by reducing heat losses.[29,30] With the advance of flat-panel display technology, indium tin oxide (ITO) became the most commonly used TCO material for transparent electrodes.[31] Nowadays, ITO is one of the commonly used commercial TCOs, with a lowest resistivity of the order of $1–2 \times 10^{-4}\,\Omega$ cm, for applications in flat-panel displays and solar cells.[32]

Considerable enhancement in the overall performance of OSCs, including efficiency, durability, and cost competitiveness, is needed if this technology is to become a viable option for sustainable energy. Although organic photovoltaic technology offers an attractive option for achieving alternative clean energy sources, the use of an ITO transparent electrode is not a long-term solution for large area OSCs at low cost. This is because indium is not abundant on Earth. There remains a need for the development of alternatives to ITO for the application in OSCs. Varieties of ITO replacements have been investigated. Among them, ZnO has received significant attention by taking advantages of the nontoxicity and abundance in nature. The electrical conductivity of ZnO films can be easily modified by the introduction of extrinsic dopants such as B, Al, In, Ga, Si, Sn, F, and Cl. Among these impurity-doped ZnO films, aluminum-doped zinc oxide (AZO) is a possible alternative to ITO due to its unique optical and electrical characteristics. AZO is much cheaper compared to ITO and has good potential for the application in OSCs.[33]

2.3.2 ITO-Free Contacts for UV-Durable OSCs

In 2013, Zhu et al.[34] reported an efficient and UV-durable inverted OSC based on an AZO front transparent cathode. This work yielded AZO-based OSCs with a promising PCE of 6.15% using PTB7:$PC_{70}BM$ as an active layer shown in Figure 2.2(a), which was slightly lower than 6.57% of a control ITO-based OSC; however, a significant enhancement in the stability of AZO-based OSCs was observed under a UV-assisted acceleration aging test. The wavelength-dependent transmittance $T(\lambda)$ of AZO/ glass and ITO/glass used in this work were measured, as shown in Figure 2.2(b). It can be seen obviously that $T(\lambda)$ of AZO exhibits a cutoff at a short wavelength of around 380 nm. That is caused by UV absorption of ZnO (band gap ~3.0 eV), which is the main component of AZO. As the plot of AM1.5G solar spectrum in Figure 2.2(b), the photons with wavelength below 380 nm are able to enter into an ITO-based control OSC contributing to the photocurrent generation, while the AZO front cathode in an AZO-based OSC behaves like a natural UV filter, blocking short-wavelength light from penetrating into OSCs. The UV-assisted acceleration aging test was conducted to manifest this advantage. The variation in cell parameters including PCE, fill factor (FF), short circuit current density (J_{SC}), and open circuit voltage (V_{OC}) measured for the inverted OSCs with AZO and ITO (control) under different UV exposure times was recorded as shown in Figure 2.2(c). After a 20-min continuous UV exposure, V_{OC}, FF, and J_{SC} of the control OSC dropped from 0.74 V, 67%, and 13.2 mA/cm^2 to 0.58 V, 49%, and 11.5 mA/cm^2, resulting in a falling of PCE from 6.53% to 3.3%. For an AZO-based OSC, however, it only experienced a moderate decrease in the cell performance under the same UV-assisted acceleration aging test. The results reveal that V_{OC} changed from 0.75 V to 0.72 V, FF decreased slightly from 63% to 59%, while J_{SC} remained almost unchanged, from 12.4 mA/cm^2 to 12.3 mA/cm^2, resulting in that an overall PCE of AZO-based OSCs only had a minor decrease from 5.86% to 5.2% after a 20-min continuous UV exposure. The distinctive enhancement in the stability of AZO-based OSCs arises from

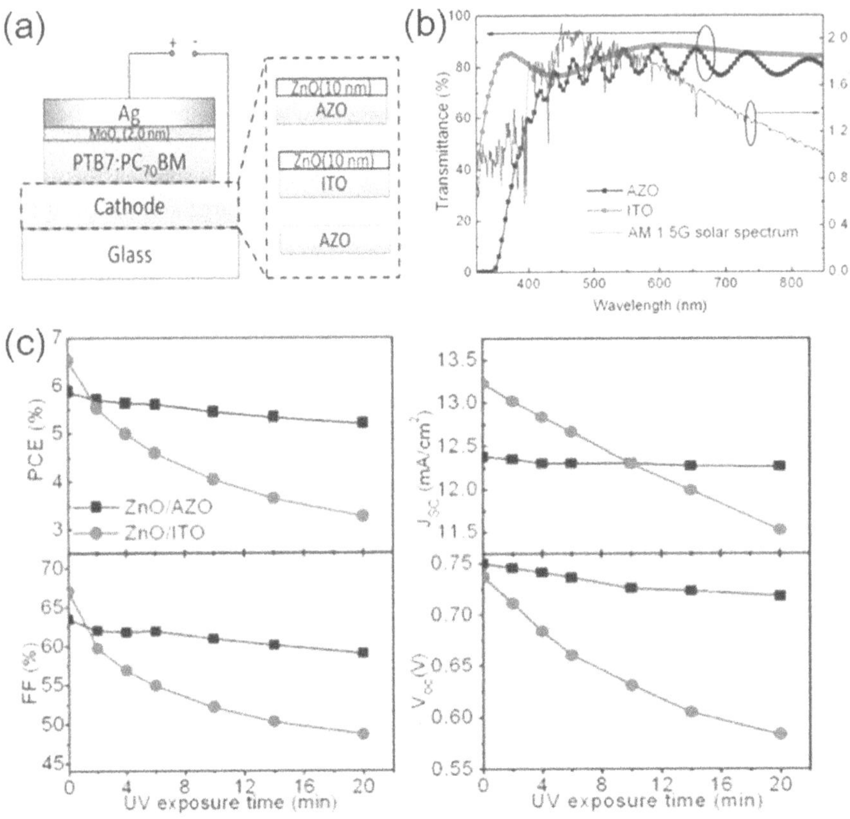

FIGURE 2.2 (a) The cross-sectional view of an inverted OSC made with a front transparent cathode. (b) Transparency measured for an AZO/glass and an ITO/glass over the wavelength range from 320 to 850 nm, spectrum of AM1.5G solar irradiance is also plotted. (c) Recorded PCE, FF, J_{SC}, and V_{OC} of inverted OSCs made with ITO and AZO front cathodes versus the UV exposure time. (Reproduced with permission.[34] Copyright 2013, AIP Publishing.)

the tailored absorption of AZO electrode in wavelength <380 nm, serving as a UV filter to inhibit an inevitable degradation in ITO-based OSCs caused by the UV exposure.

The UV stability of the GZO/AgTi/AZO (GATG) multilayer transparent conducting films was investigated by Chen et al and Luo et al.[35,36] The sheet resistance of 5 Ω/sq and a maximum optical transmittance of 86% were obtained for the GATG films, prepared by magnetron sputtering. In addition, the transparency of GATG film is relatively poor in UV range which is expected to play the role of UCF. PTB7-TH:PC_{71}BM-based inverted OSCs with GATG electrode gave PCE of 9.20%, which is comparable to PCE (9.23%) of the control OSCs with ITO electrode. The PCE of OSCs with the GATG and ITO electrodes, respectively, remain 59% and 23% of the original PCE values after UV exposure for 20 min with a relativize humidity of 68% in air, indicating that OSCs with GATG show better UV durability. These results suggest the use of the TCOs with a good UV absorption serving as an internal built-in UV filter for improving the UV stability of OSCs, removing the need of having a separate UCF to reduce the cost and simplifying the fabrication process.

2.3.3 OSCs with a Built-in UV Filter

The photo-induced degradation in chemical properties of the organic functional materials can be effectively hindered by the introduction of a UV-blocking layer in the cell encapsulation. However, cutting out the UV part of the electromagnetic spectrum also leads to a loss in energy conversion due to the missing photocurrent generation. To avoid the problem of decreasing device performance by stabilization with a UV-blocking layer, down conversion of a part of the blocked UV light enables recycling of the photons which is otherwise dismissed. Engmann et al.[37] employed a combination of titanium and silicon oxide to form a UV-blocking layer on the top of OSCs, as shown in Figure 2.3(a). It shows that the increase in the PCE of OSCs is due to the photon recycling through the introduction of a luminescent layer on the top of a UV-blocking layer. Figure 2.3(b) shows that the transmission of high-energy photons through the substrate with a UV-blocking layer is drastically reduced compared to the reference substrate. Meanwhile, the absorption and the corresponding photocurrent loss due to the use of the UV-blocking layer can be compensated by the application of an anti-reflection and luminescent layer, if the luminescence is efficiently matched to the absorption of the photo-active layer. Using this approach, more than 10% increase in the current output of OSCs was obtained. Luminescent layers based on a fluorophore inside a transparent matrix can easily be produced by roll-to-roll techniques. Such a fluorophore acts as the additive in the luminescence substrates, which has a higher conversion efficiency than that of the photon down-conversion, leading to an increase in the performance of the flexible solar cells.

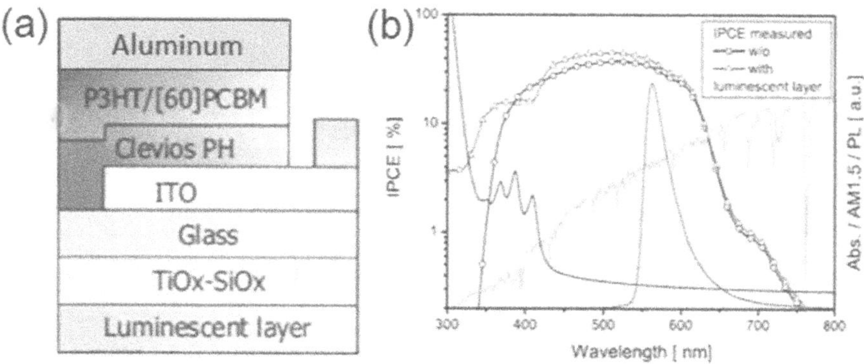

FIGURE 2.3 (a) The cross-sectional view of an OSC with a front UV photoluminescent down-conversion layer to recycle the UV photons. (b) IPCE spectrum measured for the P3HT:PCBM-based OSCs with and without a UV photoluminescent down-conversion layer for photon recycling. For comparison, the absorption spectrum, the photoluminescence spectrum of the UV photoluminescent down-conversion layer/glass, and the spectrum of the AM1.5G solar irradiance are shown. (Reproduced with permission.[37] Copyright 2012, AIP Publishing.)

Kimura et al.[38] developed a 1.0-μm-thick ultrathin transparent polyimide substrate with an almost complete UV (less than 350 nm) filtering properties for the application in flexible OSCs. The ultrathin flexible OSCs, based on a 2,5-bis(3-(2-ethylhexyl-5-(trimethylstannyl)thiophen-2-yl)thiazolo[5,4-d]thiazole-2-butyloctyl (PTzNTz-BOBO) and $PC_{71}BM$ blend system, had a PCE of 9.0%. The PCE was maintained at 90% after 3 h in a maximum power point tracking test, indicating much better operational stability than the reference OSCs prepared using the rigid glass substrate.

In order to overcome the UV stability of OSCs and to fully utilize sunlight, another approach was developed using an inorganic-perovskite/organic four-terminal tandem solar cell based on a semitransparent inorganic $CsPbBr_3$ perovskite solar cell (PSC) as the top cell and an OSC as the bottom cell, as illustrated in Figure 2.4.[39] The high-quality $CsPbBr_3$ photoactive layer of the planar PSC is prepared with a dual-source vacuum co-evaporation method, using stoichiometric precursors of CsBr and $PbBr_2$ with a low evaporation rate. The related semitransparent PSC can almost completely filter UV light and well maintain photovoltaic performance; it additionally shows an extremely high average visible transmittance. The resulting tandem solar cells with a bottom OSC based on PBDB-T-SF:IT-4F exhibited the best PCE of 14.03% and excellent photostability retention after 120 h of strong UV irradiation. Such a strategy can not only improve the photovoltaic

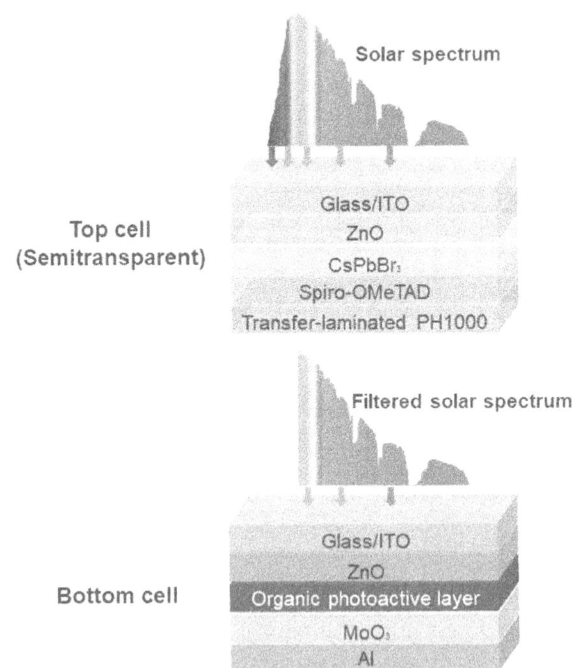

FIGURE 2.4 The cross-sectional view of a four-terminal tandem solar cell comprising a semitransparent top cell of glass/ITO/ZnO/$CsPbBr_3$/spiro-OMeTAD/transfer laminated PH1000 and a bottom OSC of glass/ITO/ZnO/organic photoactive layer/MoO_3/Al.[38]

performance but also overcome the UV-light stability problem of OSCs, with the PSC acting as a UV filter.

2.4 METAL NANOSTRUCTURED TRANSPARENT ELECTRODES

2.4.1 An Overview of Metal Nanostructured Transparent Electrodes

Transparent electrodes are crucial for various optoelectronic devices including OSCs, liquid-crystal displays, organic light-emitting diodes. Currently, the mainstream of transparent electrodes relies on the techniques of high vacuum processes, which are considered to be well compatible with the existing silicon-based device fabrication, and the materials of TCOs, including ITO and AZO. With the development of the next generation of optoelectronic devices, there is an increased and huge demand for the function of flexibility. As a necessary component, flexible transparent electrodes (FTEs) have thus shown potential and fascinating applications in future flexible optoelectronics due to the advantages of flexibility, low-cost, and large-scale manufacturing. Figure 2.5(a) shows the publications on FTEs from Web of Science during the past few decades.[40] Before 2009, there were very few reports on FTEs, and there has been a significant increase in the number of publications since 2011. A similar publication trend of FTEs used in organic electronic devices can also be seen from Figure 2.5(b).

FTEs are critical for high-performance flexible organic electronic devices. However, the development of flexible electronic devices still lags far behind that of the organic electronic devices due to the absence of ideal FTEs, which require excellent electrical conductivity, high transmittance, good mechanical flexibility, and long-term stability. A variety of FTEs were proposed to replace ITO such as metallic nanostructures (*e.g.*, nanowires, grids),[41,42] carbon-based materials (*e.g.*, nanotubes, graphene),[43,44] and conducting polymers.[45] Among them, a layer of silver nanowires (AgNWs) is one of the most promising FTE candidates for use in flexible optoelectronic devices because of its superior optical and electric properties.[46] However, AgNW-based FTEs still face some technical challenges: first, the

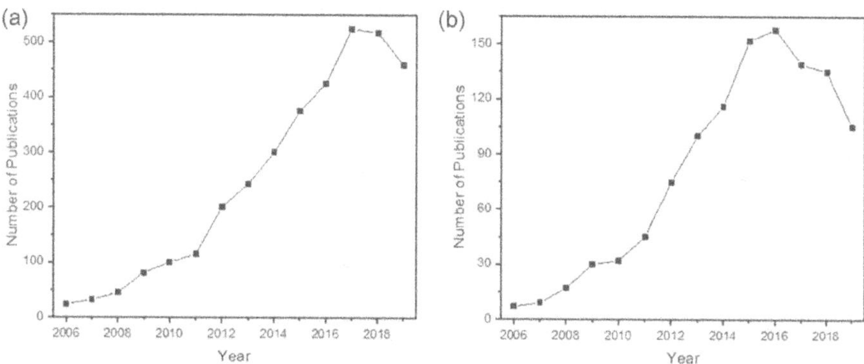

FIGURE 2.5 Numbers of the research publications on (a) FTEs and (b) applications of FTEs in organic electronic devices since 2016. (From Web of Science.)

randomly distributed AgNWs formed during the solution process result in a large surface roughness that may penetrate through the active layer, leading to electric shorts; second, percolation of charges through junctions between AgNWs results in large contact resistance. The tensile deformation also leads to the irreversible detachment of the NWs at the junctions. As a result, the production of a highly conductive and flexible AgNW-based electrode is still one of the open challenges for efficient OSCs. In parallel to enhance the operation stability of OSCs, the UV durability is another critical factor for applications in flexible OSCs. It is well known that OSCs are unstable under UV irradiation,[47,48] however, the development of UV-durable flexible nonfullerene OSCs has not yet been systemically studied.

2.4.2 Plasmonic Nano-Antennas

Since UV radiations may degrade the PCE and overall lifetime of solar cells, plasmonic band-stop filters with a periodic array of plasmonic cross-shaped nano-antennas have been designed, as it is shown in Figure 2.6.[49] Such filters are polarization-independent which make them suitable for photovoltaic utilization and their optical properties can be tuned by varying the size and periodicity of the antennas. By adjusting the antenna parameters, UV and IR cut-off filters with rigid and flexible substrates have been obtained. Plasmonic crossed-shaped nano-antenna arrays with SiO_2 substrate have blocked 79.6 and 65.2% of the incident wavelength from 300–400 nm and 1100-1800 nm, respectively. In order to investigate the proposed filters for flexible solar cell application, plasmonic filters with PET substrate have been designed. In this case, about 70.8 and 56.2% of the light in solar radiation, over the wavelength ranges from 300 to 400 nm and that from 650 to 1800 nm, are filtered. The proposed filters on rigid and flexible substrates could not only improve the power efficiency of the silicon and organic solar cells for long-term applications but also increase the lifetime due to the removal of heat and UV light.

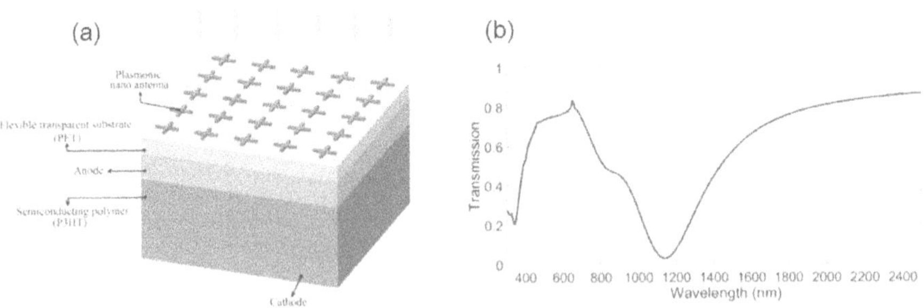

FIGURE 2.6 (a) The schematic diagram illustrating the incorporation of a front plasmonic UV cut-off filter in an OSC. (b) Transmission spectrum of the plasmonic UV cut-off filter from UV to IR wavelength range. (Reproduced with permission.[48] Copyright 2018, Elsevier.)

2.4.3 HYBRID NANOSTRUCTURED TRANSPARENT ELECTRODES

Recently, a novel solution-processable hybrid nanostructured FTE, comprising a mixture of 0D AgNPs, 1D AgNWs, and 2D exfoliated graphene (EG) sheets, has been demonstrated by Xu et al.[50,51] The schematic fabrication process of the hybrid nanostructured FTEs is shown in Figure 2.7(a). A layer of AgNWs:AgNPs was first coated on the poly(methyl methacrylate) (PMMA)-coated polyethylene terephthalate (PET) substrates, forming a network of AgNWs:AgNPs embedded in the PMMA, which was flattened by a post-mechanical pressing process. A layer of EG sheets was then overlaid on the surface of the AgNWs:AgNPs matrix forming a hybrid nanostructured FTE. AgNPs were added to fill in the empty spaces between the AgNWs,

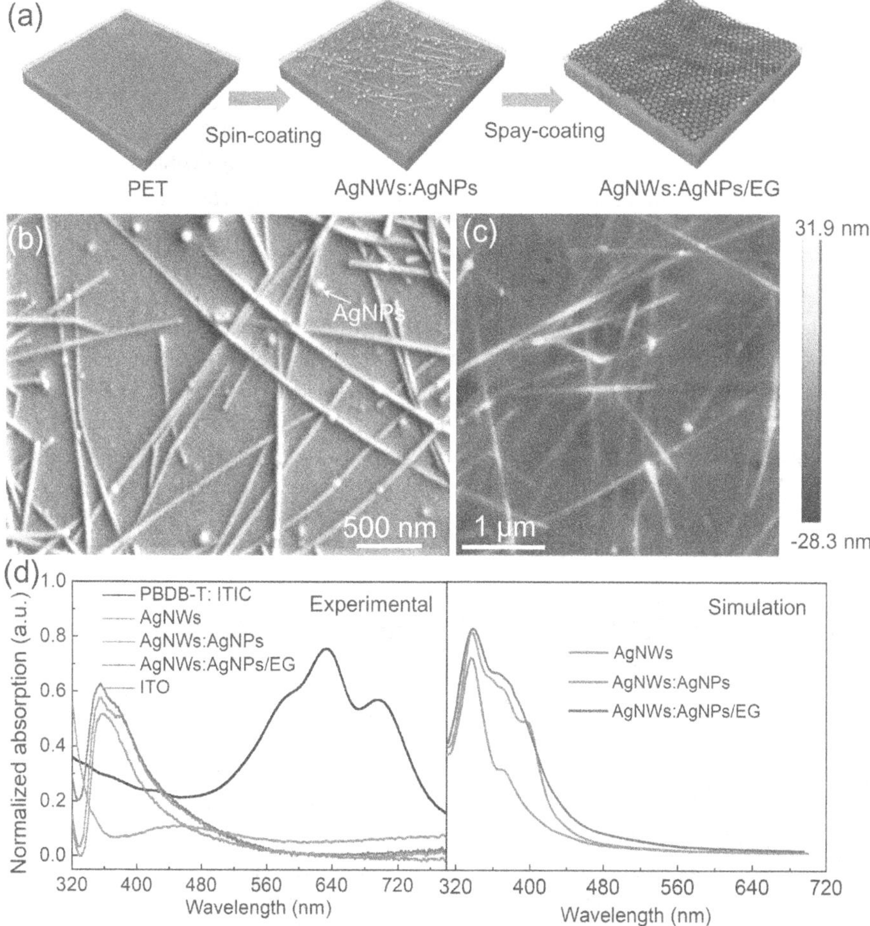

FIGURE 2.7 (a) Schematic fabrication process of AgNWs:AgNPs/EG FTE. (b) SEM and (c) AFM images measured for the surface of AgNWs:AgNPs/EG FTE. (d) Absorption spectra measured and calculated for different layers of AgNWs, AgNWs:AgNPs, AgNWs: AgNPs/EG.

leading to a decrease in the sheet resistances (R_s),[52] while the use of graphene sheets helps to reduce the surface roughness and improves the mechanical stability, taking advantage of the 2D graphene pathways for enhancing charge transfer and stretchability of graphene.[53,54] The scanning electron microscopy (SEM) and atomic force microscopy (AFM) images measured for the surface of the hybrid nanostructured FTEs are shown in Figure 2.7(b) and (c). The AFM and SEM measurements clearly revealed that the EG sheets were uniformly coated on the surface of the AgNWs:AgNPs interpenetrating networks. Additionally, the hybrid nanostructured FTE had a low sheet resistance of 23 Ω sq^{-1}, a high optical transparency of 82% over the visible light wavelength range, and a superior mechanical flexibility.

The absorption spectra measured for the different layers of AgNWs, AgNWs:AgNPs, AgNWs:AgNPs/EG, PBDB-T:ITIC, and ITO are shown in Figure 2.7(d). It becomes clear that the hybrid nanostructured AgNWs:AgNPs/EG FTE has a strong absorption over a short wavelength range with a peak located at ~360 nm, as compared with that of the ITO/glass substrate. The AgNWs:AgNPs/EG FTE has a weak absorption over the visible light wavelength range from 480 to 720 nm. The absorption profile of the AgNWs:AgNPs/EG FTE matches well with the absorption of the PBDB-T:ITIC blend layer. It allows the visible light passing through for maximum absorption in the PBDB-T:ITIC bulk BHJ layer, and prevents the UV part of the incident light from entering into the organic active layer. The UV absorption behaviors of the pristine AgNW, AgNW:AgNP, and AgNW:AgNPs/EG structures were analyzed using Comsol Multiphysics, as is shown in Figure 2.7(d). The calculated absorptions reveal that there is a slight increase in the absorption of the AgNW:AgNPs/EG structure, particularly over the wavelength range from 300 to 600 nm. The simulation supports the experimental results showing an improved UV stability of the flexible OSCs with an AgNW:AgNPs/EG-based FTE, due to the improved UV filtering effect as compared with the ones with the AgNW- and AgNW:AgNP-based FTEs. As a result, the use of the AgNWs:AgNPs/EG FTE could have an additional advantage, serving as a built-in UV filter for improving the UV durability of the flexible OSCs.

2.5 UV-DURABLE FLEXIBLE ORGANIC SOLAR CELLS

To investigate the feasibility of hybrid nanostructured FTEs for application in flexible OSCs, a set of flexible OSCs with a blend system of poly[(2,6-(4, 8-bis(5-(2-ethylhexyl)thiophen-2-yl)-benzo[1,2-b:4,5-b′]dithiophene))-alt-(5,5-(1′,3′-di-2-thienyl-5′,7′-bis(2-ethylhexyl)benzo[1′,2′-c:4′,5′-c′]dithiophene-4,8-dione)](PBDB-T):3,9-bis(2-methylene-(3-(1,1-dicyanomethylene)-indanone))-5,5,11,11-tetrakis(4-hexylphenyl)-dithieno[2,3-d:2′,3′-d′]-s-indaceno[1,2-b:5,6-b′] dithiophene (ITIC) was fabricated. An average PCE of 8.15% was obtained for the flexible OSCs with an AgNWs:AgNPs/EG FTE, along with a J_{sc} of 13.36 mA cm^{-2}, a V_{oc} of 0.89 V, and an FF of 68.55%. The mechanical flexibility of the flexible OSCs made with the AgNWs:AgNPs/EG FTE was studied by a bending test with a radius of 4.0 mm. The V_{oc}, J_{sc}, and FF of the flexible OSCs with an AgNWs:AgNPs/EG FTE can retain >96% of their initial values after a test of 1000 bends, which is much higher than that of the ones made with an AgNWs:AgNPs FTE (<60%). The flexible OSCs, made with an AgNWs:AgNPs/EG FTE, having an average PCE of >8% are clearly demonstrated.[50]

Most importantly, the flexible OSCs also possess an excellent UV durability as compared with the control OSC made with an ITO anode, taking advantage of the tailored FTE absorption in wavelength <380 nm, as shown in Figure 2.7(d). The UV durability of the flexible OSCs and a control OSC made with an ITO anode was analyzed by exposing the cells to a UV lamp, with a peak wavelength of 360 nm and an intensity of 200 W/m^2, under different UV exposure times. To avoid the possible cell degradation due the moisture and oxygen encroachment, the accelerated UV-durable tests were conducted in the N_2-purged glove box with O_2 and H_2O levels <0.1 ppm. The variation in PCE measured for the flexible OSCs under different UV exposure times is compared to that of the control cell, and the results are plotted in Figure 2.8.

Although the ITO-based control cell had a higher initial PCE, a fast decrease in its V_{oc}, J_{sc}, FF, and PCE was observed after a 60-min UV exposure. For example, a loss in V_{oc} from 0.89 V to 0.84 V, a decrease in J_{sc} from 15.45 mA cm^{-2} to 10.37 mA cm^{-2}, and a reduction in FF from 68.1% to 57.5 % were observed, leading to a 47% drop in PCE from 9.7% to 5%. A slower degradation in the performance of the flexible OSCs with an AgNWs:AgNPs/EG FTE, aged under the same UV exposure condition, has been observed, for example, a mild drop in V_{oc} from 0.89 V to 0.87 V, a smaller decrease in J_{sc} from 13.4 mA cm^{-2} to 11.65 mA cm^{-2}, and a lesser reduction in FF from 67.7% to 61.2%, resulting in a smaller decrease of 23% in PCE from 8.15% to 6.20%. It is clear that the flexible OSCs have a superior UV durability as compared with that of a control OSC made with an ITO contact. The enhancement in the UV durability of the flexible OSCs is attributed to the tailored

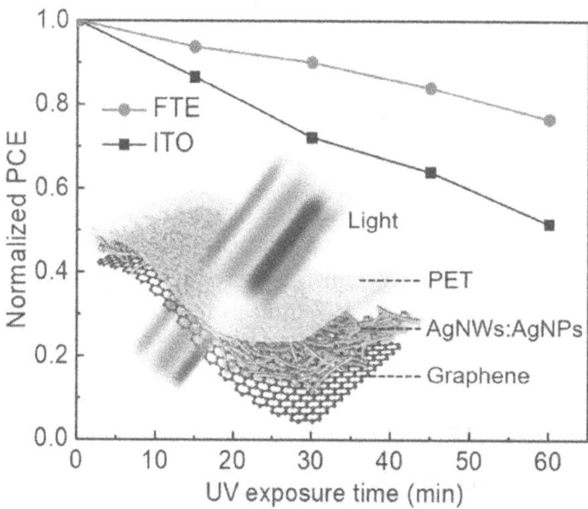

FIGURE 2.8 Evolution of PCE measured for a flexible OSC with an AgNWs:AgNPs/EG FTE and a control cell with an ITO anode as a function of the UV exposure time. Inset: schematic view of the FTE filtering the UV portion of the incoming light.

absorption of the AgNWs:AgNPs/EG FTE over the short wavelength range. Apart from the superior mechanical flexibility, electric conductivity, and solution fabrication capability, the use of the solution-processable FTE aids in further benefit on its tailored absorption in wavelength <380 nm, offering an exciting option for the application in low-cost large area UV-durable flexible OSCs.

2.6 CONCLUSIONS

OSCs have attracted worldwide attention due to their potential of large area solution-fabrication capability at low-cost. ITO-coated glass is one of the widely used front electrodes for application in OSCs and other optoelectronic devices. However, the ITO layer is very brittle and is not a perfect solution for use in flexible OSCs. This aside, indium is also not an abundant element on Earth. The relatively high cost of the ITO/glass substrate is not an ultimate choice for the commercialization of the flexible OSCs at a low cost. The ITO layer does not have a good UV filtering effect, making the ITO-based OSCs have a relatively poorer stability upon UV exposure. UV-absorbing metal nanostructured transparent electrode and AZO have relatively lower transmission over the UV light wavelength range, a promising ITO alternative due to the advantages of high electric conductivity, optical transparency, nontoxicity, and low cost. In this chapter, we have discussed different approaches developed for enhancing the UV stability of OSCs, including the use of UV-cut filter, metal oxide electrodes, and nanostructured electrodes that absorb UV light serving as both ITO-free contact and internal built-in UV filter.

The properties of the UV-filtering layer or UV-absorbing visible light transparent electrodes including optical transparency, sheet resistance, surface morphology, and surface work function are discussed. It shows that that OSCs with a built-in UV filter have a superior UV stability than the ones with conventional ITO/glass-based OSCs. The presence of the built-in UV filter allows the suppressing of the undesired UV absorption in the functional organic layers in OSCs. Excellent UV-durable ITO-free flexible OSCs are realized by incorporating a hybrid nanostructured flexible transparent electrode with a tailored absorption in wavelength <380 nm. A hybrid nanostructured FTE, comprising a mixture of AgNWs and exfoliated graphene sheets, has been developed. The FTE possesses high optical transparency and electric conductivity, good air stability, full-solution fabrication capability at a low processing temperature. The FTE provides a promising alternative to ITO for use in UV-durable flexible OSCs, serving as a built-in UV filter to impede an inevitable UV-induced degradation in ITO-based OSCs.

ACKNOWLEDGMENTS

This work was supported by National Natural Science Foundation of China (12174244), Natural Scientific Foundation of Shanghai (19ZR1419500), and Research Grants Council, University Grants Committee, Hong Kong Special Administrative Region, China (12302419, C5037-18GF, and N_HKBU201/19).

REFERENCES

1. Q. D. Ou, Y. Q. Li, and J. X. Tang. "Light manipulation in organic photovoltaics." *Adv. Sci.*, 3 (2016): 1600123.
2. J. D. Chen, C. H. Cui, Y. Q. Li, L. Zhou, and Q. D. Ou. "Single-junction polymer solar cells exceeding 10% power conversion efficiency." *Adv. Mater.*, 7 (2014): 1035–1041.
3. C. C. Chen, L. Dou, R. Zhu, C. H. Chung, and T. B. Song. "Visibly transparent polymer solar cells produced by solution processing." *ACS Nano*, 6 (2012): 7185–7190.
4. J. Zhang, H. S. Tan, X. Guo, A. Facchetti, and H. Yan. "Material insights and challenges for non-fullerene organic solar cells based on small molecular acceptors." *Nat. Energy*, 3 (2018) : 720–731.
5. T. Ameri, P. Khoram, J. Min, and C. J. Brabec. "Organic ternary solar cells: a review." *Adv. Mater.*, 25 (2013): 4245–4266.
6. L. Duan, Y. Zhang, H. Yi, F. Haque, R. Deng, H. Guan, Y. Zou, and et al. "Tradeoff between exciton dissociation and carrier recombination and dielectric properties in Y6-sensitized nonfullerene ternary organic solar cells." *Energy Technol*, 8 (2019): 1900924.
7. Q. D. Ou, H. J. Xie, J. D. Chen, and L. Zhou. "Enhanced light harvesting in flexible polymer solar cells: synergistic simulation of a plasmonic meta-mirror and a transparent silver mesowire electrode." *J. Mater. Chem. A*, 4 (2016): 18952–18962.
8. J. Yuan, Y. Zhang, L. Zhou, G. Zhang, H. L. Yip, T. K. Lau, X. Lu, and et al. "Single-junction rganic solar cell with over 15% efficiency using fused-ring acceptor with electron-deficient core." *Joule*, 3 (2019): 1140–1151.
9. L. Hong, H. Yao, Z. Wu, Y. Cui, T. Zhang, Y. Xu, R. Yu, and et al. "Eco-compatible solvent-processed organic photovoltaic cells with over 16% efficiency." *Adv. Mater.*, 31 (2019): 1903441.
10. L. Meng, Y. Zhang, X. Wan, C. Li, X. Zhang, Y. Wang, X. Ke, and et al. "Organic and solution-processed tandem solar cells with 17.3% efficiency." *Science*, 361 (2018): 1094–1098.
11. Y. Cui, H. Yao, J. Zhang, K. Xian, T. Zhang, L. Hong, Y. Wang, and et al. "Single-junction organic photovoltaic cells with approaching 18% efficiency." *Adv. Mater.*, 32 (2020): 1908205.
12. P. Bi, S. Zhang, Z. Chen, Y. Xu, Y. Cui, T. Zhang, J. Ren, and et al. "Reduced non-radiative charge recombination enables organic photovoltaic cell approaching 19% efficiency." *Joule*, ISSN (2021): 2542–4351.
13. H. Kang, G. Kim, J. Kim, S. Kwon, H. Kim, and K. Lee. "Bulk-eterojunction organic solar cells: Five core technologies for their commercialization." *Adv. Mater.*, 28 (2016): 7821–7861.
14. P. Cheng, and X. Zhan, Chem. "Stability of organic solar cells: challenges and strategies." *Soc. Rev.*, 45 (2016): 2544–2588.
15. W. Lan, J. Gu, S. Wu, Y. Peng, M. Zhao, Y. Liao, T. Xu, and et al. "Toward improved stability of nonfullerene organic solar cells: impact of interlayer and built-in potential." *EcoMat.*, 3 (2021): e12134.
16. S. W. Wu, Y. L. Zhao, C. Wang, S. Li, R. Bachelot, B. Wei, and T. Xu. "Sub-nanometer atomic layer deposited Al_2O_3 barrier layer for improving stability of nonfullerene organic solar cells." *Org. Electron.*, 99 (2021): 106351.
17. L. Duan, N. K. Elumalai, Y. Zhang, and A. Uddin. "Progress in non-fullerene acceptor based organic solar cells." *Sol. Energy Mater. Sol. Cells*, 193 (2019): 22–65.
18. H. Hintz, C. Sessler, H. Peisert, H. J. Egelhaaf, and T. Chasse. "Wavelength-dependent pathways of poly-3-hexylthiophene photo-oxidation." *Chem. Mater.*, 24 (2012): 2739–2743.

19. A. Tournebize, P. O. Bussiere, A. Rivaton, J. L. Gardette, H. Medlej, R. C. Hiorns, C. Dagron-Lartigau, and et al. "New insights into the mechanisms of photodegradation/stabilization of P3HT:PCBM active layers using Poly(3-hexyl-d13-Thiophene)." *Chem. Mater.*, 25 (2013): 4522–4528.
20. H. Kimura, K. Fukuda, H. Jinno, S. Park, M. Saito, I. Osaka, K. Takimiya, and et al. "High operation stability of ultraflexible organic solar cells with ultraviolet-filtering substrates." *Adv. Mater.*, 31 (2019): 1808033.
21. Q. Wan, X. Guo, Z. Y. Wang, W. Li, B. Guo, W. Ma, M. J. Zhang, and et al. "10.8% efficiency polymer solar cells based on PTB7—Th and PC71BM via binary solvent additives treatment." *Adv. Mater.*, 26 (2015): 6635–6640.
22. J. Jeong, J. Seo, S. Nam, H. Han, and H Kim. "Significant stability enhancement in high-efficiency polymer:fullerene bulk heterojunction solar cells by blocking ultraviolet photons from solar light." *Adv. Sci.*, 3 (2015): 1500269.
23. N. Y. Doumon, G. Wang, X. Qiu, A. J. Minnaard, R. C. Chiechi, and L. J. A. Koster. "1, 8-diiodooctane acts as a photo-acid in organic solar cells." *Sci. Rep.*, 9 (2019): 4350.
24. Q. Liu, J. Toudert, F. Liu, P. Mantilla-Perez, M. M. Bajo, T. P. Russell, and J. Martorell. "Circumventing UV light induced nanomorphology disorder to achieve long lifetime PTB7-Th: PCBM based solar cells." *Adv. Energy Mater.*, 7 (2017): 1701201.
25. J. Jeong, J. Seo, S. Nam, H. Han, H. Kim, and T. D. Anthopoulos. "Bradley and Youngkyoo Kim. Significant stability enhancement in high-efficiency polymer:fullerene bulk heterojunction solar cells by blocking ultraviolet photons from solar light." *Adv. Sci.*, 3 (2016): 1500269.
26. A. Larkin, J. D. Haigh, and S. Djavidnia. "The effect of solar UV irradiance variations on the Earth's atmosphere." *Space Sci. Rev.*, 94 (2000): 199214.
27. C. N. Weng, H. C. Yang, C. Y. Tsai, S. H. Chen, Y. S. Chen, C. H. Chen, K. M. Huang, and et al. "The influence of UV filter and Al/Ag moisture barrier layer on the outdoor stability of polymer solar cells." *Solar Energy*, 199 (2020): 308–316.
28. D. S. Ginley, H. Hosono, and D. C. Paine. *Handbook of Transparent Conductors*. Springer, 2010.
29. K. Bädeke. "Über die elektrische Leitfähigkeit und die thermoelektrische Kraft einiger Schwermetallverbindungen." *Ann. Phys.*, 22 (1907): 749–766.
30. H. J. Vanboort, and R. Groth. "Low-pressure sodium lamps with indium oxide filters." *Phil. Tech. Rev.*, 29 (1968): 47–48.
31. H. Köstlin, R. Jost, W. Lems. "Optical and electrical properties of doped In_2O_3 films." *Phys. Stat. Sol. A*, 29 (1975): 87–93.
32. D. L. White, and M. Feldman. "Liquid-crystal light valves." *Electron. Lett.*, 6 (1970): 837–839.
33. M. Katayama. "TFT-LCD technology." *Thin Solid Films*, 341 (1999): 140–147.
34. H. X. Liu, Z. H. Wu, J. Q. Hu, Q. L. Song, B. Wu, H. L. Tam, Q. Y. Yang, and et al. "Efficient and ultraviolet durable inverted organic solar cells based on an aluminum-doped zinc oxide transparent cathode." *Appl. Phys. Lett.*, 103 (2013): 043309.
35. T. L. Chen, R. Betancur, D. S. Ghosh, J. Martorell, and V. Pruneri. "Efficient polymer solar cell employing an oxidized Ni capped Al:ZnO anode without the need of additional hole-transporting-layer." *Appl. Phys. Lett.*, 100 (2012): 013310.
36. H. Luo, X. Jia, J. Wang, J. Zhou, Z. Jiang, L. Pan, S. Huang, and et al. "Efficient and ultraviolet durable inverted polymer solar cells using thermal stable GZO-AgTi-GZO multilayers as a transparent electrode." *Org. Electron.*, 39 (2016): 177–183.
37. S. Engmann, M, Machalett, V. Turkovic, R. Rosch, E. Radlein, G. Gobsch, and H. Hoppe. "Photon recycling across a ultraviolet-blocking layer by luminescence in polymer solar cells" *J. Appl. Phys.*, 112 (2012): 034517.

38. H. Kimura, K. Fukuda, H. Jinno, S. Park, M. Saito, I. Osaka, K. Takimiya, S. Umezu, and T. Someya. "High operation stability of ultraflexible organic solar cells with ultraviolet-filtering substrates." *Adv. Mater.*, 32 (2019): 1808033.
39. W. J. Chen, J. W. Zhang, G. Y. Xu, R. M. Xue, Y. W. Li, Y. H. Zhou, J. H. Hou, and et al. "A semitransparent inorganic perovskite film for overcoming ultraviolet light instability of organic solar cells and achieving 14.03% efficiency." *Adv. Mater.*, 30 (2018): 1800855.
40. W. T. Li, H. Zhang, S. W. Shi, J. X. Xu, X. Qin, Q. Q. He, K. Yang, and et al. "Silva and Mats Fahlman. Recent progress in silver nanowire networks for flexible organic electronics." *J. Mater. Chem. C*, 8 (2020): 4636–4674.
41. J. Y. Lee, S. T. Connor, Y. Cui, and P. Peumans. "Solution-processed metal nanowire mesh transparent electrodes." *Nano Lett.*, 8 (2008): 689–692.
42. P. Lee, J. Lee, H. Lee, J. Yeo, S. Hong, K. H. Nam, D. Lee, and et al. "Highly stretchable and highly conductive metal electrode by very long metal nanowire percolation network." *Adv. Mater.*, 24 (2012): 3326–3332.
43. D. S. Hecht, L. Hu, and G. Irvin. "Emerging transparent electrodes based on thin films of carbon nanotubes, graphene, and metallic nanostructures." *Adv. Mater.*, 23 (2011): 1482–1513.
44. G. D. M. R. Dabera, K. D. G. I. Jayawardena, M. R. R. Prabhath, I. Yahya, Y. Y. Tan, N. A. Nismy, H. Shiozawa, and et al. "Hybrid carbon nanotube networks as efficient hole extraction layers for organic photovoltaics." *ACS Nano*, 7 (2013): 556–565.
45. S. Wang, Y. Zhao, H. Lian, C. Peng, X. Yang, Y. Gao, Y. Peng, and et al. "Towards all-solution-processed top-illuminated flexible organic solar cells using ultrathin Ag-modified graphite-coated polyethylene terephthalate substrates." *Nanophotonics*, 8 (2019): 297.
46. Y. Xu, and J. Liu. "Graphene as transparent electrodes: fabrication and new emerging applications." *Small*, 12 (2016): 1400–1419.
47. H. X. Liu, Z. H. Wu, J. Q. Hu, Q. L. Song, B. Wu, H. L. Tam, Q. Y. Yang, and et al. "Efficient and ultraviolet durable inverted organic solar cells based on an aluminum-doped zinc oxide transparent cathode." *Appl. Phys. Lett.*, 103 (2013): 043309.
48. X. Z. Wang, C. X. Zhao, G. Xu, Z. K. Chen, and F. R. Zhu. "Degradation mechanisms in organic solar cells: Localized moisture encroachment and cathode reaction." *Sol. Energy Mater. Sol. Cells*, 104 (2012): 1–6.
49. V. Khoshdel, M. Joodaki, and M. Shokooh-Saremi. "UV and IR cut-off filters based on plasmonic crossed-shaped nano-antennas for solar cell applications." *Opt. Commun.*, 433 (2019): 275–282.
50. T. Xu, C. L. Gong, S. L. Wang, H. Lian, W. X. Lan, G. Lévêque, B. Grandidier, and et al. "Ultraviolet-durable flexible nonfullerene organic solar cells realized by a hybrid nanostructured transparent electrode." *Sol. RRL*, 4 (2020): 1900522.
51. S. Wu, Y. Li, H. Lian, G. Lévêque, B. Grandidier, P. Adam, D. Gérard, and et al. "Hybrid nanostructured plasmonic electrodes for flexible organic light-emitting diodes." *Nanotechnology.*, 31 (2020): 375203.
52. T. Kim, S. Kang, J. Heo, S. Cho, J. Kim, A. Choe, B. Walker, and et al. "Nanoparticle-enhanced silver-nanowire plasmonic electrodes for high-performance organic optoelectronic devices." *Adv. Mater.*, 30 (2018): 1800659.
53. A. G. Ricciardulli, S. Yang, G. A. H Wetzelaer, X. Feng, and P. W. M. Blom. "Hybrid silver nanowire and graphene-based solution-processed transparent electrode for organic optoelectronics." *Adv. Funct. Mater.*, 28 (2018): 1706010.
54. N. Ye, T. Liang, L. Zhan, Y. Kong, S. Xie, X. Ma, H. Chen, and et al. "High-performance bendable organic solar cells with silver nanowire-graphene hybrid electrode." *IEEE J. Photovolt.*, 9 (2019): 214–219.

3 Absorption Enhancement in Photonic-Structured Organic Solar Cells

Weixia Lan[a] and Furong Zhu[b]
[a]Shanghai University, Shanghai, People's Republic of China
[b]Hong Kong Baptist University, Hong Kong, People's Republic of China

CONTENTS

3.1 INTRODUCTION

The energy consumption is growing tremendously to meet the demand for the development of industrialization, technology innovation, and improvement of our life quality. Renewable energy resources play an important role in meeting the growing energy demands and achieving sustainable development of the society. The deployment of renewable energy technologies has attracted interests for the sustainable

DOI: 10.1201/9781003141358-3

development of our society, as the natural resources, e.g., fossil fuel, are not abundant on the earth. Solar energy, as one of the inexhaustible and environmentally friendly energy sources, is attracting increasingly more attention.

Conjugated polymer-based organic solar cells (OSCs) are attractive alternative to the traditional inorganic semiconductor-based solar cells due to the low-cost solution fabrication processes.[1–3] However, despite the rapid progresses made in the development of new donor and accepter materials, the relatively poor efficiency and stability of OSCs are still limitations on the potential applications. The performance of OSCs is primarily hampered by the limited light absorption, caused by the mismatch between light absorption depth and carrier transport scale, low carrier mobility, and unstable electrode/organic interfacial properties.[4,5] There are two major research approaches toward improving the performance of OSCs. One attention has been paid on the development of low bandgap conjugated organic polymers and nonfullerene acceptors whereby to increase power conversion efficiency (PCE) of OSCs through improved spectral response over the long wavelength region. The other is the device approach in which the enhanced absorption in OSCs is achieved using light-trapping effects. Light absorption enhancement in OSCs based on different light-trapping features has been proposed. For example, metallic nanostructures inducing surface plasmon polaritons (SPPs) and localized surface plasmons (LSPs) are demonstrated for absorption enhancement without increasing thickness of active layer. Incorporation of photonic structures in the organic photoactive layers often results in absorption enhancement in OSCs. Theoretical simulation predicts that enhanced absorption in the photonic structures OSCs can be realized via photonic waveguide modes in the organic photoactive layer.[6,7] The unique design flexibility and features of the photonic structures also add a decorative and aesthetic dimension to the OSCs that can be prepared using curved and flexible substrates, which cannot be done using traditional rigid silicon solar cells.

The introduction of photonic structures could effectively solve these two problems. Photonic structure was first proposed in 1987, which was formed by periodic arrangement of materials with different refractive index in space. Due to the influence of Bragg scattering, the propagation of photons in photonic structures is limited.[8] The band limited by photon propagation is called photonic band gap, or photonic band gap. The existence of photonic band gap limits the propagation of photons in a specific band.[9] The optical bandgap of the photonic-structured materials can be adjusted artificially through the designing of the photonic structures.[10] This realizes the on-demand regulation of photons and light fields in photonic-structured devices using one-, two-, or three-dimension photonic structures, as shown in Figure 3.1. One-dimensional (1-D) photonic structures can be made of layers deposited or stuck together. Two-dimensional (2-D) ones can be made by photolithography, or by drilling holes in a suitable substrate. Fabrication methods for three-dimensional (3-D) ones include drilling under different angles, stacking multiple 2-D layers on top of each other, direct laser writing, or, for example, instigating self-assembly of spheres in a matrix and dissolving the spheres.

One-dimensional (1-D) photonic structures can be made of layers deposited or stuck together. Two-dimensional (2-D) ones can be made by photolithography, or by drilling holes in a suitable substrate. Fabrication methods for three-dimensional

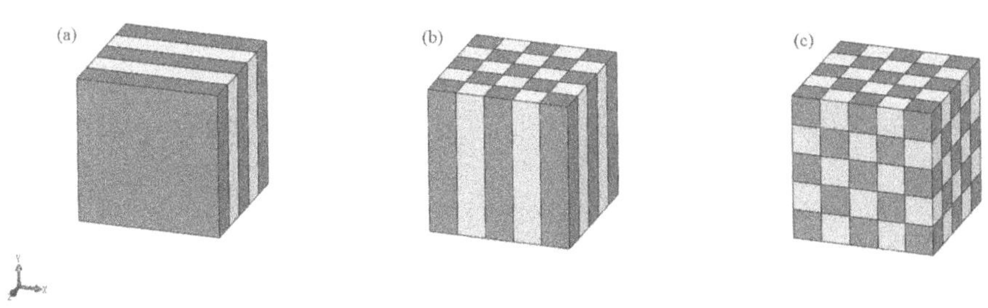

FIGURE 3.1 Schematic diagram of (a) one-dimensional, (b) two-dimensional, (c) three-dimensional photonic structures.

(3-D) ones include drilling under different angles, stacking multiple 2-D layers on top of each other, direct laser writing, or, for example, instigating self-assembly of spheres in a matrix and dissolving the spheres.

3.2 APPLICATION OF 1-D PHOTONIC STRUCTURES IN ORGANIC SOLAR CELLS

3.2.1 Light Absorption in Planar Organic Solar Cells

The optical performance of the planar OSCs with a layer configuration of indium tin oxide (ITO)/ZnO/poly[4,8-bis[(2-ethylhexyl)oxy]benzo[1,2-b:4,5-bA] dithiophene-2, 6-diyl][3-fluoro2-[(2- ethylhexyl) carbonyl]thieno[3,4-b]-thiophenediyl] (PTB7): [6,6]-phenyl-C71-butyric acid methyl ester ($PC_{71}BM$)/MoO_3/Ag was analyzed. The ITO/glass substrates, with a sheet resistance of 10 Ω/square, were cleaned by ultrasonication sequentially with detergent, deionized water, acetone, and isopropanol each for 20 min. A 10-nm-thick ZnO electron extraction layer (EEL) was then fabricated on ITO/glass by spin coating inside a N_2-purged glovebox with O_2 and H_2O levels less than 0.1 ppm. A PTB7:$PC_{71}BM$ bulk heterojunction layer was then deposited on ZnO-modified ITO/glass substrates by spin coating inside the glovebox. Samples were finally transferred to an adjacent vacuum evaporator, with a base pressure of 5.0×10^{-5} Pa, for the deposition of a 2-nm-thick MoO_3 hole extraction layer (HEL) and a 100-nm-thick Ag top contact.

To achieve the best performance, the integrated absorbance of the active layer is first calculated using the following equation:[11,12]

$$\bar{A} = \frac{\int A(\lambda) \cdot \Phi(\lambda)\, d\lambda}{\int \Phi(\lambda)\, d\lambda}, \tag{3.1}$$

where $\Phi(\lambda)$ is the flux of AM 1.5G solar irradiation (in $W{\cdot}m^{-2}{\cdot}nm^{-1}$), $A(\lambda)$ is the light absorption in the active layer calculated using finite-difference time-domain (FDTD) simulation. The integrated absorbance of the active layer as a function of its layer thickness over the active layer thickness range from 10 to 150 nm is shown in

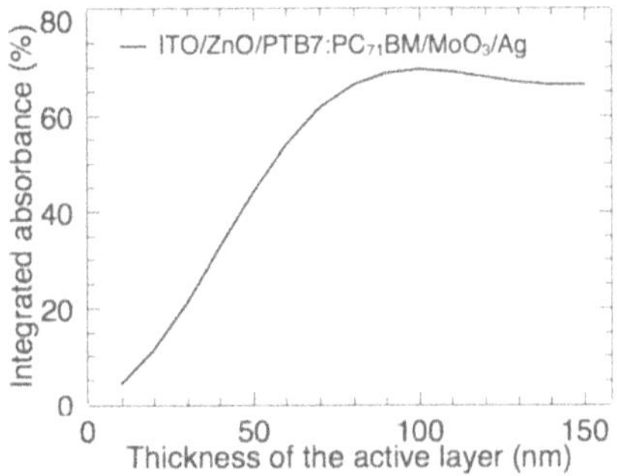

FIGURE 3.2 Relationship between the integrated absorbance of the PTB7:PC_{71}BM active layer as a function of its layer thickness in a planar OSC.

Figure 3.2. The results allow us to optimize the thickness of the active layer through maximizing the integrated absorbance. Figure 3.2 indicates the oscillation behavior relationship between the integrated absorbance and the active layer thickness. It is clear that the maximum absorption occurs at the active layer thickness of 90–100 nm for the PTB7:PC_{71}BM-based planar OSC.

The electric field distribution $|E|$ in the planar device is analyzed using the FDTD simulation. The $|E|$ profiles of the planar device at different wavelengths of 379 nm, 455 nm, and 842 nm are illustrated in Figure 3.3. It can be seen that the electric field distributions in the active layer vary with the different wavelengths of incident light. As the intensity of light is in direct proportion to $|E|^2$, higher electric field distribution indicates higher light intensity. It would be beneficial for efficient light absorption in OSCs if the highest electric field distribution locates exactly within the active layer. The results in Figure 3.3 reveal clearly that the electric field distribution in the active layer is relatively low for the planar device. The location of highest electric field distributions is outside of the active layer at the wavelengths of 379 nm and 455 nm, expanding to the ZnO EEL and the ITO layer that would not contribute to the absorption enhancement in the OSCs.

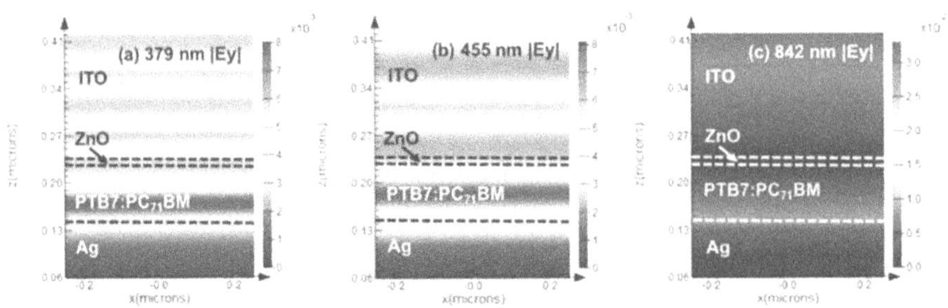

FIGURE 3.3 Simulated electric ($|E|$) field distributions in the PTB7:PC_{71}BM-based planar OSCs at different wavelengths of (a) 379 nm, (b) 455nm, and (c) 842 nm.

3.2.2 Light Trapping in 1-D Photonic-Structured Organic Solar Cells

1-D photonic structures are widely used in OSCs, especially for semi-transparent devices, due to the flexible design and the adaptivity to the large area manufacturability.[13] By tuning the photonic band gap, 1-D photonic structures also enable to improve the light absorption over a desired wavelength range, thus enhancing the optical absorption capacity of the active layer for a high *PCE* and a high short circuit current density (J_{SC}).

1-D photonic structures with a layer-by-layer structure of $(WO_3/LiF)^N$, where N is the periodic number, were proposed.[11] When applied into OSCs, the center wavelength of $(WO_3/LiF)^N$ was set to match the absorption spectrum of the active layer. In devices with poly[(4,4′-bis(2-ethylhexyl) dithieno[3,2-b:2′,3′-d]silole)-2,6-diyl-alt-(2,1,3 benzothiadiazole)-4,7-diyl] (PSBTBT):$PC_{71}BM$ bulk heterojunction, the absorption peak is about 580 nm to 780 nm; thus, the thickness of WO_3 and LiF in each layer was calculated to be 81.6 nm and 118.3 nm, respectively, based on equation (3.1) using the following equation:[11]

$$d = n_A d_A + n_B d_B = \frac{\lambda_0}{2} X, \tag{3.2}$$

where n_A and n_B are the refractive index of materials A and B, d_A and d_B are the corresponding material thickness, λ_0 is the central wavelength of the photonic band gap; X is an adjustable parameter and generally taken as 1 during the calculation.

As shown in Figure 3.4(b), the reflectivity of $(WO_3/LiF)^N$ increases with the number of the WO_3/LiF bilayer, and the transmittance decreases with the increase of N. When $N = 8$, the reflectivity of $(WO_3/LiF)^8$ is close to 100% in the photonic band gap of 580 nm to 780 nm, which is consistent with the absorption spectrum of the active layer. A summary of the results of ST-OSCs prepared without and with $(WO_3/LiF)^N$ is listed in Table 3.1. The ST-OSCs with $(WO_3/LiF)^8$ achieved the highest *PCE* and J_{SC}, which are 28.1% and 31.67% higher than those without using 1-D photonic structures.

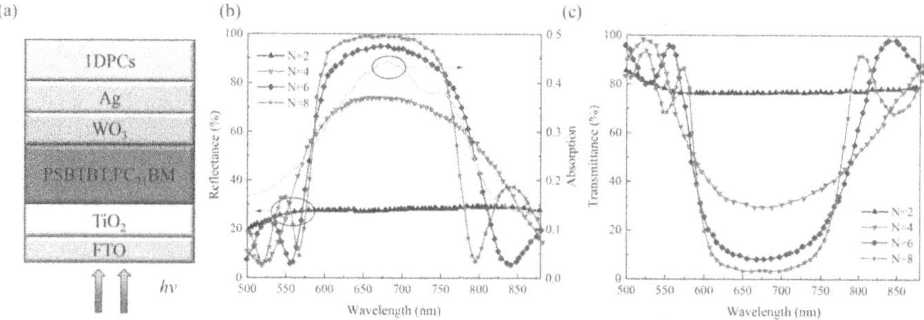

FIGURE 3.4 (a) The structure of ST-OSC with $(WO_3/LiF)^N$, (b) reflection spectrum and (c) transmission spectrum of $(WO_3/LiF)^N$[11].

TABLE 3.1
A Summary of the Performance of the Control ST-OSCs and ST-OSCs with the $(WO_3/LiF)^{N}$[11] Photonic Structures

Device Structure	J_{SC} (mA/cm²)	V_{OC} (V)	*FF* (%)	*PCE* (%)
Control cell	6.00	0.64	50.0	1.92
$(WO_3/LiF)^2$	6.39	0.64	50.1	2.05
$(WO_3/LiF)^4$	7.01	0.64	50.4	2.26
$(WO_3/LiF)^6$	7.51	0.64	48.7	2.34
$(WO_3/LiF)^8$	7.90	0.64	48.7	2.46

The improvement in the performance of the OSCs with the 1-D $(WO_3/LiF)^N$ photonic structure is due to an enhanced absorption that matches with the photoactive layer used in the device. With the photonic band gap designed to match the PSBTBT:PC_{71}BM absorption spectrum, the propagation of incident light from 580 nm to 780 nm is limited and reflected back into the active layer, enabled by the 1-D $(WO_3/LiF)^N$ photonic structure. The reabsorption of photons enhances the light absorption ability of the active layer, thus increasing the *PCE* and J_{SC}. 1-D photonic structures can also be applied to tune the transmission spectrum of ST-OSC and enhance the color diversity of the integrated photovoltaic industry. However, the see-though colors of ST-OSC can only be adjusted in the range of red and neutral colors for single-layer 1-D photonic structures prepared by using traditional materials, such as MoO_3, WO_3, LiF, and TiO_2.[14,15] Liang et al. successfully obtained ST-OSC devices with green transmission color by using stacked 1-D photonic structures without reducing other color ranges.[16] The structure of ST-OSC with 1-D photonic structures is shown in Figure 3.5. The photonic structures consist of two parts: the top 1-D photonic structures and the bottom 1-D photonic structures. Both of them are prepared by TiO_2 and SiO_2. The central wavelength of the photonic band gap of the two kinds of photonic structures is complementary. With these calculated data as inputs, the photovoltaic conversion efficiency and the see-though colors are calculated.

When the center wavelengths of the top and bottom photonic structures are set at 360 nm and 580 nm respectively, ST-OSC displays a blue color. ST-OSC exhibits a green color when the center wavelengths of the top and bottom photonic structures are selected at 400 nm and 650 nm, respectively. ST-OSCs can also have a red color

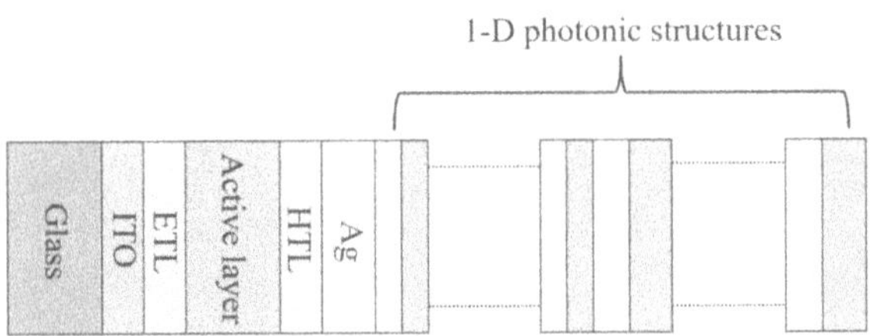

FIGURE 3.5 ST-OSC with 1-D photonic structures.

when the central wavelengths of the top and bottom photonic structures are chosen at 470 nm and 760 nm, respectively.

3.3 REALIZATION OF ABSORPTION ENHANCEMENT IN ORGANIC SOLAR CELLS WITH 2-D PHOTONIC STRUCTURES

3.3.1 LIGHT TRAPPING IN 2-D PHOTONIC-STRUCTURED ORGANIC SOLAR CELLS

PTB7:$PC_{71}BM$-based OSCs incorporating 2-D photonic structures in the active layer, prepared by nanoimprinting using a PDMS mold, have been reported.[6] The height of the photonic structures in the active layer was optimized by controlling the mold pressure during the imprinting process for achieving high efficiency. The imprinting process was carried out at room temperature with a duration of 5 min. After the mold was removed, samples were then transferred to an adjacent vacuum evaporator for the deposition of the 2-nm-thick MoO_3 anode interlayer and 100-nm-thick Ag top contact.

The schematic 3-D drawing of the photonic-structured OSCs is shown in Figure 3.6(a). Figure 3.6(b) reveals the scanning electron microscopy (SEM) image measured for the nano-structured active layer. The SEM result confirms the creation of the hexagonally arranged periodic structures in the PTB7:$PC_{71}BM$ active layer, formed by the nano-imprinting process. Considering the wettability and the capillary effect during the cell fabrication via the one-step imprint process, a mold with a periodicity of ~480 nm was used. Apart from the periodicity of the pattern, the structure depth was found to be another important factor on the device performance. For the same mold, photonic structure with different depths, which can be controlled by the pressing pressure, had a great impact on the performance of the cells. The photonic-structured OSCs with different structure heights in the active layer were made, along with the control planar cells for comparison studies.

Atomic force microscopy (AFM) images measured for the top surface of the PTB7:$PC_{71}BM$ layer, formed by the imprinting process with a pressure of 9.68 kPa, and the Ag cathode surface of the photonic-structured cells are shown in

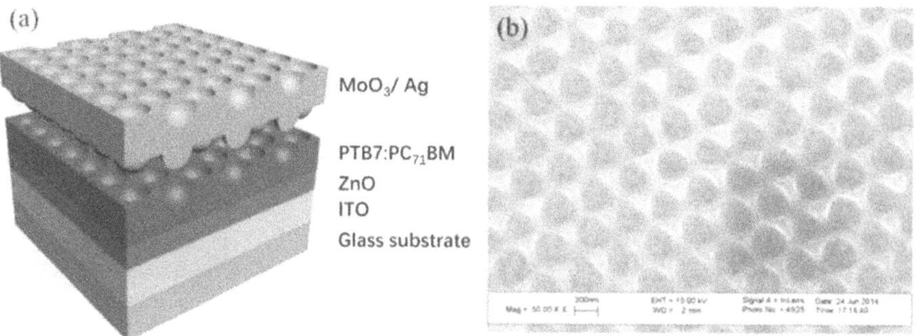

FIGURE 3.6 (a) Schematic diagram of a 2-D photonic-structured OSC, (b) SEM image measured for the nano-imprinted PTB7:$PC_{71}BM$ layer.[6]

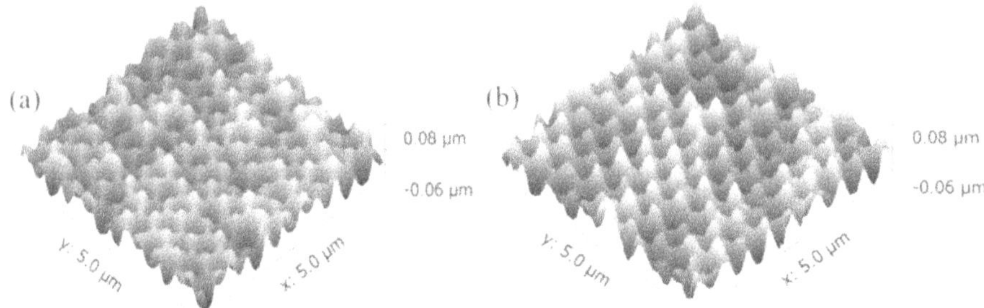

FIGURE 3.7 3-D AFM images measured for (a) the PTB7:$PC_{71}BM$ active layer imprinted with a pressure of 9.68 kPa, and (b) the Ag cathode surface of the corresponding photonic-structured OSC.[6]

Figure 3.7(a) and (b), respectively. The AFM images further confirm the periodic structure in the active layer and also in the silver layer being conformal with the imprint pattern having a period of ~480 nm and a height of ~51 nm. In this work, the periodic pattern with different structure heights in the active layer was controlled by adjusting the imprinting pressure over the range from 1.94 to 19.36 kPa, with the corresponding average height of the periodic structure in the active layer ranging from 42 to 64 nm. The modulation parameters of the photonic structure obtained in the AFM measurements were then used in the simulation.

Figure 3.8 presents an example of the $|E|$ field distributions calculated for the imprinted devices with different structure heights in the active layer at the wavelength of 379 nm. It reveals that the electric field distributions in the device can be controlled by varying the structure height in the active layer. For devices with 51 nm structure height, the electric field in the active layer is stronger than that with 42 nm structure height, indicating better light trapping in the active layer. While for that with 64 nm structure height, although a much stronger electric field distribution is observed, part of the electric field concentration locates outside of the active layer which does not contribute to the light absorption of the active layer.

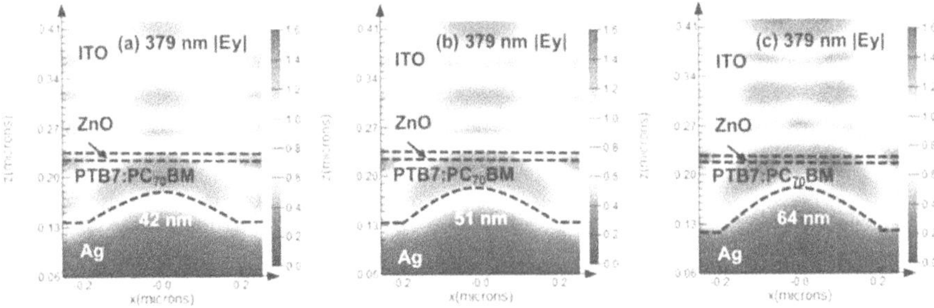

FIGURE 3.8 Simulated electric ($|E|$) field distributions in photonic-structured cells with different structure heights of (a) 42 nm, (b) 51 nm, and (c) 64 nm for plane wave incident light at wavelength of 379 nm.[6]

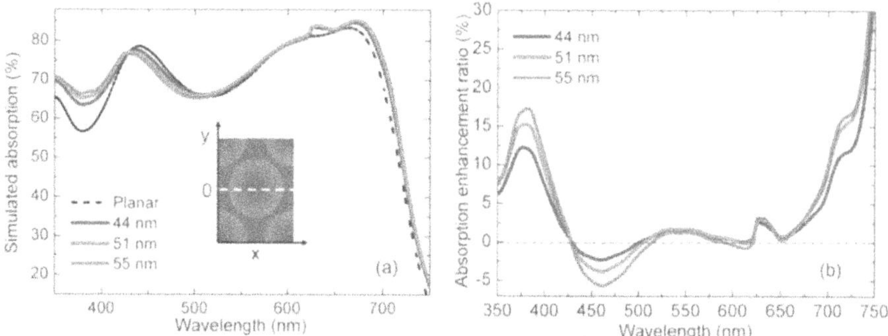

FIGURE 3.9 (a) Calculated absorption spectra of photonic-structured and control planar OSCs (inset: the analog cell used in the FDTD simulation. The dash line refers to the cross section of the field distribution); (b) Enhancement factor on absorption of the imprinted OSCs over the control planar device.[6]

The simulated absorption spectra and the corresponding enhancement factor of the active layer at normal incidence for different OSCs are plotted in Figure 3.9(a) and (b). The unit cell in the x–y plane used in FDTD simulation is shown in the inset in Figure 3.9(a), and the cross section of the field distribution is set at $y = 0$ in the FDTD simulation. The results, shown in Figure 3.9(b), reveal that the enhancement factor on absorption calculated for the imprinted OSCs over the planar cell is wavelength dependent and occurs at specific wavelengths.

3.3.2 Performance Enhancement of Organic Solar Cells with 2-D Photonic Structures

The current density–voltage (*J–V*) characteristics measured for a series of the photonic-structured and planar OSCs are shown in Figure 3.10(a). The statistical analyses of J_{SC} and *PCE* with the corresponding measurement errors are summarized in

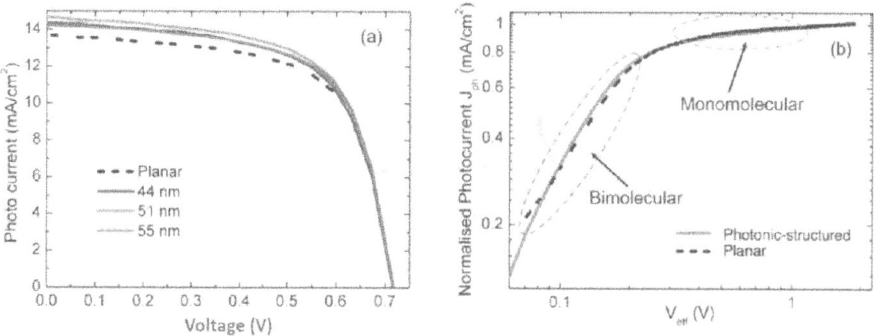

FIGURE 3.10 (a) *J–V* characteristics measured for the best planar cell and photonic-structured OSCs having different structure heights in the active layer, (b) J_{ph}–V_{eff} characteristics of the optimized photonic-structured OSCs and the control planar cells.[6]

TABLE 3.2
Summary of the Performance of the Optimized Planar Cell and Photonic-Structured OSCs Having Different Structure Heights in the Active Layer

Structure Height (nm)	V_{OC} (V)	FF (%)	J_{SC} (mA/cm^2)	PCE (%)	R_S (Ω·cm^2)	R_{SH} (Ω·cm^2)
64	0.71	63.3	15.0 ± 0.2	6.7 ± 0.2	6.4	591.2
55	0.72	65.2	14.7 ± 0.1	6.9 ± 0.1	3.6	739.1
51	0.72	64.3	15.2 ± 0.1	7.0 ± 0.2	3.9	688.4
44	0.72	64.0	14.9 ± 0.2	6.9 ± 0.1	3.9	620.0
42	0.71	62.7	15.0 ± 0.1	6.7 ± 0.1	4.1	597.6
Planar	0.72	65.1	14.1 ± 0.2	6.6 ± 0.1	3.7	710.4

Table 3.2, revealing that there is an obvious enhancement in the performance of the photonic-structured cells, as compared to the best performing (optimized) planar cells. Regardless of the variation in the structure of the active layer, these OSCs yielded a consistent V_{OC} of 0.71 V to 0.72 V, which are in good agreement with the reported values. The J_{SC} of 15.2 ± 0.1 mA/cm^2 was obtained for imprinted cells having an average structure height of 51 nm in the active layer, showing a 7.8% increase in J_{SC} compared to that of a structurally identical control planar cell (14.1 ± 0.2 mA/cm^2). 7.8% enhancement in J_{SC} is an improvement averaged from more than 10 cells, not the result from the champion device.

A 7.8% enhancement in J_{SC} can be quite impactful as it is realized via a very simple one-step imprinting process without acquiring any post annealing treatment. There is a small reduction in the FF of the imprinted devices, induced by a slight increase in the series resistance (R_S) and a small decrease in the shunt resistance (R_{SH}). For example, compared to the control planar cell, 5% increase in R_S and 3% decrease in R_{SH} were observed in the imprinted OSCs with a 51 nm height pattern in the active layer. The optimized photonic-structured OSCs with a PCE of 7.0 ± 0.2% were obtained, showing a 6.1% increase in PCE compared to that of a structurally identical best performing control planar cell (6.6 ± 0.1%).

In order to better understand the origin of the enhancement in the performance of the photonic-structured OSCs compared to the control planar cells, the recombination characteristics in the OSCs were analyzed. Figure 3.10(b) shows the double logarithmic plot of the net photocurrent density J_{ph} ($J_{ph} = J_l - J_d$, where J_l and J_d are the photocurrent and dark currents), generated as a function of the effective voltage V_{eff} ($V_{eff} = V_0 - V_b$, where V_0 is the built-in voltage measured at $J_{ph} = 0$, and V_b is the applied bias), measured for an imprinted OSC and a planar cell.[17] As V_{eff} decreases, charge recombination would increase and not all the photo-generated carriers could be collected by the electrodes. Thus, under specific V_{eff}, charge extraction efficiency P can be expressed as[18]

$$P(I, V_{eff}) = {J_{ph}(I, V_{eff})} \big/ {J_{ph,sat}(I)} \tag{3.3}$$

P approaches unity at a high V_{eff}, corresponding to the complete collection of the photo-generated charges.[19] In this regime, recombination is negligible. The recombination becomes increasingly important at low V_{eff} as P decreases with V_{eff}. As shown in Figure 3.10(b), monomolecular recombination is the dominant recombination mechanism at the high V_{eff}, while bimolecular recombination is the dominant recombination process at the low V_{eff}. $J_{ph} - V_{eff}$ characteristics of the photonic-structured OSCs are almost identical to that measured for the control planar cell over the V_{eff} range from 1 V to approximately 0.25 V that is close to the maximum power point of the cells. This suggests that both photonic-structured and control planar OSCs possess the same charge collection efficiency in this V_{eff} region. The value of P reflects an overall measure of the loss in the photo-generated charges in the OSCs. The results shown in Figure 3.10(b) reveal clearly that the photonic-structured OSCs had similar charge collection properties to that of the control planar cell, suggesting that the creation of the periodic structure in the active layer does not affect the charge recombination process and the charge collection properties.

Incident photon to current efficiency (IPCE) spectra measured for the photonic-structured OSCs and the planar control cells are shown in Figure 3.11(a). Comparing the IPCE spectra measured for the imprinted and the control planar OSCs, the enhancement factor on IPCE due to the photonic structure at different wavelengths ((IPCE of photonic-structured device)/(IPCE of planar device)–1) is shown in Figure 3.11(b). It can be observed that the enhancement factor on IPCE is wavelength dependent. The enhancement occurs at specific wavelengths, e.g., at 379 nm and 544 nm. There is a spectral region with decrease in absorption, e.g., at 455 nm. For OSCs with periodic photonic structures, light diffraction and light scattering play an important role in contributing to the absorption enhancement in the cells, although the improved spectral response due to light scattering effect is not wavelength dependent. The results in Figure 3.11 imply that absorption enhancement in the photonic-structured OSCs is mainly attributed to the 2-D periodic grating effect, as the absorption enhancement is wavelength dependent.

For photonic-structured OSCs with an optimized structure height of 51 nm in the active layer, $|E|$ and $|H|$ field distributions, at x–z plane when $y = 0$ at incident

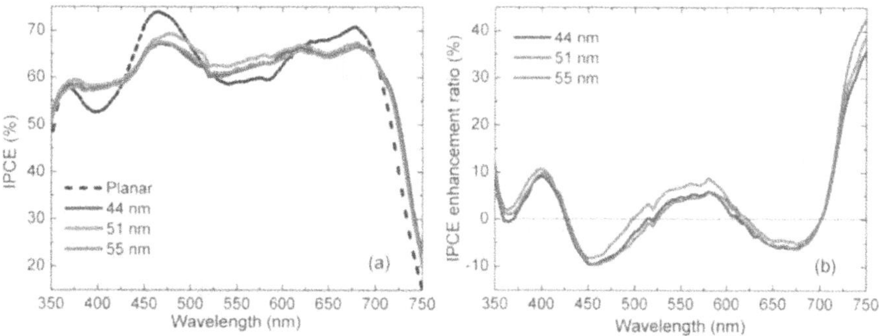

FIGURE 3.11 (a) IPCE spectra measured for the photonic-structured and the planar OSCs. (b) IPCE enhancement factor of photonic-structured cell over the planar cell.[6]

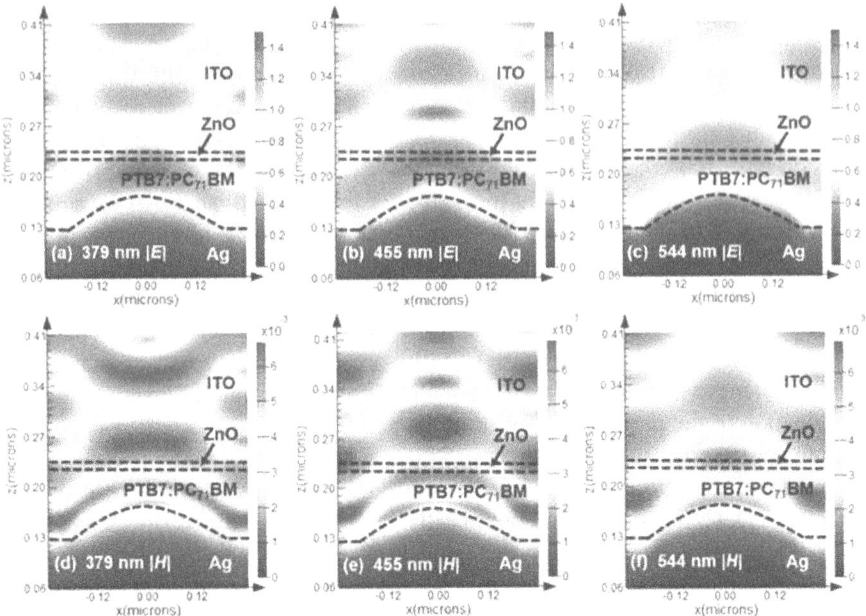

FIGURE 3.12 Simulated electric (|*E*|) and magnetic (|*H*|) field distributions in photonic-structured cells with structure height of 51 nm for plane wave incident light at different wavelengths of 379 nm, 455 nm, and 544 nm.[6]

wavelengths of 379 nm, 455 nm, and 544 nm, are illustrated in Figure 3.12. It can be seen that the *E*| and |*H*| field distributions in the active layer of the structured cells are clearly modified at these wavelengths. The profile of the field distributions arises from the distorted cavity hybridization modes, caused by the distinct optical phenomena in periodic photonic structures and the multilayer interference effect.[20] At 379 nm, the resonant optical modes of the electric field distribution cover exactly within the active layer, explaining the observed 15.2 ± 0.1% enhancement in the absorption. However, at longer wavelength (e.g., 544 nm), part of the electric field concentration locates outside the active layer, resulting in only ~2% increase in absorption. It shows that the absorption enhancement in photonic-structured OSCs occurs at the specific wavelengths, e.g., at 379 nm and 544 nm.

A decrease in absorption over a spectral region is also observed, e.g., at 455 nm, as compared to a planar cell, observed in both experiments and the simulation. The behavior of such a wavelength-dependent absorption enhancement is mainly due to the distinct optical phenomena in the OSCs containing periodic patterns. The electric and magnetic field distributions calculated for the photonic-structured and the control planar OSCs at 455 nm are shown in Figure 3.12(b) and (c). FDTD simulation reveals that a large portion of the field enhancement at 455 nm is distributed outside the active region in photonic-structured cells. This imposes a limitation in the absorption over a spectral region near 455 nm. However, PTB7 does not have a strong absorption below 500 nm. Photonic-structured cells still benefit from an overall absorption enhancement due to the 2-D periodic grating effect over the wavelength region from 500 to 750 nm, attaining a 7.8% increase in J_{SC} as compared to the control planar cell.

3.4 EFFECT OF ANGULAR DEPENDENCY ON LIGHT ABSORPTION IN ORGANIC SOLAR CELLS WITH PHOTONIC STRUCTURES

3.4.1 Optical Phenomena in 2-D Photonic Structured Organic Solar Cells

The results reveal that absorption, reflection, and diffraction of the incident light in the photonic structured OSCs strongly depend on the period of nanostructures.[20–22] To better investigate the optical characteristics of OSCs, the absorption in the active layer and reflection of the OSCs with different structure periodicity as a function of the angle of incident light are calculated.[7] The 2-D square-arranged structures are introduced into the active layer, as shown in the inset of Figure 3.13(a). For cells with different structure periodicity, the duty ratio (structure size/period) used in the calculations is kept unchanged, using a constant structure height of 30 nm. Figure 3.13(a) displays the absorption spectrum of the active layer in structured OSCs with different structure periods; the corresponding enhancement factor compared with the planar device is shown in Figure 3.13(b).

The relationship between structure period (p) varied from 200 nm to1000 nm, and the absorption in the active layer, and the correlation between p and the reflectivity of the cells under normal incidence of light are calculated; the results are plotted in Figure 3.14. It can be seen that the absorption in the active layer, integrated from 380 nm to 780 nm, increases slightly with decrease in the structure period, while the reflectivity of the cells decreases with the structure period for periodicity of the nanostructure below ~400 nm. The integrated absorption decreases slightly from 70.1% to 68.7% and the device reflection increases from 17.7% to 19% as the

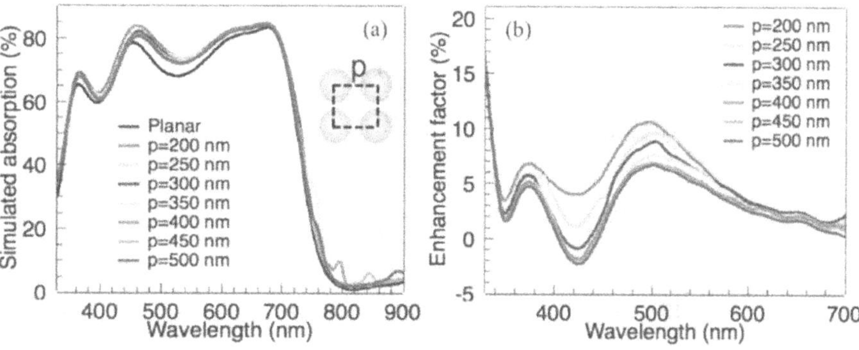

FIGURE 3.13 (a) Absorption spectra calculated for photonic-structured OSCs with different structure periods in the active layer; (b) Enhancement factor on absorption of the photonic-structured OSCs over the planar device. Reproduced with permission.[7] Copyright 2018, American Chemical Society.

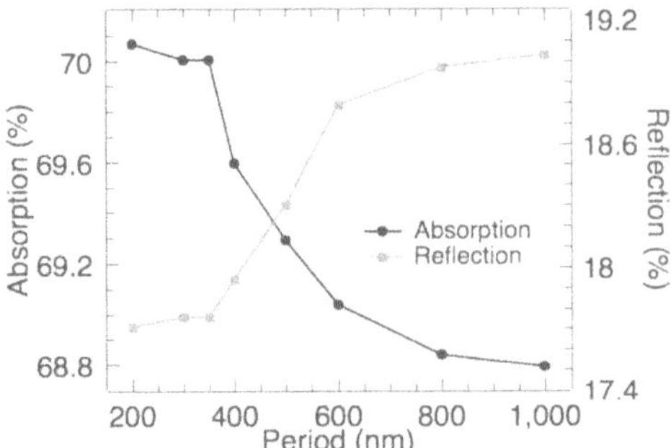

FIGURE 3.14 Relationship between the structure period and the integrated absorption in the active layer, and the reflection of the device. Reproduced with permission.[7] Copyright 2018, American Chemical Society.

structure period increases from 200 nm to 1000 nm. The corresponding absorption and reflection calculated for the planar cell under normal incidence are 66.2% and 20.3%, respectively. It indicates that photonic-structured OSCs with a smaller structure period favor for more effective light-harvesting; it shows that OSCs with a period of 200 nm have the highest absorption in the active layer. While it would be more complicated and more expensive to design periodic structure with such a small period, so there should be a tradeoff between the absorption enhancement and the structure period.

3.4.2 Angular-Dependent Absorption Profiles

To further investigate the oblique illumination performance of the planar and photonic-structured OSCs, angular-dependent absorption and reflection spectra as a function of the wavelength were calculated. Figure 3.15 shows the integrated absorption and reflection of the cells, calculated over the wavelength range from 380 nm to

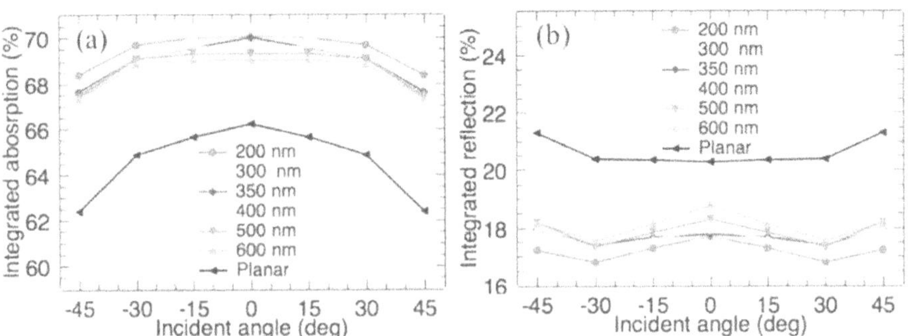

FIGURE 3.15 Integrated (a) absorption and (b) reflection calculated for the planar control cells, and those obtained for the photonic-structured OSCs with different structure periods under incident angle range from −45 to +45 deg. Reproduced with permission.[7] Copyright 2018, American Chemical Society.

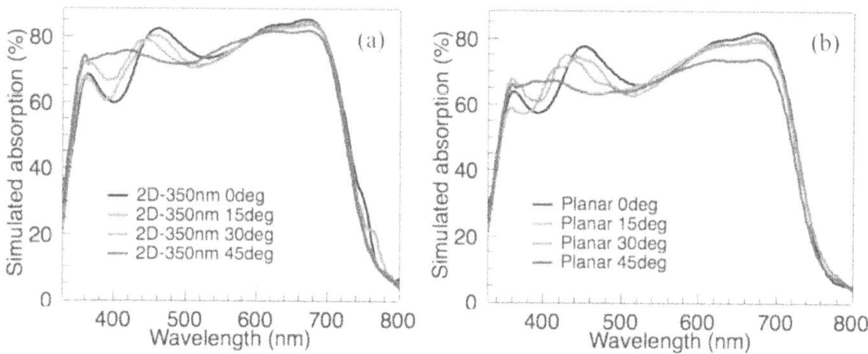

FIGURE 3.16 Absorption spectra calculated for (a) a 350 nm period photonic-structured device and (b) planar control cell under different angles of the incident light. (Reproduced with permission.[7] Copyright 2018, American Chemical Society.)

780 nm, under different angles of incident light from −45 deg and 45 deg. Though with different structure periods, both absorption and reflection of photonic-structured OSCs show similar variation trends. The absorptions of photonic-structured cell with different structure periods are all larger and also less angle-dependent than the planar cell in our case. Considering the light absorption of the active layer and the cost to design the periodic structure, a period of 350 nm for the OSCs was chosen for further study.

Figure 3.16 reveals the absorption spectra calculated for a 350 nm period photonic-structured device and the planar control cell under different angles of the incident light. It can be seen the absorption spectra of both the photonic-structured cell and the planar cell change with the angle of the incident light.

3.4.3 Broadband and Omnidirectional Absorption Enhancement in 2-D Photonic-Structured Organic Solar Cells

OSCs were fabricated with a structure of glass substrate/ITO/ZnO (10 nm)/PTB7:$PC_{71}BM$ (photonic-structured)/MoO_3 (2 nm)/Ag (100 nm), as plotted in Figure 3.17(a). The 2-D periodic structures with a structure period of 350 nm were transferred onto the active layer by nano-imprinting method using a PDMS mold. The imprint process was carried out at room temperature with a duration of 10 min. Figure 3.17(b) shows the AFM image of photonic-structured PTB7:$PC_{71}BM$ active layer, indicating well-defined 2-D square-arranged periodic structures in the active layer with a period of ~350 nm and a height of ~30 nm. We find that Ag cathode forms a conformal 2-D periodic structure on the active layer.

The J–V characteristics measured for the photonic-structured and planar cells are shown in Figure 3.18(a). A summary of the parameters is listed in Table 3.3. Figure 3.18(b) reveals the dependence of the normalized photocurrent density J_{ph} as a function of the effective voltage V_{eff}, recorded for the photonic-structured and planar cells under illumination at 100 mW/cm^2. The perfectly overlapped curves reveal clearly that the photonic-structured OSCs share similar charge collection properties

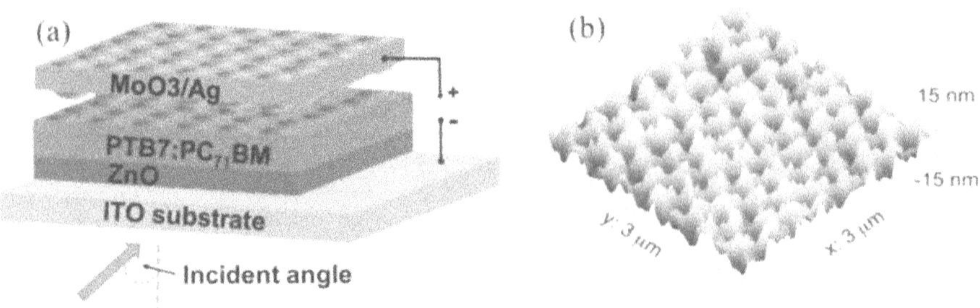

FIGURE 3.17 (a) Schematic cross-sectional view of the photonic-structured OSC; (b) AFM image measured for the surface of nano-imprinted PTB7:$PC_{71}BM$ layer. (Reproduced with permission.[7] Copyright 2018, American Chemical Society.)

compared with that of the planar control cell, suggesting that the creation of the periodic structure in the active layer does not affect the charge recombination process and the charge collection properties. This is similar to the charge recombination characteristics of the 480 nm periodic hexagonal-structured OSCs discussed in the previous chapter, indicating the imprinted OSCs with good charge collection efficiency.

Figure 3.19(a) shows the IPCE spectra measured for the photonic-structured and planar OSCs at normal incidence and the corresponding IPCE enhancement factor, the ratio of the IPCE of photonic-structured device to the (IPCE of the planar control device. Interestingly, there is an obvious broadband enhancement in IPCE of the photonic-structured devices. The broadband enhancement suggests that the increase of J_{SC} and *PCE* of the OSCs are attributed to the improved light absorption in the active layer. IPCE spectra measured for the photonic-structured and planar devices over different angles of incident light from normal incident (0 deg) to 45 deg are displayed in Figure 19(b) and (c), respectively. Figure 3.19(d) shows J_{SC} calculated using IPCE spectra measured for the photonic-structured and planar control cells under different angles of incident light (see in Figure 3.17(a)).

As shown in Figure 3.19(d), the integrated J_{SC} of the photonic-structured devices shows little dependence on the incident angle, which is largely different from the

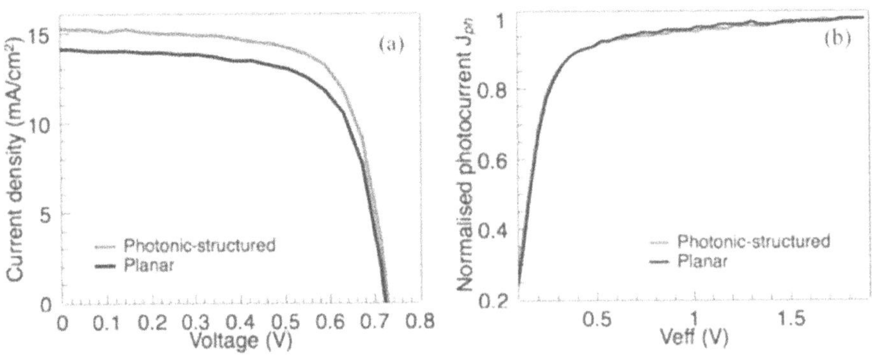

FIGURE 3.18 (a) *J–V* and (b) normalized J_{ph}–V_{eff} characteristics of photonic-structured and planar cell. (Reproduced with permission.[7] Copyright 2018, American Chemical Society.)

TABLE 3.3
Summary of the Device Parameters Obtained for the Photonic-Structured and Planar Cells[7]

Devices	V_{OC} (V)	J_{SC} (mA/cm²)	*FF* (%)	*PCE* (%)	R_S (Ω·cm²)
Structured	0.73 ± 0.01	15.17 ± 0.20	70.0 ± 1.2	7.74 ± 0.20	3.7 ± 0.5
Planar	0.73 ± 0.01	14.01 ± 0.15	68.1 ± 1.1	6.94 ± 0.15	3.8 ± 0.6

situation for the planar devices. As the incident angle increases normal (0 deg) to 45 deg, J_{SC} of photonic-structured cell is 14.76 mA/cm², which is 97.3% of the J_{SC} obtained at normal incident (15.17 mA/cm²). While for the planar cell, J_{SC} decreases from 14.02 mA/cm² (normal incident) to 13.15 mA/cm² (45 deg) and remains only 93.8%, indicating a faster decrease in the absorption of the planar cells as the angle of incident light increases. The results in Figure 3.19 reveal clearly that the use of

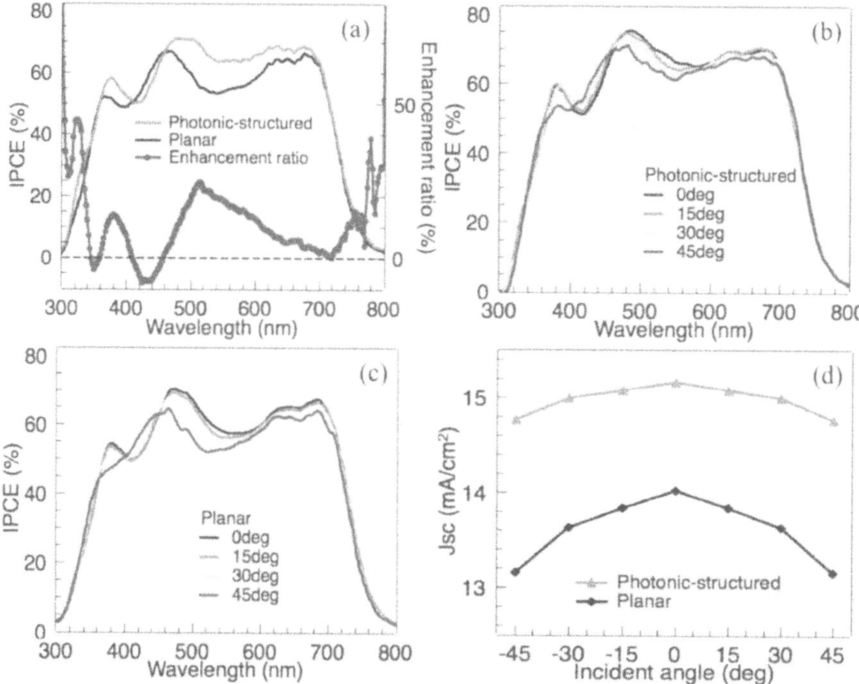

FIGURE 3.19 (a) IPCE spectra measured for the photonic-structured and planar control cells at normal incidence and the corresponding IPCE enhancement factor over the visible light wavelength range; (b) IPCE spectra measured for the photonic-structured and (c) planar devices over different angles of incident light from normal incident (0 deg) to 45 deg. (d) J_{SC} calculated using IPCE spectra measured for the nano-structured and planar control cells under different angles of incident light. Reproduced with permission.[7] Copyright 2018, American Chemical Society.

periodic nano-structures not only significantly enhances J_{SC}, but also greatly reduces the angular dependency on light absorption in the OSCs at oblique incident angles as compared to the planar control OSCs, achieved by omnidirectional and broadband absorption enhancement. The feature of weak angular dependency on light absorption in photonic-structured OSCs is of great importance in practical application. With different incidence angle of solar irradiation, light trapping can be mutated in the active layer of the photonic-structured device. The incorporation of the periodical structures enables the effective light trapping for attaining omnidirectional and broadband absorption enhancement in OSCs.

Figure 3.20(a) and (b) shows the contour plots of the absorption in the active layer calculated for the photonic-structured and planar OSCs. As can be seen, the high absorption region (i.e., red part) of the photonic-structured cell over the incident angle range from normal (0 deg) to 45 deg is broader than that of planar control cell over the wavelength region from 400 nm to 720 nm.

This is attributed to the enhanced light trapping in the active layer due to the patterned structure. Therefore, the enhanced light absorption in the nano-structured OSCs under different angles of the incident light is achieved as compared to that

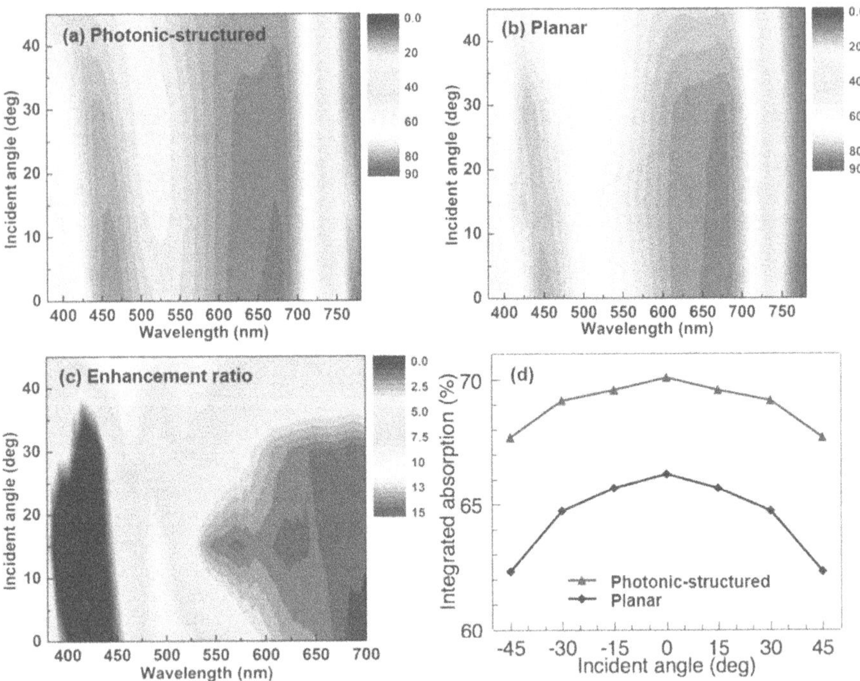

FIGURE 3.20 The contour maps of (a) absorption in photonic-structured OSCs, (b) absorption in planar control cells, (c) the ratio of absorption in the photonic-structured OSCs to the planar OSCs over range of angles of incident light from normal (0 deg) to 45 deg, and a wavelength range from 380 nm to 780 nm. (d) The integrated absorption in the photonic-structured and planar control cells as a function of the angle of incident light. Reproduced with permission.[7] Copyright 2018, American Chemical Society.

of the planar control cell over a broad wavelength range. In order to further illustrate the point, we have calculated the ratio of absorption in the photonic-structured OSCs to that in the planar control cells for a range of angles of incident light from normal (0 deg) to 45 deg, and a wavelength range from 380 nm to 780 nm. Results are shown in Figure 3.20(c), where the contour map of the absorption ratio of the photonic-structured OSCs to that of planar OSCs, absorption of photonic-structured device/(absorption of planar device), is plotted. It can be observed that the broadband absorption in the photonic structured OSCs is achieved over different angles of the incident light. In particular, at wavelengths around 450 nm to 570 nm, the higher absorption occurs in the active layer of photonic-structured OSCs, which are in good agreement with the enhancement in IPCE as shown in Figure 3.19(a).

Absorption in the active layer of the nano-structured and planar control cells was calculated using AM 1.5G solar spectrum over the wavelength range from 380 nm to 780 nm, the results are shown in Figure 3.20(d). As the angle of the incident light increases from 0 deg to 45 deg, the absorption of the active layer in photonic-structured OSC changes slightly from 70.1% to 67.7%, remaining 96.6% of its absorption obtained at the normal incidence. While for the planar OSC, the absorption in the active layer at an oblique incident angle of 45 deg remains only 94% of its absorption obtained at normal incidence, e.g., a large decrease in the absorption from 66.2% to 62.2%. In addition to the broadband absorption enhancement, the photonic-structured OSCs also have the advantages in generating high photocurrent at the oblique angle of incident light, a very unique feature for practice applications.

3.5 SUMMARY

There are two major approaches for improving the performance of OSCs. One effort has been paid on the development of narrow-bandgap conjugated polymers and non-fullerene acceptors whereby to increase *PCE* of OSCs through improved spectral response over the long wavelength region. Another is the device approach in which the enhanced absorption in OSCs is achieved using light-trapping effects. Various approaches have been reported including incorporating metal nanoparticles, surface plasmonic structures, and a textured substrate template in OSCs to boost light absorption. However, each approach has its own technical limitation for a particular device design. This chapter presents a comprehensive analysis of performance of the photonic-structured OSCs. Incorporation of the photonic structures is applicable to all OSC designs with broadband and omnidirectional light absorption enhancement. The selection of the photoactive materials and the photonic structures has effects on modulation of light absorption in OSCs. The use of the photonic structures in the OSCs gives rise to the distinct optical phenomena for absorption enhancement, creating a profound impact on the development of high-performance OSCs.

ACKNOWLEDGMENTS

This work was supported by the National Natural Science Foundation of China (62005152) and the Research Grants Council, Hong Kong Special Administrative Region, China (12302419, C5037-18GF, and N_HKBU201/19).

REFERENCES

1. Peumans, P.; Uchida, S.; Forrest, S. R., Efficient bulk heterojunction photovoltaic cells using small-molecular-weight organic thin films. *Nature* 2003, *425* (6954), 158–162.
2. Li, G.; Shrotriya, V.; Huang, J.; Yao, Y.; Moriarty, T.; Emery, K.; Yang, Y., High-efficiency solution processable polymer photovoltaic cells by self-organization of polymer blends. *Nat. Mater.* 2005, *4* (11), 864–868.
3. Kim, J. Y.; Lee, K.; Coates, N. E.; Moses, D.; Nguyen, T.; Dante, M.; Heeger, A. J., Efficient tandem polymer solar cells fabricated by all-solution processing. *Science* 2007, *317* (5835), 222–225.
4. Halls, J. J. M.; Walsh, C. A.; Greenham, N. C.; Marseglia, E. A.; Friend, R. H.; Moratti, S. C.; Holmes, A. B., Efficient photodiodes from interpenetrating polymer networks. *Nature* 1995, *376* (6540), 498–500.
5. Yu, G.; Gao, J.; Hummelen, J. C.; Wudl, F.; Heeger, A. J., Polymer photovoltaic cells: enhanced efficiencies via a network of internal donor-acceptor heterojunctions. *Science* 1995, *270* (5243), 1789–1791.
6. Lan, W.; Cui, Y.; Yang, Q.; Lo, M.-F.; Lee, C.-S.; Zhu, F., Broadband light absorption enhancement in moth's eye nanostructured organic solar cells. *AIP Advances* 2015, *5* (5), 057164.
7. Lan, W.; Wang, Y.; Singh, J.; Zhu, F., Omnidirectional and broadband light absorption enhancement in 2-D photonic-structured organic solar cells. *ACS Photonics* 2018, *5* (3), 1144–1150.
8. Liang, Z.; Sun, J.; Jiang, Y.; Jiang, L.; Chen, X., Plasmonic enhanced optoelectronic devices. *Plasmonics* 2014, *9*(4), 1–8.
9. Villeneuve, P. R.; Piché, M., Photonic Band Gaps in Two-Dimensional Square and Triangular Lattices. In *Photonic Band Gaps and Localization*, Soukoulis, C. M., Ed. Springer US: Boston, MA, 1993; 283–288.
10. Weickert, J.; Dunbar, R. B.; Hesse, H. C.; Wiedemann, W.; Schmidt-Mende, L., Nanostructured organic and hybrid solar cells. *Adv. Mater.* 2011, *23* (16), 1810–1828.
11. Furong, Z.; Jai, S. *In* Optical design of thin film amorphous silicon solar cells, Photovoltaic Specialists Conference, 1993., Conference Record of the Twenty Third IEEE, 10–14 May 1993; 1047–1050.
12. Liu, Y.; Kim, J., Polarization-diverse broadband absorption enhancement in thin-film photovoltaic devices using long-pitch metallic gratings. *J. Opt. Soc. Am. B* 2011, *28* (8), 1934–1939.
13. Yu, W.; Shen, L.; Shen, P.; Meng, F.; Long, Y.; Wang, Y.; Lv, T.; Ruan, S.; Chen, G., Simultaneous improvement in efficiency and transmittance of low bandgap semitransparent polymer solar cells with one-dimensional photonic crystals. *Sol. Energy Mater. Sol. Cells* 2013, *117*, 198–202.
14. Liu, F.; Zhou, Z.; Zhang, C.; Zhang, J.; Hu, Q.; Vergote, T.; Liu, F.; Russell, T. P.; Zhu, X., Efficient semitransparent solar cells with high NIR responsiveness enabled by a small-bandgap electron acceptor. *Adv. Mater.* 2017, *29* (21), 1606574.
15. Xu, G.; Shen, L.; Cui, C.; Wen, S.; Xue, R.; Chen, W.; Chen, H.; Zhang, J.; Li, H.; Li, Y.; Li, Y., High-performance colorful semitransparent polymer solar cells with ultrathin hybrid-metal electrodes and fine-tuned dielectric mirrors. *Adv. Funct. Mater.* 2017, *27* (15), 1605908.
16. Liang, W.; Zhong, J.; Xu, H.; Deng, H.; Wang, Q.; Long, Y., Tailoring the performance of semitransparent organic solar cells by tandem one-dimensional photonic crystals. *Acta Photonica Sinica* 2018, *47* (8), 823003–0823003.
17. Kyaw, A. K. K.; Wang, D. H.; Gupta, V.; Leong, W. L.; Ke, L.; Bazan, G. C.; Heeger, A. J., Intensity dependence of current–voltage characteristics and recombination in high-efficiency solution-processed small-molecule solar cells. *ACS Nano* 2013, *7* (5), 4569–4577.

18. Cowan, S. R.; Street, R. A.; Cho, S.; Heeger, A. J., Transient photoconductivity in polymer bulk heterojunction solar cells: Competition between sweep-out and recombination. *Phys. Rev. B* 2011, *83* (3), 035205.
19. Tian, X.; Wang, W.; Hao, Y.; Lin, Y.; Cui, Y.; Zhang, Y.; Wang, H.; Wei, B.; Xu, B., Omnidirectional and polarization-insensitive light absorption enhancement in an organic photovoltaic device using a one-dimensional nanograting. *J. Mod. Opt.* 2014, *61* (21), 1–9.
20. Leem, J. W.; Kim, S.; Park, C.; Kim, E.; Yu, J. S., Strong photocurrent enhancements in plasmonic organic photovoltaics by biomimetic nanoarchitectures with efficient light harvesting. *ACS Appl. Mater. Interfaces* 2015, *7* (12), 6706–6715.
21. Jin, Y.; Feng, J.; Xu, M.; Zhang, X.L.; Wang, L.; Chen, Q.D.; Wang, H.Y.; Sun, H.-B., Matching photocurrents of sub-cells in double-junction organic solar cells via coupling between surface plasmon polaritons and microcavity modes. *Adv. Opt. Mater.* 2013, *1* (11), 809–813.
22. Kim, M.S.; Kim, J.S.; Cho, J. C.; Shtein, M.; Kim, J.; Guo, L. J.; Kim, J., Flexible conjugated polymer photovoltaic cells with controlled heterojunctions fabricated using nanoimprint lithography. *Appl. Phys. Lett.* 2007, *90* (12), 123113.

4 Recent Advances in Gas Sensors for Device Applications

Bharat Sharma[a] *and Ashutosh Sharma*[b]
[a]Institute of Microstructure Technology, Karlsruhe Institute of Technology, Eggenstein-Leopoldshafen, Germany
[b]Department of Materials Science and Engineering, Ajou University, Gyeonggi-do, Republic of Korea

CONTENTS

DOI: 10.1201/9781003141358-4

4.1 INTRODUCTION AND BACKGROUND

Gas sensors have made an important contribution to mankind in detecting various gases and pollutants and made life comfortable. Various sensors technologies have been developed in the past. Earlier developments include the temperature sensitivity of materials subject to a change in resistance due to current flow in the eighteenth century. Later, Wilhelm von Siemens developed a copper temperature sensor in 1860 [1–4]. Piezoelectric materials like single-crystal quartz or ceramics have made possibility to design various sensors and pressure actuators for several engineering applications, affordable and economically viable for industries and defense architectures. Gas sensors have played a key role in common men's life. Modern sensor developments in science and technology were ushered in by the fabrication and research on large-scale silicon wafers, and exploiting the silicon chemistry to create new methods and phenomena into an electrical response to be read by computer units. Due to the development of advanced sensor materials, transistors, multilayered coatings, conducting polymers, and novel process, current sensor technologies are expected to have better control of microstructures, response, and new possibilities in materials device world. Interesting features have been included and appreciated by electronic manufacturers such as low cost, freedom reliability, and industrial scalability [5, 6].

Gas sensors are crucial for many important applications in our daily lives. Most attractive applications are detection of toxic gases and pollutants in air atmosphere for public security, industries, and monitoring air quality and climate [7–12]. Various developments in metal oxide sensing layer for sensor applications are being done to improve the various sensor features such as sensitivity, selectivity, sensor speed, recovery, and stability, so coined as "4s" attribute. Modern developments combined with novel materials contributed by the nanoscience and nanotechnology have achieved significant progress in the sensor world and are expected to have unlimited opportunities in this area. However, before proceeding further we would like to have a brief background of this advanced technology [13–16].

4.1.1 Classification

Gas sensors can be classified into various categories for diverse applications in both industry and academia. Notable applications include the following areas: (1) Gaseous detections, CH_4, SF_6, CO, etc. in mining [17–20], (2) pollution from automotive vehicles [21, 22], (3) electronic noses, glucose sensing, etc. in biomedical applications [23–25], (4) greenhouse gas monitoring in environment [26–31]. Common classification of various sensors includes methods of variation in electrical response and characteristics as shown in Figure 4.1.

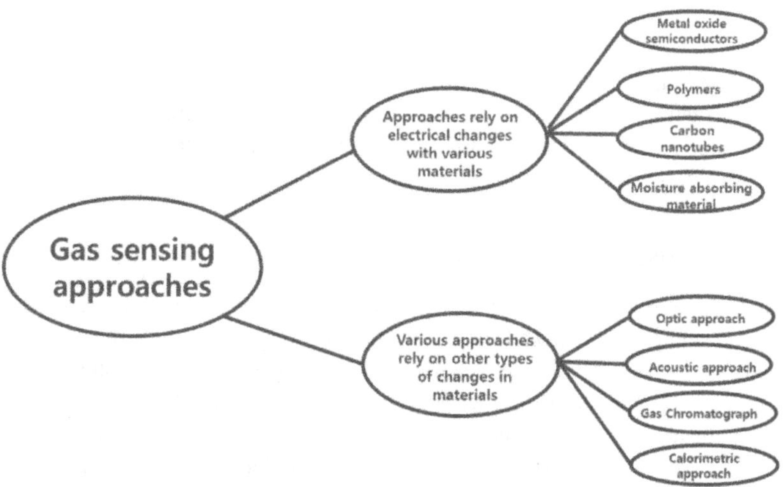

FIGURE 4.1 Classification of various gas sensing technologies.

TABLE 4.1
Physical and Chemical Response Signals

Mechanical	**Gravity**	**Pressure**		**Stress**	
Thermal	Heating	Thermal resistance		Cooling.	Thermal conductivity
Electrical	Piezoelectricity.	Hall effect	.	Johnsons noise	Electrolysis
Magnetic	Magnetostriction	Biot-Savart's Law	Faraday effect	Galvano-magnetic effects	
Radiation	Photo-electric	Photovoltaic	Refractive effects	Luminescence	Photo-synthesis
Chemical	Hygrometer	Potentiometric	Amperometric	Photo-chemical	
Thermo-electric	Seebeck effect.	Pyroelectricity	Peltier effect		

All these methods rely on the various interaction among the sensor components such as gas and sensor elements. For example, the common parts of the sensor device consist of various portions [1].

1. Sensor element: The part or material that functions to convert the input signal into an output gas response. "Sensor element" can be of more than one component known as compound sensors.
2. Sensor: The "sensor" combines the sensor element and its physical packaging with external environment.
3. Sensor system: "Sensor system" is the entire signal processing hardware either in same or in different platforms.

In the beginning, Lion presented a system of sensor responses based on the various forms of energies received and generated as shown in Table 4.1 [3, 4].

Gopel and his coworkers summarized Table 4.1 in 1989, which contains the most common sensing and actuating principles except biological and nuclear effects [3]. Accordingly, the terms sensors and actuators can be used according to the intent of the application rather than principles of physics.

4.1.2 Environmental Pollutants

The area of sensor technology is extremely wide due to the changing chemistry of toxic gases and pollutants. The future development of gas sensors will surely depend upon the interaction of various disciplines including electronics, materials science and engineering, mechanical, chemical, or biological areas. Environmental pollutions have brought one of the most critical concerns among global issues worldwide.

Figure 4.2 shows the various forms of pollution and its effect on human surroundings and wildlife. Wildlife has been affected severely due to the various chemical disposal in water resources over decades. Various biological tools have been devised

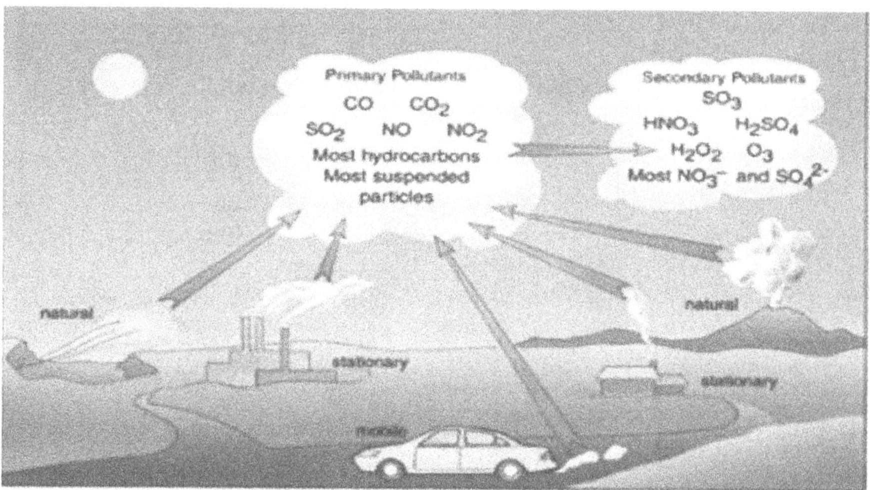

FIGURE 4.2 Types and sources of air pollutants.

to purify water and check contamination by novel water sensor elements. Recent biomedical tools and traditional design allow us to sense microorganisms, degrade hazardous materials from environment, and prevent chemical deposits in the seawater [6–16, 32, 33]. To improve agriculture, water contamination should be avoided by using novel gas sensors.

Most of the gas contaminants are particulate matter in the environment coming from the industrial wastes and smoke such as oil refineries, fly ash, cement dust, tobacco, and smog (Table 4.2). We now make a survey of the various pollutants and contaminants in our human environment in the following sections.

4.1.2.1 Carbon Monoxide (CO)

CO is the major gas pollutant in our environment which comes from the combustion of fuels and carbon emissions. The CO poisoning includes headache, dizziness, nausea, and can lead to final death over long-term exposure. CO has a high affinity to hemoglobin levels and reduces the oxygen level in the body. Loss of the oxygen in veins causes hypoxia and serious cardiovascular issues in human beings. CO is the main greenhouse gas responsible for worldwide global warming and climate changes [36].

4.1.2.2 Nitrogen Oxide (NO_2)

NO_2 is mainly present among the pollution emitting from the automotives [37, 38]. The presence of NO_2 irritates the respiratory tracts and may cause coughing, bronchitis, and severe heart problems when inhaled at high doses. According to reports, 0.2 ppm (parts per million) levels of NO_2 can cause adverse effects in humans. Higher concentrations can affect T-cells, CD^{8+} cells, and NK cells and may destroy the immunity system [39]. Long-term exposure may cause chronic lung problems, tuberculosis, sensory organs, and respiration [39].

TABLE 4.2
Various Types of Particulate Matter (PM) [34, 35]

Type	PM diameter (μm)	
Particulates	Smog	0.01–1
	Soot	0.01–0.8
	Tobacco smoke	0.01–1
	Fly ash	1–100
	Cement	8–100
Bio-contaminants	Bacteria	0.7–10
	Viruses	0.01–1
	Fungus and mold	2–12
	Allergens (pollen dust)	0.1–100
Dust	Air dust	0.01–1
	Heavy dust	100–1000
	Settling dust	1–100
Gases	Pollutants	0.0001–0.01

4.1.2.3 Sulfur Dioxide (SO_2)

SO_2 is a very fatal chemical in our surroundings which comes mainly from industrial plants and burning of fossil deposits. The industrial standard of SO_2 is 0.03 ppm. SO_2 affects human beings and wildlife severely causing lung failure and heart issues. The kids are high-risk damage categories affected by respiratory issues, bronchitis, loss of smell and taste. Also, skin inflammation, eyes blurring, damage to mucosa, and cardiovascular health have been observed [39]. SO_2 has been a major factor in soil acidity and acid rain in our environment.

4.1.2.4 Lead

Pb is a heavy metal and used in many electronic gadgets and consumer appliances as an interconnect material. After the discard of electronic devices, there is a high risk of lead poisoning in water resources and inhalation of lead vapors in environment. It is also present in the motor engines, batteries, radiators, wastewaters, ores, and engine aircraft [40]. Pb has been associated with harmful effects on nervous system, brain, and heart. Pb poisoning occurs when it is inhaled or accumulated in blood, tissues, liver, bones, and reproductive organs [40, 41].

4.1.2.5 Hydrocarbons

Hydrocarbons such as aromatic, acrylic (PAHs) are ubiquitous in our surroundings. Such polyacrylic hydrocarbons are mostly seen in coal tar deposits. The examples of these pollutants include benzopyrene, naphthalene, anthracene, and fluoranthene recognized as carcinogenic and mutation-causing products. They have been reported to cause lung cancer in many cases [42].

4.1.2.6 Volatile Organics

Several volatile organic compounds (VOCs) such as toluene, xylenes, benzene, and its derivatives, present in our environment, may cause severe issues on human beings [43]. Due to the setting up of modern industries and vehicles, new organic products are increasing in human environments [44]. Short-term exposure causes eye irritation, nose, and throat problems while long-term issues include fatal toxicity to human organs [43, 44].

4.1.2.7 Dioxins

Dioxin compounds come from industrial and natural causes like fires and volcanoes. Their accumulation happens in food sources such as meats and milk products, seafood, and mainly in the animal tissues [45]. The exposure to dioxins causes short freckles and blemishes on the skin and foreheads. Other long-term effects are congenital disease, weakening of endocrine, nervous and immune systems, infertility, and skin cancer [45].

4.1.3 Various Sensing Materials

The last few decades have seen enormous development in the sensing materials and processes for diverse applications. Different types of sensing materials are metal oxides, composites, nanomaterials, polymers, CNTs, graphene, and their derivatives.

4.1.3.1 Metal Oxides

A lot of developments in this area have been available in literature as shown in Figure 4.3. Metal oxides are advantageous due to their high thermal stability for extreme environment, high inertness to various reducing or oxidizing atmospheres by a change in their electrical resistance in the gaseous environment.

Different types of metal oxides have different gas sensing mechanisms for different gases. The most popular metal oxides for gas sensing include SnO_2, In_2O_3, WO_3, ZnO, etc. These metal oxides have been used to detect various gases such as H_2, NH_3, CO, C_2H_5OH. Figure 4.3 shows the various metal oxides used in sensing [46–51].

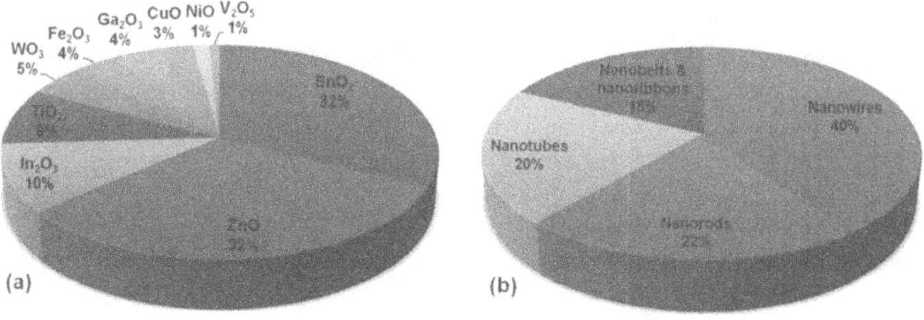

FIGURE 4.3 (a, b) Trends in application of some important metal oxides for various gas sensing applications [46].

4.1.3.2 Carbon-Based Materials

Nanotechnology has given several alternative and novel materials that can replace the traditional metal oxides and have higher sensitivity toward a gas species. Examples include carbon nanotubes (CNTs), graphene, nanoparticles, nanorods, and nanodots. Nanomaterials have unique properties such as high thermomechanical stability, high specific area, and excellent electrical and thermal properties. CNTs have been used widely in gas sensing. There are two types of CNTs – single-walled CNTs and multi-walled CNTs. Single-walled CNTs are simply the roll graphene sheets in cylindrical form (diameter: 0.4 to 2 nm). Multiwalled CNTs contain multilayered structure with many rings of graphene rolled into it. The various synthesis methods of these carbon nanomaterials include laser ablation, plasma methods, and chemical vapor deposition. There is a very strong sp^2 hybridized C-C bond that imparts high stability. The narrow diameter and hollow tubes are best suited for the gas adsorption and detection. Any defects (vacancies, line defects) inside the hollow structure increase or decrease the sensing response of the device [52].

4.1.3.3 CNTs-Based Nanocomposites

Literature shows that composite materials are more feasible for sensing response and detection. It has been found that CNT composites have been more efficient toward NH_3 as compared to CO. An increase in conductivity is noted for CO and vice versa for NH_3 [53]. Nanocomposites are materials that contain secondary nano-reinforcements to improve the desired microstructural and thermodynamic properties of the matrix. Nanocomposite reinforcements include metal oxides, polymers, CNTs, graphene for promising tuning of sensing properties. CNT/polymer composites can be used to higher stability and mechanical strength of polymer matrix at high temperatures [54]. However, the dispersion of CNTs in a composite matrix is a problem and should be controlled for better realization of sensing properties [54].

4.1.3.4 Conducting Polymers

Electron conducting polymers are found to be exceptionally superior candidates for gas sensing applications. Various researchers have added CNT, graphene to increase the sensitivity and response to manifolds due to the inability of CO sensing at room temperature (RT) using pristine single-walled CNTs [55]. Choi and his coworkers worked together to examine the electrical properties of PANI/CNT composite for CO and NH_3 detection. More recently, CNTs and metal nanoparticles have been used for sensing. Mohsen Asad et al. [56] used SWCNT nanotubes decorated with Cu nanoparticles as a sensing layer produced by drop-casting. They tested the sensing layer for a variety of gases, such as CH_3COCH_3, C_2H_5OH, and H_2S. The highest selective response was observed for H_2S among other gases.

4.1.3.5 Graphene

Graphene is a 2D carbon nanostructure with extraordinary electrical and thermal properties. Lower resistance and large surface area contribute to an attractive sensing material [57]. There are various derivatives of graphene (graphene oxide and reduced graphene oxide) used for various gas sensing, e.g., H_2O, NO_2, CO, and NH_3. The sensing output is quantified in terms of the variation in graphene sheet resistance.

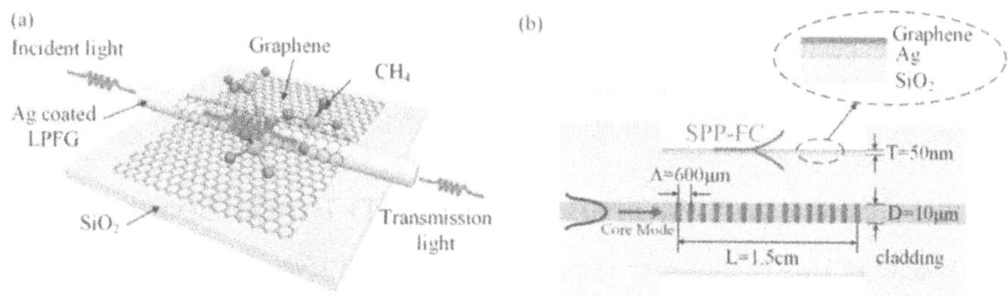

FIGURE 4.4 Graphene sensor for CH_4 gas sensing [58].

Figure 4.4(a) and (b) shows a graphene sensor made of graphene sheets by Wei et al. [58]. They used it for sensing humid environment. They reported that for a relative humidity at lower levels lesser than 50%, a decrease in 1.38 kHz frequency is due to the mass of water vapor on the graphene sheets.

4.2 TYPES OF GAS SENSORS

Various types of sensors are shown in Table 4.3. The most popular input signals are based on the variations in current, voltage, electric field or impedance, etc. [59].

We now describe these sensors, their mechanism, and application fields in the following sections.

4.2.1 Calorimetric, Catalytic Bead Type Gas Sensor

The calorimetric gas sensor measures the heat of reaction on sensor. These are the earliest portable gas sensors made of Pt wire developed in the 1950s. The current applied can heat the Pt wire to 900°C to 1000°C which can cause combustion on the Pt surface exposed to a gas mixed with air. The resistance change due to heat was calibrated to temperature changes of wire. Therefore, it is like a heater, catalytic surface for combustion, and thermometer. The oxide-supported catalyst has higher catalytic activity and can operate at lower temperatures (300°C to 500°C) which in turn enhance the lifetime and reduce dramatically the drift problems for the Pt wire

TABLE 4.3
Various Types of Gas Sensors and Materials

Types of sensor	Sensing materials	Principle of detection
Calorimetric	Pellistor	Thermo-chemical
Catalytic field-effect sensors	Catalytic metals	Electric field
Polymer sensors	Conducting polymers	Resistance
Electrochemical sensors	Solid/liquid electrolytes	Current or voltage
MOX (Metal oxide semiconductors)	Doped SnO_2, GaO	Resistance

heater. Units are commonly calibrated in terms of lower explosive limit (LEL) of the gas that is expected to be measured in the application. For example, a reading of 10% LEL for methane corresponds to a concentration of methane in air that is 10% LEL, with a net concentration of 0.5% by volume [60–62].

4.2.2 Catalytic Field-Effect Sensors (MOSFET) Type Gas Sensor

MOX sensors have high sensor response (sub-ppm/ppb levels) and react to various oxidizing materials such as zinc oxide (ZnO), iron oxide (Fe_2O_3), and various reducing materials, such as nickel oxide (NiO) or cobalt oxide (Co_2O_3) [63, 64]. MOSFET was originally stated by Lundström et al. [65] in 1975 which relied on propensity of amount of metals to adsorb and desorb hydrogen (H_2) [66]. FET-based gas sensors have exciting characteristics that include: (1) to work at RT, (2) diversity of sensing nanomaterials thus improve sensing surface for sensitive and selective detection; and (3) small magnitudes and well-suited with nanofabrication technology, made into sensors for high sensing characteristics [67]. Recently Y. Hong et al. studied the capacitive-based FET (CFET) gas sensor that relies on Si FET. Five photomasks are utilized to construct CFET sensor and ZnO thin film is implemented as sensing material [68]. NO_2 gas sensing properties are attained by utilizing pulse pre-bias technique.

4.2.3 Conducting Polymer Sensors

Conductive polymers have fascinated attention in usage as electronic noses, mainly due to their high sensor response, fast response times, simple fabrication processes [69, 70]. Conductive polymer-based gas sensors contain a substrate, generally silicon, gold electrodes, and conducting polymer as sensing material [71]. Conducting polymers are typically produced by chemical oxidizing of equivalent monomers. Most broadly utilized sensor monomers are polyaniline, polypyrrole, polythiophene [72], but polyacetylene, poly(phenylenevinylene) and have also been studied [73]. The communal characteristic of conductive polymer is the existence of conjugated pi-electron structure which spreads over polymer. C. H.A. Esteves at al. studied the alteration in steady pair of conductive polymer dopant that is proposed, substituting metallated porphyrins. Doping of the porphyrin rings with a similar conductive polymer leads to enormous changes in sensor response to nominated analytes [74]. Two dissimilar arrays (doped with free-base and metallized porphyrins) utilizing these sensors were unable to recognize four VOCs, such as ethyl acetate, toluene, propanone, and ethanol. One of the foremost drawbacks of conductive polymers is high susceptibility to ambient conditions such as environmental humidity, though sorption of water inside a polymer might play a significant part in the mechanism of gas response [75]. Sensor response of conducting polymer is usually less than MOS, yet, at sub-ppm level, has been studied for certain analytes with appropriate electronic circuit [76].

4.2.4 Electrochemical Gas Sensor

Electrochemical gas sensors can operate at room temperatures, low power consumption, and robust, nonetheless large [77, 78]. Their detection method relies on electrochemical reduction/oxidation of gas molecules at sensing electrode [78]. This methodology has decent significance when functional to the detection of electrochemically gases species, nevertheless not sensitive to an extensive variety of materials, specifically aromatic hydrocarbons [79]. It was examined that nanocomposite materials show good results, and in certain situations, a stable prearrangement is united with solid electrolyte. Q. Lin et al. studied the fast response-recovery rates of YSZ electrochemical sensor through the laser ablation method. Sensor showed high selectivity to other reducing gases at low working temperatures [80].

4.2.5 Metal Oxides Semiconductor (MOS) Gas Sensor

Detection of flammable, and exhaust gases is significant for energy saving along with environmental protection [81–83]. Gas sensors have been utilized for detecting flammable and toxic gases for industrial applications [83]. The inexpensive, dependable, trivial, and lower power consumption gas sensors are in excessive mandate because of wide series of applications. MOS gas sensors are creating attention as this material accomplishes the condition of a perfect sensor to a great extent. MOX is appropriate for sensing reducing/oxidizing gases by conductive studies. Some examples of MOS that show a decent gas response are Mn_2O_3, Cr_2O_3, SnO_2, ZnO, TiO_2, etc. [84]. MOS nominated for gas sensors can be known by their electronic structure. Variety of MOS based is divided into two categories [85]: (1) Transition-metal oxides (2) Non-transition-metal oxides. The precise essential gas mechanisms of these sensors are yet debatable, which mostly includes band bending tempted by charged molecules and variation in conductivity.

4.3 THIN-FILM DEPOSITION METHODS FOR GAS SENSORS

4.3.1 Physical Vapor Deposition (PVD)

PVD is mostly functional to deposit thin film of dissimilar materials on substrate. The choice substrate could be glass or silicon. Most communal approaches of PVD for metals are e-beam evaporation, thermal evaporation, plasma deposition, and sputtering. A short-term study of various thin-film deposition methods is enlightened as follows [86, 87].

4.3.1.1 e-Beam Evaporation

e-beam evaporation is a vital means of PVD way for depositing thin film of metals, and metal oxides in vacuum environment. High purity material is positioned inside vacuum chamber, characteristically as a pellet in crucible. The electron energy is utilized to heat pellets, instigating depositing material to go in gas [88]. Because of vacuum atmosphere, evaporated particles can be transportable to substrate deprived of colliding with other particles. High melting point depositing materials could be coated at high coating rates, producing a preferred procedure for ceramic materials. Presently accessible materials library are metals (Al, Cr), oxides (SiO_2, Al_2O_3, ITO), fluorides (MgF_2), and semiconductors (Si, Ge) [89].

4.3.1.2 Thermal Evaporation

Thermal evaporation is another method of PVD that could be utilized to coat metals, organic/inorganic polymers. In this process, electrical energy is utilized to heat filament, which in turn melts and evaporates the material [90, 91]. Simplest bases utilized in thermal evaporation are wire-based filaments of high melting point, which are heated by passing flow of electrical current through them.

4.3.1.3 Sputtering

Sputtering can be utilized to deposit a wide range of thin films that include metals, metal oxides, and alloys. The positive particles (ions) in plasma are quickened to target. Sputtered atoms would diffuse through plasma area and some will impose onto substrate [92]. Uncertainty sputtered atoms contain high sticking coefficient (low vapor pressure), which are adsorbed onto the wafers. In case of ion-beam sputtering, ions are quicker to target and intrude on its surface [93].

4.3.1.4 Molecular Beam Epitaxy (MBE)

MBE is also a PVD method without any vapor-phase deposition. Its benefits over epitaxial methods are high preciseness management of doping and the prospect of heterostructures creation. It includes reactions of more than one beam sustained in high vacuum (10^{-8} Pa). Due to high vacuum, growth rate is sluggish as compared to

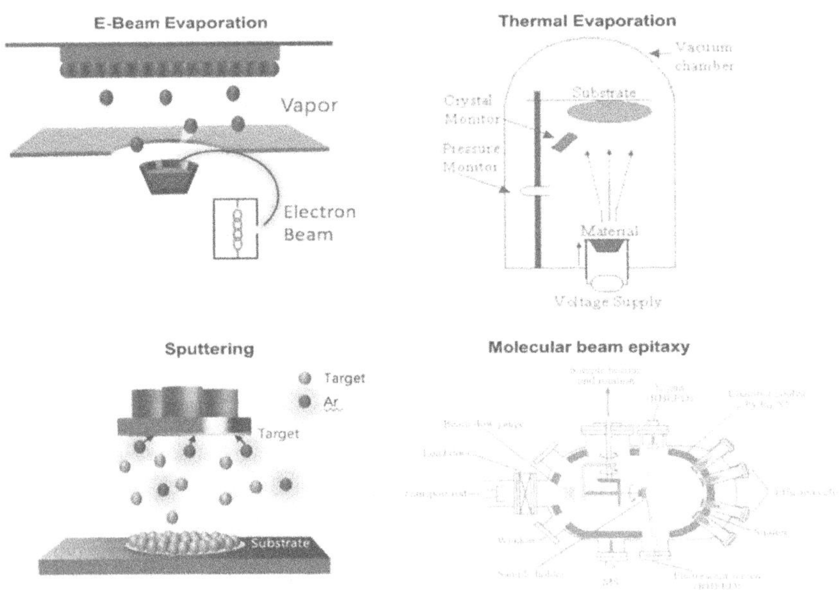

FIGURE 4.5 Types of physical vapor deposition (PVD); e-beam evaporation, thermal evaporation, sputtering, molecular beam epitaxy.

other epitaxial deposition methods; for this reason the MBE method is not an appropriate industrial application [94, 95]. Nonetheless, exactly control chemical composition of device can be achieved. Single crystal multilayer design at sub-atomic level is probable with MBE in thin-film deposition.

4.3.2 Chemical Vapor Deposition (CVD)

CVD is a striking method due to conformal coatings, good step coverage, and hefty number of wafers at the same time. In general CVD setup is similar as utilized for coating of polysilicon and dielectrics [96, 97]. Low-pressure CVD is proficient in making homogenous deposition over extensive series of structural outlines, regularly with lesser electrical resistivity as compared to PVD.

4.3.2.1 Atmospheric Pressure CVD (APCVD)

APCVD epitaxial films are single-layer films coated by the CVD technique. It is extremely utilized in industrial procedures and proposals for low cost and high throughput [98]. Vapor-phase chemicals are evaporated and flow in reaction cavity with partial vapor pressure for reactions. Different types of vapor deposition methods are given in Figure 4.6.

4.3.2.2 Plasma-Enhanced CVD (PECVD)

PECVD is a superior method of CVD method where glow discharge plasma perseveres in reaction cavity. Deposition temperature could be condensed to amount of 500°C by employing plasma than furnace structure to attain dielectric layer of dissimilar materials. The most exceptional technique to excite plasma is RF electric field. PECVD is typically utilized to coat insulating layer/dielectrics. The frequency series utilized is usually from 200 kHz to 35 MHz [96].

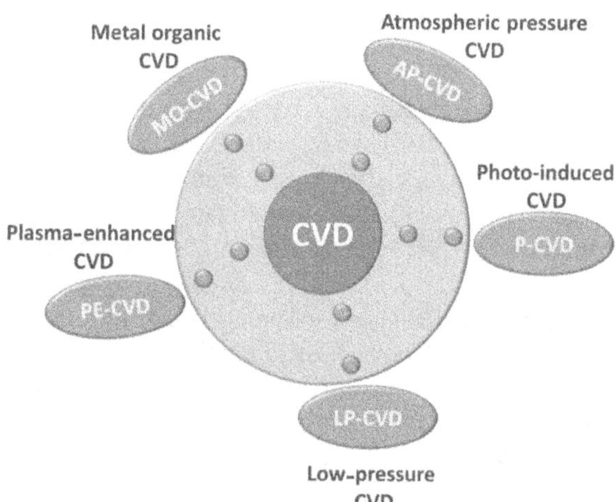

FIGURE 4.6 Various types of chemical vapor deposition; plasma-enhanced CVD, atmospheric pressure CVD, metal organic CVD, photo-induced and low pressure CVD.

4.3.2.3 Metal Organic Chemical Vapor Deposition (MOCVD)

MOCVD is a method for the coating of epitaxial thin film. MOCVD is broadly utilized in semiconductors but not appropriate for elemental semiconductors. The semiconductor is mainly utilized in the assembly of heterostructure devices, e.g. FET. The focal benefit of the MOCVD method is easy to eliminate volatile compounds to evade contaminations for next procedure stage. It is difficult to eliminate solid or liquid contaminants [99]. The chief drawback of the MOCVD procedure is carbon contaminations and the configuration is the same as CVD.

4.3.2.4 Low-Pressure Chemical Vapor Deposition (LPCVD)

LPCVD is a CVD method that utilizes heat to start reactions of precursor gas on a solid substrate. The reaction at the surface takes place in the form of solid phase. Low pressure is utilized to reduce slightly undesirable gas-phase reactions and rises homogeneousness across substrate. LPCVD is utilized to coat extensive series of thin-film configurations with respectable conformal step-coverages [100]. This thin-film comprises variability of materials with polysilicon (pSi) for electrical contacts, oxides utilized for isolation, nitrides, and dielectrics.

4.3.2.5 Photo-Induced Chemical Vapor Deposition (PCVD)

Photo-CVD has fascinated an excessive deal of attention. In the deposition technique, gases are utilized by way of sources, photon energy is used to tempt reaction leads to coating of thin film on substrates. In photoinduced approaches, light is an important part and all types of light sources have been studied [101]. Consequently, traditional lights and lasers are utilized. Lasers are significant with their properties like high power, monochromaticity, and coherence.

4.3.3 Miscellaneous Methods

4.3.3.1 Additional Approaches Contain

1. **Metallization:** Metal layers are coated onto wafer to form conductive paths. Common metals utilized are particularly noble metals that include aluminum (Al), gold (Au), silver (Ag), platinum (Pt), and copper (Cu) [102, 103]. First is ohmic-contact metallization that is contact electrode produces ohmic contacts by surface layer, and second is Schottky contacts metallization.
2. **Filament evaporation:** Filament evaporation, as well termed as thermal evaporation, simplest technique of metallization. It is proficient by resistive heating of filament that is held inside chamber. The current flow by filament is amplified, while the heat produced by filament heats entire metal till it evaporates fully. Metal vapor shrinks once it is in contact with substrate, establishing conformal metal layers [104, 105].
3. **Electron beam evaporation:** Beams of electron evaporate metal that is held in the crucible and unmask for certain time in which state initiating it to evaporate and deposit on wafers [106].

4. **Induction evaporation:** Induction evaporation utilizes radio frequency (RF) radiations to evaporate metal held in crucible. Metal is at that time heated and coated onto wafers [107].
5. **Sputtering:** Sputtering takes place in inert gas surroundings. Ions of inert gases are acquainted inside chamber of low vacuum conditions. Ions are made by applications of electric field to atoms. After ions attack target, metal atoms are sputtered onto silicon substrate [93].
6. **Wire bonding:** Wire bonding is used for electrical connections utilizing thin wire and with reaction of pressure, heat, and ultrasonic energy. Ultrasonic energy is utilized to improve surface smoothness. There are three categories of wire bonding methods that are dependent on sorts of energy utilized [108]. (1) Thermocompression, (2) Thermosonic, and (3) Ultrasonic wire bonding.

4.4 SENSING PARAMETER

4.4.1 Operating Temperature

Various sensing parameters affect the sensing response in gas sensors. Based on the operating temperature there are two groups of materials: (1) 400-600°C operating temperature and (2) greater than 700°C. The first group of materials includes SnO_2 and ZnO known as surface conductance materials, while the second group of materials are bulk conductance materials. TiO_2, CeO_2, Nb_2O_5 are few known bulk conductance materials [109, 110].

4.4.2 Sensor Response

The sensor response can be optimized by several methods. The most important one is microstructures tailoring. For example, large grain size D (grain size) $\gg$ 2L (space charge width), limits conductance by forming Schottky barrier at the grain boundaries. For instance $D = 2L$, conductance is restricted by necks among grains and for instance $D < 2L$, conductance is affected by every grain or summary of porous microstructure with advanced surface area and sensor response [111, 112].

4.4.3 Selectivity

Selectivity in gas sensors is a major challenge in sensing applications. Dopants addition to metal oxides has been shown to improve selectivity of sensors to different gases. In advanced applications, both sensitivity and selectivity of a gas sensor are prime factors in design. The selectivity is generally defined as *Ra/Rg* for reducing gases, where *Ra* is the reference gas resistance and *Rg* is resistance in target gas [113–116].

Table 4.4 presents the complete summaries for related methods for enhancing sensor and selectivity as stated above. There are approaches that adjust sensing materials composition such as doping or decorating with catalytic nanoparticles, which might advance gas sensor response and selectivity.

TABLE 4.4
Summary of Approaches Improving Sensitivity and Selectivity [46, 113]

Methods	Sensor response	Selectivity
Dielectric resonator	Higher surface area Conductivity as a function of gas concentration	NA
Thermostatic	Sensor response of target gases Gases with different operating temperatures	Sensor response of target gases Gases with different operating temperatures
Pre-concentrator	Concentrations of gases	Selective
Photoacoustic	Optic and acoustic merits	NA
Sensor array	NA	Multidimensional Multi-target gases in various situations

Figure 4.7 displays selectivity of gas sensors to blend various gases. Various levels of sensor response are essential under dissimilar conditions, for instance, ppth (parts per thousand) is sufficient for industrial purposes, while ppm or ppb (parts per million or billion) are desirable for accurate detection in laboratory scale [114, 117]. Chemical characteristics of sensing material, physical structure, temperature, and humidity of the surroundings are the probable influences that people can examine to enhance sensor response.

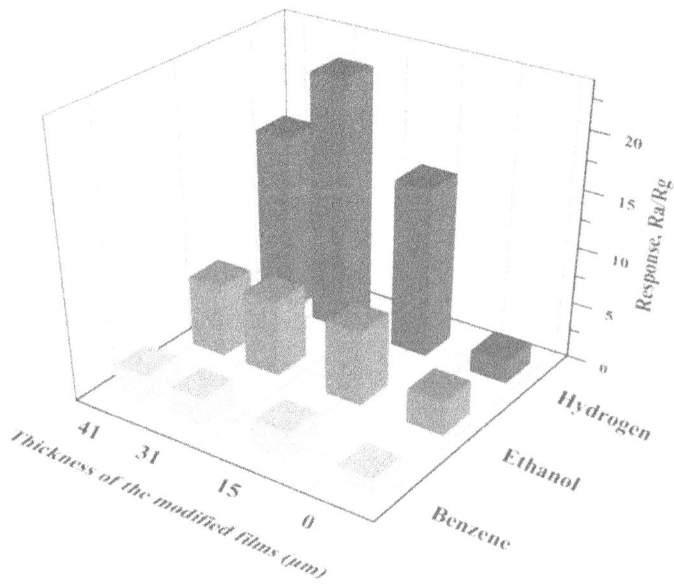

FIGURE 4.7 The schematic diagram for selectivity of gas sensors toward different gases [114].

4.4.4 Stability

Stability is a crucial parameter in the expansion of gas sensors for market as sensors would yield stability in addition to recycled electrical signals as a minimum for 1-2 years which resembles to 8,000 h–16,000 h of operations [117]. Stability of sensor is associated with recyclability of sensor properties in a definite period at working situations with high operating temperature and existence of identified analytes. Other stability is related to retentive the sensor response and selectivity with certain time at regular loading circumstances like RT and relative humidity. According to Korotcenkov et al. features that may be in control for variability are phase transformation, structural transformation, degradation of contacts, bulk diffusion, variation in humidity, variations of operating temperature in surroundings, and interfering effect. Stability could be amplified to some amount by calcination or annealing employing postprocessing dealing and reducing operating temperature of sensing materials. Doping of MOS with metals or synthesis of nanocomposites intensifies stability of sensor materials.

4.5 APPLICATIONS OF GAS SENSORS

Gas sensors are being applied to several industrial applications for detecting gasses and pollutants. In addition, gas sensors have many applications in biomedical, MEMS, mechanical, and electronics. A summary of various application sectors can be given in Figure 4.8 [114, 116–118].

Table 4.5 gives an overview of advantages and applications of various gas sensors.

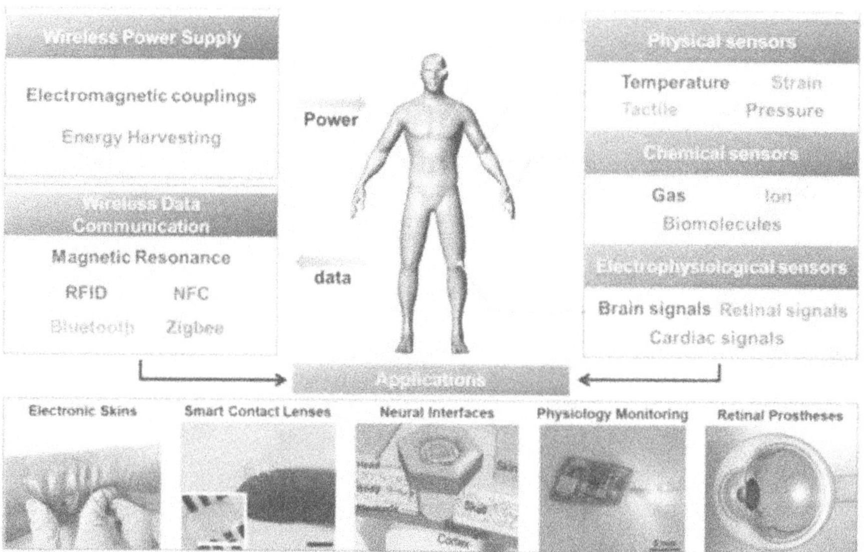

FIGURE 4.8 Summary and comparison for different gas sensor applications [118].

TABLE 4.5
An Overview of Gas Sensing Materials and Applications [113]

	Merits and demerits of materials for target gas applications		
MOX	(1) Cheaper (2) Fast response	(1) Low sensitivity (2) Sensitive to environment	(1) Industries and public use.
Polymer	(1) High sensitivity; (2) Fast response (3) Cheap (4) Portable, light; (5) Low power.	(1) Long-time instability; (1) Irreversibility (2) Poor selectivity	(1) Indoor air monitoring; (1) Paints, wax or fuels storage (2) Chemical industries.
CNTs	(1) Ultra-sensitive; (2) Great adsorption (3) High surface to volume (4) Fast response (5) Light	(1) Processing issues (2) Expensive.	(1) Partial discharge
Moisture Absorbing Material	(1) Cheaper; (2) Lightweight; (3) High selectivity to water	(1) Vulnerable to friction; (2) High irreversibility	(1) Humidity monitoring
Optical Methods	(1) High sensitivity and selectivity (2) Insensitive to environment (3) Durable	(1) Difficulty in miniaturization; (2) High cost	(1) Air quality monitoring; (2) Gas leakage (3) Accuracy and safety
Calorimetric Methods	(1) High stability at ambient temperature; (2) Cheaper (3) ppth range sensitivity	(1) Catalyst poisoning and explosion (2) Selectivity issues	(1) Combustible gases in industries (2) Chemical plants; (3) Mining (4) Kitchen.
Gas Chromatograph	(1) Excellent separation (2) High sensitivity and selectivity	(1) High cost; (2) Difficulty in miniaturization.	Lab analysis.
Acoustics	(1) Highly durable; (2) No secondary pollution.	(1) Low sensitivity; (2) Sensitive to environment	(1) Wireless networks

4.5.1 Bio-medical Applications

Biomedical application of sensors is very important for human life. In such a sensor the target species of the gas or analyte concentration brings about a change in the electrical signal [119–121]. Various electrochemical cell variations are chosen for electrical characterization of the sensors such as potentiometric, amperometric, or conductometric sensors [122]. The working operation of a particular sensor depends upon the analyte concentration, sensitivity, and selectivity of the overall device. Mostly field effect (FET) type sensors have been utilized for these configurations in potentiometric or amperometric format. The examples include FET-based electrochemical cells with several electrodes, e.g., reference, counter, and working electrode where the reactions occur [121, 122].

4.5.1.1 pH Sensors

Other important sensors include pH sensors in biomedical applications for the determination of body pH and health monitoring and diagnosing various serious diseases due to environmental pollutants and other causes. Most of the therapies use pH sensors for monitoring body fluids in critical cases. The pH sensors are also utilized for the control of biochemical processes and treatments such as tumor cells, blood plasma levels, bone marrow applications for which indication of pH is known as the onset of the cell division [123, 124]. The use of pH sensors for various media can also be exploited for the understanding of the various life processes such as endocytosis and phagocytosis [125, 126].

4.5.1.2 Urea Sensors

The urea sensors are generally known for indicators in eliminating the levels of urea in blood plasma. Biosensors have been developed for the detection of urea in the joints and bones. Most of these are based on the FETs such as enzymatic FETs [127, 128]. The applications include medical diagnostics in pathology, such as kidney, blood cancer, sugar levels, and enlargement of the thyroid glands.

4.5.1.3 Glucose Sensors

Glucose sensors are the need of the hour in common men's life especially old people due to their easy and busy lifestyle. The blood glucose levels are an important indication to control diabetes and functioning of the pancreas. Clark and Lyons in 1962 discovered the first biosensor for the measurement of glucose levels in bloodstream [129]. The working principle depends upon the production of gluconate and H_2O_2 with glucose oxidase as a catalyst in glucose solution. The sensing response is measured by the dissociation reaction of H_2O_2 which produces a change in the membrane potential usually made of zinc oxide.

Various nanostructured sensing membranes are available in the past such as micro-, nano-ZnO wires, tubes, and nails manufactured at <100°C or 650 and 960°C for glucose detection. The sensitivity varies from 99% and greater according to the type of nanostructure [130]. Besides, a change in morphology of the nanostructure also affects the sensing behavior of the glucose sensors.

4.5.1.4 Calcium Ion Sensors

Ca ions play a key role in our body for various life processes, e.g., cell division, gene expression, and apoptosis. They interact with the proteins with varying degrees of affinity and trigger the biochemical processes [131, 132]. Various types of nanostructured materials used for sensing mostly based on semiconductors offer good biocompatibility, response, and miniaturization. Most widely used materials for sensing membranes are specially made of zinc oxides nanorods to measure the concentration of intra- and extracellular Ca ions.

4.5.1.5 DNA Sensors

DNA or nucleic acid biosensors are very important for a variety of applications, e.g., determination of pathogens, screening, and diagnosis of genetic disorders, virus, and tumor cell growth prediction. The operating mechanism of DNA sensors involves the recognition of a single sequence of the nucleic acids hybridized with a complementary DNA/RNA strand (Figure 4.9).

The detection system consists of a series of enzyme–DNA nanostructures for target recognition, signaling, and visual detection. In the presence of complementary target DNA, the complex dissociates to activate the polymerase activity. Modified deoxynucleotides are incorporated to immobilize horseradish peroxidase onto the signaling nanostructures. Upon the addition of optical substrate, visual signals can be enzymatically enhanced, detected by the naked eye, and quantified with a smartphone camera. The common cartridge houses the universal signaling nanostructures,

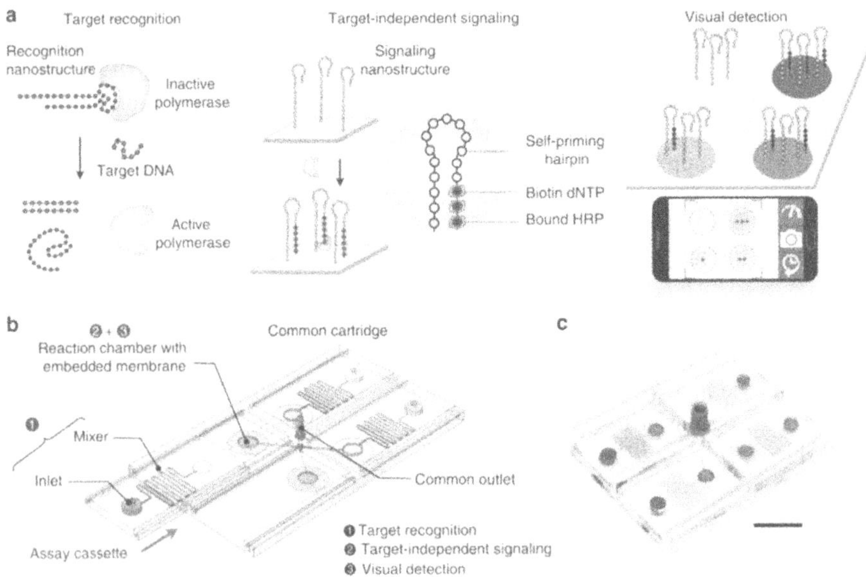

FIGURE 4.9 Visual and modular detection of pathogen nucleic acids [133]. (a) Target recognition, target-independent signaling and visual detection, (b) Microfluidic reaction chamber, and (c) Microfluidic prototype.

which are immobilized on embedded membranes, for target independent signaling and visual detection [133]. Various FET sensors have been used for DNA sensing where FET acts as sensing membranes that can amplify the sequence of nucleic acid hybridization [133].

4.5.1.6 Immunosensors

Immunosensors are usually based on the interaction of the antigen and antibody in body fluids in clinical diagnostics. Mostly they are used in case of environmental pollution and microbes' contaminants such as bacteria, viruses, or insecticides. More recently, immunosensors have been fabricated from FETs, such as separate extended gate SEGFET technology. It can detect the dengue virus protein [134]. The extended gate terminal is made of Au electrode decorated with an anti-NS1 antibody. Such type of device can measure less than 1 g/mL with an accuracy of 0.25 g/mL.

4.5.1.7 Wearable Sensors

With the advancement of modern technology and sensors in various fields of applications, wearable sensors find a top place in sensor applications. Such wearable sensors can be divided according to their function and applications, such as motion sensors, biosensors, and environmental sensors. As already discussed, biosensors are used for a variety of diseases and monitoring of patients, such as glucose sensors, blood pressure and ECG monitor, and temperature [135, 136]. Wearable and flexible sensors mostly comprise PDMS polymer with secondary reinforcements such as CNT and graphene for transparent and flexible NH_3 gas sensors [137].

They reported a considerable ECG signal stability even after a week of wearing which showed promising applications in sensor healthcare area. Therefore, more research in these directions needed to promote the development of wearable devices for gas sensing and monitoring of various diseases and diagnostics in practice [138].

4.5.2 Detection of Gases and Pollutants

Most important application of gas sensors belongs to the identification of air pollution levels. The air pollutants are invisible particulates and therefore we cannot realize their effect immediately. For example, CH_4 gas is a clean source of energy but its leakage can cause severe global warming after CO_2. Various nanoparticles such as Au and Ag, quantum dots, magnetic nanoparticles, and CNTs have highly selective and sensitive target detection [56, 86].

4.5.3 Water Contamination

Water quality monitoring is done by sampling of individual points in time and space, and analysis in laboratory-scale within days. Therefore, there is a little probability of catching periodic pollution even [139] and follow-up chance is even smaller. In recent years, advanced machine learning tools have been used to monitor massive data related to water quality as shown in Figure 4.10.

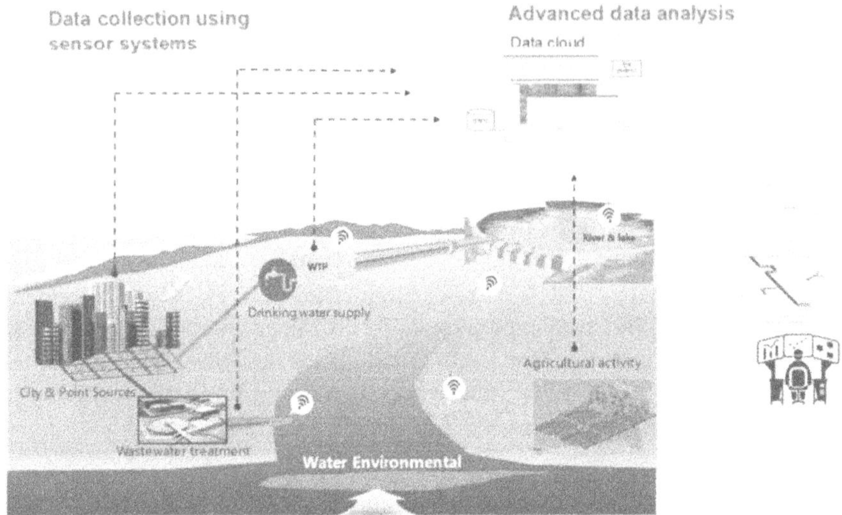

FIGURE 4.10 Gas sensor analysis for detecting water contamination level [140].

Figure 4.11 shows the gas sensor data for detecting water contamination level. Various fast methods such as measuring ionic conductivity, oxygen level, pH, and turbidity are already well established in various systems [140].

4.6 LATEST DEVELOPMENTS IN GAS SENSORS

Rapid development in smart, intelligent materials and internet of things (IoT) technology has taken place in modern research world. Therefore, sensors technology has already gripped people's lives across the globe. The most famous flexible electronic skin (E-skin) is used as a pressure sensor made of polymer switching matrices for display panels, robotics, etc. [141, 142]. Simulating human skin with high-resolution sensing response to temperature and pressure is quite interesting though it has some hurdles for intelligent health care monitoring [143, 144].

4.6.1 MOX/MOSFET Sensors

Figure 4.11 shows the MOSFET gas sensor based on the change in the work potential of gas-sensitive gate electrode when exposed to gases. The construction includes a metal–SiO_2–Si structure where Si regions serve as source and drain. The source being grounded, the charge carriers that move flow between source and drain can be controlled by applied gate voltage. The sensing response depends on the gate voltage change in a gas atmosphere. Recent studies are devoted to the MOSFET for hydrogen sensing where H_2 atoms are diffused to the interface of metal oxide layer and form a dipole layer inducing the change in gate voltage. Consequently, gate voltage

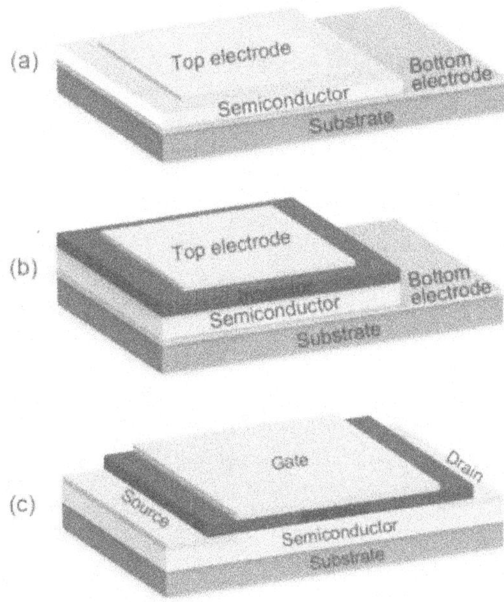

FIGURE 4.11 Semiconductor hydrogen sensor [145]. (a) Schottky type, (b) MOS type, and (c) MOSFET type.

leads to the change in measured output sensing response concerning different hydrogen concentrations [145].

Common materials used for hydrogen sensing include Pd, Pt, Au, Ag, Cu, etc. Yamamoti et al. gave an alternative mechanism of hydrogen sensing of Pd/TiO2 Schottky diode [146]. Their observation suggests that reduced work potential of sensitive Pd layer could be related to the adsorption reaction of O_2 ions on Pd with the H_2 molecule. Zemel et al. also commented that hydrogen-induced interface states result in sensing response [147].

4.6.2 Nanostructured Sensor

Nanostructured sensors play a great role in sensing applications. Various nanomaterials have been investigated for various applications including structural, electronic packaging, as well as sensing technology for many years [148, 149]. Nanostructured materials possess a higher surface-to-volume ratio which can be exploited for higher sensitivity, response, lower operating temperatures, and recovery times. In the last few years, an enormous amount of research has been done to examine hydrogen sensing of thin films. These parameters include particle diameter, pores, grain orientation, doping levels, metal additives, and geometry of electrodes, etc. Various metal oxide nanoparticles are SnO_2, TiO_2, ZnO, In_2O_3, WO_3, CuO, NiO thin films [111]. Table 4.6 shows the recent literature on hydrogen gas sensors.

TABLE 4.6
Hydrogen Sensors Materials and Processes [145]

Materials	Synthesis Method	Working Temp. (°C)	Hydrogen response time	Ref.
SnO_2	Sol-gel annealing	100–300	<10 s	[150]
CNT-doped SnO_2	Sol-gel annealing	150–300	<5s	[151]
Pd-doped SnO_2	Reactive sputtering	50–300	Several minutes	[152]
Al-doped ZnO	Magnetron sputtering	40–100	10 min	[153]
Nanoporous TiO_2	Thermal oxidation	500	10 s	[154]
CNT-doped WO_3	Evaporation	200–400	NA	[155]
Pd-doped WO_3	Sol-gel annealing	20–350	<100 s	[156]
Pt-doped WO_3	RF sputtering	95–220	0.7 min	[157]
CuO	Thermal oxidation	300–800	5 min	[158]
NiO	sputtering	300–650	5 min	[159]

4.6.3 Graphene Sensors

Graphene is an excellent carbon nanomaterial which composed of single layer of graphite atoms with a thickness 0.35 nm and 97.7% transparency. Graphene is known to be a material with greatest strength known at present. Graphene is a 2-D structure that can be further transformed into novel zero-dimensional fullerenes, 1 D CNTs or can be stacked into 3-D graphite [57]. All these great properties of graphene make an excellent candidate for photocatalysts, sensing, supercapacitors, energy storage, and so on [15, 57, 58, 95].

4.7 FUTURE DIRECTIONS

Gas sensors are the need of the hour in modern technology and development. The field of gas sensors is very diverse and multidisciplinary with many technologies of expertise. Examples of various applications include biomedical, transport, health monitoring, climate control, and agriculture. This field is expected to grow more rapidly with the modern Internet of Things. However, various guidelines can be though based on this chapter can be summarized as:

- In-depth understanding of basic mechanisms associated with the sensor and response process.
- Ability to tune the sensitivity to a wide variety of gases and increasing levels of pollutants in the environment to assure appropriate use.
- Tailoring of microstructure using novel nanomaterials and control of material properties for high response and selectivity.
- Use of lightweight materials and technology for higher-level integration of sensor devices and robotic feasibility.
- Multidisciplinary approach and design enabling present technology and providing new avenues for versatile gas sensors.

REFERENCES

1. Error, N.G., Coppersmith, S.N., Hill, M., Dean, P.D., Murray, R.W., Peercy, P.S., Rogers, C.A., Sadoway, D.R., Thome, J.R., Wagner, J.W., 1995. Expanding the Vision of Sensor Materials, National Academic Press, Washington, DC, USA.
2. Gimzewski, J.K., Gerber, C., Meyer, E. and Schlittler, R.R., 1994. Observation of a chemical reaction using a micromechanical sensor. Chemical Physics Letters, 217(5–6), pp. 589–594.
3. Gopel, W., Hesse, J. and Zemel, J.N., 1989. Sensors a Comprehensive Survey. vol. 1, Fundamentals and General Aspects, John Wiley & Sons, Inc., NY, USA.
4. Lion, K.S., 1969. Transducers: Problems and prospects. IEEE Transactions on Industrial Electronics and Control Instrumentation, (1), pp. 2–5.
5. Middelhoek, S. and Noorlag, D.J.W., 1981. Three-dimensional representation of input and output transducers. Sensors and Actuators, 2, pp.29–41.
6. Albrecht, M. and van Koten, G., 1999. Gas sensor materials based on metallodendrimers. Advanced Materials, 11(2), pp.171–174.
7. Brunet, J., Dubois, M., Pauly, A., Spinelle, L., Ndiaye, A., Guérin, K., Varenne, C. and Lauron, B., 2012. An innovative gas sensor system designed from a sensitive organic semiconductor downstream a nanocarbonaceous chemical filter for the selective detection of NO_2 in an environmental context: Part I: Development of a nanocarbon filter for the removal of ozone. Sensors and Actuators B: Chemical, 173, pp.659–667.
8. Li, X., Wang, Y., Yang, X., Chen, J., Fu, H. and Cheng, T., 2012. Conducting polymers in environmental analysis. TrAC Trends in Analytical Chemistry, 39, pp.163–179.
9. Long, Y.Z., Li, M.M., Gu, C., Wan, M., Duvail, J.L., Liu, Z. and Fan, Z., 2011. Recent advances in synthesis, physical properties and applications of conducting polymer nanotubes and nanofibers. Progress in Polymer Science, 36(10), pp.1415–1442.
10. Zimmer, M., Burgmair, M., Scharnagl, K., Karthigeyan, A., Doll, T. and Eisele, I., 2001. Gold and platinum as ozone sensitive layer in work-function gas sensors. Sensors and Actuators B: Chemical, 80(3), pp.174–178.
11. Huang, Y., Wieck, L. and Tao, S., 2013. Development and evaluation of optical fiber NH_3 sensors for application in air quality monitoring. Atmospheric Environment, 66, pp.1–7.
12. Daza, L., Dassy, S. and Delmon, B., 1993. Chemical sensors based on SnO_2 and WO_3 for the detection of formaldehyde: cooperative effects. Sensors and Actuators B: Chemical, 10(2), pp.99–105.
13. Takada, T., Fukunaga, T. and Maekawa, T., 2000. New method for gas identification using a single semiconductor sensor. Sensors and Actuators B: Chemical, 66(1-3), pp.22–24.
14. Zhao, B., Liu, Z., Liu, Z., Liu, G., Li, Z., Wang, J. and Dong, X., 2009. Silver microspheres for application as hydrogen peroxide sensor. Electrochemistry Communications, 11(8), pp.1707–1710.
15. Ohno, Y., Maehashi, K. and Matsumoto, K., 2010. Chemical and biological sensing applications based on graphene field-effect transistors. Biosensors and Bioelectronics, 26(4), pp.1727–1730.
16. Upadhyayula, V.K., 2012. Functionalized gold nanoparticle supported sensory mechanisms applied in detection of chemical and biological threat agents: a review. Analytica Chimica Acta, 715, pp.1–18.
17. Yi, W., Ke-Jia, W., Qi, W. and Feng, T., 2009, August. Measurement of CH4 by differential infrared optical absorption spectroscopy. In 2009 9th International Conference on Electronic Measurement & Instruments (pp. 1–761). IEEE.
18. Zhu, C., Shan, M., Zhang, H., Zhang, X., Zhen, P. and Tao, Y., 2006. A precise sensor for SF 6 based on piezoelectric ultrasound. Xiyou Jinshu Cailiao yu Gongcheng(Rare Metal Materials and Engineering), 35, pp.157–158.

19. Changping, Z., Minglei, S., Yongfu, L. and Hongping, Z., 2005. Micro-concentration Detector for SF6 Based on CPLD. Chinese Journal of Scientific Instrument, 26(8), pp. 448.
20. Endres, H.E., Göttler, W., Hartinger, R., Drost, S., Hellmich, W., Müller, G., Braunmühl, C.B.V., Krenkow, A., Perego, C. and Sberveglieri, G., 1996. A thin-film SnO_2 sensor system for simultaneous detection of CO and NO_2 with neural signal evaluation. Sensors and Actuators B: Chemical, 36(1-3), pp.353–357.
21. Frodl, R. and Tille, T., 2006. A High-Precision NDIR CO_2 Gas Sensor for Automotive Applications. IEEE Sensors Journal, 6(6), pp.1697–1705.
22. Billi, E., Viricelle, J.P., Montanaro, L. and Pijolat, C., 2002. Development of a protected gas sensor for exhaust automotive applications. IEEE Sensors Journal, 2(4), pp.342–348.
23. Song, K., Wang, Q., Liu, Q., Zhang, H. and Cheng, Y., 2011. A wireless electronic nose system using a Fe_2O_3 gas sensing array and least squares support vector regression. Sensors, 11(1), pp.485–505.
24. Lilienthal, A.J., Loutfi, A. and Duckett, T., 2006. Airborne chemical sensing with mobile robots. Sensors, 6(11), pp.1616–1678.
25. Tian, F., Yang, S.X. and Dong, K., 2005. Circuit and noise analysis of odorant gas sensors in an E-nose. Sensors, 5(1), pp.85–96.
26. Hulko, M., Hospach, I., Krasteva, N. and Nelles, G., 2011. Cytochrome C biosensor—A model for gas sensing. Sensors, 11(6), pp.5968–5980.
27. Lazik, D., Ebert, S., Leuthold, M., Hagenau, J. and Geistlinger, H., 2009. Membrane based measurement technology for in situ monitoring of gases in soil. Sensors, 9(2), pp.756–767.
28. Fine, G.F., Cavanagh, L.M., Afonja, A. and Binions, R., 2010. Metal oxide semiconductor gas sensors in environmental monitoring. Sensors, 10(6), pp.5469–5502.
29. Jimenez, I., Vilà, A.M., Calveras, A.C. and Morante, J.R., 2005. Gas-sensing properties of catalytically modified WO_3 with copper and vanadium for NH_3 detection. IEEE Sensors Journal, 5(3), pp.385–391.
30. Kanan, S.M., El-Kadri, O.M., Abu-Yousef, I.A. and Kanan, M.C., 2009. Semiconducting metal oxide based sensors for selective gas pollutant detection. Sensors, 9(10), pp.8158–8196.
31. Yang, L., Zhang, R., Staiculescu, D., Wong, C.P. and Tentzeris, M.M., 2009. A novel conformal RFID-enabled module utilizing inkjet-printed antennas and carbon nanotubes for gas-detection applications. IEEE Antennas and Wireless Propagation Letters, 8, pp.653–656.
32. Phillips, M., Gleeson, K., Hughes, J.M.B., Greenberg, J., Cataneo, R.N., Baker, L. and McVay, W.P., 1999. Volatile organic compounds in breath as markers of lung cancer: a cross-sectional study. The Lancet, 353(9168), pp.1930–1933.
33. Phillips, M., Cataneo, R.N., Ditkoff, B.A., Fisher, P., Greenberg, J., Gunawardena, R., Kwon, C.S., Rahbari-Oskoui, F. and Wong, C., 2003. Volatile markers of breast cancer in the breath. The Breast Journal, 9(3), pp.184–191.
34. Heal, M.R., Kumar, P. and Harrison, R.M., 2012. Particles, air quality, policy and health. Chemical Society Reviews, 41(19), pp.6606–6630.
35. Manisalidis, I., Stavropoulou, E., Stavropoulos, A. and Bezirtzoglou, E., 2020. Environmental and health impacts of air pollution: A review. Frontiers in Public Health, 8, pp. 14.
36. Robinson, A.B., Baliunas, S.L., Soon, W. and Robinson, Z.W., 1998. Environmental effects of increased atmospheric carbon dioxide. Medical Sentinel, 3(5), pp.171–178.
37. Richmond-Bryant, J., Owen, R.C., Graham, S., Snyder, M., McDow, S., Oakes, M. and Kimbrough, S., 2017. Estimation of on-road NO_2 concentrations, NO_2/NO_X ratios, and related roadway gradients from near-road monitoring data. Air Quality, Atmosphere & Health, 10(5), pp.611–625.

38. Hesterberg, T.W., Bunn, W.B., McClellan, R.O., Hamade, A.K., Long, C.M. and Valberg, P.A., 2009. Critical review of the human data on short-term nitrogen dioxide (NO_2) exposures: evidence for NO2 no-effect levels. Critical Reviews in Toxicology, 39(9), pp.743–781.
39. Chen, T.M., Kuschner, W.G., Gokhale, J. and Shofer, S., 2007. Outdoor air pollution: nitrogen dioxide, sulfur dioxide, and carbon monoxide health effects. The American Journal of the Medical Sciences, 333(4), pp.249–256.
40. Ezzati, M., Lopez, A.D., Rodgers, A. and Murray, C.J., 2004. Comparative quantification of health risks. Global and regional burden of disease attributable to selected major risk factors. Geneva: World Health Organization, pp.1987–97.
41. Goyer, R.A., 1990. Transplacental transport of lead Environ Health Perspect 89: 101–105. Find this article online.
42. Abdel-Shafy, H.I. and Mansour, M.S., 2016. A review on polycyclic aromatic hydrocarbons: source, environmental impact, effect on human health and remediation. Egyptian Journal of Petroleum, 25(1), pp.107–123.
43. Kumar, A., Singh, B.P., Punia, M., Singh, D., Kumar, K. and Jain, V.K., 2014. Assessment of indoor air concentrations of VOCs and their associated health risks in the library of Jawaharlal Nehru University, New Delhi., Environmental Science and Pollution Research, 21(3), pp. 2240.
44. E Mølhave, L., Clausen, G., Berglund, B., De Ceaurriz, J., Kettrup, A., Lindvall, T., Maroni, M., Pickering, A.C., Risse, U., Rothweiler, H. and Seifert, B., 1997. Total volatile organic compounds (TVOC) in indoor air quality investigations. Indoor Air, 7(4), pp.225–2248.
45. Manisalidis, I., Stavropoulou, E., Stavropoulos, A. and Bezirtzoglou, E., 2020. Environmental and health impacts of air pollution: A review. Frontiers Public Health, 8, pp. 14.
46. Choi, K.J. and Jang, H.W., 2010. One-dimensional oxide nanostructures as gas-sensing materials: review and issues. Sensors, 10(4), pp.4083–4099.
47. Raj, V.B., Singh, H., Nimal, A.T., Tomar, M., Sharma, M.U. and Gupta, V., 2013. Effect of metal oxide sensing layers on the distinct detection of ammonia using surface acoustic wave (SAW) sensors. Sensors and Actuators B: Chemical, 187, pp.563–573.
48. Phan, D.T. and Chung, G.S., 2012. Surface acoustic wave hydrogen sensors based on ZnO nanoparticles incorporated with a Pt catalyst. Sensors and Actuators B: Chemical, 161(1), pp.341–348.
49. Sadek, A.Z., Wlodarski, W., Shin, K., Kaner, R.B. and Kalantar-Zadeh, K., 2008. A polyaniline/WO3 nanofiber composite-based ZnO/64 YX LiNbO3 SAW hydrogen gas sensor. Synthetic Metals, 158(1-2), pp.29–32.
50. Tang, Y.L., Li, Z.J., Ma, J.Y., Su, H.Q., Guo, Y.J., Wang, L., Du, B., Chen, J.J., Zhou, W., Yu, Q.K. and Zu, X.T., 2014. Highly sensitive room-temperature surface acoustic wave (SAW) ammonia sensors based on Co_3O_4/SiO_2 composite films. Journal of Hazardous Materials, 280, pp.127–133.
51. Tang, Y.L., Li, Z.J., Ma, J.Y., Guo, Y.J., Fu, Y.Q. and Zu, X.T., 2014. Ammonia gas sensors based on ZnO/SiO_2 bi-layer nanofilms on ST-cut quartz surface acoustic wave devices. Sensors and Actuators B: Chemical, 201, pp.114–121.
52. Reich, S., Thomsen, C. and Maultzsch, J., 2008. Carbon Nanotubes: Basic Concepts and Physical Properties. John Wiley & Sons.
53. Zhao, L., Choi, M., Kim, H.S. and Hong, S.H., 2007. The effect of multiwalled carbon nanotube doping on the CO gas sensitivity of SnO2-based nanomaterials. Nanotechnology, 18(44), p.445501.
54. Zhang, W.D. and Zhang, W.H., 2009. Carbon nanotubes as active components for gas sensors. Journal of Sensors, 2009, pp. 160698.

55. Sayago, I., Fernández, M.J., Fontecha, J.L., Horrillo, M.C., Vera, C., Obieta, I. and Bustero, I., 2011. Surface acoustic wave gas sensors based on polyisobutylene and carbon nanotube composites. Sensors and Actuators B: Chemical, 156(1), pp.1–5.
56. Sayago, I., Fernández, M.J., Fontecha, J.L., Horrillo, M.C., Vera, C., Obieta, I. and Bustero, I., 2012. New sensitive layers for surface acoustic wave gas sensors based on polymer and carbon nanotube composites. Sensors and Actuators B: Chemical, 175, pp.67–72.
57. Nazarpour, S. and Waite, S.R. eds. Graphene Technology: From Laboratory to Fabrication. John Wiley & Sons, New York, NY, USA, 2016.
58. Wei, W., Nong, J., Zhang, G., Tang, L., Jiang, X., Chen, N., Luo, S., Lan, G., Zhu, Y. 2017. Graphene-based long-period fiber grating surface plasmon resonance sensor for high-sensitivity gas sensing. Sensors, 17(1), pp.2.
59. Sharma, B., Sharma, A. and Kim, J.S., 2018. Recent advances on H2 sensor technologies based on MOX and FET devices: A review. Sensors and Actuators B: Chemical, 262, pp.758–770.
60. Korotcenkov, G., 2008. The role of morphology and crystallographic structure of metal oxides in response of conductometric-type gas sensors. Materials Science and Engineering: R: Reports, 61(1-6), pp.1–39.
61. Schierbaum, K.D., Weimar, U., Göpel, W. and Kowalkowski, R., 1991. Conductance, work function and catalytic activity of SnO2-based gas sensors. Sensors and Actuators B: Chemical, 3(3), pp.205–214.
62. Han, C.H., Han, S.D., Singh, I. and Toupance, T., 2005. Micro-bead of nano-crystalline F-doped SnO2 as a sensitive hydrogen gas sensor. Sensors and Actuators B: Chemical, 109(2), pp.264–269.
63. Zeng, Z., Wang, K., Zhang, Z., Chen, J. and Zhou, W., 2008. The detection of H2S at room temperature by using individual indium oxide nanowire transistors. Nanotechnology, 20(4), p.045503.
64. Zhang, D., Li, C., Liu, X., Han, S., Tang, T. and Zhou, C., 2003. Doping dependent NH 3 sensing of indium oxide nanowires. Applied Physics Letters, 83(9), pp.1845–1847.
65. Lundström, I., Shivaraman, S., Svensson, C. and Lundkvist, L., 1975. A hydrogen–sensitive MOS field– effect transistor. Applied Physics Letters, 26(2), pp.55–57.
66. Lundström, I. and Söderberg, D., 1981. Hydrogen sensitive mos-structures part 2: characterization. Sensors and Actuators, 2, pp.105–138.
67. Morita, Y., Nakamura, K.I. and Kim, C., 1996. Langmuir analysis on hydrogen gas response of palladium-gate FET. Sensors and Actuators B: Chemical, 33(1-3), pp.96–99.
68. Hong, Y., Wu, M., Bae, J.H., Hong, S., Jeong, Y., Jang, D., Kim, J.S., Hwang, C.S., Park, B.G. and Lee, J.H., 2020. A new sensing mechanism of Si FET-based gas sensor using pre-bias. Sensors and Actuators B: Chemical, 302, p.127147.
69. Miasik, J.J., Hooper, A. and Tofield, B.C., 1986. Conducting polymer gas sensors. Journal of the Chemical Society, Faraday Transactions 1: Physical Chemistry in Condensed Phases, 82(4), pp.1117–1126.
70. Barisci, J.N., Conn, C. and Wallace, G.G., 1996. Conducting polymer sensors. Trends in Polymer Science, 9(4), pp.307–311.
71. Matsui, J., Akamatsu, K., Nishiguchi, S., Miyoshi, D., Nawafune, H., Tamaki, K. and Sugimoto, N., 2004. Composite of Au nanoparticles and molecularly imprinted polymer as a sensing material. Analytical Chemistry, 76(5), pp.1310–1315.
72. Ram, M.K., Yavuz, Ö., Lahsangah, V. and Aldissi, M., 2005. CO gas sensing from ultrathin nano-composite conducting polymer film. Sensors and Actuators B: Chemical, 106(2), pp.750–757.
73. Liu, Y., Mills, R.C., Boncella, J.M. and Schanze, K.S., 2001. Fluorescent polyacetylene thin film sensor for nitroaromatics. Langmuir, 17(24), pp.7452–7455.

74. Esteves, C.H., Iglesias, B.A., Li, R.W., Ogawa, T., Araki, K. and Gruber, J., 2014. New composite porphyrin-conductive polymer gas sensors for application in electronic noses. Sensors and Actuators B: Chemical, 193, pp.136–141.
75. Sergeyeva, T.A., Piletsky, S.A., Brovko, A.A., Slinchenko, E.A., Sergeeva, L.M. and El'Skaya, A.V., 1999. Selective recognition of atrazine by molecularly imprinted polymer membranes. Development of conductometric sensor for herbicides detection. Analytica Chimica Acta, 392(2-3), pp.105–111.
76. Zee, F. and Judy, J.W., 2001. Micromachined polymer-based chemical gas sensor array. Sensors and Actuators B: Chemical, 72(2), pp.120–128.
77. Weppner, W., 1987. Solid-state electrochemical gas sensors. Sensors and Actuators, 12(2), pp. 107–119.
78. Mao, L., Yamamoto, K., Zhou, W. and Jin, L., 2000. Electrochemical nitric oxide sensors based on electropolymerized film of M (salen) with central ions of Fe, Co, Cu, and Mn. Electroanalysis: An International Journal Devoted to Fundamental and Practical Aspects of Electroanalysis, 12(1), pp.72–77.
79. Matuszewski, W. and Meyerhoff, M.E., 1991. Continuous monitoring of gas-phase species at trace levels with electrochemical detectors: Part 1. Direct amperometric measurement of hydrogen peroxide and enzyme-based detection of alcohols and sulfur dioxide. Analytica Chimica Acta, 248(2), pp.379–389.
80. Lin, Q., Cheng, C., Zou, J., Kane, N., Jin, H., Zhang, X., Gao, W., Jin, Q. and Jian, J., 2020. Study of response and recovery rate of YSZ-based electrochemical sensor by laser ablation method. Ionics, 4163, pp. 26.
81. Kang, B.S., Ren, F., Gila, B.P., Abernathy, C.R. and Pearton, S.J., 2004. AlGaN/GaN-based metal–oxide–semiconductor diode-based hydrogen gas sensor. Applied Physics Letters, 84(7), pp.1123–1125.
82. Kim, Y.S., Ha, S.C., Kim, K., Yang, H., Choi, S.Y., Kim, Y.T., Park, J.T., Lee, C.H., Choi, J., Paek, J. and Lee, K., 2005. Room-temperature semiconductor gas sensor based on nonstoichiometric tungsten oxide nanorod film. Applied Physics Letters, 86(21), p.213105.
83. Utriainen, M., Kärpänoja, E. and Paakkanen, H., 2003. Combining miniaturized ion mobility spectrometer and metal oxide gas sensor for the fast detection of toxic chemical vapors. Sensors and Actuators B: Chemical, 93(1-3), pp.17–24.
84. Rai, P., Majhi, S.M., Yu, Y.T. and Lee, J.H., 2015. Noble metal@ metal oxide semiconductor core@ shell nano-architectures as a new platform for gas sensor applications. RSC Advances, 5(93), pp.76229–76248.
85. de Mos, M., De Bruijn, A.G.J., Huygen, F.J.P.M., Dieleman, J.P., Stricker, B.C. and Sturkenboom, M.C.J.M., 2007. The incidence of complex regional pain syndrome: a population-based study. Pain, 129(1-2), pp.12–20.
86. Iizuka, K., Kambara, M. and Yoshida, T., 2012. Highly sensitive SnO_2 porous film gas sensors fabricated by plasma spray physical vapor deposition. Sensors and Actuators B: Chemical, 173, pp.455–461.
87. Park, S.H., Son, Y.C., Willis, W.S., Suib, S.L. and Creasy, K.E., 1998. Tin oxide films made by physical vapor deposition-thermal oxidation and spray pyrolysis. Chemistry of Materials, 10(9), pp.2389–2398.
88. Punetha, D. and Pandey, S.K., 2018. CO gas sensor based on e-beam evaporated ZnO, MgZnO, and CdZnO thin films: a comparative study. IEEE Sensors Journal, 19(7), pp.2450–2457.
89. Wisitsoraat, A., Tuantranont, A., Patthanasettakul, V., Lomas, T. and Chindaudom, P., 2005. Ion-assisted e-beam evaporated gas sensor for environmental monitoring. Science and Technology of Advanced Materials, 6(3-4), p.261.
90. Van Hieu, N., Van Vuong, H., Van Duy, N. and Hoa, N.D., 2012. A morphological control of tungsten oxide nanowires by thermal evaporation method for sub-ppm NO2 gas sensor application. Sensors and Actuators B: Chemical, 171, pp.760–768.

91. Van Hieu, N., Khoang, N.D., Minh, N.T., Trung, T. and Chien, N.D., 2010. A facile thermal evaporation route for large-area synthesis of tin oxide nanowires: characterizations and their use for liquid petroleum gas sensor. Current Applied Physics, 10(2), pp.636–641.
92. Samarasekara, P., Kumara, N.T.R.N. and Yapa, N.U.S., 2006. Sputtered copper oxide (CuO) thin films for gas sensor devices. Journal of Physics: Condensed Matter, 18(8), p.2417.
93. Ferroni, M., Guidi, V., Martinelli, G., Nelli, P., Sacerdoti, M. and Sberveglieri, G., 1997. Characterization of a molybdenum oxide sputtered thin film as a gas sensor. Thin Solid Films, 307(1-2), pp.148–151.
94. Heo, Y.W., Varadarajan, V., Kaufman, M., Kim, K., Norton, D.P., Ren, F. and Fleming, P.H., 2002. Site-specific growth of ZnO nanorods using catalysis-driven molecular-beam epitaxy. Applied Physics Letters, 81(16), pp.3046–3048.
95. Zhong, A. and Hane, K., 2012. Growth of GaN nanowall network on Si (111) substrate by molecular beam epitaxy. Nanoscale Research Letters, 7(1), pp.1–7.
96. Huang, H., Tan, O.K., Lee, Y.C., Tran, T.D., Tse, M.S. and Yao, X., 2005. Semiconductor gas sensor based on tin oxide nanorods prepared by plasma-enhanced chemical vapor deposition with postplasma treatment. Applied Physics Letters, 87(16), p.163123.
97. Abdullah, Q.N., Yam, F.K., Hassan, J.J., Chin, C.W., Hassan, Z. and Bououdina, M., 2013. High performance room temperature GaN-nanowires hydrogen gas sensor fabricated by chemical vapor deposition (CVD) technique. International Journal of Hydrogen Energy, 38(32), pp.14085–14101.
98. Yu, K., Bo, Z., Lu, G., Mao, S., Cui, S., Zhu, Y., Chen, X., Ruoff, R.S. and Chen, J., 2011. Growth of carbon nanowalls at atmospheric pressure for one-step gas sensor fabrication. Nanoscale Research Letters, 6(1), p.202.
99. Wu, M.R., Li, W.Z., Tung, C.Y., Huang, C.Y., Chiang, Y.H., Liu, P.L. and Horng, R.H., 2019. NO gas sensor based on $ZnGa_2O_4$ epilayer grown by metalorganic chemical vapor deposition. Scientific Reports, 9(1), pp.1–9.
100. Chan, P.C., Yan, G.Z., Sheng, L.Y., Sharma, R.K., Tang, Z., Sin, J.K., Hsing, I.M. and Wang, Y., 2002. An integrated gas sensor technology using surface micro-machining. Sensors and Actuators B: Chemical, 82(2-3), pp.277–283.
101. Yu-Sheng, S. and Tian-Shu, Z., 1993. Preparation, structure and gas-sensing properties of ultramicro $ZnSnO_3$ powder. Sensors and Actuators B: Chemical, 12(1), pp.5–9.
102. Yan, G., Tang, Z., Chan, P.C., Sin, J.K., Hsing, I.M. and Wang, Y., 2002. An experimental study on high-temperature metallization for micro-hotplate-based integrated gas sensors. Sensors and Actuators B: Chemical, 86(1), pp.1–11.
103. Talazac, L., Barbarin, F., Mazet, L. and Varenne, C., 2004. Improvement in sensitivity and selectivity of InP-based gas sensors: Pseudo-Schottky diodes with palladium metallizations. IEEE Sensors Journal, 4(1), pp.45–51.
104. Meng, D., Yamazaki, T., Shen, Y., Liu, Z. and Kikuta, T., 2009. Preparation of WO_3 nanoparticles and application to NO_2 sensor. Applied Surface Science, 256(4), pp.1050–1053.
105. Salehi, A., 2002. Selectivity enhancement of indium-doped SnO_2 gas sensors. Thin Solid Films, 416(1-2), pp.260–263.
106. Tamaki, J., Naruo, C., Yamamoto, Y. and Matsuoka, M., 2002. Sensing properties to dilute chlorine gas of indium oxide based thin film sensors prepared by electron beam evaporation. Sensors and Actuators B: Chemical, 83(1-3), pp.190–194.
107. Li, X., Peng, K., Dou, Y., Chen, J., Zhang, Y. and An, G., 2018. Facile synthesis of wormhole-like mesoporous tin oxide via evaporation-induced self-assembly and the enhanced gas-sensing properties. Nanoscale Research Letters, 13(1), p.14.
108. Seals, L., Gole, J.L., Tse, L.A. and Hesketh, P.J., 2002. Rapid, reversible, sensitive porous silicon gas sensor. Journal of Applied Physics, 91(4), pp.2519–2523.

109. Wang, C., Yin, L., Zhang, L., Xiang, D. and Gao, R., 2010. Metal oxide gas sensors: sensitivity and influencing factors. Sensors, 10(3), pp.2088–2106.
110. Barsan, N.; Koziej, D.; Weimar, U. Metal oxide-based gas sensor research: How to? Sensors and Actuators B 2007, *121*, 18–35.
111. Lee, J.M., Park, J.E., Kim, S., Kim, S., Lee, E., Kim, S.J. and Lee, W., 2010. Ultra-sensitive hydrogen gas sensors based on Pd-decorated tin dioxide nanostructures: Room temperature operating sensors. International Journal of Hydrogen Energy, 35(22), pp.12568–12573.
112. Kumar, M.K., Tan, L.K., Gosvami, N.N. and Gao, H., 2009. Conduction-atomic force microscopy study of H_2 sensing mechanism in Pd nanoparticles decorated TiO_2 nano-film. Journal of Applied Physics, 106(4), p.044308.
113. Liu, X., Cheng, S., Liu, H., Hu, S., Zhang, D. and Ning, H., 2012. A survey on gas sensing technology. Sensors, 12(7), pp.9635–9665.
114. Xue, N., Zhang, Q., Zhang, S., Zong, P. and Yang, F., 2017. Highly sensitive and selective hydrogen gas sensor using the mesoporous SnO_2 modified layers. Sensors, 17(10), p.2351.
115. Dey, A., 2018. Semiconductor metal oxide gas sensors: A review. Materials Science and Engineering: B, 229, pp.206–217.
116. Wang, C., Wang, Y., Zhang, S.Y., Fan, L. and Shui, X.J., 2012. Characteristics of SAW hydrogen sensors based on InOx/128° YX-LiNbO3 structures at room temperature. Sensors and Actuators B: Chemical, 173, pp.710–715.
117. Karadurmuz, L., Kurbanoglu, S., Uslu, B. and A Ozkan, S., 2017. Electrochemical DNA biosensors in drug analysis. Current Pharmaceutical Analysis, 13(3), pp.195–207.
118. Park, Y.G., Lee, S., Park, J.U., 2019. Recent progress in wireless sensors for wearable electronics. Sensors, 19(20), pp. 4353.
119. Guth, U., Vonau, W. and Zosel, J., 2009. Recent developments in electrochemical sensor application and technology—a review. Measurement Science and Technology, 20(4), p.042002.
120. Harguindey, S., Pedraz, J.L., Canero, R.G., de Diego, J.P. and Cragoe Jr, E.J., 1995. Hydrogen ion-dependent oncogenesis and parallel new avenues to cancer prevention and treatment using a H+-mediated unifying approach: pH-related and pH-unrelated mechanisms. Critical Reviews™ in Oncogenesis, 6(1).
121. Gottlieb, R.A. and Dosanjh, A., 1996. Mutant cystic fibrosis transmembrane conductance regulator inhibits acidification and apoptosis in C127 cells: possible relevance to cystic fibrosis. Proceedings of the National Academy of Sciences, 93(8), pp.3587–3591.
122. Miksa, M., Komura, H., Wu, R., Shah, K.G. and Wang, P., 2009. A novel method to determine the engulfment of apoptotic cells by macrophages using pHrodo succinimidyl ester. Journal of Immunological Methods, 342(1-2), pp.71–77.
123. Richter, A., Paschew, G., Klatt, S., Lienig, J., Arndt, K.F. and Adler, H.J.P., 2008. Review on hydrogel-based pH sensors and microsensors. Sensors, 8(1), pp.561–581.
124. Fog, A. and Buck, R.P., 1984. Electronic semiconducting oxides as pH sensors. Sensors and Actuators, 5(2), pp.137–146.
125. Talley, C.E., Jusinski, L., Hollars, C.W., Lane, S.M. and Huser, T., 2004. Intracellular pH sensors based on surface-enhanced Raman scattering. Analytical Chemistry, 76(23), pp.7064–7068.
126. Kreft, O., Javier, A.M., Sukhorukov, G.B. and Parak, W.J., 2007. Polymer microcapsules as mobile local pH-sensors. Journal of Materials Chemistry, 17(42), pp.4471–4476.
127. Lue, C.E., Yu, T.C., Yang, C.M., Pijanowska, D.G. and Lai, C.S., 2011. Optimization of urea-EnFET based on Ta2O5 layer with post annealing. Sensors, 11(5), pp.4562–4571.

128. Abd-Alghafour, N.M., Ahmed, N.M., Hassan, Z., Almessiere, M.A., Bououdina, M. and Al-Hardan, N.H., 2017. High sensitivity extended gate effect transistor based on V_2O_5 nanorods. Journal of Materials Science: Materials in Electronics, 28(2), pp.1364–1369.
129. Clark Jr, L.C., Noyes, L.K., Spokane, R.B., Sudan, R. and Miller, M.L., 1988. [6] Long-term implantation of voltammetric oxidase/peroxide glucose sensors in the rat peritoneum. In Methods in Enzymology (Vol. 137, pp. 68–89). Academic Press.
130. Clark, L.C., Spokane, R.B., Sudan, R. and Stroup, T.L., 1987. Long-lived implanted silastic drum glucose sensors. ASAIO Journal, 33(3), pp.323–328.
131. Suzuki, K., Tohda, K., Tanda, Y., Ohzora, H., Nishihama, S., Inoue, H. and Shirai, T., 1989. Fiber-optic magnesium and calcium ion sensor based on a natural carboxylic polyether antibiotic. Analytical Chemistry, 61(4), pp.382–384.
132. Craggs, A., Moody, G.J. and Thomas, J.D.R., 1979. Evaluation of calcium ion-selective electrodes based on di (n-alkylphenyl)-phosphate sensors and their calibration with ion buffers. Analyst, 104(1238), pp.412–418.
133. Ho, N.R., Lim, G.S., Sundah, N.R., Lim, D., Loh, T.P. and Shao, H., 2018. Visual and modular detection of pathogen nucleic acids with enzyme–DNA molecular complexes. Nature Communications, 9(1), pp.1–11.
134. Prodromidis, M.I., 2010. Impedimetric immunosensors—A review. Electrochimica Acta, 55(14), pp.4227–4233.
135. Heikenfeld, J., Jajack, A., Rogers, J., Gutruf, P., Tian, L., Pan, T., Li, R., Khine, M., Kim, J. and Wang, J., 2018. Wearable sensors: modalities, challenges, and prospects. Lab on a Chip, 18(2), pp.217–248.
136. Mukhopadhyay, S.C., 2014. Wearable sensors for human activity monitoring: A review. IEEE Sensors Journal, 15(3), pp.1321–1330.
137. Xue, L., Wang, W., Guo, Y., Liu, G. and Wan, P., 2017. Flexible polyaniline/carbon nanotube nanocomposite film-based electronic gas sensors. Sensors and Actuators B: Chemical, 244, pp.47–53.
138. Khan, S., Ali, S. and Bermak, A., 2019. Recent developments in printing flexible and wearable sensing electronics for healthcare applications. Sensors, 19(5), p.1230.
139. Vaseashta, A., Vaclavikova, M., Vaseashta, S., Gallios, G., Roy, P. and Pummakarnchana, O., 2007. Nanostructures in environmental pollution detection, monitoring, and remediation. Science and Technology of Advanced Materials, 8(1-2), p.47.
140. Park, J., Kim, K.T., Lee, W.H., 2020. Recent advances in information communications technology (ICT) and sensor technology for monitoring water quality. Water, 12(2), pp.510.
141. Choi, S.J. and Kim, I.D., 2018. Recent developments in 2D nanomaterials for chemiresistive-type gas sensors. Electronic Materials Letters, 14(3), pp.221–260.
142. Moseley, P.T., 2017. Progress in the development of semiconducting metal oxide gas sensors: a review. Measurement Science and Technology, 28(8), p.082001.
143. Zhang, J., Qin, Z., Zeng, D. and Xie, C., 2017. Metal-oxide-semiconductor based gas sensors: screening, preparation, and integration. Physical Chemistry Chemical Physics, 19(9), pp. 6313–6329.
144. Lin, Y. and Fan, Z., 2020. Compositing strategies to enhance the performance of chemiresistive CO2 gas sensors. Materials Science in Semiconductor Processing, 107, p. 104820.
145. Gu, H., Wang, Z. and Hu, Y., 2012. Hydrogen gas sensors based on semiconductor oxide nanostructures. Sensors, 12(5), pp. 5517–5550.
146. Yamamoto, N., Tonomura, S., Matsuoka, T. and Tsubomura, H., 1980. A study on a palladium-titanium oxide Schottky diode as a detector for gaseous components. Surface Science, 92(2-3), pp. 400–406.

147. Keramati, B. and Zemel, J.N., 1982. Pd–thin-SiO2–Si diode. I. Isothermal variation of H2-induced interfacial trapping states. Journal of Applied Physics, 53(2), pp. 1091–1099.
148. Sharma, A., Lim, D.U., Jung, J.P., 2016. Microstructure and brazeability of SiC nanoparticles reinforced Al–9Si–20Cu produced by induction melting. Materials Science and Technology, 32(8), pp. 773–779.
149. Sharma, A., Das, S., Das, K., 2016. Pulse electrodeposition of Lead-Free Tin-Based Composites for Microelectronic Packaging. In Electrodeposition of Composite Materials, Adel M. A. Mohamed and Teresa D. Golden (Eds.), IntechOpen, DOI: 10.5772/62036.
150. Adamyan, A., Adamyan, Z., Aroutiounian, V., Arakelyan, A., Touryan, K., Turner, J., 2007. Sol-gel derived thin-film semiconductor hydrogen gas sensor. International Journal Hydrogen Energy, 32, pp. 4101–4108.
151. Gong, J., Sun, J., Chen, Q., 2008. Micromachined sol-gel carbon nanotube/SnO2 nanocomposite hydrogen sensor. Sensors and Actuators B, 130, pp. 829–835.
152. Shen, Y., Yamazaki, T., Liu, Z., Meng, D., Kikuta, T., 2009. Hydrogen sensing properties of Pd-doped SnO2 sputtered films with columnar nanostructures. Thin Solid Films, 517, pp. 6119–6123.
153. Galstyan, V.E., Aroutiounian, V.M., Arakelyan, V.M., Shahnazaryan, G.E., 2009. Investigation of hydrogen sensor made of ZnO thin film. Armenian Journal of Physics, 1, pp. 242–246.
154. Lu, C., Chen, Z., 2009. High-temperature resistive hydrogen sensor based on thin nanoporous rutile TiO2 film on anodic aluminum oxide. Sensors and Actuators B, 140, pp. 109–115.
155. Wongchoosuk, C., Wisitsoraat, A., Phokharatkul, D., Tuantranont, A., Kerdcharoen, T., 2010. Multi-walled carbon nanotube-doped tungsten oxide thin films for hydrogen gas sensing. Sensors, 10, pp. 7705–7715.
156. Fardindoost, S., Iraji Zad, A., Rahimi, F., Ghasempour, R., 2010. Pd doped WO3 films prepared by sol-gel process for hydrogen sensing. International Journal Hydrogen Energy, 35, pp. 854–860.
157. Patil, L.A., Shinde, M.D., Bari, A.R., Deo, V.V., 2009. Highly sensitive and quickly responding ultrasonically sprayed nanostructured SnO2 thin films for hydrogen gas sensing. Sensors and Actuators B, 143, pp. 270–277.
158. Hoa, N.D., An, S.Y., Dung, N.Q., Van Quy, N., Kim, D., 2010. Synthesis of p-type semiconducting cupric oxide thin films and their application to hydrogen detection. Sensors and Actuators B, 146, pp. 239–244.
159. Steinebach, H., Kannan, S., Rieth, L., Solzbacher, F., 2010. H2 gas sensor performance of NiO at high temperatures in gas mixtures. Sensors and Actuators B, 151, pp. 162–168.

5 The Sputtered Thin Films as the Sensing Materials for the MEMS Gas Sensors

Hairong Wang, Xin Tian, and Yankun Tang
Xi'an Jiaotong University, Xi'an, People's Republic of China

CONTENTS

DOI: 10.1201/9781003141358-5

5.1 INTRODUCTION

5.1.1 Application Requirements of Gas Sensors

With rapid development of the world economy, the increasing gas emissions cause serious air pollution. The gases may come from various sources such as byproducts of the factory, automobile exhaust. Most of them are harmful or toxic gases, which pose a great threat to the people in their daily lives. To keep a clear and safe atmosphere for healthy living, it is a must to monitor the air pollution and adopt measures to improve air quality.

Some flammable and explosive gases (e.g. methane, carbon monoxide, hydrogen) are often found in coal mining and chemical production. When the concentration of one of these gases increases to a certain value without being perceived, serious accidents may occur and pose a threat to properties and lives. Hence, it is necessary to monitor the flammable and explosive gases and alarm online in production.

Various toxic gases can be encountered on many industrial and living sites, while the hydrogen sulfide, formaldehyde, and hydrogen chloride are the classical ones.

People will feel sick, headache, or uncomfortable in an environment where these toxic gases are even at very low concentrations, and what is worsen they can be ill or die in some incidents. Therefore, detection of these gases and early alarming are efficient ways of ensuring the safety of people lives.

The internet industry and the real physical industry have permeated and merged with each other, leading to booming of internet of things (IoT). However, the development of IoT, especially the industrial IoT, is not as fast as expected. One primary reason is the lack of a large number of reliable low-cost sensors to meet the massive demand of the industrial IoT applications. As the sensing elements of acquiring environmental information, gas sensors, just like other types of sensors, are developing toward miniaturization, low power consumption, high integration and intelligence [1], and the miniaturization is the foundation of realizing the low power consumption and high integration, while the high integration is beneficial to ensure the high intelligence. Therefore, to develop miniaturized gas sensors based on MEMS is of great significance for the IoT development.

5.1.2 Classifications and Working Principles of Gas Sensors

5.1.2.1 General Classifications of Gas Sensors

The gas sensor is a device to detect the gas concentration by converting variation of the gas concentration into intensity variation of the electrical signal. Based on the working principles, the gas sensors usually can be classified into the electrochemical, infrared absorption, semiconductor, catalytic combustion, photo ionization detector (PID), and other types [2].

5.1.2.2 Electrochemical Gas Sensors

The electrochemical gas sensors usually include ion-selective electrode type, constant potential type, galvanic cell type, and limiting current type. In a traditional electrochemical sensor, the electrolyte is liquid, which makes the sensor difficult to miniaturize. The solid-state electrochemical gas sensor usually consists of a solid electrolyte, a sensing electrode, and a reference electrode. The gas sensors based on the solid electrolyte may be potentiometric type, mixed potential type, or amperometric type according to the testing methods.

For a potentiometric gas sensor, the target gas concentration is to be measured by the potential difference between the sensing electrode and the reference electrode. For example, in a potentiometric gas sensor based on Li_3PO_4 thin film, the sensitive electrode is Li_2SO_4/V_2O_5 prepared by screen-printing on the Au pattern [3], as shown in Figure 5.1, and the target gas SO_2 is adsorbed on the sensing electrode and generates SO_3 as equation (5.1) shows. SO_3 continues to diffuse inside of the Li_3PO_4 film and chemically reacts with Li^+ (5.2), reducing the potential of the sensing electrode. Hence, the gas concentration is calculated by the potential difference between the sensing electrode and the reference electrode.

$$SO_2 + \frac{1}{2}O_2 \Leftrightarrow SO_3 \tag{5.1}$$

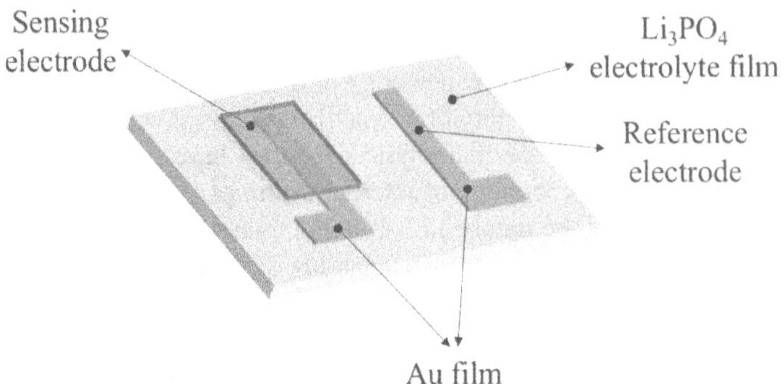

FIGURE 5.1 A potentiometric gas sensor based on Li_3PO_4 film [3].

$$2Li^{+} + 2e^{-} + SO_3 + \frac{1}{2}O_2 \Leftrightarrow Li_2SO_4 \tag{5.2}$$

Besides the potentiometric type, the gas sensor of limiting current type belongs to the amperometric solid-electrolyte ones. For a limiting-current-type gas sensor, a voltage is applied between the sensing electrode and the reference electrode to make the sensing electrode as cathode and the reference electrode as anode. When the voltage is greater than a certain value, the loop current between the two electrodes is linearly dependent on the concentration of the target gas. For example, the limiting-current-type O_2 sensor based on ZrO_2, has been often used for detecting oxygen partial pressure in the wide range.

With advantages of high accuracy, good selectivity, high resolution, the electrochemical gas sensors have been widely used in industrial fields, and the detectable gases include O_2, CO_2, CO, H_2, H_2S, HCl, Cl_2, and NO.

5.1.2.3 NDIR Gas Sensors

The NDIR (non-dispersive infrared) gas sensor, which consists of an infrared light source, a filter, and an infrared detector, is to identify the target gas composition based on the infrared absorption characteristics of different gas molecules, and determine its concentration according to the relationship between gas concentration and absorption intensity (Lambert-Bill law).

As illustrated in Figure 5.2, the light from the infrared light source is divided into two beams, which are led into the target gas chamber and the reference gas chamber, respectively. The reference gas chamber is filled with air free of the target gas. After passing through the target and reference gas chambers, the intensities of the two beams are expressed as follows:

$$I_1(\lambda) = I_0(\lambda)\exp(-KCL) \tag{5.3}$$

$$I_2(\lambda) = I_0'(\lambda)\exp(-KC'L) \tag{5.4}$$

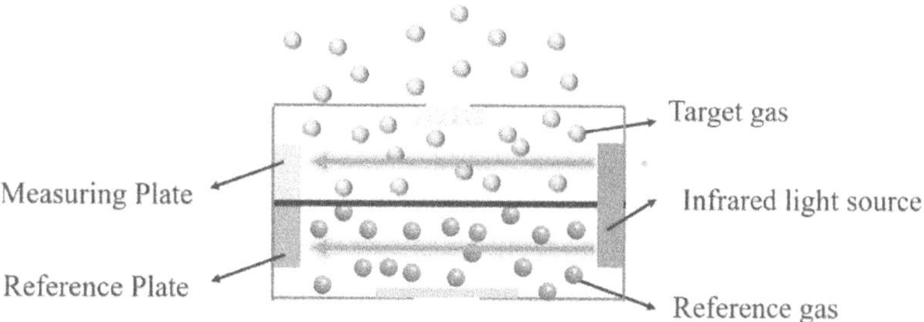

FIGURE 5.2 Schematic diagram of the NDIR gas sensor.

where $I_0(\lambda)$ and $I_0'(\lambda)$ are intensities of the beams before going through the target gas chamber and the reference chamber, K is absorption coefficient, L is absorption length, while C and C' are the gas concentrations in the target and reference chambers. The gas to be measured is free in the reference gas chamber, thus letting C'=0. From equations (5.3) and (5.4), one may derive

$$\exp(-KCL) = \frac{I_1(\lambda)I_0'(\lambda)}{I_2(\lambda)I_0(\lambda)} \tag{5.5}$$

Gas concentration C can be calculated by equation (5.5). Since the two beams are emitted from the same source, their intensities can be regarded as the same, namely

$$I_0(\lambda) = I_0'(\lambda) \tag{5.6}$$

The NDIR gas sensors can detect a variety of gases, such as CO, CO_2, CH_4, SO_2, NO, and NH_3, which have strong characteristic absorption peaks in the infrared band, and meanwhile there are appropriate filters according to the absorption peaks.

5.1.2.4 Semiconductor Gas Sensors

The semiconductor gas sensors mainly consist of metals, semiconductors, and/or metal oxides, and have advantages of simple structure, and compatibility with the thin-film process. When a semiconductor gas sensor is placed in the atmosphere containing the target gas, some electrical properties of the sensors will change, so as to detect the concentration of the target gas. The semiconductor gas sensors can be divided into the resistive gas sensors and non-resistive gas sensors, and the non-resistive gas sensors can be subdivided into diode type, field-effect transistor type, capacitance type, and other types according to the transducers used.

5.1.2.4.1 Resistive Gas Sensors

The resistive gas sensor is usually made of the SMOs such as SnO_2, Zn_0, TiO_2, Ni_0. As a sensing element, the SMO can be porous sintered bulk, thick film, or thin film. Since most of the SMOs react with the target gas only at enhanced temperatures, the resistive SMO gas sensors are usually equipped with heating wire and sensing electrodes, of which the heating wire is to heat the sensing element to the operating

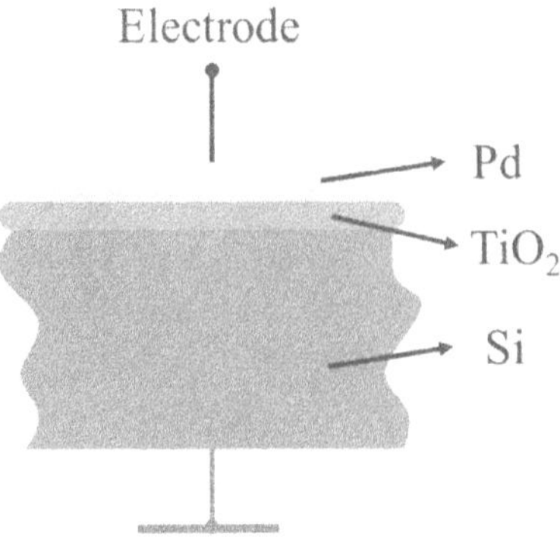

FIGURE 5.3 Schottky barrier SMO gas sensor.

temperature, and the testing electrodes are to detect the conductivity change of the sensing element. In a resistive SMO gas sensor, the sensing element with the adsorbed oxygen contacts with the target gas (such as H_2 and CO), releasing the electrons bound by oxygen and changing the surface conductance, thus reducing the device resistance. The gas concentration can be obtained according to the resistance change of the sensing element.

5.1.2.4.2 Non-resistive Gas Sensors

As a typical non-resistive type, the diode gas sensor is mainly composed of an n-type semiconductor and a p-type semiconductor or a semiconductor and a metal, which form the p–n junction or Schottky junction. As shown in Figure 5.3, in the Schottky barrier formed by the metal Pd and the SMO TiO_2, when applying a forward voltage to the Pd electrode, electrons in TiO_2 will flow to Pd. At this time, the p–n junction is conductive. In this case, when the diode sensor is placed in the target gas atmosphere, the Pd has a strong adsorption effect on H_2, leading to a decrease in the work function of Pd and resulting in the Schottky barrier height reduction at the Pd–TiO_2 interface which causes the increase in forward current. Therefore, the concentration of the adsorbed target gas can be reflected by the change in the forward current through the junction.

5.2 PRINCIPLES AND PROPERTIES OF SMO THIN-FILM GAS SENSORS

5.2.1 Principles of the SMO Thin-Film Gas Sensors

5.2.1.1 Structure of the SMO Thin-Film Gas Sensors

As a kind of semiconductor, a metal oxide whose majority carriers are electrons is n-type SMO, and an SMO whose majority carriers are holes is p-type SMO.

In the resistive gas sensors, the SMO gas sensing materials develop from sintered body to thick film, and then to today's nanostructured thin film. The nanostructured thin films have a large specific surface area, a high surface activity, and a high chemical adsorption at low temperature, and some of them can be prepared by the

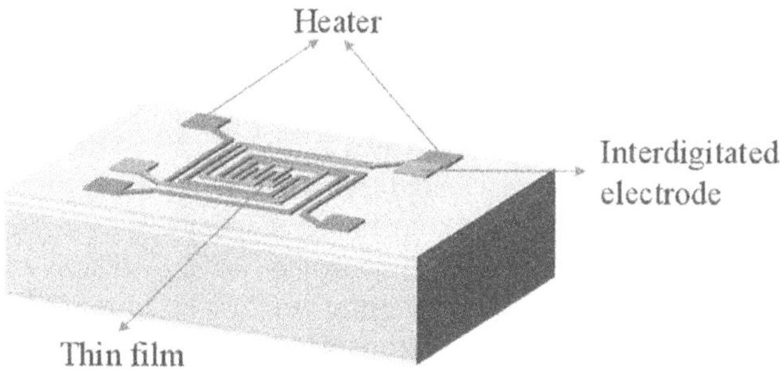

FIGURE 5.4 Schematic of an SMO thin-film gas sensor.

MEMS-compatible processes such as sputtering, CVD. Due to these advantages, the nanostructured thin films have become the important gas sensing materials for the MEMS gas sensors with good performance. For example, Sharma, A. et al. indicate the porous rough SnO_2 thin films prepared by RF (radio frequency) sputtering show high sensitivity, good stability, and long life to NO_2 gas [4].

A resistive gas sensor based on SMO thin film is basically composed of an SMO sensing material in the form of thin film, a pair of interdigitated electrodes, and a heater to provide an enhanced operating temperature, as shown in Figure 5.4. Sensing performance of a resistive gas sensor is closely related to the characteristics of the SMO. In the following text of this chapter, all gas sensors refer to the resistive gas sensors.

5.2.1.2 Gas Sensing Mechanism of SMO Thin Films

The gas sensing mechanism of SMO thin film can be explained mainly from two perspectives: adsorption/desorption and electron conduction.

As the mainstream gas sensing mechanism, the oxygen adsorption model may be applicable to most SMOs. Here, the n-type SnO_2 and the p-type CuO are used to elucidate the oxygen adsorption gas sensing mechanism.

When SnO_2 is exposed to air, oxygen will be adsorbed on the surface of the SnO_2. Due to the high electron affinity of oxygen molecules, electrons on the sensing material surface are captured into the form of anion adsorption, depending on the temperature of oxygen adsorption, which is either physical adsorption or chemical adsorption. Equation (5.7) describes the physical adsorption when the temperature is below 150°C. When the temperature rises, the reaction shown in equation (5.8) occurs, and when the temperature continues to rise above 400°C, the reaction shown in equation (5.9) occurs. Both equations (5.8) and (5.9) are chemical adsorption. Most of the SMOs work at 200–400°C, within which oxygen ions normally exist in the form of O^- ions.

$$O_2 + e^- \leftrightarrow O_2^- \tag{5.7}$$

$$O_2^- + e^- \leftrightarrow 2O^- \tag{5.8}$$

$$O^- + e^- \leftrightarrow O^{2-} \tag{5.9}$$

Majority carriers in the SnO_2 are electrons, and its conductivity decreases as the electrons are captured by adsorbed oxygen that reduces the concentration of the electrons, and then its conductivity decreases. While majority carriers in the CuO are holes, the concentration of the holes increases as a result of the minority carrier electrons captured by adsorbed oxygen, and then its conductivity increases.

For an SMO thin film, at a certain operating temperature, an electron depletion layer is formed on its surface due to the adsorbed oxygen, resulting in a built-in potential barrier between grains. Taking SnO_2 thin film as an example, under heating condition, oxygen in the air will capture electrons from the conduction band of the SnO_2 and leave the adsorbed negative oxygen ions on the SnO_2 surface, which will increase the surface potential and thus hinder the movement of free electrons. Therefore, the SnO_2 thin-film gas sensor has higher resistance in air. When the sensor contacts with the reducing gas, the redox reaction takes place between the reducing gas and the oxygen anion adsorbed on the SnO_2 surface, producing desorption of the adsorbed oxygen. In this case, the electrons are released into the SnO_2 again, which results in both decreases in the surface potential and resistance. Redox reaction is the essence of gas sensing mechanisms for most SMOs. For the n-type SMO thin film, its resistance decreases in reducing gases, while in oxidizing gases, its resistance increases. For the p-type SMO thin film, the reaction between reducing gas and oxygen negative ions will release excess electrons into the film and reduce the concentration of the majority carrier, thus increasing the thin-film resistance.

The n-type SMO thin film (where film thickness $Z \gg$ Debye length L_D) and reducing gas CO are used to illustrate the working principle of the thin-film gas sensor. As shown in Figure 5.5, free electrons move from the grain to the surface captured by the adsorbed oxygen, forming the electron depletion layer on the film surface with thickness of Z_0. The depletion layer causes bending of energy band on the surface and creates a Schottky barrier $eV_{surface}$ on the surface of grains. In the target gas CO, the reaction of equation (5.10) occurs, in which electrons are released to conduction band of the film. Due to the large film thickness, the conduction happens mostly

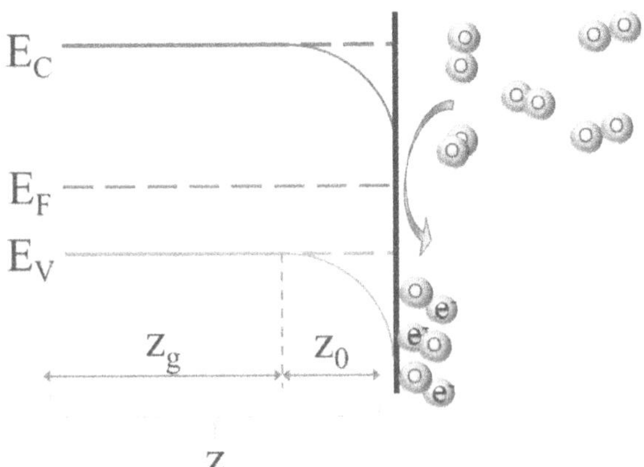

FIGURE 5.5 Energy band bending of the oxygen adsorbed surface. Redrawn figure.[5]

within the layer thickness Z_g, and the conductivity is expressed as equation (5.11). The relationship between Z_0 and barrier voltage V_{surface} is written as equation (5.12), and the relationship between barrier voltage V_{surface} and concentration of the target gas CO as equation (5.13) [5]. Combining equations (5.12) and (5.13) can get equation (5.14), which describes the relationship between Z_0 and the CO concentration. The direct relationship (5.15) between the film conductance and the CO concentration can be given by combining (5.14) and (5.11).

$$CO + O^- \rightarrow CO_2 + e^- \tag{5.10}$$

$$G = const1 \cdot (Z - Z_0) \tag{5.11}$$

$$Z_0 = \sqrt{\frac{2\varepsilon\varepsilon_0}{en_e} V_{\text{surface}}} \tag{5.12}$$

$$const2 \cdot P_{co} = n_e \cdot \exp\left(-\frac{eV_{surface}}{k_B T} \right) \tag{5.13}$$

$$Z_0 = \frac{1}{e}\sqrt{2\varepsilon\varepsilon_0 k_B T \ln \frac{n_e}{const2} - \frac{k_B T}{1+\delta} \ln P_{co}} \tag{5.14}$$

$$G = const1 - \sqrt{const3 - const4 \ln P_{co}} \tag{5.15}$$

where L_D is Debye length/nm, ε_0 vacuum dielectric constant, ε dielectric constant, k_B Boltzmann constant, T temperature/K, e charge/C, n_e carrier concentration/m^{-3}, G conductance/S, V_{surface} barrier voltage/V, Z film thickness/m, Z_0 depletion layer thickness/m, P_{CO} CO concentration/ppm, $\delta = 0$–0.2.

Equation (5.15) shows the measured conductance G increases as the CO concentration increases. Since the film thickness $Z \gg L_D$, the conductivity G is little influenced by the concentration change of target gas CO [5].

5.2.2 Gas Sensing Properties of Resistive SMO Gas Sensors

5.2.2.1 Response

Response, usually denoted with the symbol S, reflects the reaction of the sensing element to the target gas after it reaches its equilibrium state. Obviously, when a sensor is placed in the air without the target gas, the sensor has a resistance R_a when it reaches stability. Then placing the sensor in the atmosphere with target gas, the sensor output changes and has a resistance R_g when it reaches stability. Then, response of the gas sensor can be described by R_a and R_g, and it is defined in different expressions for the different types of SMO, as shown in Table 5.1.

Obviously, the response of a gas sensor is different for a certain target gas at different concentrations. Being the most important parameter of a gas sensor, response is corresponding to a specific concentration of the target gas without considering other interference factors.

TABLE 5.1
Responses Defined for The SMO to Different Types of Gases

SMO Type	Reducing Gas	Oxidizing Gas
n-Type	$S = \frac{R_a}{R_g}$	$S = \frac{R_g}{R_a}$
p-Type	$S = \frac{R_g}{R_a}$	$S = \frac{R_a}{R_g}$

5.2.2.2 Selectivity

Selectivity indicates that the sensor has a high response to the target gas, but negligible or small responses to other interfering gases. In application, other kinds of gases also exist in the measured atmosphere. In this case, the sensor is required to respond only to the target gas or has obviously a high response to the target gas over other gases. Otherwise, the sensor cannot accurately detect the target gas or output a wrong result. Selectivity of a gas sensor is usually defined by the ratio of response of the target gas to responses of the interfering gases. The larger the ratio, the better the selectivity. As an example, Katsuki, A. et al. prepared a gas sensor based on SnO_2 [6], which has significantly higher response to H_2 than other interfering gases such as methanol, ethanol, methane, CO.

5.2.2.3 Response and Recovery Time

Response and recovery time are the parameters to describe the sensor's response rate to the target gas and the recovery rate from the target gas to the air. When the gas sensor is transferred from the air to the target gas environment, resistance of the sensor increases from R_a in the air to equilibrium value R_g in the target gas, and resistance variation is $(R_g - R_a)$. Response time refers to the time from initial rising of the sensor resistance R_a to it first attaining to 90% of the resistance variation. When the gas sensor is transferred from the target gas environment to the air, response drops from R_g to R_a, and recovery time refers to the time from initial decreasing of the sensor resistance R_g to it first attaining to 90% of the resistance variation.

5.2.2.4 Repeatability

Repeatability refers to the results of cycle tests on the same sample by the same operator in the same conditions. In the cycling environment with a clean air and an air including a certain concentration of target gas, the continuous reversible cycles of response and recovery curves reflect repeatability of the gas sensor. It can be described by the distribution error of the response characteristics for testing a certain number of cycles. For example, the H_2 sensor developed by Mourya, S. et al. shows a good repeatability [7].

5.2.2.5 Stability

After working a period of time, response of a gas sensor will decrease, and the response and recovery time will become longer, which will greatly affect the accuracy of the sensor, and reduce the reliability of the sensor for long-time working. Stability is the ability of the sensor maintaining its performance of the original design after being used for a period. Normally, the initial resistance of the sensor in air is affected by the external environment and may vary with time and environmental factors. The stability may be defined by the response variation of the sensor used over a certain time, e.g. ΔS/month. For instance, an SnO_2-based ethanol gas sensor shows good stability after 5 months of working [8].

5.2.3 Methodology of Improving Gas Sensing Properties of SMO Thin Films

5.2.3.1 Noble Metal Doping

A pure SMO thin film is often unable to meet the high requirements for a gas sensor. To improve gas sensing properties of an SMO thin film, small amounts of noble metals such as Ag, Pd, Au, and Pt are often introduced. The doping of noble metal into an SMO thin film can effectively improve the response and shorten the response time of the thin-film gas sensor. Noble metal with catalytic activity can reduce adsorption and activation energy of gas on the surface of the SMO thin film. Moreover, the surface doping of noble metals can generate more active sites on the surface, which helps to preferentially adsorb the target gas and increase its concentration on the surface. At the same time, noble metals provide a reaction pathway to reduce activation energy, thereby increasing the reaction rate, response, selectivity, and stability. Additionally, since the Fermi level of the noble metal is usually lower than that of the SMO, electrons will transfer from the SMO to the noble metal after carrier redistribution until the Fermi levels of the two materials become identical. In the process, the noble metal becomes negatively charged, while the SMO becomes positively charged. At the interface, the energy band bending forms the Schottky barrier, which can effectively prevent electron–hole pairs from recombination and improve the response to the target gas. In another case, the noble metal atoms enter the crystal lattice of the SMO thin film and replace some metal atoms. Take TiO_2 as an example, if the noble metal replaces some Ti ions in the lattice, then lattice parameters and bond length will change. Therefore, the centers of positive and negative charges in the octahedron will shift and form a dipole moment inside. Then, a local electric field is built throughout the crystal, and facilitate in the separation of the electron–hole pairs.

In addition, spillover effect plays an important role in noble metal-doped SMO thin films. Spillover is a process in which the gases adsorb and dissociate in a certain phase. The dissociated material migrates to the second phase and is activated. The adsorption of these activated substances in the second phase cannot occur under the same conditions. In the gas sensing process, this means that the target gas molecule first reacts with the noble metal in the sensing layer and generates the active substances. The substances are then adsorbed onto the SMO surface and affect its sensing properties. For example, Figure 5.6 shows the spillover of a Pd-doped hydrogen sensor schematically, in which Pd can dissociate H_2 to H, and then the H spills and adsorbs on the surface of the SMO. Here, the function of Pd is to reduce activation

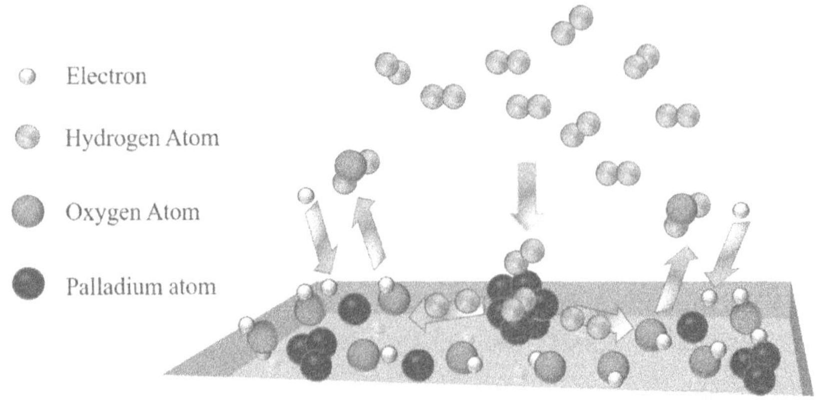

FIGURE 5.6 Schematic diagram of precious metal spillover effect. Redrawn figure. [9]

energy of the reaction. As a result, response-recovery time and operating temperature decrease [9].

Xiao, L. et al. prepared an alcohol sensor based on Pd-doped SnO_2 [10], which has high response due to the chemical adsorption and dissociation of gas molecules caused by Pd doping. In particular, the SnO_2 nanoparticles enhance local conductivity and possibly speed up phase transfer reactions, resulting in fast response/recovery properties. Ivanov, P. et al. prepared a very sensitive alcohol sensor by using the Pt-doped SnO_2 [11]. The Pt atoms are distributed uniformly throughout the SnO_2 layer, and the 3 wt% Pt doping sample has the best response properties at the optimal operating temperature 300°C. Hastir, A. et al. used Ag-doped ZnO as the gas sensing material [12]. Compared with pure ZnO, the Ag-doped ZnO sensor shows a significantly enhanced response to ethanol at a concentration of about 10 ppm. Xu, X. et al. investigated gas sensing properties of Au-loaded In_2O_3 nanofibers to ethanol [13]. The sensor based on 0.2 wt % Au-loaded In_2O_3 has high response and fast response/recovery speed.

5.2.3.2 Heterojunction

Gas sensing properties of an SMO thin film, such as response, operating temperature, selectivity, and response-recovery time, can be improved by combining it with another SMO to form heterojunction. Interface of the heterojunction is the most important factor in analyzing gas sensing properties of the SMO thin films.

According to the types of the SMOs, heterojunctions can be divided into isotype heterojunctions (n–n or p–p) and an isotype heterojunction (p–n heterojunction). Due to the poor oxygen adsorption of p–p heterojunction formed by two p-type SMOs, which results in poor gas sensing properties, the p–p heterojunction composites are rarely used. In the following, principle and application of the p–n and n–n heterojunctions are briefly described.

Figure 5.7(a) and (c) show the energy bands of p-type Co_3O_4/n-type ZnO and n-type SnO_2/n-type TiO_2 before contact, where Fermi levels of ZnO and TiO_2 are higher than those of Co_3O_4 and SnO_2 [14]. When two types of SMO come in contact,

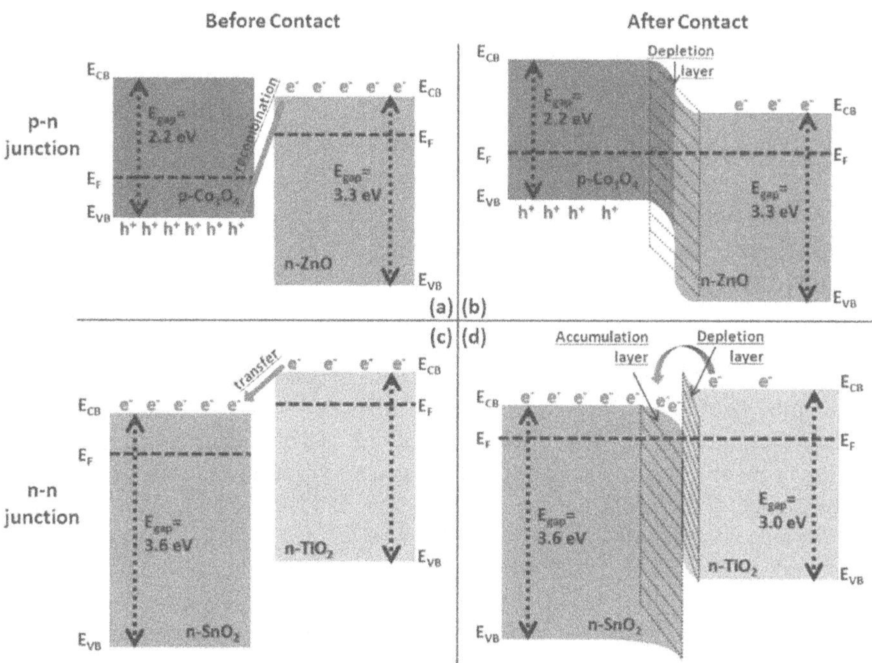

FIGURE 5.7 Heterojunction band bending of SMO, (a) p–n junction formed between the p-type Co_3O_4 and n-type ZnO, (b) depletion layer and potential barrier formed between p-type Co_3O_4 and n-type ZnO, (c) n–n junction formed between n-type SnO_2 and n-type TiO_2, (d) depletion layer and potential barrier formed between n-type SnO_2 and n-type TiO_2 [14]. (Note: E_{CB}: conduction band edge energy, E_{VB}: valence band edge energy, E_{gap}: band gap energy, E_F: Fermi energy).

the built-in electric field is created at their interface, and electrons in the conduction band of the SMO with high Fermi level flow to the conduction band of other SMO with low Fermi level. It continues to flow until the Fermi levels of the two materials reach a new balance, as illustrated in Figure 5.7(b) and (d). After that, the interface of n-type ZnO and n-type TiO_2 will form an electron depletion layer due to the electron loss, and the band will bend upward. In the p-type Co_3O_4 interface, donated electrons neutralizing holes generate hole depletion layer and lead to band bending downward, while in the n-type SnO_2 interface an electron accumulation layer that is created due to the obtained electrons makes energy bending downward. As the carrier concentrations of both p-type SMO and n-type SMO decrease with the formation of p–n junction, the resistance of p–n junction composite increases compared with those that are without composite. The effect of n–n junction on the resistance of SMO is complex, and the presence of electron accumulation layer also affects the gas adsorption and reaction process.

Heterojunction composite has obviously enhanced gas sensing properties. The n–n heterojunction of ZnO/In_2O_3 prepared by Zhang, K. et al. can detect 1 ppm of ethanol gas at 240°C, and it has high selectivity, long-term stability, and fast response and recovery speed [15]. Wang, Q. et al. adopted the p–n heterojunction of CoO/SnO_2 and found that it has improved response and selectivity to ethanol as compared with pure SnO_2 [16].

5.2.3.3 Ultraviolet Light

The UV (ultraviolet) photon energy is similar to the forbidden gaps of many SMOs. Thus UV radiation can be effectively absorbed by the SMO, to produce a series of physical and chemical processes inside and on the surface of the SMO, and eventually affect the gas sensing properties of the SMO.

UV radiation can excite electron–hole pairs inside the grain. On the one hand, it increases carrier concentration. On the other hand, photogenic carriers can migrate to the surface under the effect of the built-in electric field, and recombine with the surface states or adsorbed species on the surface. This will reduce interface barrier and thickness of the depletion layer, increase the tunneling probability of the barrier during carrier transport, and thus increase the conductance of the SMO.

The response-recovery time of an SMO is related to the adsorption and desorption process of gas molecules on the surface. The reaction between molecules and the SMO surface (namely the adsorption process) is slow at room temperature, while the holes generated by UV light in the grains move to the surface under the effect of electric field and combine with the adsorbed negative oxygen ions, which can accelerate the adsorption rate by reducing the concentration of the reaction product. When the target gas is pumped away, UV light can accelerate desorption of adsorbed molecules and reduce the recovery time.

Zhang, C. et al. tested the response properties of a WO_3 sensor to 160 ppb and 320 ppb NO_2 at room temperature under visible/ultraviolet light [17]. Compared with no light radiation, gas sensing properties of the WO_3 sensor are largely improved under the UV light radiation.

5.2.4 Influential Parameters of the Gas Sensing Thin Films

5.2.4.1 Grain Size

Grain size is an important factor affecting the gas sensing properties of the SMO thin film. The smaller grain size changes the energy level of the electron near the Fermi level of metal, making it into a discrete state rather than a quasi-continuous state, and then the metal has a more active catalytic property. With the decreasing of the SMO grain size, surface energy of the SMO increases, and then the SMO has a higher reactivity with gases.

According to grain size D and thickness of electron depletion layer L, three models can illustrate the effect of the grain size on gas sensing properties of an SMO thin film [18]. In Figure 5.8(a), when $D \gg 2L$, the response of the SMO will depend on the grain boundary control, of which there is the unobstructed electronic channel between the neighboring grains. In Figure 5.8(b), when $D \geq 2L$, the SMO response is attributed to the neck control, and in this case with the decrease in grain spacing, the electronic channels between the grains become narrow due to the overlapping grains of each other. In many cases, the gas sensing mechanism of

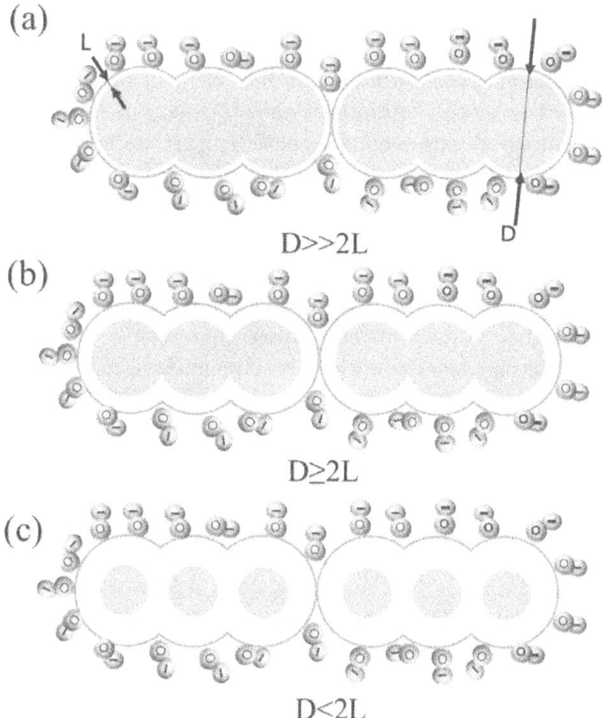

FIGURE 5.8 Effect of grain size on the SMO conductivity. (a) the grain size (D)>> 2L (double depletion layer), (b) D ≥2L, (c) D < 2L. [18]

SMO thin film is controlled jointly by the boundary barrier and the neck. If the grain size is large, the gas response is as the control of the grain boundary barrier, which is independent of particle size, so is the gas response. However, if the grain size is small, the gas response will depend on the grain size. In Figure 5.8(c), when $D < 2L$, the SMO response will be determined by the grain control. In this case, the depletion layer fills the whole grain, and the SMO surface is activated at its maximum that makes the material highly responsive at lower optimal operating temperature.

The above analysis shows that precisely controlling grain size of the SMO at about twice of the depletion layer thickness is helpful to improve its response and provides a useful theoretical basis for preparing the SMO gas sensing materials. Therefore, methods such as refining grain size, shortening grain spacing, and increasing active surface area of sensing materials can significantly improve the sensing properties of the SMO thin films.

5.2.4.2 Surface Morphology

Complex morphology of the SMO with a large specific surface area provides more active sites in reacting with the target gas. Due to these active sites, the target gas molecules are more likely to react on the surface of the SMO, improving the reaction degree and response.

The SMO thin films in different nanostructures have quite different electron conductivities. The zero-dimensional hollow spherical structure not only makes the outer surface of the material participate in the reaction, but also fully utilizes the interior surface active sites. One-dimensional SMO nanostructures, such as nanowires, nanorods, nanofibers, and nanotubes, have high crystallinity and a large specific surface area. Electrons can flow conveniently through the ordered one-dimensional SMO nanostructures. Two-dimensional nanosheets and nanoribbons also own the large specific surface area and are extremely thin, which greatly increase the number of the surface active sites. Three-dimensional SMO nanostructures are composed of many low-dimensional nanostructures. In a porous form, the three-dimensional SMOs have large surface areas, which expose more active sites and facilitate the adsorption, transmission, and diffusion of reaction molecules.

In gas sensors, the SMO thin films should be the surface-sensitive type, and porous loose structure will make the gas molecules rapidly spread to the surface of the thin films and go deep into the materials. Hence, the gas molecules can contact with more catalytic active sites, accelerating response-recovery speed, and improving the response at the same time.

5.2.4.3 Defect

The gas sensing properties of an SMO thin film mainly depend on the physical and chemical reactions on its surface, and the surface defects play an important role in the reactions. The surface defects cause the carriers to scatter that are generated by gas adsorption and be captured at the grain boundary during the conduction process inside the SMO, which affects the transport of carriers and thus affects the gas sensing properties of the thin film. Surface defects can be divided into intrinsic defects and extrinsic defects, which have very different effects on the gas sensing properties of the SMO thin films.

For example, ZnO has six intrinsic defects: interstitial zinc (Zni), interstitial oxygen (O), zinc vacancy (Vzn), oxygen vacancy (Vo), antisite zinc (ZnO), and antisite oxygen (Ozm). Among them, the defect of antisite zinc is generally not considered because it is impossible to exist stably in the equilibrium state. Zni and Vo are used as electron donors, while the remaining three defects are electron acceptors. Higher the proportion of Zni and Vo defects, better the gas response. The two donor defects can release electrons and enhance the adsorption of oxygen species on the surface, thus improving the gas sensing properties.

Extrinsic defects are introduced by doping other metals or metal oxides in an SMO thin film. For the SMO thin film sensor, these defects possibly increase the active sites on the SMO surface, resulting in reducing operating temperature, shortening response time, increasing response, and improving stability and selectivity.

5.3 SPUTTERED SMO THIN FILMS FOR GAS SENSING

5.3.1 Types of Sputtering

5.3.1.1 General of Sputtering

Sputtering is a process in a vacuum environment to deposit thin film on the surface of the substrate by bombarding the target material surface with the charged ions. Sputtering techniques are divided into magnetron sputtering, radio frequency (RF) sputtering, direct current (DC) sputtering, reactive sputtering, and other types. According to the type and the requirement of the deposited material, appropriate sputtering techniques can be selected. Moreover, one can get a new sputtering method by combining these techniques, e.g., the RF reactive sputtering is the combination of the RF sputtering and the reactive sputtering.

5.3.1.2 DC Sputtering

DC sputtering is also called bipolar sputtering, and its basic principle and mode is the foundation of all sputtering deposition techniques. As shown in Figure 5.9(a), the DC sputtering equipment adopts a pair of parallel-plate electrodes, of which the electrode supporting the target is the cathode, and the electrode supporting the substrate is the anode. After being evacuated to 10^{-3}–10^{-4} Pa, the chamber is introduced with inert gas (e.g. Ar) having pressure of 0.1–1 Pa, and a high voltage of thousands of volts is applied between the electrodes to produce a glow discharge and establish a plasma area. Ion bombardment on the target causes the target atoms to detach from the target and deposit on the substrate.

DC sputtering is simple in structure, easy to operate, and suitable for long-term working. However, the technique has the following disadvantages. (1) The deposition rate is relatively low, (2) it is only utilized for depositing metal target, (3) the secondary electrons produced by ion bombarding the cathode directly strike the substrate, so that the substrate needs to withstand high temperature, and (4) high working pressure and residual gases in the background vacuum may pollute the film, and also affect the deposition rate. To overcome these disadvantages, triode sputtering and quadrupole sputtering with the addition of auxiliary electrodes are developed to keep high plasma density and improve deposition rate after reducing the working pressure. However, these two techniques cannot solve the problem of high temperature caused by secondary electron bombardment on the substrate. Hence, the DC sputtering can be used for preparing electrodes and heating wire in the gas sensors, but not for the SMO gas sensing thin films.

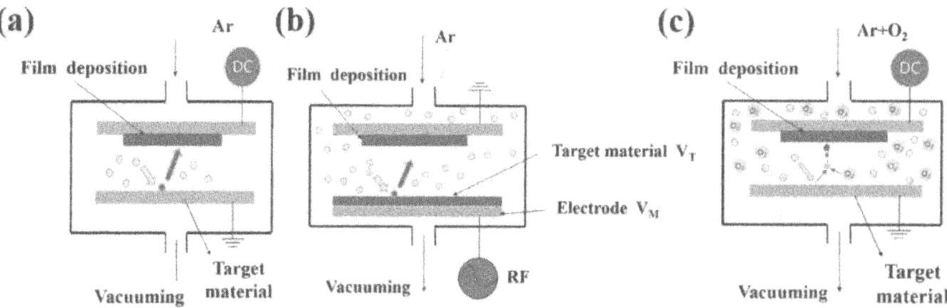

FIGURE 5.9 Schematic diagram of the (a) DC sputtering, (b) RF sputtering, and (c) reactive sputtering.

5.3.1.3 RF Sputtering

While in depositing a nonmetallic material with poor electrical conductance, RF sputtering is one of the candidate methods. In the RF sputtering, a metal electrode is installed on the back of the insulation target and a high frequency (usually 13.56 MHz) electric field is applied to make sputtering continuously. Figure 5.9(b) shows the RF sputtering schematic, in which the substrate is grounded with zero potential, the target voltage is V_T, and the AC voltage of V_M is applied on the target metal electrode. Supposing the AC voltage on the target is a sinusoidal wave, in the positive half cycle, because the electrons move easily, V_T and V_M are rapidly charged. In the negative half cycle, the ions move much more slowly than the electrons, so the electrically charged capacitor begins to delay discharge. If the number of electrons arriving at the substrate when the substrate in positive potential, equals the number of ions arriving at the substrate when the substrate in negative potential, the target will be negative for most of the time, so that the target can be sputtered under the bombardments of positive ions.

Unlike DC sputtering that is only capable of depositing metals, the RF sputtering technique can deposit various metal and nonmetal materials, including high or low melting point metals, and multi-component compounds. The amount of sputtering yield can be enhanced by increasing the cathode surface. Therefore, RF sputtering can be used to deposit the nanostructured SMO thin films by controlling the process parameters.

5.3.1.4 Reactive Sputtering

Reactive sputtering is an effective technique to deposit various compounds thin films. A multicomponent compound film can be deposited either by direct sputtering on a target made of the compound directly, or by reactive sputtering. When depositing a compound thin film by the reactive sputtering, the pure metal or alloy targets are sputtered, while at the same time the reaction gases are introduced, such as oxygen and nitrogen, as shown in Figure 5.9(c). Through reactive sputtering, various oxide and nitride thin films can be prepared by using the high purity metal target and the reactive gas. Therefore, reactive sputtering is often used to prepare various SMO thin films.

5.3.1.5 Magnetron Sputtering

Magnetron sputtering is an improvement of DC or RF sputtering by controlling the secondary electrons with the magnetic field. As shown in Figure 5.10, secondary electron in the perpendicular electromagnetic fields is bound near the target surface and moves with rolling along the "runway" around magnetic field lines. In this way, the gas ionization rate is enhanced, and even if the working pressure is as low as 10^{-1}–10^{-2} Pa, the plasma density still will be increased, thus improving the incident ion density, reducing sputtering voltage, and increasing the deposition rate. Moreover, secondary electrons can escape from the target surface and fall on the anode only after their energy is exhausted. Therefore, the substrate is free of the bombardments of the secondary electrons, that makes the substrate have low temperature rising and little damaging.

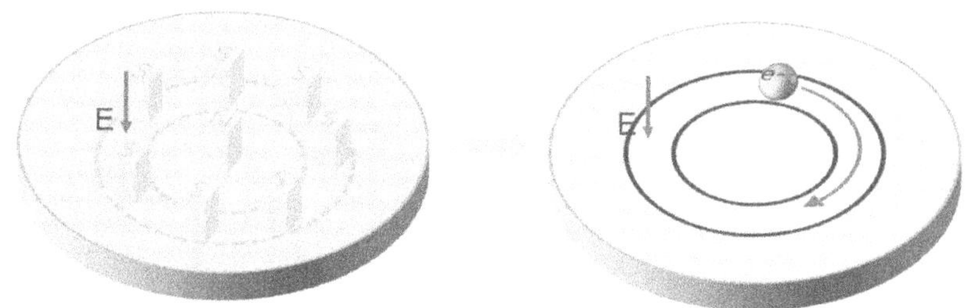

FIGURE 5.10 Schematic of magnetron sputtering.

In magnetron sputtering, plasma area is strongly bound within the region near the target surface about 60 mm due to the magnetic field effect [19]. Therefore, to ensure that the substrate can be bombarded by ions, the substrate should be placed within the plasma region. Magnetron sputtering can be used to deposit metals, as well as SMOs.

The sputtering techniques have been applied to deposit SMO gas sensing thin films. Zoolfakar, A. S. et al. used RF sputtering to prepare CuO thin films, which have grain size of about 30 nm and large ratio of surface area to volume [20]. The film-based gas sensor can detect ethanol at ppm level, and it has good repeatability, high response, and fast response-recovery properties. Dong, Y. F. et al. applied reaction sputtering of CdIn target in Ar and O_2 atmosphere to prepare $CdIn_2O_4$ thin film, and then sputtered the thin discontinuous Pt layer on the thin-film surface [21]. The film has high response, low operating temperature, and short response and recovery time (about 10 s) to detect ethanol.

5.3.2 Improving Gas Responses of Sputtered SMO Thin Films

5.3.2.1 Co-sputtering

Sputtering is a MEMS-compatible process and can deposit the SMO film with good uniformity over the whole wafer. Due to these advantages, the sputtered SMO films are very attractive in the batch fabrication of the SMO gas sensors. However, the SMO films sputtered under normal conditions usually cannot meet the requirements of the gas sensors. To improve gas sensing properties of the sputtered SMO thin films, various measures may be taken, e.g., controlling surface morphology of thin film by adjusting sputtering parameters, adding noble metals, creating heterojunctions. The co-sputtering can effectively improve the gas sensing properties of the SMO films. According to the microstructure of the deposited film, the co-sputtering methods can be simply classified into three types: (1) co-sputtering of two targets to different layers, (2) co-sputtering of two targets to one layer, and (3) sputtering single target made of powder mixed by different materials.

The thin film prepared by co-sputtering of two targets layer by layer may have good gas sensing properties. Wang, H. et al. deposited a TiO_2/SnO_2 multi-layer gas sensing film by co-sputtering TiO_2 target and SnO_2 target alternatively. The sensor based on the co-sputtering multi-layer film has a response of 20–200 ppm ethanol at 300°C and other gas sensing properties are quite good [22].

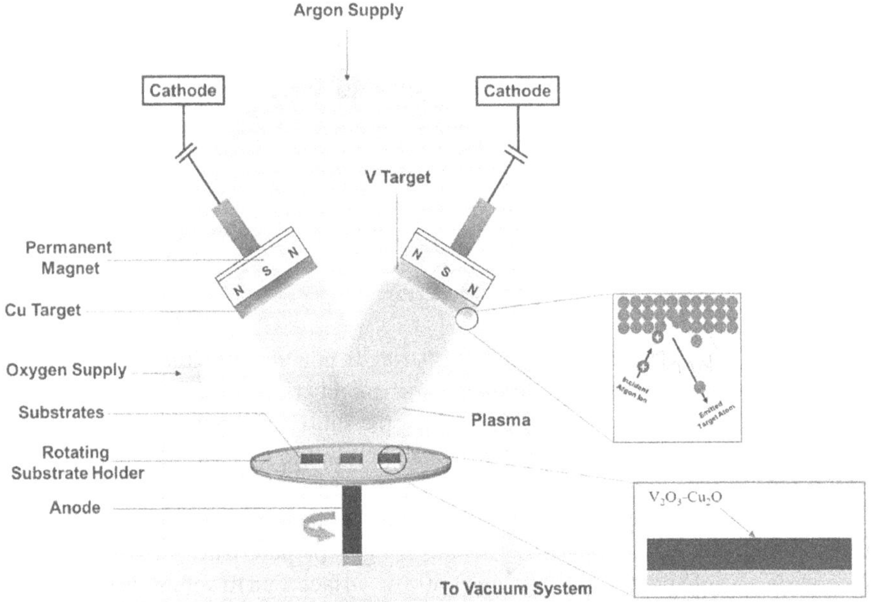

FIGURE 5.11 Co-sputtering with two targets to deposit SMO composite film [23].

In a co-sputtering, the sputtering of two targets begins and ends at the same time, and a uniform single composite film of the two target materials forms, which is composed of numerous micro heterojunctions. Mounasamy, V. et al. [23] applied sputtering powers of 30 W and 100 W to Cu and V targets, respectively, at the working pressure at 9.9×10^{-3} mbar, and oxygen was injected into the chamber for a total reaction of 60 minutes, as shown in Figure 5.11. The deposited V_2O_3-Cu_2O composite film has a response of 1.08 to 3 ppm trimethylamine.

Before sputtering a single target of mixed powder, two or more materials in powders are mixed first, and then the mixture is pressed into the target material for sputtering. Girija, K. G. et al. synthesized doped ZnO films by RF magnetron sputtering of composite target prepared with powders of ZnO and copper. The doped films are highly selective to H_2S, 2 wt% Cu-doped film has highest sensitivity at an operating temperature of 250°C [24].

5.3.2.2 Adjustment of Sputtering Parameters

The sputtered film's grain size, defect, microstructure, and surface morphology are the important factors affecting its gas sensing properties. These characteristics can be changed by adjusting the sputtering parameters. In the following text, the effects of sputtering power, substrate temperature, atmosphere, and other parameters on the characteristics of the sputtered films are discussed.

5.3.2.2.1 Sputtering Power

The SMO film deposition rate is greatly influenced by the sputtering power. The sputtering power is increased by increasing the voltage between the target and the substrate, which will increase the ionic current formed by ionized gas, causing more particles of the target to deposit and then enhancing the deposition rate. In the early stage of deposition, a high deposition rate will generate more nucleating centers on the substrate, to easily grow fine grains, which makes the SMO film more compact, smooth, and better orientated. However, if the rate is too high, then the film crystal orientation becomes worse, the grain size grows larger, surface is rougher, and the crystalline order of the SMO film declines and even mixed orientation appears.

5.3.2.2.2 Substrate Temperature

The appropriate substrate temperature in the sputtering process is beneficial to facilitate grain crystallization, enhance adhesion between the thin film and the substrate, and improve electrical properties of the film. Åbom, A. E. et al. [25] found that during sputtering Pt, with the increasing of substrate temperature, grain size of the thin film gradually increases. The grain size of the thin film deposited at 600°C of substrate is much larger than that at room temperature. However, the gas response of the thin film deposited on the substrate at room temperature is lower than that at 600°C.

5.3.2.2.3 Protective Gas Ratio

During sputtering, a certain proportion of Ar and O_2 is usually injected as protective gases to protect the sputtered SMO or metal from the influence of other interfering gases. Different proportions of protective gases affect the properties of the sputtered films greatly. Hien, V. X. et al. [26] sputtered Cu_2O for 40 min in accordance with Ar/O_2 at different ratios, and obtained SMO films with different structures. The results indicated that the Cu_2O film with an Ar/O_2 ratio of 97/3 has the best response to 1–10 ppm H_2S, and the proportions of protective gases are vital to alter microstructures of the films and then affect their responses.

5.3.2.2.4 Sputtering Material Composition Ratio

The greater the sputtering power, the larger the sputtering yield, and there is a rough linear relationship between them within a certain range. In the co-sputtering process, different powers are applied to the different targets, that is, the deposited film is a composition with a certain proportion of the target materials. Wang, Y. et al. prepared the Pd-doped SnO_2: NiO film on the micro-hotplate with co-sputtering technique [27], and changed the Pd content by adjusting the corresponding RF sputtering power from 0 W to 30 W. The optimal response of Pd-doped SnO_2: NiO films was obtained with sputtering power of 20 W to deposit Pd.

5.3.2.2.5 Film Thickness

If the deposition time is too short, and only a very thin film is formed, the sputtering particles have no time to diffuse, migrate, and nucleate, resulting in uneven surface distribution and incomplete crystal. When the film thickness increases gradually, the diffusion and migration of sputtering particles make the surface gradually uniform, which can cover the substrate well to form a smooth surface. However, if the film is too thick, the nucleation growth of particles in the film becomes larger, thus affecting the surface flatness. Prajapati, C. S. et al. prepared V_2O_5 films of different thicknesses by reactive-sputtering, and concluded that the optimal film thickness of the V_2O_5 films is 61 nm to detect SO_2 [28].

5.3.3 Advantages and Drawbacks of Sputtered SMO Thin Films

5.3.3.1 Advantages

5.3.3.1.1 Compatibility

A sputtered SMO thin film can be deposited directly on the IDEs. As compared to the "dispense coating," in which the SMO is prepared as paste or suspension and then coated or sprayed on the IDEs to form the gas sensing layer, the sputtering technique can deposit an SMO film directly like the CVD, plasma spray, EBPVD, which can be integrated with MEMS technology. Moreover, thickness of the sputtered film can be accurately controlled over the whole wafer. These advantages make it possible to achieve massive production of gas sensors.

5.3.3.1.2 Consistency

The sputtered SMO thin films have a good consistency. Sputtering is to make the atoms on the target surface splash in all directions by means of high-speed plasma bombardment, and finally deposit on the substrate to form a film. In sputtering the parameters such as sputtering power and substrate temperature can be properly adjusted, which ensures the good consistency of the sputtered thin films in different batches, thus ensuring good yield of the gas sensors produced by wafer-level mass production.

5.3.3.1.3 Low Power Consumption

A MEMS gas sensor, in which the sputtered SMO thin film is used as the gas sensing material, will have lower power consumption. To fabricate a MEMS gas sensor, the common method (dispense coating) is to coat the suspension droplet or smear the paste of the SMO on the micro-hotplate, and then heat it for sintering. The gas sensor fabricated in this way possibly has low power consumption due to the micro-hotplate. However, it is difficult to further reduce the power consumption since the SMO sensing layer is relatively thick and the sensing pattern is relatively large. As a MEMS-compatible technology, the sputtering process can deposit thinner SMO film and define the SMO film pattern in a smaller dimension. Thinner the film and smaller the horizontal dimension, smaller the lower power consumption.

5.3.3.2 Drawbacks

5.3.3.2.1 Front Protection

To fabricate a MEMS gas sensor based on the sputtered SMO film, the front side is often protected to ensure the defined SMO pattern free of damage by the following process. In the process of fabricating the micro-hotplate for the SMO gas sensor, the adiabatic groove is etched on the back side of the silicon wafer, by wet or dry etching. Therefore, front side protection prevents the patterns of SMO film and electrodes from destroying during etching that are already defined on the front side. The improper front protection is one main reason for the sputtered SMO gas sensor failure in the fabrication, while the micro inkjet method usually does not need a front protection.

5.3.3.2.2 Complex Process

There are about five steps to fabricate a MEMS gas sensor: preparation of support layer, preparation of sensitive layer, preparation of electrode, preparation of adiabatic groove, slicing and packaging. A variety of MEMS processes are applied, such as resist spinning, lithography, lift-off, deposition, sputtering, dry etching, and wet etching. Since all of the processes are arranged in series, any mistake in one process will result in the failure of the gas sensor fabrication.

5.4 THE MEMS GAS SENSORS BASED ON SPUTTERED SMO THIN FILMS

In this section, methodology of fabricating MEMS gas sensors is introduced, and sputtered SMO thin films for detecting hydrogen and ethanol are briefly reviewed. Two examples are given to develop MEMS gas sensors based on the sputtered thin films in detail.

5.4.1 METHODOLOGY OF FABRICATING THE MEMS GAS SENSORS

Various methods can prepare the nanostructured gas sensing SMO thin films with large specific area, such as sol-gel method, hydrothermal method, template method, and electrostatic spinning method. However, these methods usually are not compatible with the MEMS processes, and they cannot be directly adopted in MEMS to massively fabricate gas sensors based on these SMO thin films at wafer level. At present, the common practice is to prepare suspension of gas sensing nanomaterials, and then apply "dispense coating" to integrate the gas sensing nanomaterials with the fabricated micro-hotplates, as shown in Figure 5.12(a). The micro-printing or inkjet, as the very common methods to fabricate the MEMS gas sensor, can be

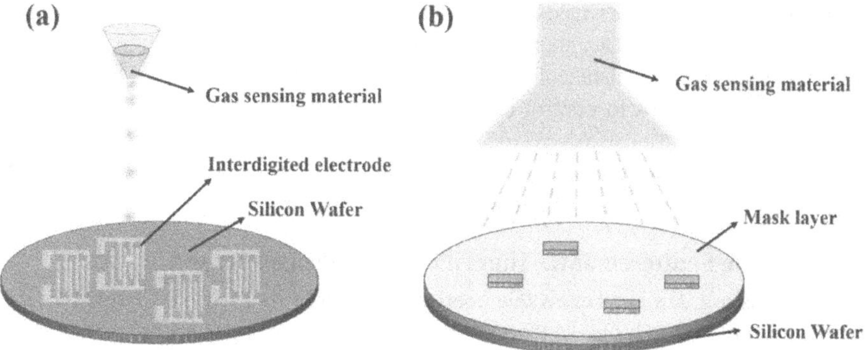

FIGURE 5.12 Method of integrating the gas sensing nanomaterials and the interdigitated electrodes, (a) dispense coating, and (b) large area coating.

categorized as a kind of "dispense coating." Actually, the micro-printing or inkjet is to fabricate the gas sensor one by one, and in production it proposes high requirements for the equipment. So far it still cannot overcome the disadvantages of poor process consistency and low efficiency. In essence, the current process of fabricating a micro-hotplate first and then preparing a small area of sensing film is not an integration method at wafer level. A "top-down" and "bottom-up" combination strategy was proposed to manufacture wafer-scale miniaturized gas sensors by in-situ growth of $Ni(OH)_2$ nanowalls at specific locations [29]. To integrate the gas sensing material and IDEs and micro-hotplate at wafer level, the gas sensing thin films should be prepared at wafer level and can also be compatible with the MEMS process. This wafer-level integration method can be schematically illustrated by "large area coating," as shown in Figure 5.12(b). In the "large area coating," any thin-film preparation process can be used if it can realize uniform gas sensing layer over the wafer, and the mask may be photoresist, hard mask, or other types if it can define the pattern of sensing material and can also be removed easily after use. For example, with hard mask, Barborini, E. et al. fabricated the NO_2 gas sensors based on thin film in batch [30]. The method theoretically can solve the problem of integrating the nanostructured gas sensing material with IDEs and micro-hotplate. However, in application it needs higher accuracy of alignment between the hard mask and the micro-hotplate, and at the same time, size of sensing area defined by the hard mask cannot be very small.

In the "dispense coating," coating gas sensing SMO is usually arranged as the last step of the whole fabrication process. However, in "large area coating," deposition of sensing SMO thin film usually is not designed as the last step. These analyses indicate that an MEMS-compatible SMO thin film that has good gas sensing properties is precondition of fabricating gas sensors at wafer level by "large area coating."

So far, some gas sensors based on the sputtered SMO thin film exhibit high response, high reliability, high anti-interference, and stable properties under harsh environments. Since the thickness of the sputtered thin film can be controlled uniformly on the wafer, the sensor size may be smaller and the power consumption of the sensor may be lower. More importantly, sputtered SMO thin films are well compatible with MEMS technology, suitable for batch production. These advantages will make the gas sensors based on the sputtered thin films with better consistency than sensors based on films prepared by "dispense coating." Moreover, as mentioned above, gas sensing properties of sputtered thin films could be improved by adjusting the sputtering process to control crystal size, surface morphology, microstructure, and so on.

5.4.2 MEMS Hydrogen Sensors

5.4.2.1 The Sputtered SMO Thin Films for Hydrogen Sensor

Hydrogen is a green and renewable energy that is used in a variety of applications, including fuel cell vehicle, spacecraft, automobile, electric generator, and aircraft. This light, odorless, highly flammable gas, if leaking, will cause serious accidents. Therefore, there is a great demand for the gas sensors that can effectively monitor hydrogen concentration and alarm its leaking.

The sputtered SMO thin films usually are dense and unable to respond with good properties to the low concentration of H_2. On the basis of the gas sensing mechanism, how to improve gas sensing responses of the sputtered thin films is an important topic. Presently works showed that the sputtered SMO thin films have good response properties to hydrogen.

Yamazaki, T. et al. [31] studied H_2 gas sensing properties of SnO_2 thin films deposited by DC magnetron sputtering. The SnO_2 films were prepared with varied Pd doping and densities by changing the deposition parameters like discharge pressure and substrate temperature. The film with low density and Pd doping shows high sensitivity and rapid response.

Imawan, C. et al. [32] prepared TiOx-modified NiO thin film by sputtering TiOx-overlayers in pure Ar gas, and found it has the enhanced sensitivity to H_2, as compared with the pure NiO thin film. The maximum sensitivity of the film occurs at 250°C while the cross-sensitivity is relatively small for other gases such as NO_2 and NH_3.

Vijayalakshmi, K. et al. [33] deposited entirely (101)-oriented ZnO thin films by DC sputtering. These films are composed of tightly packed nanoparticles with a size range of 40–70 nm. The rise in band gap of the films from 3.33 to 3.38 eV is attributed to the decrease in particle size. At an operating temperature of 200°C, the ZnO film with a thickness of 44 nm has the highest response of 145–500 ppm H_2.

Panahi, N. et al. [34] studied hydrogen sensing properties of SnO_2 thin films prepared by DC magnetron sputtering. The crystallinity degree of the SnO_2 thin films greatly depends on the deposition time. With increase in deposition time, sizes of nanoparticles or agglomerates increase, and then the surface roughness also increases. The nanostructured SnO_2 films show acceptable responses to hydrogen.

Shen, Y. B. et al. [35] prepared SnO_2 different nanostructures by sputtering and thermal evaporation. The nanostructured films, nanorods, and nanowire-based sensors have highest response to H_2 at operating temperatures 250°C, 200°C, and 150°C, respectively. The activation energy of H_2 at the surface of the material decreases in the sequence of nanofilms, nanorods, and nanowires, thus progressively improving the gas sensing properties.

Bhati, V. S. et al. [36] investigated undoped and Ni-doped ZnO nanostructures, which were deposited by RF magnetron sputtering. Both the undoped and Ni-doped ZnO are in hexagonal wurtzite structure. The surface morphology varies with Ni doping amount. The 4% Ni-doped ZnO sensor has an outstanding response, due to its large surface sites for oxygen adsorption and lowest activation energy.

Lee, J. H. et al. [37] deposited anatase TiO_2 sensing films with thickness 200 nm and then a Pd layer with a thickness ranging from 3 nm to 13 nm by RF sputtering. After annealing, the Pd layer turns into PdO NPs on TiO_2 films. The thickness of PdO NPs has a great effect on the surface morphology, which determines gas sensing properties of the films.

Drmosh, Q. A. et al. [38] prepared ZnO thin films decorated with Pt nanoparticles (PtNPs@ZnO) by sequential sputtering and annealing at 600°C. Responses of the PtNPs@ZnO thin films to H_2 are greatly enhanced and the Pt nanoparticle density influences the response of the thin film-based gas sensors.

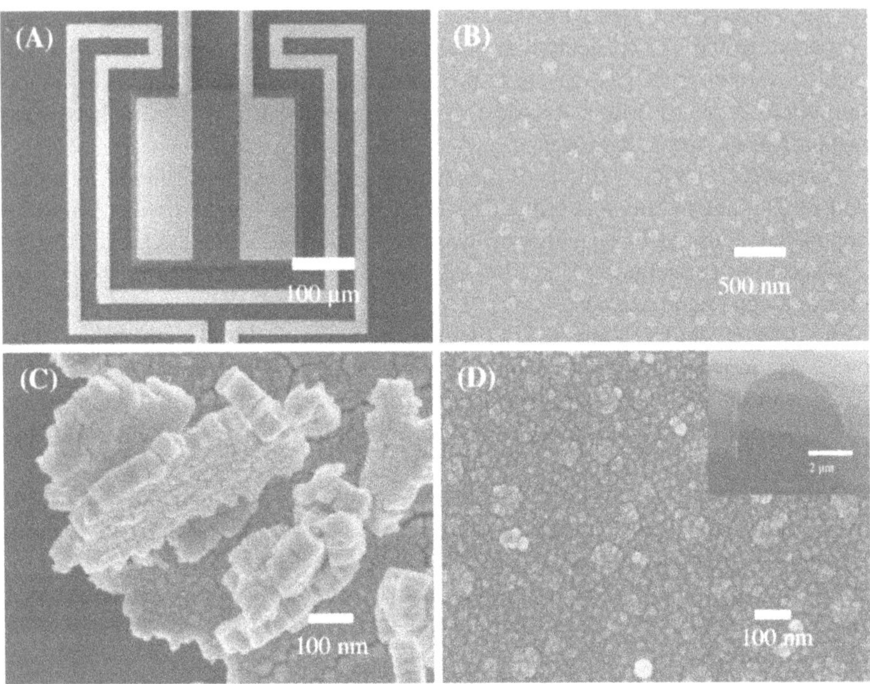

FIGURE 5.13 SEM images: (a) sensor chip, (b) SnO_2 film, (c) cross-section, (d) SnO_2-Pd [40].

Horprathum. M. et al. [39] developed ultrasensitive hydrogen gas sensors by decorating sputtered Pt nanoparticles on vertically aligned WO_3 nanorods prepared by DC magnetron sputtering with the glancing angle deposition (GLAD) technique. The optimal Pt-decorated WO_3 nanorods sensor has a remarkably high response of 2.2×10^5 to 3000 ppm of H_2 at 200°C. The ultrasensitive property may be attributed to dispersion of fine Pt nanoparticles on WO_3 nanorods with a very large effective surface area, leading to highly effective spillover.

Toan, N. V. et al. [40] prepared SnO_2 thin films of different thickness by reactive sputtering. On the SnO_2 film surface, a sputtered Pd layer of 10 nm significantly improves the gas sensing performance of the film. The Pd particles are uniformly distributed on the SnO_2 film, as shown in Figure 5.13(b). Figure 5.13(c) shows that the film thickness is about 80 nm. In Figure 5.13(d), the Pd islands have a diameter of about 5 μm, while the diameter of the nanoparticles that make up the SnO_2 film is less than 10 nm.

5.4.2.2 MEMS Hydrogen Sensor Based on the Sputtered Thin Film

The Pd modified SnO_2 thin film mentioned above has a good hydrogen response. When the film is used as the gas sensing material, the H_2 gas sensor chip can be fabricated by the MEMS process as shown in Figure 5.14, which is described briefly as follows [40].

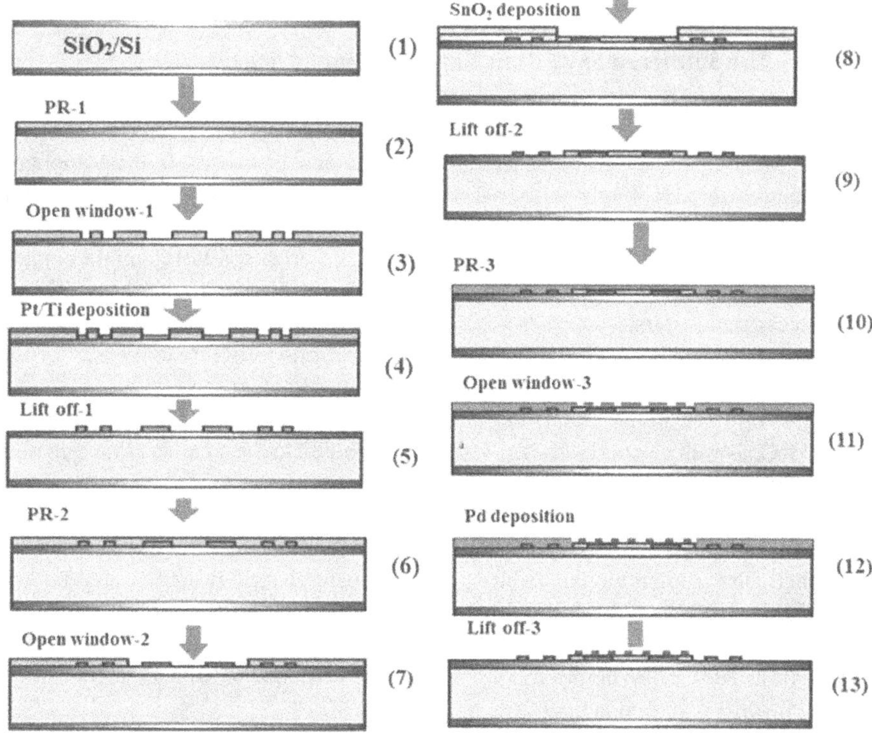

FIGURE 5.14 Fabrication process of an H_2 gas sensor [40].

A silicon wafer with double-sided oxidation was prepared and one side was selected as the front side, on which a layer of photoresist was evenly coated. Under the shelter of mask 1, a specific window 1 was defined by photolithography, and the layers of Pt/Ti were deposited and then the heating wire and electrodes were fabricated by lift off process 1. A photoresist was coated evenly on the front side and a specific window 2 was formed by photolithography through mask 2. SnO_2 film was deposited and its pattern was defined by lift off process 2. The photoresist was continuously evenly coated on the front side, and a specific window 3 was opened by photolithography through the mask 3, and Pd was deposited and then defined by lift off process 3. After slicing, sensor chips were obtained.

In step 8, the SnO_2 was prepared by RF sputtering with an Ar/O_2 ratio of 5: 5 to make the working pressure at 5×10^{-3} torr. The thicknesses of SnO_2 films are 20, 40, 60, and 80 nm and the thicknesses of sputtered Pd films are 5, 10, 25, and 40 nm. The Pd-modified SnO_2 films were treated for 2 h in air at 400 °C.

The sputtered SMO thin films usually have good sensing properties only at an enhanced temperature, instead of room temperature. The H_2 sensor chip mentioned above has no micro heater. To ensure good sensing properties, an SMO thin-film-based gas sensor should be integrated with a micro-hotplate, and an adiabatic groove is required to reduce power consumption of the sensor.

5.4.3 MEMS Ethanol Sensors

5.4.3.1 The Sputtered SMO Thin Films for Ethanol Sensor

Ethanol, as one of the most common VOCs (volatile organic compounds), is widely used in drunk driving tests, safe work in the food industry, brewing process control, medical and clinical applications, and bio-technological processes. In these applications, the sputtered SMO thin-film-based ethanol sensors are expected to have high response, good selectivity, and rapid response/recovery. For this purpose, sputtering techniques such as DC sputtering, RF sputtering, reactive sputtering, and co-sputtering have been attempted to prepare the gas sensing SMO thin films by optimizing the process parameters.

Micocci, G. prepared the vanadium oxide thin films by RF reactive sputtering in an Ar/O_2 atmosphere (O_2 from 5 to 20%) [41]. The film deposited with 15% O_2/Ar ratio has high sensitivity and selectivity at operating temperature between 280°C and 300°C, as well as good stability, which is attributed to the smaller grain size and content of V_2O_5.

Mg-doped ZnO thin films were deposited by RF magnetron sputtering for gas-sensing application [42]. The film with 10 mol% of Mg has better response toward ethanol, methanol, and ammonia, and its surface-to-volume ratio is higher than others.

NiO thin films were deposited directly by RF magnetron sputtering using a NiO target [43]. The film deposited at 200°C has the smallest crystal size and accordingly the highest surface-to-volume ratio. The highest response to ethanol of the film-based sensor is recorded at operating temperature 400°C.

The WO_3 thin films composed of WO_3 nanorods were prepared by RF sputtering via GLAD [44]. At operating temperature 300°C, the optimized thin-film sensor has the maximum response of 10–200 ppm ethanol. The good sensing properties are attributed to the surface nanostructures with very high surface-to-volume ratio of the thin films.

The nanocrystalline SnO_2 thin films with noble metals and 3*d*-transition metal additives in the bulk and on the surface have been studied [45]. Thin SnO_2 films were deposited in oxygen–argon plasma by magnetron DC sputtering of a tin–antimony alloy target. The thin-film sensors can detect ethanol and acetone vapors at 1 ppm level, and their responses to ethanol are higher than to acetone at the operating temperature less than 650 K.

NiO thin films were prepared by DC-reactive magnetron sputtering. NiO-based sensors with thin film of thickness 50 nm have relatively higher response in the detection of the ethanol [46], which can be attributed to the 50 nm NiO films formed by nanocrystals with a lateral size of about 30 nm homogenously covering the Pt/alumina surface.

ITO thin films were deposited on the Si substrate by RF magnetron sputtering and the RF power strongly affected the sensitivity of detection [47]. The ITO films prepared at 200 W are more sensitive as compared with the films prepared at 300 W. ITO films sputtered at 200 W are composed of the uniformly distributed particles of 5–10 nm.

The $MgIn_2O_4$ thin films were deposited by RF sputtering by altering the parameters as sputtering power and substrate temperature [48], which have obvious effect on the sensing responses of the films to ethanol. With increasing RF power there is an obvious improvement in the sensitivity of the film-based sensor.

Compositionally gradient tin oxide thin films were prepared by co-sputtering from tin and SnO_2 targets [49]. The coexisting p-type SnO and n-type SnO_2 affect the direction of electrical resistance and the response during gas reaction and recovery. SnO and SnO_2 respond in the opposite direction of electrical resistance and show different responses for each volatile gas because of their different semiconductor properties.

ZnO-NiO p-n composite thin films were grown through RF co-sputtering [50]. The Ni content varies with the NiO sputtering power. The surface grain size of the films increases and its morphology turns into a large pyramidal stereo-geometry when the NiO sputtering power is greater than 90 W. An optimal composition for the film shows superior ethanol gas-sensing properties, which are attributed to the p–n junctions in the composite films.

Cross-linked SnO_2:NiO networks were fabricated by sputtering SnO_2:NiO target onto the etched self-assembled triangle polystyrene (PS) microsphere arrays [51]. The cross-linked films show high response, low detection limit, high selectivity to ethanol, which can be attributed to the more active adsorption sites in the cross-linked SnO_2:NiO network.

Nanocrystalline CuO-$CuFe_2O_4$ composite thin films were developed from the $CuFeO_2$ ceramic target by using the RF sputtering method followed by a thermal oxidation process [52]. The thin film of thickness 25 nm exhibits 128% response toward 500 ppm of ethanol with 90 seconds response time at the operating temperature of 400°C.

A heterojunction thin film is composed of NiO deposited by electron beam evaporation and SnO_2 prepared by RF magnetron sputtering [53]. The 5 nm NiO/100 nm SnO_2 film-based gas sensor has a high response of 7.9 at 250°C for 100 ppm ethanol, good reproducibility and linearity. In addition, it can detect the ultra-low concentration ethanol as low as 100 ppb.

5.4.3.2 Sputtered SMO Thin-Film-Based MEMS Ethanol Sensor

To show how to develop the MEMS gas sensor based on the sputtered SMO film, an ethanol MEMS sensor is taken as example, in which the SMO composite thin film is used as the gas sensing material [22]. With similar lattice parameters, the SMOs such as TiO_2 and SnO_2 were selected to prepare the SnO_2/TiO_2 multi-layer composite thin film by magnetron sputtering.

Before depositing the SMO thin film, a silicon wafer was thermally oxidized on both sides, and Si_3N_4 thin film was deposited on both the sides by LPCVD. Then, the TiO_2 and SnO_2 targets (with purity 99.99%) were used to sputter the multi-layer composite thin film with parameters as shown in Table 5.2. The layers of SnO_2 and TiO_2 were sputtered alternately for a total of six times, and the deposition rates of SnO_2 and TiO_2 are 2.8 nm/min and 1.7 nm/min, respectively. Finally, the SnO_2/TiO_2 multi-layer composite thin film, as shown schematically in Figure 5.15, was heat-treated at 500 °C for 3 h.

5.4.3.3 Characterization

As the XRD pattern in Figure 5.16(a) shows, there are only TiO_2 and SnO_2 in the multi-layer composite thin film without other impurity peaks. The diffraction peaks

TABLE 5.2
Parameters of Sputtering the Multi-Layer Nanocomposite Thin Film [22]

Material	Power (W)	Time (min)
SnO_2	100	25
TiO_2	150	15
SnO_2	100	25
TiO_2	150	15
SnO_2	100	25
TiO_2	150	15

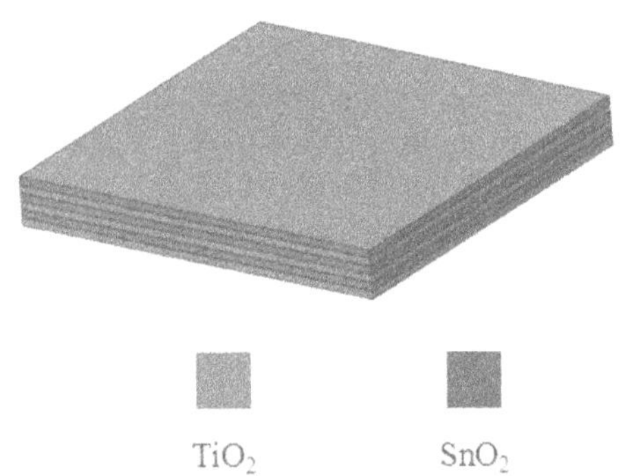

FIGURE 5.15 Multi-layer nanocomposite thin film.

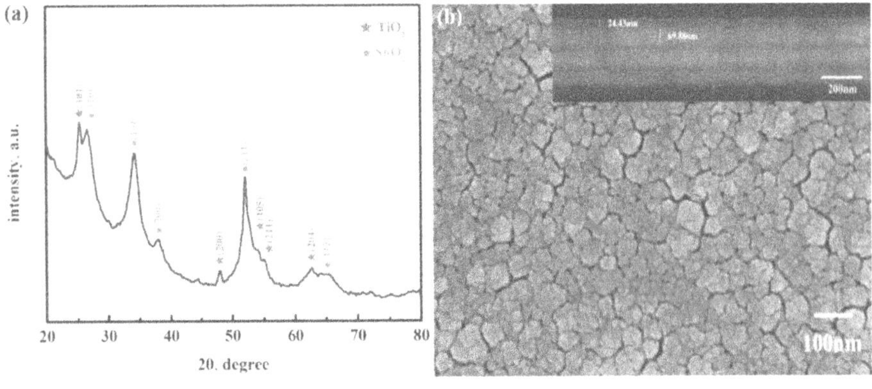

FIGURE 5.16 (a) XRD pattern and (b) FE-SEM image of the SnO_2/TiO_2 nanocomposite thin film [22].

of TiO_2 are at different 2θ angles, corresponding to (101), (200), (105), (211), and (204) crystal planes of anatase TiO_2, respectively. The diffraction peaks of SnO_2 are corresponding to (110), (101), (200), (211), and (112), respectively. As shown in Figure 5.16(b), on the surface of the composite thin film, there are many irregular nanoclusters in different shapes and sizes, and each cluster is composed of many tiny nanoparticles. Scattering cracks between the nanoclusters will improve the specific surface area of the film and the adsorption sites of gas molecules. The embedding image on the upper right corner in Figure 5.16(b) is the cross-section of the composite thin film, from which one may clearly see that the thin film consists of six layers of two SMOs, and the thicknesses of the SnO_2 and TiO_2 layers are about 70 nm and 24 nm, respectively.

5.4.3.4 Design of the MEMS Gas Sensor

To realize low power consumption, mass production, and high sensitivity, the SMO thin-film-based MEMS gas sensors are designed as shown schematically in Figure 5.17. From bottom to top, the gas sensor successively consists of a silicon substrate, a supporting element, a sensing material, a heating wire, and a pair of interdigitated electrodes. The sensing material is the SnO_2/TiO_2 multi-layer thin film mentioned above. The heating wire provides the operating temperature by applying a certain voltage, and the interdigitated electrodes transmit changes in the resistance of the sensing material. The supporting element conducts heat from the heating wire to the gas sensing material, and it also acts as an insulator.

Due to the high thermal conductivity of silicon, to reduce power consumption, the silicon substrate below the heating area in the sensor should be etched to form an adiabatic groove, which can efficiently reduce the heat transfer between the heating wire and the substrate. Generally, silicon removal can be classified into two types. One is to form a concave part under sensitive materials and heating wires by top-down etching (wet or dry etching), which leads to the suspended membrane, as shown in Figure 5.18(a). The other is to remove silicon (e.g. wet etching) from bottom-top to form the trapezoidal adiabatic groove on the wafer backside, and the front side is a closed membrane to support the sensitive material, as shown in Figure 5.18(b). In the suspended membrane, the heating wire is

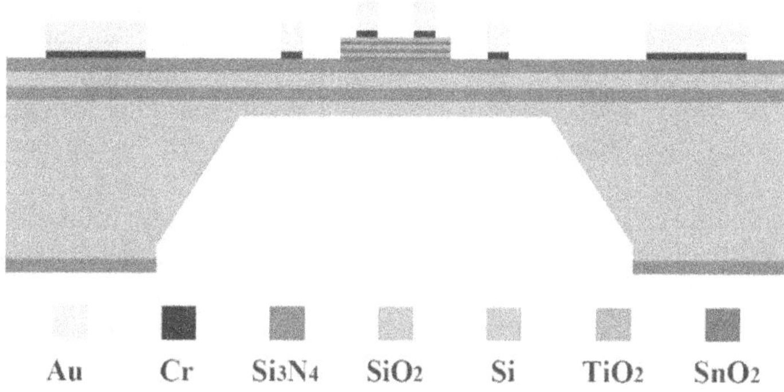

FIGURE 5.17 MEMS gas sensor based on the sputtered SMO thin film [22].

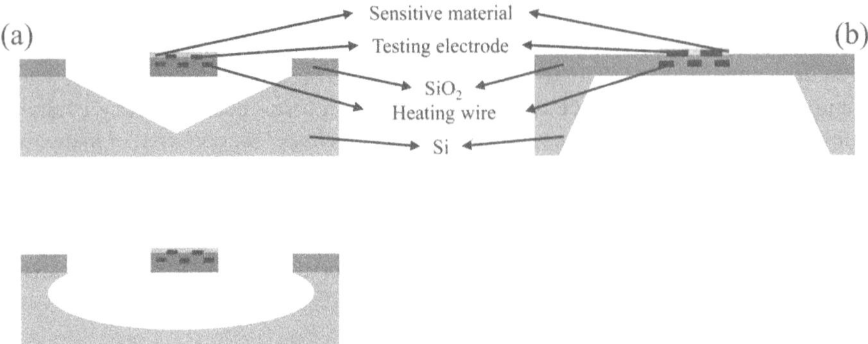

FIGURE 5.18 The two types of the adiabatic grooves for MEMS gas sensors (a) cantilever beam, (b) closed membrane. [54].

suspended, and the heat dissipation area and heat transfer are greatly reduced, which can realize very low power consumption. However, its poor structural stability may reduce the yield of devices. By contrast, the closed membrane is more mechanically stable, and its power consumption may also be very low because of the adiabatic groove. Therefore, the closed membrane structure is selected for the MEMS gas sensor in our case [54].

The supporting layer must have enough mechanical strength, and residual stress in the layer should be as small as possible. For this purpose, the supporting layer is designed as the membrane successively composed of four layers of $SiO_2/Si_3N_4/SiO_2/Si_3N_4$, in which the thickness of the SiO_2 layer is 500 nm and the thickness of the Si_3N_4 layer is 150 nm. The residual stress in the SiO_2 thin films is usually in a compressive state, whereas that in the Si_3N_4 thin films is in a tensile state. The contrary stress states of the different films can compensate for each other, thus reducing the stress of the supporting layer.

The SnO_2/TiO_2 nanocomposite thin film is designed as a square of 100 μm × 100 μm, locating in the upside surface center of the gas sensor, as Figure 5.19 shows. A double-helix heating wire with a central symmetry is used to provide a uniform and symmetrical temperature field in the central sensing material area and reduce the thermal stress caused by temperature inhomogeneity or asymmetry. The interdigitated electrodes cover on the surface of the gas sensing material, and are spirally wound with the heating wires, showing a central symmetrical structure. Near the heating wires, two resistors are designed as temperature sensors, which can be used for measuring the operating temperature on the applications with higher accuracy requirements.

Both Pt and Au are usually selected as electrode materials. Pt is used as the heating wire due to its antioxidant properties, chemical and thermal stability at high temperature, and can also be used as the temperature sensing material. With the advantages of high electric conductivity, chemical and thermal stability, suitability for subsequent wiring bonding, Au is mostly used in the electrodes. In the example of this chapter, the operating temperature is not measured for simplification, so Au is used as the material for the heating wire and sensing electrodes.

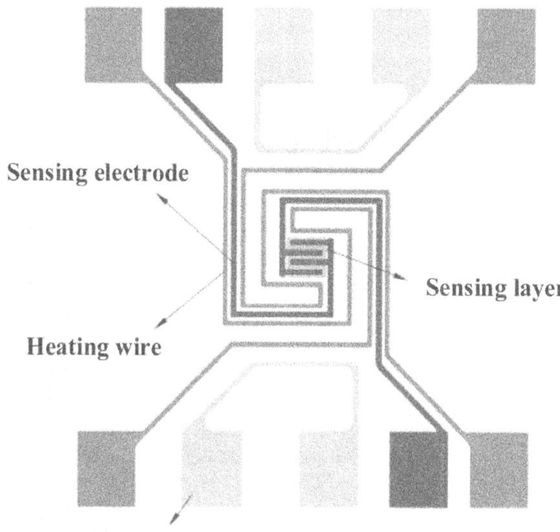

FIGURE 5.19 Layout of the electrodes and sensing material [22].

5.4.3.5 Fabrication Process

Fabrication of the MEMS ethanol gas sensor mainly includes four processes: (1) fabrication of the supporting layer, (2) definition of the sensitive pattern, (3) fabrication of the electrodes, and (4) etching of the adiabatic groove. Various MEMS technologies, such as CVD, lithography, magnetron sputtering, ICP etching, and wet etching, are involved. In Table 5.3, the process flow of 15 steps was used for fabricating the MEMS ethanol gas sensor.

TABLE 5.3
Process Flow of Fabricating the Low-Power Integrated MEMS Gas Sensor

Fabrication Process	Schematic Diagram	Key Parameters
1. Silicon substrate		N-type doping, 4 inches (100) wafer, thickness 400 ± 10μm, resistance 3–4.8 Ω·cm
2. SiO_2 by thermal oxidization		Double-side thermal oxidation of SiO_2 film (thickness 500 nm)
3. Si_3N_4 thin film by LPCVD		Si_3N_4 film of 150 nm by LPCVD, substrate temperature 810°C
4. SiO_2/Si_3N_4 muilti-layers by PECVD		SiO_2 film of 500 nm by PECVD, substrate temperature 350°C; Si_3N_4 film of 150 nm by PECVD, substrate temperature 350°C

(Continued)

TABLE 5.3 (*Continued*)
Process Flow of Fabricating the Low-Power Integrated MEMS Gas Sensor

Fabrication Process	Schematic Diagram	Key Parameters
5. Resist pattern by lithography		EPG535 resist of 2 μm by spinning, prebaking, exposure, developing, and baking
6. SnO_2/TiO_2 composite thin-film deposition	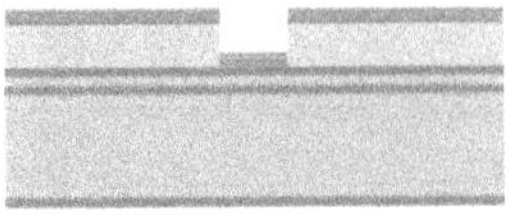	Magnetic sputtering 70 nm TiO_2 and 20 nm SnO_2 in turn three time
7. Lift-off and heat treatment		Lift-off in acetone, and then in ethyl alcohol, deionized water; cleaning and baking; heat treatment at 500°C for 3 h
8. Resist pattern of the electrodes		EPG535 resist of 2 μm by spinning, prebaking, exposure, developing, and baking
9. Cr/Au films by electron beam		Cr of 50 nm and Au of 50 nm by electron beam deposition
10. Lift-off, heat treatment		Lift-off in acetone, cleaning, baking, and heat treatment at 300°C, 10 min
11. Open window on back side		Cleaning, baking, spinning 2 μm of EPG535, prebaking, exposure, developing, and baking
12. Dry etching	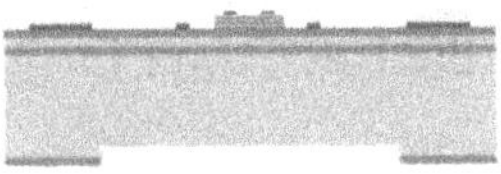	ICP etching to remove the 150 nm Si_3N_4 and 500 nm SiO_2 films
13. Upside protection		Spinning 4 μm EPG535 resist, drop coating Dow Corning 184, baking; Adhering to a glass wafer, secondary sealing, and drying
14. Backside wet etching	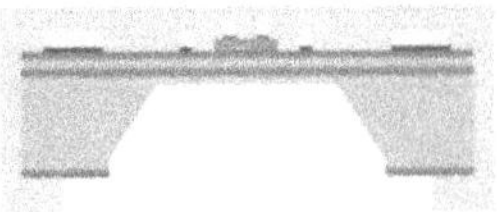	Wet etching in 25% TMAH, at temperature 85°C, 17 h
15. Removing photoresist and slicing	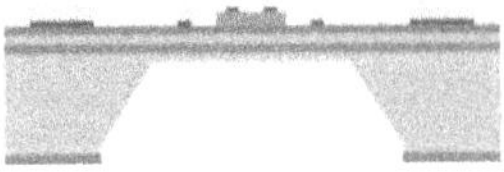	Cutting PDMS edge, separating PDMS and glass wafer, cleaning and baking; slicing wafer

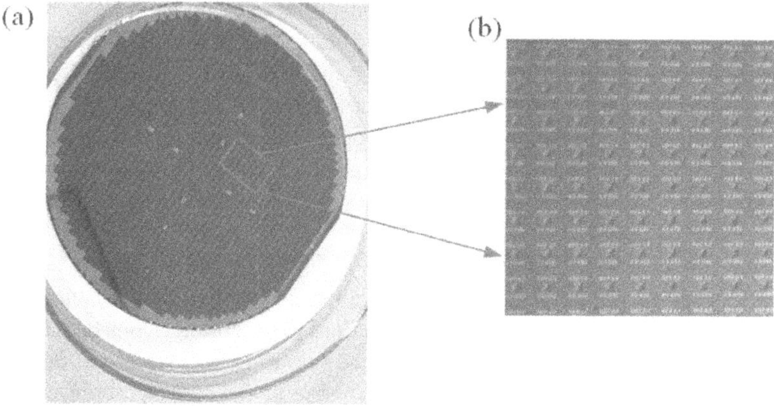

FIGURE 5.20 The fabricated MEMS gas sensor chip on (a) the wafer and (b) the partial enlarged drawing.

The MEMS gas sensor chips are successfully fabricated at wafer level. Figure 5.20(a) is the unsliced sensor wafer and (b) is the magnified local image in which the MEMS gas sensor chips can be clearly seen. The fabrication process is highly repeatable. In the lab, the yield is over 85%, namely for the total number of 1704 sensor chips in the wafer, there are more than 1400 chips which have perfect structure and can work properly.

5.4.3.6 Sensing Properties of the MEMS Gas Sensor

5.4.3.6.1 Operating Temperature

First, the sensor chip's responses to 200 ppm ethanol gas were tested at operating temperature from 148.8 °C to 320 °C. The background gas is synthetic air. As shown in Figure 5.21(a), the sensor response to ethanol gas increases with the increase in

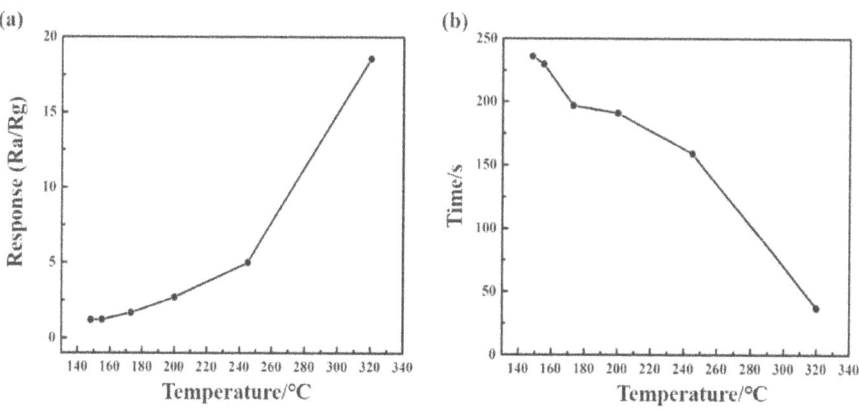

FIGURE 5.21 (a) Response of the sensor at different operating temperatures to 200 ppm ethanol, (b) response time of the sensor at different operating temperatures to 200 ppm ethanol.

operating temperature, and the sensor response can reach up to 18.57 at 320 °C. At different operating temperatures, response time of the sensor to 200 ppm ethanol decrease sharply as shown in Figure 5.21(b), and it is only 37 s at 320°C. It can be predicted that the sensor would have a better response if the operating temperature is higher than 320°C. In our application, the operating temperature is set at 320°C, at which the voltage applied on the heating wire is 1.3 V, and the power consumption is 39 mW.

5.4.3.6.2 Responses to Different Ethanol Concentrations

At the operating temperature 320°C, response properties of the sensor to different ethanol gas concentrations were studied. The ethanol concentration is in the range of 50–600 ppm, and the resistance variation is collected at an interval of 1 s. As shown in Figure 5.22(a), the response of the sensor increases significantly with the increase in ethanol concentration and presents an obvious step-like variation. According to the response-recovery histogram in Figure 5.22(b), with ethanol concentration increasing, the response time of the sensor decreases, whereas the recovery time decreases first and then increases when the concentration is greater than 200 ppm. The response time at each concentration is far less than the recovery time. The built-in picture of Figure 5.22(b) shows variation of the sensor response time vs. the concentration more clearly. When the ethanol concentration varies from 50 ppm to 600 ppm, the response time drops from 102 s to 13 s, while the recovery time becomes much longer, varying from 193 s to 900 s.

Gas sensing properties of the multi-chips randomly taken from the same wafer were investigated at an operating temperature of 148°C–320°C by measuring their resistances to 200 ppm ethanol gas. As shown in Figure 5.23, three sensor chips exhibit noticeable increasing in response with rising the temperature up to 320°C.

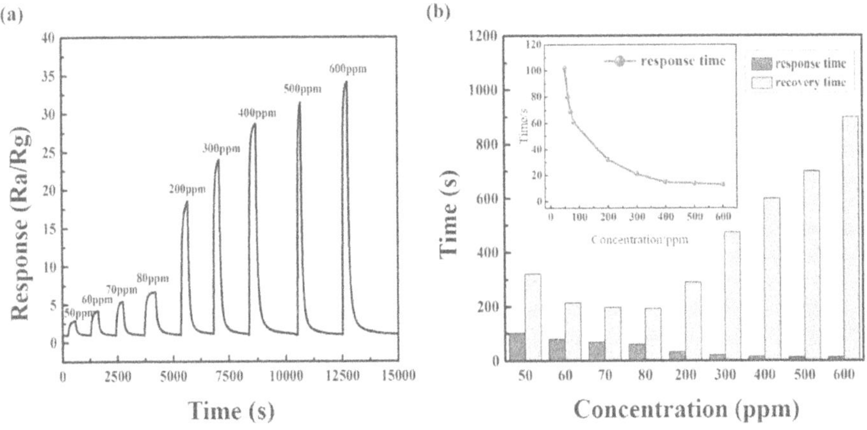

FIGURE 5.22 Response and response/recovery time of the gas sensor at optimal operating temperature 320°C [22].

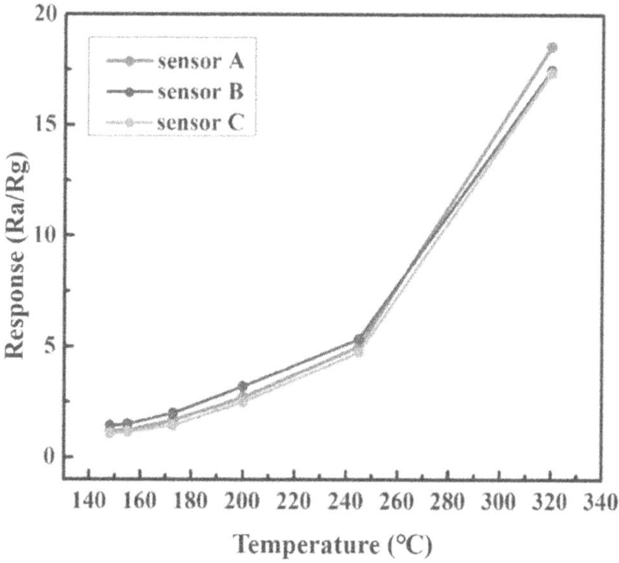

FIGURE 5.23 Responses vs. temperature in 200 ppm alcohol of three sensors [22].

Since the chips are selected randomly from the same wafer after slicing, the result verifies that the MEMS sensor chips are in good consistency of responses. For one of sensors, eight cycles of tests were carried out at an operating temperature of 320°C to 50 ppm ethanol. The results indicate the eight response curves are almost consistent, and the highest and lowest responses are 3.8 and 3.6 respectively, verifying good repeatability of the sensor.

In the above MEMS sensor, the gas sensing material is the sputtered multi-layer composite thin film. If other sputtered SMO film with good sensing properties to the target gas replaces it, then a new MEMS gas sensor for detecting the target gas can also be fabricated with the same process.

5.5 SUMMARY

To meet requirements of IoT development, gas sensors with small sizes, low cost, low power consumption, and good responses are needed badly. The MEMS gas sensors based on the SMOs are in accordant with all these characteristics. If the SMO gas sensing materials can be prepared with a MEMS-compatible process, it is possible to fabricate the MEMS gas sensors in a large batch with good consistency. In this chapter, we discussed the sputtering techniques, and analyzed the factors, which may influence gas sensing properties of the sputtered SMO thin films. e.g., microstructures, composition, and surface topography. The sputtering SMO thin films as the sensing materials for detecting ethanol and hydrogen are briefly reviewed. The examples of the MEMS gas sensors based on the sputtered SMO thin films are given in detail.

REFERENCES

1. Lakkis, S., Younes, R., Alayli, Y., and Sawan, M. 2014. Review of recent trends in gas sensing technologies and their miniaturization potential. *Sensor Review* 34:24–35. 10.1108/SR-11-2012-724
2. Fergus, J. 2007. Materials for high temperature electrochemical NOx gas sensors. *Sensors and Actuators B: Chemical* 121:652–663. 10.1016/j.snb.2006.04.077
3. Wang, H., Liu, Z., Chen, D., and Jiang, Z. 2015. A new potentiometric SO_2 sensor based on Li_3PO_4 electrolyte film and its response characteristics. *Review of Scientific Instrument* 86:075007. 10.1063/1.4927173
4. Sharma, A., Tomar, M., and Gupta, V. 2011. SnO_2 thin film sensor with enhanced response for NO_2 gas at lower temperatures. *Sensors and Actuators B: Chemical* 156:743–752. 10.1016/j.snb.2011.02.033
5. Barsan, N., and Weima, U. 2001. Conduction model of metal oxide gas sensors. *Journal of Electroceramics* 7:143–167. 10.1023/A:1014405811371
6. Katsuki, A., and Fukui, K. 1998. H_2 selective gas sensor based on SnO_2. *Sensors and Actuators B: Chemical* 52:30–37. 10.1016/S0925-4005(98)00252-4
7. Mourya, S., Kumar, A., Jaiswal, J., Malik, G., Kumar, B., and Chandra, R. 2019. Development of Pd-Pt functionalized high performance H_2 gas sensor based on silicon carbide coated porous silicon for extreme environment applications. *Sensors and Actuators B: Chemical* 283:373–383. 10.1016/j.snb.2018.12.042
8. Tan, W., Yu, Q., Ruan, X., and Huang, X. 2015. Design of SnO_2-based highly sensitive ethanol gas sensor based on quasi molecular-cluster imprinting mechanism. *Sensors and Actuators B: Chemical* 212:47–54. 10.1016/j.snb.2015.01.035
9. Luo, Y., Zhang, C., Zheng, B., Geng, X., and Debliquy, M. 2017. Hydrogen sensors based on noble metal doped metal-oxide semiconductor: A review. *International Journal of Hydrogen Energy* 42:20386–20397. 10.1016/j.ijhydene.2017.06.066
10. Xiao, L., Xu, S., Yu, G., and Liu, S. 2018. Efficient hierarchical mixed Pd/SnO_2 porous architecture deposited microheater for low power ethanol gas sensor. *Sensors and Actuators B: Chemical* 255:2002–2010. 10.1016/j.snb.2017.08.216
11. Ivanov, P., Llobet, E., Vilanova, X., Brezmes, J., Hubalek, J., and Correig, X. 2004. Development of high sensitivity ethanol gas sensors based on Pt-doped SnO_2 surfaces. *Sensors and Actuators B: Chemical* 99:201–206. 10.1016/j.snb.2003.11.012
12. Hastir, A., Kohli, N., and Singh, R. C. 2017. Ag doped ZnO nanowires as highly sensitive ethanol gas sensor. *Materials Today: Proceedings* 4:9476–9480. 10.1016/j.matpr.2017.06.207
13. Xu, X., Fan, H., Liu, Y., Wang, L., and Zhang, T. 2011. Au-loaded In_2O_3 nanofibers-based ethanol micro gas sensor with low power consumption. *Sensors and Actuators B: Chemical* 160:713–719. 10.1016/j.snb.2011.08.053
14. Miller, D. R., Akbar, S. A., and Morris, P. A. 2014. Nanoscale metal oxide-based heterojunctions for gas sensing: A review. *Sensors and Actuators B: Chemical* 204:250–272. 10.1016/j.snb.2014.07.074
15. Zhang, K., Qin, S., Tang, P., Feng, Y., and Li, D. 2020. Ultra-sensitive ethanol gas sensors based on nanosheet-assembled hierarchical ZnO-In_2O_3 heterostructures. *Journal of Hazardous Materials* 391:122191. 10.1016/j.jhazmat.2020.122191
16. Wang, Q., Kou, X., Liu, C., and et al. 2018. Hydrothermal synthesis of hierarchical CoO/SnO_2 nanostructures for ethanol gas sensor. *Journal of Colloid and Interface Science* 513:760–766. 10.1016/j.jcis.2017.11.073
17. Zhang, C., Boudiba, A., Marco, P. D., Snyders, R., Olivier, M., and Debliquy, M. 2013. Room temperature responses of visible-light illuminated WO_3 sensors to NO_2 in sub-ppm range. *Sensors and Actuators B: Chemical* 181:395–401. 10.1016/j.snb.2013.01.082

18. C.N. Xu, J. Tamaki, N. Miura, N. Yamazoe.1991. Grain-size effects on gas sensitivity of porous SnO_2-based elements. *Sensors and Actuators B: Chemical* 3:147–155. 10.1016/0925-4005(91)80207-Z
19. Kelly, P. J., and Arnell, R. D. 2013. Magnetron sputtering: a review of recent developments and applications. *Vacuum* 56:159–172. 10.1016/S0042-207X(99)00189-X
20. Zoolfakar, A. S., Ahmad, M. Z., Rani, R. A., and et al. 2013. Nanostructured copper oxides as ethanol vapour sensors. *Sensors and Actuators B: Chemical* 185:620–627. 10.1016/j.snb.2013.05.042
21. Dong, Y. F., Wang, W. L., and Liao, K. J. 2000. Ethanol-sensing characteristics of pure and Pt-activated $CdIn_2O_4$ films prepared by r.f. reactive sputtering. *Sensors and Actuators B: Chemical* 67:254–257. 10.1016/S0925-4005(00)00515-3
22. Wang, H., Wang, M., Lei, W., Chen, X., Huang, H., and Wang, J. 2019. Wafer-level fabrication of low power consumption integrated alcohol micro sensor. *Micro & Nano letters* 14:11–16. 10.1049/mnl.2018.5183
23. Mounasamy, V., Mani, G. K., Ponnusamy, D., Tsuchiya, K., Prasad, A. K., and Madanagurusamy, S. 2019. Sub-ppm level detection of trimethylamine using V_2O_3 -Cu_2O mixed oxide thin films. *Ceramics International* 45:19528–19533. 10.1016/j.ceramint.2019.06.074
24. Girija, K. G., Somasundaram, K., Topkar, A., and Vatsa, R. K. 2016. Highly selective H_2S gas sensor based on Cu-doped ZnO nanocrystalline films deposited by RF magnetron sputtering of powder target. *Journal of Alloys and Compounds* 684:15–20. 10.1016/j.jallcom.2016.05.125
25. Åbom, A. E., Persson, P., Hultman, L., and Eriksson, M. 2002. Influence of gate metal film growth parameters on the properties of gas sensing field-effect devices. *Thin Solid Films* 409:233–242. 10.1016/S0040-6090(02)00135-9.
26. Hien, V. X., You, J., Jo, K., and et al. 2014. H_2S-sensing properties of Cu_2O submicron-sized rods and trees synthesized by radio-frequency magnetron sputtering. *Sensors and Actuators B: Chemical* 202:330–338. 10.1016/j.snb.2014.05.070
27. Wang, Y., Tong, W. G., and Han, N. 2020. Co-sputtered Pd/SnO_2:NiO heterostructured sensing films for MEMS-based ethanol sensors. *Materials Letters* 273:127924. 10.1016/j.matlet.2020.127924
28. Prajapati, C. S., and Bhat, N. 2018. Growth optimization, morphological, electrical and sensing characterization of V_2O_5 films for SO_2 sensor chip. *IEEE Sensors Conference* 1:1–4. 10.1109/ICSENS.2018.8589650
29. Liu, L., Wang, Y., Sun, F., and et al. 2020. "Top-down" and "bottom-up" strategies for wafer-scaled miniaturized gas sensors design and fabrication. *Microsystems & Nanoengineering* 6:1–10. 10.1038/s41378-020-0144-4
30. Barborini, E., Vinati, S., Leccardi, M., and et al. 2008. Batch fabrication of metal oxide sensors on micro-hotplates. *Journal of Micromechanics and Microengineering* 18:055015. 10.1088/0960-1317/18/5/055015
31. Yamazaki, T., Okumura, H., Jin, C., Nakayama, A., Kikuta, T., and Nakatani, N. 2005. Effect of density and thickness on H_2 -gas sensing property of sputtered SnO_2 films. *Vacuum* 77:237–243. 10.1016/j.vacuum.2004.09.024
32. Imawan, C., Solzbacher, F., Steffes, H., and Obermeier, E. 2000. TiOx -modified NiO thin films for H_2 gas sensors: effects of TiOx -overlayer sputtering parameters. *Sensors and Actuators B: Chemical* 68:184–188. 10.1016/S0925-4005(00)00427-5
33. Vijayalakshmi, K., Karthick, K., and Tamilarasan, K. 2013. Enhanced H_2 sensing properties of a-plane ZnO prepared on c-cut sapphire substrate by sputtering. *Journal of Materials Science* 24:1325–1331. 10.1007/s10854-012-0927-y
34. Panahi, N., Hosseinnejad, M. T., Shirazi, M., and Ghoranneviss, M. 2016. Optimization of Gas Sensing Performance of Nanocrystalline SnO_2 Thin Films Synthesized by Magnetron Sputtering. *Chinese Physics Letters* 33:103–107. 10.1088/0256-307x/33/6/066802

35. Shen, Y., Wang, W., Fan, A., and et al. 2015. Highly sensitive hydrogen sensors based on SnO_2 nanomaterials with different morphologies. *International Journal of Hydrogen Energy* 40:15773–15779. 10.1016/j.ijhydene.2015.09.077
36. Bhati, V. S., Ranwa, S., Fanetti, M., Valant, M., and Kumar, M. 2018. Efficient hydrogen sensor based on Ni-doped ZnO nanostructures by RF sputtering. *Sensors and Actuators B: Chemical* 255:588–597. 10.1016/j.snb.2017.08.106
37. Lee, J. H., Kwak, S., Lee, J., and et al. 2018. Sputtered PdO decorated TiO_2 sensing layer for a hydrogen gas sensor. *Journal of Nanomaterials* 2018:1–8. 10.1155/2018/8678519
38. Drmosh, Q. A., and Yamani, Z. H. 2016. Hydrogen sensing properties of sputtered ZnO films decorated with Pt nanoparticles. *Ceramics International* 42:12378–12384. 10.1016/j.ceramint.2016.05.011
39. Horprathum, M., Srichaiyaperk, T., Samransuksamer, B., and et al. 2014. Ultrasensitive hydrogen sensor based on Pt-decorated WO_3 nanorods prepared by glancing-angle dc magnetron sputtering. *ACS Applied Materials & Interfaces* 6:22051–60. 10.1021/am505127g
40. Toan, N. V., Chien, N. V., Duy, N. V., and et al. 2016. Fabrication of highly sensitive and selective H_2 gas sensor based on SnO_2 thin film sensitized with microsized Pd islands. *Journal of Hazardous Materials* 301:433–442. 10.1016/j.jhazmat.2015.09.013
41. Micocci, G., Serra, A., Tepore, A., Capone, S., Rella, R., and Siciliano, P. 1998. Properties of vanadium oxide thin films for ethanol sensor. *Journal of Vacuum Science & Technology A* 15:34–38. 10.1116/1.580471
42. Vinoth, E., Gowrishankar, S., and Gopalakrishnan, N. 2018. Effect of Mg doping in the gas-sensing performance of RF-sputtered ZnO thin films. *Applied Physics A-Materials Science & Processing* 124:433. 10.1007/s00339-018-1852-6
43. Kuma, R., Baratto, C., Faglia, G., Sberveglieri, G., Bontempi, E., and Borgese, L. 2015. Tailoring the textured surface of porous nanostructured NiO thin films for the detection of pollutant gases. *Thin Solid Films* 583:233–238. 10.1016/j.tsf.2015.04.004
44. Ahmad, M. Z., Wisitsoraat, A., Zoolfakar, A. S., Kadir, R. A., and Wlodarski, W. 2013. Investigation of RF sputtered tungsten trioxide nanorod thin film gas sensors prepared with a glancing angle deposition method toward reductive and oxidative analytes. *Sensors and Actuators B: Chemical* 183:364–371. 10.1016/j.snb.2013.04.027
45. Sevastyanov, E. Y., Maksimova, N. K., Khludkova, L. S., Chernikov, E. V., and Sergeychenko, N. V. 2016. Acetone and ethanol sensors based on nanocrystalline SnO_2 thin films with various catalysts. *Bionanoscience* 7:654–658. 10.1007/s12668-016-0377-8
46. Gupta, C., Sharma, S., Bhowmik, B., Sampath, K., Periasamy, C., and Sancheti, S. 2019. Development of highly sensitive and selective ethanol sensors based on RF sputtered ZnO nanoplates. *Journal of Electronic Materials* 48:3686–3691. 10.1007/s11664-019-07127-4
47. Pandya, H. J., Chandra, S., and Vyas, A. L. 2011. Fabrication and characterization of ethanol sensor based on RF sputtered ITO films. *Sensors and Transducers Journal* 10:141–150.
48. Anuradha, B., and Sanjeeviraja, C. 2012. Gas sensing properties of RF magnetron sputtered $MgIn_2O_4$ thin films. *Sensors and Actuators A: Physical* 179:98–104. 10.1016/j.sna.2012.03.039
49. Ahn, H., Park, H., Joo, J., and Kim, D. 2012. Volatile gas sensing properties of phase and composition gradient SnOx thin films by combinatorial sputter deposition. *ECS Solid State Letters* 2:11–13. 10.1149/2.001302ssl
50. Liang, Y., and Chan, Y. 2020. The effect of Ni content on gas-sensing behaviors of ZnO–NiO p–n composite thin films grown through radio-frequency cosputtering of ceramic ZnO and NiO targets. *Crystengcomm* 22:2315–2326. 10.1039/D0CE00052C

51. Tong, W., Wang, Y., Bian, Y., Wang, A., Hanl, N., and Chen, Y. 2020. Sensitive cross-linked SnO_2: NiO networks for MEMS compatible ethanol gas sensors. *Nanoscale Research Letters* 15:35. 10.1186/s11671-020-3269-3
52. De, S., Venkataramani, N., Prasad, S., and et al. 2018. Ethanol and hydrogen gas-sensing properties of CuO-$CuFe_2O_4$ nanostructured thin films. *IEEE Sensors Journal* 18:6937–6945. 10.1109/JSEN.2018.2849330
53. Fang, J., Zhu, Y., Wu, D., and et al. 2017. Gas sensing properties of NiO/SnO_2 heterojunction thin film. *Sensors and Actuators B: Chemical* 252:1163–1168. 10.1016/j.snb.2017.07.013
54. Simon, I., Bârsan, N., Bauer, M., and Weimar, U. 2001. Micromachined metal oxide gas sensors: opportunities to improve sensor performance. *Sensors and Actuators B: Chemical* 73:1–26. 10.1016/S0925-4005(00)00639-0

6 Electron Retarding Materials Making More Efficient Light-Emitting Devices

Dong-Sing Wuu
National Chi Nan University, Nantou, Taiwan, Taipei

CONTENTS

6.1 INTRODUCTION

Recently, the invention of emitting devices, such as laser diodes (LDs) and light-emitting diodes (LEDs), has brought convenience to human's life [1–3]. Moreover, the improvement in the efficiency of emitting devices can expand their practicalities. To enhance the emission efficiency, several issues involving light extraction, optical design, heat dissipation, package technique, carrier confinement, and internal quantum efficiency (IQE) have been solved for the emitting devices [4–9]. Even though the emission efficiency of devices can be improved efficiently, an essential problem in these emitting devices, i.e., the excessively large velocity (mobility) difference between electron and hole carriers, is still not overcome. For example, the velocity of the electron in a conventional blue InGaN LED is approximately 26 times faster than that of the hole. This large difference between the electron and hole velocities would increase the non-recombination rate of the device, degrading its emission

DOI: 10.1201/9781003141358-6

performance [10, 11]. Actually, to improve the recombination rate of electron–hole pairs of InGaN LEDs, the electron blocking layer (EBL) [12] and electron tunneling barrier (ETB) [13] were incorporated into the epitaxial structures of InGaN LEDs. However, the hole blocking problem and the increment of defect density in the multiple quantum well (MQW) region could be generated as the EBL and ETB were used, respectively. Most importantly, the velocity of the electron in the LED is not affected through the insertions of EBL and ETB. Thus, how to create the electrons with the slower velocity in the emitting devices becomes an exciting challenge.

In this chapter, to produce the slow electrons in the emitting device, the electron retarding n-electrode (ERN) on the n-GaN layer was proposed to enhance the optoelectronic performance of nitride LEDs. A promising ERN (on n-GaN) should satisfy these three requirements simultaneously. First, the contact characteristic between ERN and n-GaN is Ohmic, ensuring that the LEDs can be operated. Second, the ERN should be incorporated with a small amount of magnetic ions (i.e., the dilute magnetic doping) [14]. This is attributed to the fact that the electrons can be scattered because of the collisions between the magnetic ions (spin–orbit interaction) and the electrons, resulting in the decrease in the electron mobility. Is the material with heavy magnetic doping also a suitable ERN? This will result in the formation of magnetic metal-oxides, which do not help to scatter the electrons and could degrade their conductivity. Finally, the ERN with a good electrical conductivity is also required. When the ERN has good electrical conductivity, a more uniform current injection can be achieved in the LED by inserting the ERN, which induces a higher probability of encountering between electrons and magnetic ions. Furthermore, via the deposition of the ERN on n-GaN, the electron mobility becomes slower, making more efficient radiation emission on nitride LEDs. Most importantly, the problem of the huge velocity difference between the electron and hole carriers that occurred in various emitting devices can be successfully solved by introducing the slow electrons.

6.2 ELECTRON RETARDING MATERIALS FOR LIGHT-EMITTING DEVICES

As mentioned previously, a promising electron retarding material should simultaneously satisfy these three requirements. How to choose a suitable material as an ERN on the n-GaN layer becomes an important issue. In this section, we first compared different transparent conductive oxide materials including ITO, ZnO, ITO-ZnO, Ga_2O_3, Ga-doped ZnO, and Co-doped ZnO to see if the three requirements are met. Next, we compared electron mobility of patterned ITO, ZnO, 400°C, and 700°C-grown CZO on n-GaN. The band structures of ITO/n-GaN and ZnO/n-GaN isotype heterojunctions are also compared. Finally, the relation between the internal quantum efficiency and the electron retarding ratio of the conventional blue LED was simulated by using SiLENSe software.

6.2.1 Various Candidates for Electron Retarding Electrodes

In conventional LED devices, metal electrodes (such as Ti/Al) are usually directly prepared on n-GaN; however, most of the metals are not good candidates as the ERNs

TABLE 6.1
A Summary of Various TCOs Prepared on n-GaN; the Contact to n-GaN, Resistivity, Dilute Magnetic Doping, and Electron Retarding Properties of TCOs Are Compared

Material	Method	Contact to n-GaN	Resistivity (Ω-cm)	Dilute magnetic doping	Electron retarding	Ref.
ITO	Evaporation	Schottky	~10^{-4}	w/o	w/o	[15]
ITO	Sputtering and annealing	Ohmic	~10^{-4}	w/o	w/o	[16]
ZnO:Ga	Chemical vapor deposition	Ohmic	~10^{-1} ~ 10^{-2}	w/o	w/o	[17]
ITO-ZnO	Co-sputtering	Schottky	3.82×10^{-4}	w/o	w/o	[18]
Ga_2O_3	PLD	Schottky	$>10^5$	w/o	w/o	[19]
Co-doped ZnO	PLD	Ohmic	4.3×10^{-2}	w	w	This chapter

Source: From Wuu, D.S. et al. 2018. Scientific Reports 8:4865. Copyright 2018, Springer Nature.

because the electron mobilities are very high in most of the metals. Transparent conductive oxides (TCOs) are the possible candidates as the ERNs. At present, very few studies on the TCO/n-GaN have been presented. The characteristics of various TCOs (consisting of ITO, ZnO, ITO-ZnO, Ga_2O_3, and Co-doped ZnO) prepared on n-GaN are summarized in Table 6.1 [15–19]. ITO and ZnO are more often prepared on n-GaN layers. However, ITO, ZnO, ITO-ZnO, and Ga_2O_3 are all not suitable to be ERNs on n-GaN according to Table 6.1.

Moreover, their mobilities were also analyzed to investigate the electron transport behaviors of these TCO/n-GaN samples. Figure 6.1 shows the comparison of the electron mobility and the measurement method between the n-GaN and patterned-TCO/n-GaN samples (area size: 1×1 cm^2) [14]. Here, the thickness of the n-GaN layer was fixed at 2 μm. As shown in the schematic diagram of Hall measurement performed on n-GaN, the indium balls served as four electrodes were directly prepared on the corners of the sample. However, for the measurements on the patterned-TCO/n-GaN samples, TCO should be first prepared on the corners of the n-GaN layer via standard photolithography and wet etching processes, and the indium balls were formed on the TCO.

The electron mobility of the n-GaN layer was measured to be 176 cm^2/V·s. After fabricating various 120-nm-thick patterned-TCOs (400°C-grown ITO, 400°C-grown ZnO, 400°C-grown CZO, and 700°C-grown CZO) on n-GaN layers, the electron mobilities of these four samples decreased to 170, 160, 141, and 155 cm^2/V·s, respectively. Obviously, the efficient reduction in the electron mobility can both occur in the patterned-CZO films (T_s: 400 and 700°C) on n-GaN, revealing the doping of magnetic Co atoms into ZnO film is helpful to reduce the electron mobility of patterned-TCO/n-GaN. Actually, the CZO films possess a homogeneous microstructure, and

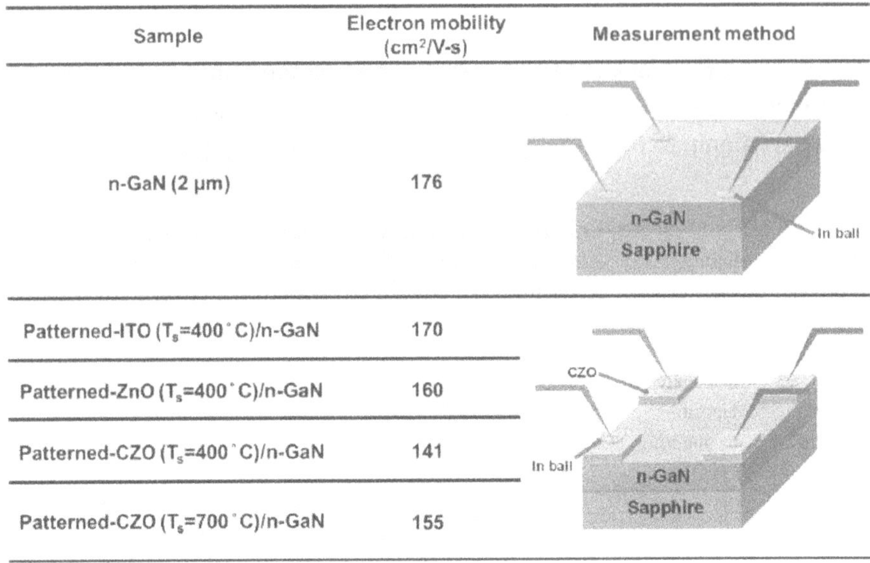

Sample	Electron mobility (cm²/V-s)	Measurement method
n-GaN (2 μm)	176	In ball, n-GaN, Sapphire
Patterned-ITO (T_s=400°C)/n-GaN	170	CZO, In ball, n-GaN, Sapphire
Patterned-ZnO (T_s=400°C)/n-GaN	160	
Patterned-CZO (T_s=400°C)/n-GaN	141	
Patterned-CZO (T_s=700°C)/n-GaN	155	

FIGURE 6.1 Comparisons of electron mobility and measurement method between n-GaN (2 μm) and patterned-TCO/n-GaN samples (area size: 1 × 1 cm²). (From Liu, H.R. et al. 2016. *Applied Physics Letters* 109:021110. Reproduced by permission of AIP Publishing.)

the direction of the magnetic field formed in the films is random. Therefore, it is deduced that the path of electron transfer would be disturbed. In other words, as the CZO film was deposited on the n-GaN layer, the electrons were scattered via the spin–orbit interaction of Co^{2+} ions, causing the reduction in the mobility of the electron carrier.

Additionally, we observe the 400°C-grown patterned-CZO/n-GaN can reach lower electron mobility than that of the 700°C-grown patterned-CZO/n-GaN. This could be attributed to the 400 (C-grown CZO having the lowest electrical resistivity (4.3 × 10^{-2} (·cm) compared to the other CZO films, as shown in Figure 6.8. As the CZO was deposited at 400°C, it had a better conductivity than that of 700°C-grown film (Figure 6.8). Thus, when the light emitter was operated, a more uniform current injection could be achieved in the 400°C-grown CZO film, and there is a higher probability of encountering between electrons and Co ions. This indicates that more electrons can meet the Co ions in the 400°C-grown CZO film. By preparing the 400°C-grown CZO film on the n-GaN layer, more electrons are scattered as the device is driven, leading to the efficient reduction of the electron mobility.

6.2.2 Band Structure and Device Simulation

Figure 6.2(a) and (b) shows the band diagrams of ITO/n-GaN and ZnO/n-GaN isotype heterojunctions, respectively [15–17, 20]. In comparison to the Ohmic contact between ZnO and n-GaN, ITO usually has the Schottky contact with the n-GaN. This is attributed to the barrier heights of 0.63–0.95 eV between ITO and n-GaN

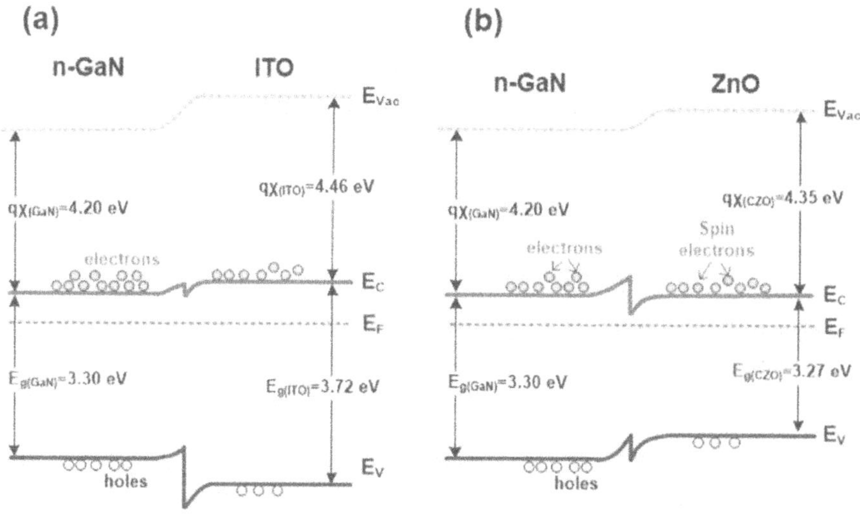

FIGURE 6.2 Band diagrams of (a) ITO/n-GaN and (b) ZnO/n-GaN isotype heterojunctions. (From Wuu, D.S. et al. 2018. *Scientific Reports* 8:4865. Copyright 2018, Springer Nature.)

[15, 16]. Thus, ZnO-based materials could be more feasible as the ERNs than ITO. However, even though the contact characteristic between ITO and n-GaN can be transferred from Schottky to Ohmic, ITO is still not a good ERN.

To investigate the effect of slow electrons on the emission property of nitride LED, the internal quantum efficiencies (IQEs) of the conventional blue LED as a function of the electron retarding ratio were simulated using SiLENSe software [21]. The electron retarding ratio is defined as $(\mu_a - \mu_b)/\mu_a$, where μ_a and μ_b are the original and reduced mobilities of n-GaN, respectively. Table 6.2 shows the details of the simulated epitaxial structure for the blue LED (including concentrations and mobilities of various epitaxial layers) and Figure 6.3 shows the simulated IQE performance with various electron retarding ratios (from 0% to 80%). Based on the simulated results, the IQE of blue LED can be increased from 60.6% to 63.67% with increasing the electron retarding ratio from 0% to 80%.

6.3 PULSED LASER DEPOSITION OF COBALT-DOPED ZNO FILMS

From the previous section, we know that the 400°C-grown CZO film on the n-GaN layer has the highest efficiency in reducing the electron mobility. The next step is to find out the best method to prepare the CZO film serving as the electron retarding material. Up to present, Co-doped ZnO (CZO) thin films and nanostructures have been prepared by several techniques including sputtering [22], sol-gel [23, 24], spin-coating [25], solvothermal [26], ultrasonic spray [27, 28], and pulsed laser deposition (PLD) [29–31]. Compared with other techniques, PLD is more useful for the growth of epitaxial ZnO-based thin films because its atomic-layer control can be achieved

TABLE 6.2
Details of the Simulated Epitaxial Structure for The Blue LED (Including Concentrations and Mobilities of Various Epitaxial Layers)

Epitaxial structure	Concentration (cm^{-3})	Electron mobility with various retarding ratio (cm^2/Vs)					
		0%	20%	40%	50%	60%	80%
n-GaN	1.00E+19	120	96	72	60	48	24
GaN barrier × 4	1.00E+18	250	200	150	125	100	50
QW × 4	5.00E+16	550	440	330	275	220	110
GaN barrier × 2	5.00E+16	550	440	330	275	220	110
QW × 2	5.00E+16	550	440	330	275	220	110
GaN barrier	5.00E+16	550	440	330	275	220	110
p-AlGaN	1.00E+19	1	0.8	0.6	0.5	0.4	0.2
p-GaN	1.00E+18	1	0.8	0.6	0.5	0.4	0.2

Source: From Wuu, D.S. et al. 2018. Scientific Reports 8:4865. Copyright 2018, Springer Nature.

through the adjustment of laser repetition rate. Moreover, during the PLD process, the source particles possess high energy, which can enhance the surface mobility of the ad-atoms.

In this chapter, CZO thin films were prepared on c-plane (001) sapphire substrates by PLD (PLD/MBE-2000, PVD products). The KrF excimer laser (λ = 248 nm) was put in the deposition system, and the pulse duration of 25 ns was employed to ablate the CZO target. Moreover, the energy fluence and the repetition rate of KrF laser are set at 600 mJ/pulse and 2 Hz, respectively. During the film's growth, a stoichiometric

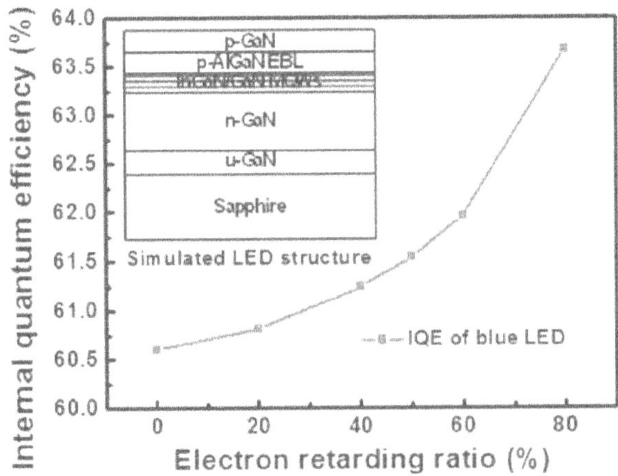

FIGURE 6.3 The simulated IQE performance with various electron retarding ratios. (From Wuu, D.S. et al. 2018. *Scientific Reports* 8:4865. Copyright 2018, Springer Nature.)

ceramic CZO target with a composition of 95 at% ZnO and 5 at% Co was used. By using a resistive heater, the T_s can be raised from 100 to 700°C. The distance between the substrate and the target was maintained at 8 cm. After pumping the growth chamber to a base pressure, the 15-sccm pure Ar gas was introduced, and the working pressure was kept at 2×10^{-3} torr. The CZO films were fixed at 120 nm in thickness.

6.4 CHARACTERIZATION OF COBALT-DOPED ZnO FILMS

At present, most researches of ZnO-based semiconductors focused on the improvement of magnetic characteristics. However, the TCO-related properties [32] such as electrical resistivity, mobility, and transmittance are usually neglected in ZnO-based semiconductors. Although several ZnO-based semiconductors with good magnetic characteristics have been proposed, the higher electrical resistivity or the lower transmittance also occurred in these materials. This will confine the optoelectronic applications of ZnO-based semiconductors. To be a promising ERN, the ZnO-based semiconductors should be incorporated with small amounts of magnetic ions (i.e., the dilute magnetic doping). The main doped metal elements consist of Mn, Co, and Fe [33–35]. Among various magnetic elements doped into ZnO, Co metal is a promising material since the Co-doped sample can exhibit a remarkable magnetization per Co ion for very low substitutions.

In this section, the structural, electrical, optical, and magnetic properties of CZO films have been analyzed in detail. By adjusting the gas atmosphere and the T_s, the optimum growth conditions of CZO films can be achieved. The PLD-CZO films in this section would possess good magnetic characteristics, excellent transmittance, low electrical resistivity, and high mobility. This suggests the CZO films can be applied for both spintronic and optoelectronic devices.

Microstructures of these films were analyzed by transmission electron microscopy (TEM). The cross-sectional images of the samples can be observed by both scanning electron microscopy (SEM) and TEM. Hall measurements were used for the electrical properties of CZO films, and the UV-Vis system was employed to obtain their transmittance spectra. The magnetic characteristics were measured by a superconducting quantum interference device (SQUID) magnetometer.

6.4.1 Structural Properties

The cross-sectional TEM images of CZO films prepared at 100 and 400°C in Ar atmosphere are displayed in Figure 6.4(a) and (b), respectively. The interfaces of CZO/sapphire in these two samples can be clearly identified. Based on our observation, the 100°C-grown CZO film exhibited a dense columnar structure [36]. When the T_s was increased, the columnar structure was transformed to the featureless structure, which was similar to the 400°C-grown CZO film, as shown in Figure 6.4(b). Figure 6.5(a)–(c) shows the cross-sectional high-resolution TEM (HR-TEM) images and the selected area electron diffraction patterns of CZO films prepared on sapphire substrates at the T_s of 100, 400, and 700°C in Ar atmosphere, respectively [37]. From our analyses, these three CZO films both had a similar d-spacing value of 2.60–2.62 Å, which was indexed to CZO(002) plane. On the other

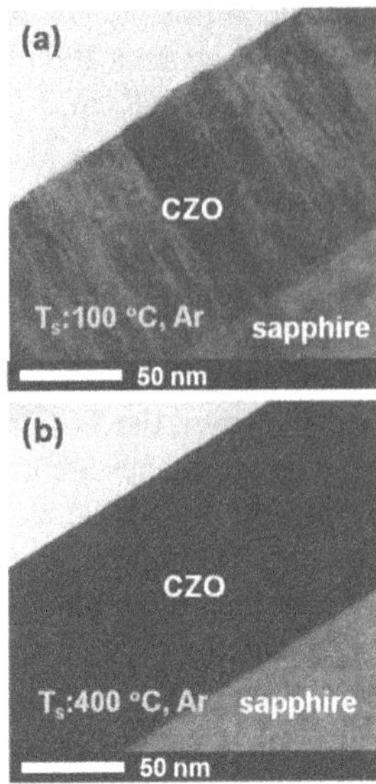

FIGURE 6.4 Cross-sectional TEM images of CZO films prepared at the T_s of (a) 100 and (b) 400°C in Ar atmosphere. (From Ou, S.L. et al. 2016. *Journal of Alloys and Compounds* 663:107–115. Copyright 2016, with permission from Elsevier.)

hand, as shown in the red (in the web version) circles of the HRTEM image of Figure 6.5(a), some lattice distortions were observed in the 100°C-grown CZO film. Meanwhile, the lattice arrangements were more regular in both the 400 and 700°C-grown CZO films. The results reveal that the phenomenon of lattice distortion in the CZO film can be gradually relaxed with increasing the T_s. This suggests that the lattice defects are easily formed in the CZO film deposited at the lower T_s. However, as the T_s was increased, the laser-ablated materials would have higher energy for the migration of ad-atoms, and the lattice distortions in the CZO film were reduced. Additionally, as observed in the electron diffraction patterns, the diffraction dots with single family possessed the zone axis of $[1\bar{0}]$ both existed in the 400 and 700°C-grown CZO films. However, it can be seen that the electron diffraction pattern of the 100°C-grown CZO film had the diffraction dots with two families (marked by the red and blue circles in the diffraction pattern, respectively), confirming again the columnar structure was generated in this film.

As discussed in Figure 6.5(a), the CZO film prepared at 100°C in Ar atmosphere possessed some lattice distortions. The formation of specific oxygen vacancies in this CZO film is the possible reason for the generation of lattice distortions. Here, we are going to explain this phenomenon via the variation of ZnO lattice structure by introducing the oxygen vacancies. Figure 6.6(a) shows the schematic illustration of ZnO lattice with an ideal wurtzite structure, where the ball and stick represent the atom and bond, respectively. After forming the oxygen vacancies with 0 and 2+ charge states (V_O^0 and V_O^{2+}), the ZnO lattice structures are displayed in Figure 6.6(b) and (c), respectively. Since the oxygen vacancy with 1+ charge state

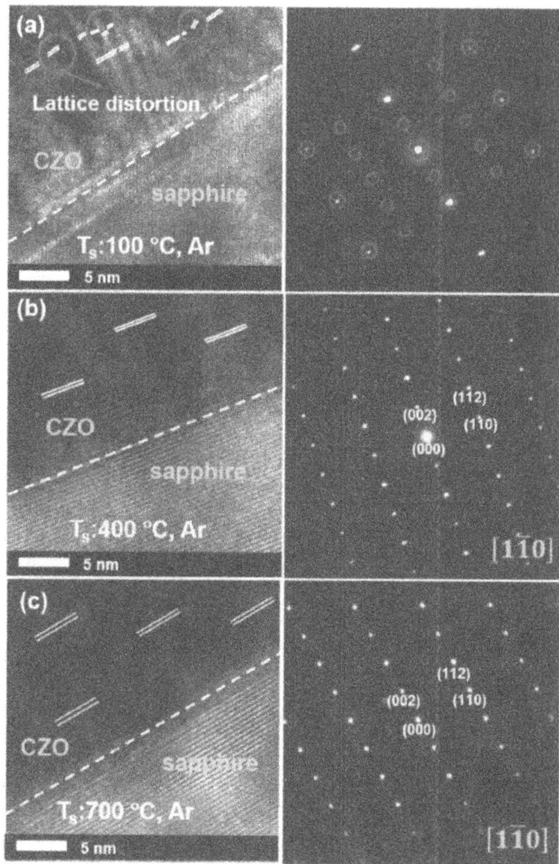

FIGURE 6.5 Cross-sectional HR-TEM images and selected area electron diffraction patterns of CZO films prepared at the T_s of (a) 100, (b) 400, and (c) 700°C in Ar atmosphere. (From Ou, S.L. et al. 2016. *Journal of Alloys and Compounds* 663:107–115. Copyright 2016, with permission from Elsevier.)

(V_O^{1+}) is very unstable in the ZnO lattice [38], we are not going to discuss this situation. The calculated results from previous research indicate the lattice relaxations of ZnO around the oxygen vacancies are large [38]. In addition, the lattice relaxations are very different as the oxygen vacancies with various charge states generate in the ZnO lattice. In comparison with the ZnO lattice with an ideal wurtzite structure (Figure 6.6(a)), the four Zn nearest neighbors exhibit the inward displacement by 12% of the equilibrium Zn–O bond length when the ZnO lattice possesses the V0 O (Figure 6.6(b)). Meanwhile, the displacement is calculated to outward by 23% for the formation of V_O^{2+} in the ZnO lattice [39], as shown in Figure 6.6(c). The origin of the abovementioned lattice relaxations lies in the electronic structure of oxygen vacancies. When an oxygen atom is removed from the ideal ZnO lattice, it will leave four Zn dangling bonds and each bond contributes 1/2 electron to a neutral vacancy. The interaction of the dangling bonds can lead to a completely symmetric a_1 state

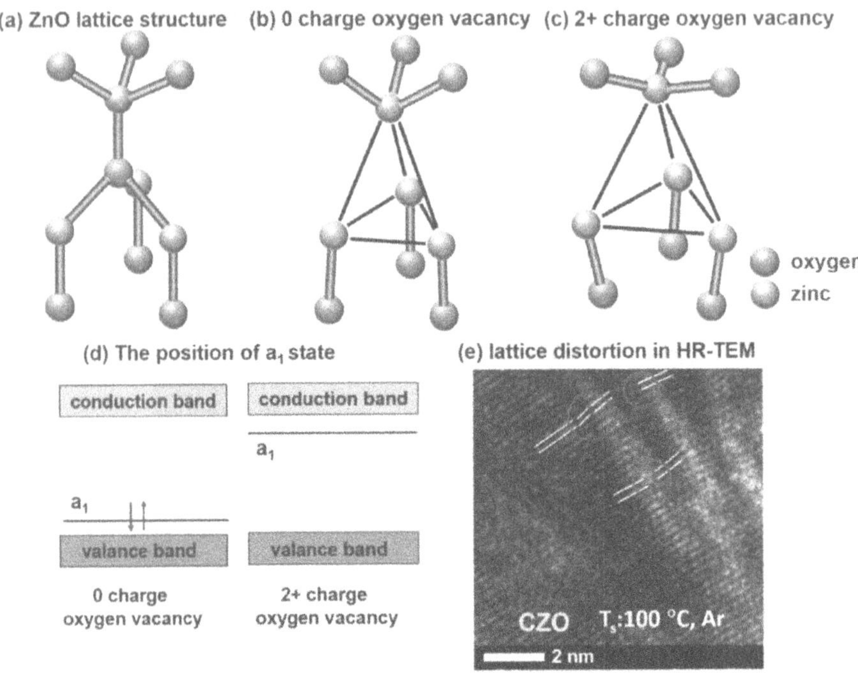

FIGURE 6.6 Schematic illustrations of ZnO lattice with (a) an ideal wurtzite structure, (b) the formation of V0 O, and (c) the formation of V_O^{2+}. (d) Positions of a_1 states for the equilibrium configurations of ZnO with V_O^0 and V_O^{2+}. (e) HR-TEM image of Figure 6.5(a) focused on the CZO region with a higher magnification. (From Ou, S.L. et al. 2016. *Journal of Alloys and Compounds* 663:107–115. Copyright 2016, with permission from Elsevier.)

and three almost degenerate higher energy states. Here, the a_1 state lies in the band gap, while the three higher energy states are resonant in the conduction band [40]. The positions of a_1 states for the equilibrium configurations of ZnO with V0 O and V2 + O are illustrated in Figure 6.6(d). For the formation of V_O^0 in the ZnO lattice, the a_1 state can be occupied by two electrons, and its energy is decreased while the four Zn atoms approach each other. This will induce the stretch of the Zn-O bonds surrounding the vacancy, as shown in Figure 6.6(b). In this case, the a_1 state lies near the top of the valence band. On the other hand, as the ZnO lattice possesses the V_O^{2+}, the a_1 state is unoccupied and there exists strongly outward relaxation in the four Zn atoms. Thus, as shown in Figure 6.6(c), the Zn-O bonds will become planar, and the empty a_1 state lies near the bottom of the conduction band. Actually, the formation of V_O^{2+} in the ZnO crystal results in a larger lattice relaxation than that of V_O^0. In other words, the lattice deformation would be more obvious when the V_O^{2+} generates in the ZnO crystal [37]. As a result, the lattice distortion occurred in the 100°C-grown CZO film can be reasonably attributed to the formation of V_O^{2+}. Figure 6.6(e) shows the HR-TEM image of Figure 6.5(a) focused on the CZO region with a higher magnification. It can be seen the lattice distortions formed in the 100°C-grown CZO film are more apparent in Figure 6.6(e).

6.4.2 Optical Properties

The transmittance spectra with wavelength ranging from 200 to 900 nm for CZO films prepared at various T_s in Ar atmosphere are shown in Figure 6.7(a). The cross-sectional SEM image of the CZO film deposited at 400°C is shown in the inset of Figure 6.7(a). In the transmittance spectra, three absorption peaks at 567, 611, and 657 nm can be found. These three absorption peaks were ascribed to $^4A_2(F)/^2A_1(G)$, $^4A_2(F)/^4T_1(P)$, and $^4A_2(F)/^2E(G)$ transitions, respectively, resulting from the crystal-field transitions in the high spin state of Co^{2+} ($3d^7$) at tetrahedral sites [41]. The results confirmed that Co existed in the ZnO lattice of the wurtzite structure when Co^{2+} ions were well substituted for Zn^{2+} at the tetrahedral sites. On the other hand, in these transmittance spectra, there was no absorption peak at around 470 nm, which was

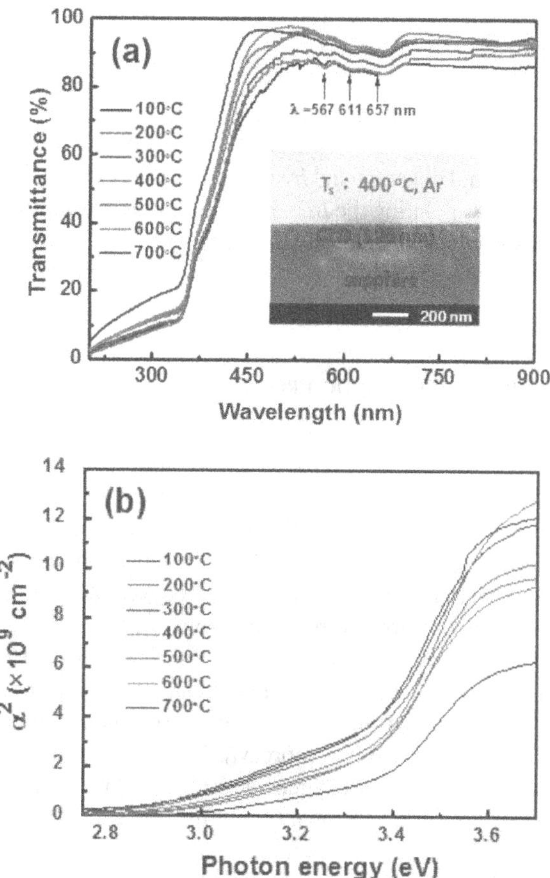

FIGURE 6.7 (a) Transmittance spectra with wavelength ranging from 200 to 900 nm with a cross-sectional TEM image of CZO films prepared at the T_s of 400°C in Ar atmosphere in the inset of the figure and (b) the square of absorption coefficient versus photon energy of CZO films prepared at various T_s in Ar atmosphere. (From Ou, S.L. et al. 2016. *Journal of Alloys and Compounds* 663:107–115. Copyright 2016, with permission from Elsevier.)

induced by the formation of the secondary Co_3O_4 spinel phase in the films [42]. This was in good agreement with the XRD result and proved again that Co^{2+} substituted for Zn^{2+} in the ZnO lattice. Additionally, at the visible region, it can be observed the transmittance of CZO film increased with increasing the T_s. The optical transmittances of CZO films can reach 73.4%, 78.4%, 80.9%, 84.1%, 88.6%, 91.1%, and 95.8% at blue wavelength (450 nm) as the T_s is increased to 100, 200, 300, 400, 500, 600, and 700°C, respectively. The absorption coefficient (α) is expressed as

$$\alpha = \frac{1}{t}\ln\left(\frac{1}{T}\right) \tag{6.1}$$

where t represents the film thickness and T is the optical transmittance. The optical band gap (E_g) values of CZO films can be determined using the following equation [43]:

$$\alpha h\nu = A\left(h\nu - E_g\right)^{1/2} \tag{6.2}$$

where A is a proportional constant and $h\nu$ is the incident photon energy, the E_g values can be obtained by extrapolating the linear portion to the photon energy axis in a plot of α^2 versus $h\nu$, as shown in Figure 6.7(b). The E_g value of the CZO film deposited in Ar atmosphere increased slightly from 3.28 to 3.33 eV with increasing the T_s from 100 to 700°C.

6.4.3 Electronic and Magnetic Properties

Figure 6.8 shows the variations of carrier concentration, electrical resistivity, and mobility of CZO films prepared in Ar atmosphere as a function of T_s. Furthermore, the cross-sectional HR-TEM image of the CZO film prepared at 400°C (Figure 6.5(b)) is shown in the inset of Figure 6.8. It was seen that the concentration of CZO film reduced from 8.10×10^{18} to 1.97×10^{18} cm^{-3} with increasing the T_s from 100 to 700°C. The decrease in the concentration of the CZO film can be attributed to the reduction of donor carriers, resulting from the enhancement in the crystal quality of CZO film with an increment of the T_s. Moreover, for the ZnO-based materials, the oxygen vacancies (V_O^0) are mainly responsible for electrons in the conduction band [44]. Thus, a lower resistivity of 4.30×10^{-2} Ω·cm can be achieved in the 400°C-grown CZO film since it had more V_O^0. In addition, when the T_s was kept at 100 or 700°C, the relatively fewer oxygen vacancies (V_O^0) were generated in the CZO film. The fewer V_O^0 in the 700°C-grown CZO film could lead to its relatively higher resistivity. Although the CZO film prepared at 100°C also had fewer V_O^0, its resistivity was unreasonably low (7.18×10^{-2} Ω·cm). This can be explained by the film's feature. As shown in Figure 6.4(a), the cross-sectional TEM image indicated that the 100°C-grown CZO film presented a dense columnar structure. Jose et al. have investigated that the existence of grain boundaries in the ZnO material can be helpful to the improvement in its conductivity [36]. This is ascribed to the trapping of electrons in the grain boundaries. Consequently, because of the formation of columnar

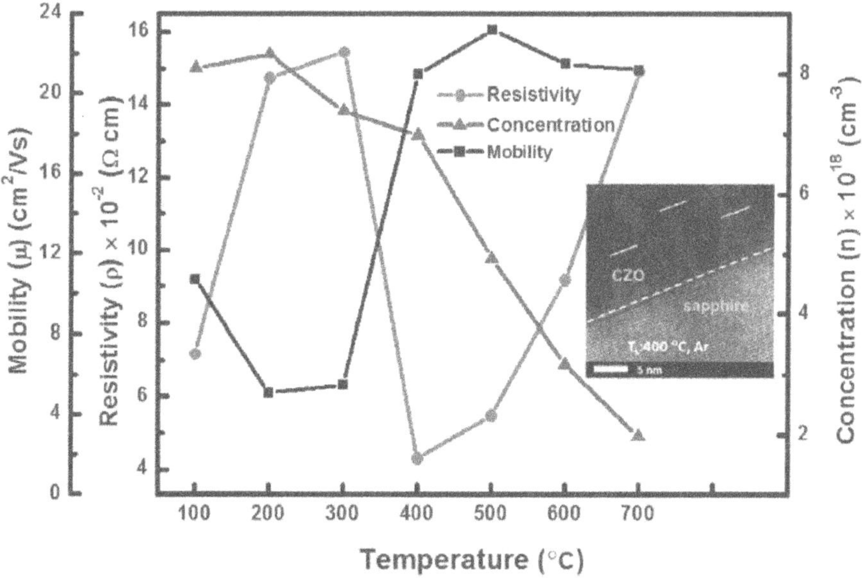

FIGURE 6.8 Variations of carrier concentration, electrical resistivity, and mobility of CZO films prepared in Ar atmosphere as a function of T_s. (From Ou, S.L. et al. 2016. *Journal of Alloys and Compounds* 663:107–115. Copyright 2016, with permission from Elsevier.)

structure, the boundaries between these columns would result in a reduction in the resistivity of the 100°C-grown CZO film.

To obtain the total mobility (μ) of ZnO-based materials, Matthiessen's rule can be employed, as expressed as follows [45]:

$$\mu = \frac{\mu_{lat}\mu_{ii}}{\mu_{lat} + \mu_{ii}} \tag{6.3}$$

where μ_{lat} is the lattice mobility of the intrinsic ZnO and the mobility μ_{ii} is limited by ionized impurity scattering. The ionized impurities consist of intrinsic lattice defects and extrinsic dopants. The expression for the mobility μ_{ii} can be obtained by the following equation:

$$\mu_{ii}^{CW} = \frac{128\sqrt{2\pi}\,(\varepsilon_r\varepsilon_0)^2\,(kT)^{3/2}}{\sqrt{m^*}\,Z^2e^3N_i} \times \left\{ ln\left[1+\left(\frac{12\pi\varepsilon_r\varepsilon_0 kT}{Ze^2N_i^{1/3}}\right)^2\right]\right\}^{-1} \tag{6.4}$$

Obviously, the mobility μ_{ii} can be determined by several parameters including the relatively static permittivity ε_r, the effective mass m^*, the charge Z of the impurity, and the concentration N_i. Moreover, the other symbols such as π, ε_0, e, k, and T possess their usual meaning. In this equation, the Z parameter can be divided into three aspects, i.e., $Z = 1$, 2, and 3. For $Z = 2$, it reveals that the intrinsic doping by oxygen vacancies occurred in ZnO materials. The case of $Z = 1$ means the extrinsic doping on the substitutional cation or anion lattice places. Additionally, the extrinsic doping on the interstitial places belongs to $Z = 3$. When the $Z = 1$ case is formed in ZnO, a higher mobility is achieved. Besides, a lowest mobility can be obtained while ZnO possesses the $Z = 3$ case. Compared with the $Z = 2$ and $Z = 3$ cases, the $Z = 1$ case has little effect on the mobility variation. As a result, we ignore this case ($Z = 1$) to discuss the mobility variation of CZO films.

During the growth in pure Ar atmosphere, Zn-rich ZnO-based films were easily formed. At the Zn-rich state, it is difficult to form the zinc vacancy and the oxygen interstitial in ZnO-based films [44, 46]. Therefore, for the mobility variation of Zn-rich ZnO-based films, the formations of oxygen vacancy ($Z = 2$) and zinc interstitial ($Z = 3$) are more dominant than those of zinc vacancy and oxygen interstitial [47]. It should be mentioned that the interstitials can cause more carrier scattering, which would reduce the mobility [45]. However, the formation energy of zinc interstitial is very high, and the high-temperature processes are recommended to generate this interstitial. Thus, for CZO films prepared in pure Ar atmosphere (Zn-rich state), the formation of various oxygen vacancies ($Z = 2$) is the main factor influencing their mobilities. As shown in Figure 6.8, the relatively low mobility of 5.1–10.7 $cm^2/V{\cdot}s$ was obtained in the 100–300°C-grown CZO films. With increasing the T_s to 400–700°C, the mobilities of these CZO films were similar to each other and reached a higher value of 21.0–23.2 $cm^2/V{\cdot}s$. At the lower T_s, these CZO films possessed fewer V_O^0. Besides, as shown in Figure 6.5(a), the HR-TEM image revealed some lattice distortions were formed in the CZO films grown at the lower T_s. In fact, the lattice distortions formed in ZnO-based films were ascribed to the formation of V_O^{2+} [47]. This is the reason why CZO films prepared at the lower T_s had relatively lower mobility. When the CZO film was prepared at 400°C, a relatively great number of oxygen vacancies (V_O^0) were created. Meanwhile, as shown in the inset of Figure 6.8, the lattice distortion phenomenon was relaxed significantly in the 400°C-grown CZO film, leading to an efficient decrease in the amount of V_O^{2+}. Based on the above-mentioned discussions, it can be deduced the amount of reduced V_O^{2+} for the 400°C-grown CZO film is considerable, which causes a slight decrease in the carrier concentration (even this film possesses more V_O^0). Thus, the increment of mobility to 21.0 $cm^2/V{\cdot}s$ in the 400°C-grown CZO film is probably attributed to the reduction of V_O^{2+}. Further increasing the T_s to 500–700°C, these CZO films possessed fewer V_O^0. Besides, the amount of V_O^{2+} in these films is also few because of the relaxation of lattice distortion. However, at the T_s of 500–700°C, the high temperature could provide enough energy to form the zinc interstitials in these CZO films, and the zinc interstitials would result in a decrease in the mobility [37]. Hence, the 500–700°C-grown CZO films had a similar mobility to that of 400°C-grown film.

The results of magnetization (M) versus magnetic field strength (H) for the CZO films grown at 100, 400, and 700ºC in Ar atmosphere are shown in Figure 6.9 [37]. These measurements were performed at room temperature. Apparently, all samples had ferromagnetism at room temperature because they exhibited well-defined hysteresis loops. The ferromagnetic behavior in these CZO films is due to the substitution of Co^{2+} ions in the ZnO lattice. If the content of Co^{2+} ions existing in the ZnO lattice is not enough, it would cause an unobvious ferromagnetic behavior in the sample. The saturation magnetization (M_s) values of CZO films prepared at 100, 400, and 700°C are determined to be 2.74×10^{-5}, 3.37×10^{-5}, and 5.33×10^{-5} emu, respectively,

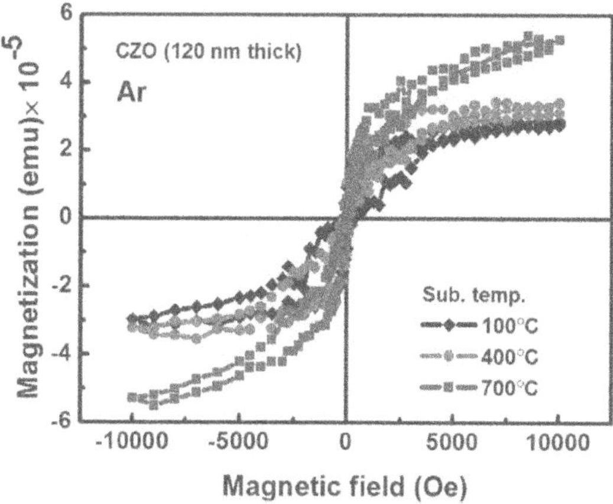

FIGURE 6.9 Magnetization versus magnetic field strength for CZO films grown at 100, 400, and 700°C in Ar atmosphere. (From Ou, S.L. et al. 2016. *Journal of Alloys and Compounds* 663:107–115. Copyright 2016, with permission from Elsevier.)

revealing that the M_s value increases with an increment of the T_s. According to our knowledge, CoO and Co_3O_4 possess antiferromagnetic and paramagnetic behaviors, respectively [29]. If CoO or Co_3O_4 phase was generated in the samples, the variation trend of M_s of CZO films with rising T_s could become irregular. Consequently, the fact that no secondary phase of Co-oxide formed in the CZO films can be verified again via the M-H results.

6.5 EFFECTS OF INSERTING THE ELECTRON RETARDING MATERIAL: COBALT-DOPED ZnO FILMS

Due to the homogeneous microstructure and random distribution of Co^{2+} ions in the CZO film, the probability of encountering electrons and Co^{2+} ions can be increased with an increment of CZO thickness. This can lead to more efficient electron scattering, improving its electron retarding effect. Figure 6.10 shows the mobilities, carrier concentrations, and retarding electron ratios of CZO/n-GaN samples as a function of CZO thickness from 0 to 240 nm [21]. The detailed measurement method for electrical properties of CZO/n-GaN samples was described in our previous research [14]. The mobilities and carrier concentrations were both analyzed by Hall measurements. Without depositing the CZO film on n-GaN, the mobility and carrier concentration of n-GaN were 176 $cm^2/V{\cdot}s$ and 1×10^{19} cm^{-3}, respectively. It should be mentioned that the n-GaN sample (without depositing the CZO layer) represents the metal Ohmic contact on n-GaN. With increasing the CZO thickness to 240 nm, the mobility of CZO/n-GaN decreased to 134 $cm^2/V{\cdot}s$, while its electron retarding ratio increased to 23.9%. The phenomenon of mobility reduction that occurred in the CZO/n-GaN samples (in comparison with the n-GaN sample) can be verified that the velocity of the electron is slowed down efficiently via the insertion of the CZO layer. Although the mobility of the CZO/n-GaN sample was reduced gradually

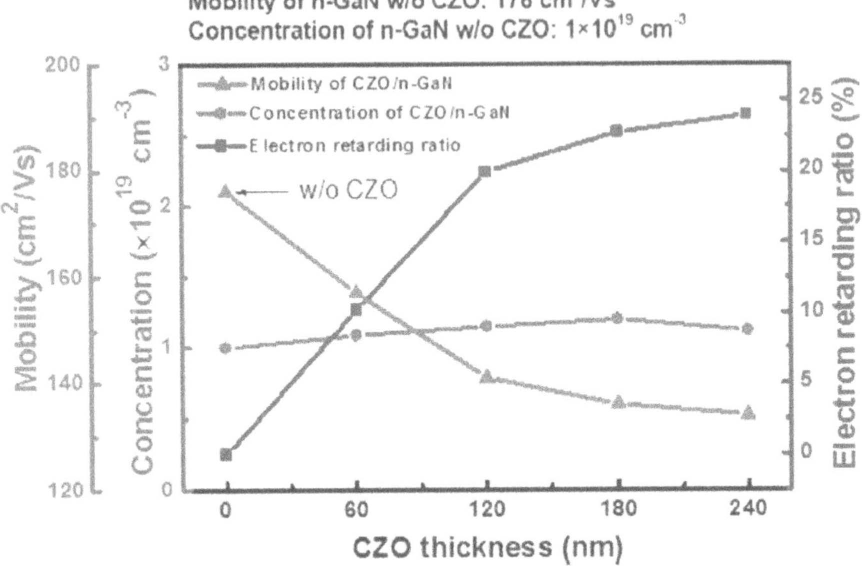

FIGURE 6.10 Mobilities, carrier concentrations, and retarding electron ratios of CZO/n-GaN samples as a function of CZO thickness from 0 to 240 nm. (From Wuu, D.S. et al. 2018. *Scientific Reports* 8:4865. Copyright 2018, Springer Nature.)

by increasing the CZO thickness, we can see that the reduction speed of mobility became obviously slow when the CZO thickness was increased to 120 nm. The mobility and the electron retarding ratio of 120-nm-thick CZO/n-GaN sample were 141 $cm^2/V{\cdot}s$ and 19.9%, respectively. This indicates that the electron retarding effect becomes saturated as the CZO thickness is larger than 120 nm. However, it can be found the carrier concentrations of all CZO/n-GaN samples are similar to that of n-GaN, revealing that the carrier concentration of n-GaN is not influenced by inserting the CZO film.

According to the results of Figure 6.10, the 120-nm-thick CZO film was selected as the ERN to fabricate the blue LED. Figure 6.11(a) shows the forward voltages and output powers as a function of the injection current of blue LEDs with and without inserting the ERN [21]. These two LEDs had almost the same forward voltages (at 0–500 mA). Based on the results of Figure 6.10 (carrier concentration of n-GaN) and Figure 6.11(a), we can confirm the electrical characteristics of the LED are not affected by inserting the ERN. As the devices were operated at 350 mA, the output powers of the blue LEDs with and without the ERN were 246.7 and 212.9 mW, respectively. Further increasing the injection current to 500 mA, the output powers of these two LEDs were measured to be 317.9 and 276.2 mW, respectively. After inserting the ERN into LED, 15.9% and 15.1% improvements can be achieved in the output powers (at 350 and 500 mA) compared with those of conventional LED. Figure 6.11(b) shows the wall-plug efficiencies (WPEs) of these two LEDs versus the injection current. The WPE of the LED with the CZO-ERN is obviously higher than

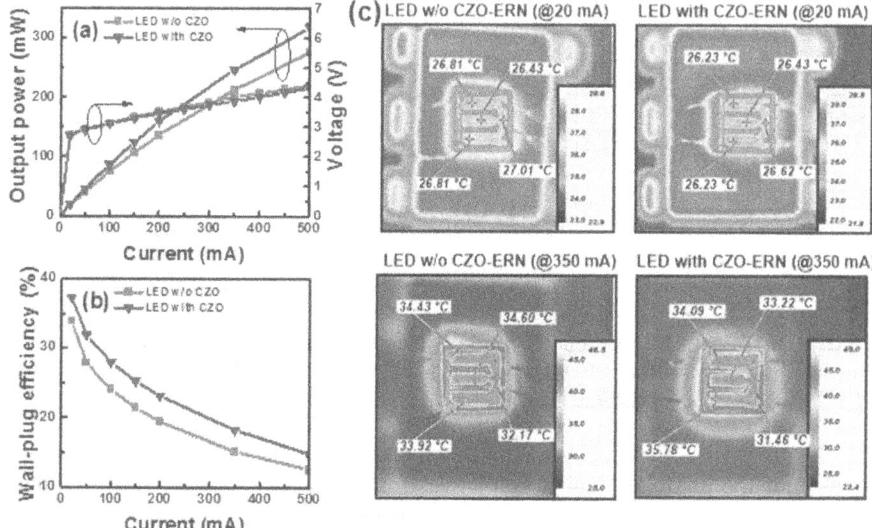

FIGURE 6.11 (a) Forward voltages, output powers and (b) wall-plug efficiencies as a function of the injection current for blue LEDs with and without inserting the ERN. (c) Surface temperature distributions of blue LEDs with and without inserting the ERN at 20 and 350 mA injection currents. (From Wuu, D.S. et al. 2018. *Scientific Reports* 8:4865. Copyright 2018, Springer Nature.)

that of a conventional LED. The WPEs of these two LEDs at 350 mA were 18.2% and 15.1%, respectively. Figure 6.11(c) displays the surface temperature distributions of LEDs with and without inserting the CZO-ERN at 20 and 350 mA injection currents. We can observe that LEDs with and without the ERN's surface temperatures (at 20 mA) are 26.23–26.62 and 26.43–27.01°C, respectively. Further increasing the driving current to 350 mA, the surface temperatures of these two LEDs are increased to 31.48–35.78 and 32.17–34.60°C, respectively. Obviously, the surface temperatures for these two LEDs are much similar to each other, indicating that the thermal characteristic of blue LED cannot be affected via the addition of the ERN.

Except for the emission improvement after inserting the CZO-ERN, the other exciting phenomenon was also induced through the formation of slow electrons. Figure 6.12(a) shows the EL spectra (at 350 mA) of blue LEDs with and without the CZO-ERN. Without inserting the ERN, the peak wavelength of EL emission was centered at 454.25 nm, as shown in the inset of Figure 6.12(a). After adding the ERN, the peak wavelength of EL emission shifted to the shorter wavelength of 452 nm (blue shift). Additionally, the full width at half maximum (FWHM) value of the EL emission peak for the blue LED with the ERN was slightly larger than that without the ERN. To explain the EL results, the carrier transport models of LEDs without and with CZO-ERN were established, as shown in Figure 6.12(b) and (c), respectively [21]. Without inserting the ERN, the velocity of the electron is much faster than that of the hole, and the electrons are difficult to fill each quantum well (Figure 6.12(b)). This will lead to the gather of electrons in the quantum

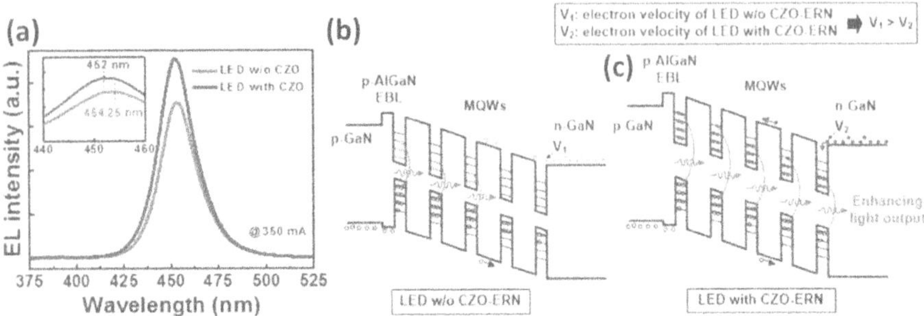

FIGURE 6.12 (a) EL spectra (at 350 mA) of blue LEDs with and without the CZO-ERN. Carrier transport models of LEDs (b) without and (c) with CZO-ERN. (From Wuu, D.S. et al. 2018. *Scientific Reports* 8:4865. Copyright 2018, Springer Nature.)

wells near the p-AlGaN EBL to radiatively recombine with holes. Thus, the non-radiative recombination rate in the MQWs is increased. However, as discussed above, the velocity of the electron can be slowed down efficiently after inserting the ERN. Therefore, as shown in Figure 6.12(c), electrons are easier to fill the quantum wells, and electrons and holes can more uniformly distribute in the MQWs, enhancing the radiative recombination rate and increasing the FWHM value of the emission peak. On the other hand, because electrons are filled in the quantum wells more effectively by inserting the ERN, electrons can transit from a high energy level to a low energy level to recombine with holes. As a result, the bandgap of MQWs can be enlarged, leading to the blue shift in the peak wavelength of EL emission.

6.6 CONCLUSIONS AND FUTURE OUTLOOK

In summary, the slow electrons were successfully generated by depositing the CZO-ERN on the n-GaN layer, reducing the mobility difference between electron and hole carriers and making more efficient radiation emission of InGaN blue LEDs. Dilute magnetic TCOs with good electrical conductivity and Ohmic contact characteristics to n-GaN have great potential for ERN applications. Via the Hall measurements on various patterned-TCO/n-GaN samples, it indicates that the 400°C-grown CZO film is more feasible for the n-electrode than the other TCOs. In comparison with the electron mobility of n-GaN (176 cm²/V·s), it can be significantly decreased to 141 cm²/V·s after depositing the patterned-CZO (T_s: 400°C) on n-GaN. This confirms the 400°C-grown CZO is indeed a promising n-electrode used to enhance the emission characteristic of the InGaN-based LED.

CZO thin films with a thickness of 120 nm were prepared by PLD. In the growth process, the T_s (100–700°C) and the gas atmosphere (Ar gas) were both adjusted. When the PLD-CZO film was grown at the T_s of 400°C in Ar atmosphere, it has more V_O^0, leading to its lowest electrical resistivity of 4.30×10^{-2} Ω·cm. When CZO films were grown at 100–300°C, the lattice distortions would be generated, leading to the formation of V_O^{2+} and relatively lower mobility of 5.1–10.7 cm²/V·s, due to the efficient relaxation of lattice distortion, there was a considerable reduction of the V_O^{2+}

amount in the 400°C-grown CZO film. As a result, an obvious increment of mobility to 21.0 $cm^2/V{\cdot}s$ can occur in this film even its V_O^0 amount was relatively large. However, although CZO films prepared at 500–700°C possessed fewer amounts of both V_O^0 and V_O^{2+}, their mobilities were similar to that of the 400°C-grown film. This can be ascribed to the generation of zinc interstitials formed through the high-temperature process, and the mobility is limited by the zinc interstitials.

Additionally, when CZO films were deposited at 100, 200, 300, 400, 500, 600, and 700°C, their transmittances (at 450 nm) were measured to be 73.4%, 78.4%, 80.9%, 84.1%, 88.6%, 91.1%, and 95.8%, respectively. Furthermore, with raising the T_s from 100 to 700°C, the M_s value was increased gradually from 2.74×10^{-5} to 5.33×10^{-5} emu. The PLD-CZO films proposed in this chapter possess lower electrical resistivity, higher mobility, good magnetic characteristic, and excellent transmittance. This reveals that PLD-CZO films are potentially useful in both spintronic and optoelectronic devices.

After depositing the 120-nm-thick CZO-ERN on n-GaN, the slow electrons with the electron retarding ratio of 19.9% can be generated. In comparison with the LED without inserting the ERN, the LED with the ERN possessed 15.9% enhancement in the output power (at 350 mA). Meanwhile, the WPEs (at 350 mA) of LEDs with and without the ERN was 18.2% and 15.1%, respectively. This indicates that the slow electrons formed via the ERN technique can improve the recombination rate of electron–hole pairs. Additionally, based on the EL results, the peak wavelengths of emission for these two devices were centered at

- LED without CZO: 454.25 nm, and
- LED with CZO: 452 nm.

The blue-shift in emission wavelength that occurred in the LED with inserting the ERN is attributed to the efficient filling of electrons in the quantum wells. The results reveal that the ERN technique is highly feasible for generating slow electrons and enhancing the optoelectronic performance of emitting devices. Via the introduction of slow electrons, the excessively large velocity difference between electron and hole carriers that occurred in various emitting devices can be reduced efficiently.

REFERENCES

1. Chi, Y.-C., Y.-F. Huang, T.-C. Wu, C.-T. Tsai, L.-Y. Chen, H.-C. Kuo, and G.-R. Lin. 2017. Violet laser diode enables lighting communication. *Scientific Reports* 7:10469.
2. Dai, W., Y. Lei, M. Xu, P. Zhao, Z. Zhang, and J. Zhou. 2017. Rare-earth free self-activated graphene quantum dots and copper-cysteamine phosphors for enhanced white light-emitting-diodes under single excitation. *Scientific Reports* 7:12872.
3. Ryu, H. Y., K. S. Jeon, M. G. Kang, H. K. Yuh, Y. H. Choi, and J. S. Lee. 2017. A comparative study of efficiency droop and internal electric field for InGaN blue lighting-emitting diodes on silicon and sapphire substrates. *Scientific Reports* 7:44814.
4. Tanaka, A., R. Chen, K. L. Jungjohann, and S. A. Dayeh. 2015. Strong geometrical effects in submillimeter selective area growth and light extraction of GaN light emitting diodes on sapphire. *Scientific Reports* 5:17314.

5. Zhang, M., Z. Chen, L. Xiao, B. Qu, and Q. Gong. 2013. Optical design for improving optical properties of top-emitting organic light emitting diodes. *Journal of Applied Physics* 113:113105.
6. Horng, R.-H., W.-C. Kao, S.-L. Ou, and D.-S. Wuu. 2012. Effect of diamond like carbon layer on heat dissipation and optoelectronic performance of vertical-type InGaN light emitting diodes. *Applied Physics Letters* 101:171102.
7. Wu, P., S. Ou, R. Horng, and D. Wuu. 2017. Enhanced light extraction of high-voltage light emitting diodes using a sidewall chamfer structure. *IEEE Photonics Journal* 9:1–9.
8. Sung-Ho, B., K. Jeom-Oh, K. Min-Ki, P. Il-Kyu, N. Seok-In, K. Ja-Yeon, K. Bongjin, and P. Seong-Ju. 2006. Enhanced carrier confinement in AlInGaN-InGaN quantum wells in near ultraviolet light-emitting diodes. *IEEE Photonics Technology Letters* 18:1276–1278.
9. Kim, G., M.-C. Sun, J. H. Kim, E. Park, and B.-G. Park. 2017. GaN-based light emitting diodes using p-type trench structure for improving internal quantum efficiency. *Applied Physics Letters* 110:021115.
10. Lin, G.-B., D. Meyaard, J. Cho, E. Fred Schubert, H. Shim, and C. Sone. 2012. Analytic model for the efficiency droop in semiconductors with asymmetric carrier-transport properties based on drift-induced reduction of injection efficiency. *Applied Physics Letters* 100:161106.
11. Dai, Q., Q. Shan, J. Cho, E. F. Schubert, M. H. Crawford, D. D. Koleske, M.-H. Kim, and Y. Park. 2011. On the symmetry of efficiency-versus-carrier-concentration curves in GaInN/GaN light-emitting diodes and relation to droop-causing mechanisms. *Applied Physics Letters* 98:033506.
12. Lee, S., C. Cho, S. Hong, S. Han, S. Yoon, S. Kim, and S. Park. 2012. Enhanced optical power of InGaN/GaN light-emitting diode by AlGaN interlayer and electron blocking layer. *IEEE Photonics Technology Letters* 24:1991–1994.
13. Kim, K. C., Y. C. Choi, D. H. Kim, T. G. Kim, S. H. Yoon, C. S. Sone, and Y. J. Park. 2004. Influence of electron tunneling barriers on the performance of InGaN–GaN ultraviolet light-emitting diodes. *Physica Status Solidi (a)* 201:2663–2667.
14. Liu, H. R., S. L. Ou, S. Y. Wang, and D. S. Wuu. 2016. On the role of diluted magnetic cobalt-doped ZnO electrodes in efficiency improvement of InGaN light emitters. *Applied Physics Letters* 109:021110.
15. Wang, R. X., S. J. Xu, A. B. Djurišić, C. D. Beling, C. K. Cheung, C. H. Cheung, S. Fung, D. G. Zhao, H. Yang, and X. M. Tao. 2006. Influence of indium-tin-oxide thin-film quality on reverse leakage current of indium-tin-oxide/n-GaN Schottky contacts. *Applied Physics Letters* 89:033503.
16. Hwang, J. D., G. H. Yang, W. T. Chang, C. C. Lin, R. W. Chuang, and S. J. Chang. 2005. A novel transparent ohmic contact of indium tin oxide to n-type GaN. *Microelectronic Engineering* 77:71–75.
17. Alivov, Y. I., J. E. Van Nostrand, D. C. Look, M. V. Chukichev, and B. M. Ataev. 2003. Observation of 430 nm electroluminescence from ZnO/GaN heterojunction light-emitting diodes. *Applied Physics Letters* 83:2943–2945.
18. Hsiao, W.-H., T.-H. Chen, L.-W. Lai, C.-T. Lee, J.-Y. Li, H.-J. Lin, N.-J. Wu, and D.-S. Liu. 2016. Investigations on the cosputtered ITO-ZnO transparent electrode ohmic contacts to n-GaN. *Applied Sciences* 6:60.
19. Lee, S.-A., J.-Y. Hwang, J.-P. Kim, S.-Y. Jeong, and C.-R. Cho. 2006. Dielectric characterization of transparent epitaxial Ga_2O_3 thin film on n-GaN/Al_2O_3 prepared by pulsed laser deposition. *Applied Physics Letters* 89:182906.
20. Soylu, M., and F. Yakuphanoglu. 2014. Properties of sol–gel synthesized n-ZnO/n-GaN (0001) isotype heterojunction. *Materials Chemistry and Physics* 143:495–502.

21. Wuu, D. S., S. L. Ou, and C. H. Tien. 2018. Slow electron making more efficient radiation emission. *Scientific Reports* 8:4865.
22. El Mir, L., Z. Ben Ayadi, H. Rahmouni, J. El Ghoul, K. Djessas, and H. J. von Bardeleben. 2009. Elaboration and characterization of Co doped, conductive ZnO thin films deposited by radio-frequency magnetron sputtering at room temperature. *Thin Solid Films* 517:6007–6011.
23. Cynthia, L.-C., M. Gabriela, C. C.-E. Juan, R.-H. Raul, P.-B. Arely, and N. A. Cristobal. 2012. Inulinase production by penicillium citrinum ESS in submerged and solid-state cultures. *American Journal of Agricultural and Biological Sciences* 7.
24. Caglar, Y. 2013. Sol–gel derived nanostructure undoped and cobalt doped ZnO: Structural, optical and electrical studies. *Journal of Alloys and Compounds* 560:181–188.
25. Gu, H., W. Zhang, Y. Xu, and M. Yan. 2012. Effect of oxygen deficiency on room temperature ferromagnetism in Co doped ZnO. *Applied Physics Letters* 100:202401.
26. Pal, B., S. Dhara, P. K. Giri, and D. Sarkar. 2014. Room temperature ferromagnetism with high magnetic moment and optical properties of Co doped ZnO nanorods synthesized by a solvothermal route. *Journal of Alloys and Compounds* 615:378–385.
27. Benramache, S., and B. Benhaoua. 2012. Influence of substrate temperature and Cobalt concentration on structural and optical properties of ZnO thin films prepared by Ultrasonic spray technique. *Superlattices and Microstructures* 52:807–815.
28. Benramache, S., B. Benhaoua, and H. Bentrah. 2013. Preparation of transparent, conductive ZnO:Co and ZnO:In thin films by ultrasonic spray method. *Journal of Nanostructure in Chemistry* 3:54.
29. Yang, S., R. Lv, C. Wang, Y. Liu, and Z. Song. 2013. Structural and magnetic properties of cobalt-doped ZnO thin films on sapphire (0001) substrate deposited by pulsed laser deposition. *Journal of Alloys and Compounds* 579:628–632.
30. Taabouche, A., A. Bouabellou, F. Kermiche, F. Hanini, Y. Bouachiba, A. Grid, and T. Kerdjac. 2014. Properties of cobalt-doped zinc oxide thin films grown by pulsed laser deposition on glass substrates. *Materials Science in Semiconductor Processing* 28:54–58.
31. Zhang, L., Z. Ye, B. Lu, J. Lu, Y. Zhang, L. Zhu, J. Huang, W. Zhang, J. Huang, J. Zhang, J. Jiang, K. Wu, and Z. Xie. 2011. Ferromagnetism induced by donor-related defects in Co-doped ZnO thin films. *Journal of Alloys and Compounds* 509:2149–2153.
32. Ravichandran, K., P. Ravikumar, and B. Sakthivel. 2013. Fabrication of protective over layer for enhanced thermal stability of zinc oxide based TCO films. *Applied Surface Science* 287:323–328.
33. Singhal, R. K., M. S. Dhawan, S. K. Gaur, S. N. Dolia, S. Kumar, T. Shripathi, U. P. Deshpande, Y. T. Xing, E. Saitovitch, and K. B. Garg. 2009. Room temperature ferromagnetism in Mn-doped dilute ZnO semiconductor: An electronic structure study using X-ray photoemission. *Journal of Alloys and Compounds* 477:379–385.
34. Fang, W., Y. Liu, B. Guo, L. Peng, Y. Zhong, J. Zhang, and Z. Zhao. 2014. Room temperature ferromagnetism and cooling effect in dilute Co-doped ZnS nanoparticles with zinc blende structure. *Journal of Alloys and Compounds* 584:240–243.
35. Kumar, S., S. Mukherjee, R. Kr. Singh, S. Chatterjee, and A. K. Ghosh. 2011. Structural and optical properties of sol-gel derived nanocrystalline Fe-doped ZnO. *Journal of Applied Physics* 110:103508.
36. Jose, J., and M. Abdul Khadar. 2001. Role of grain boundaries on the electrical conductivity of nanophase zinc oxide. *Materials Science and Engineering: A* 304-306:810–813.
37. Ou, S. L., H. R. Liu, S. Y. Wang, and D. S. Wuu. 2016. Co-doped ZnO dilute magnetic semiconductor thin films by pulsed laser deposition: excellent transmittance, low resistivity and high mobility. *Journal of Alloys and Compounds* 663:107–115.

38. Erhart, P., A. Klein, and K. Albe. 2005. First-principles study of the structure and stability of oxygen defects in zinc oxide. *Physical Review B* 72:085213.
39. Janotti, A., and C. G. Van de Walle. 2007. Native point defects in ZnO. *Physical Review B* 76:165202.
40. Janotti, A., and C. G. Van de Walle. 2005. Oxygen vacancies in ZnO. *Applied Physics Letters* 87:122102.
41. Liu, X.-C., E.-W. Shi, Z.-Z. Chen, H.-W. Zhang, L.-X. Song, H. Wang, and S.-D. Yao. 2006. Structural, optical and magnetic properties of Co-doped ZnO films. *Journal of Crystal Growth* 296:135–140.
42. Qiu, X., G. Li, X. Sun, L. Li, and X. Fu. 2008. Doping effects of Co^{2+}ions on ZnO nanorods and their photocatalytic properties. *Nanotechnology* 19:215703.
43. Pankove, J. I. 1971. Optical Processes in Semiconductors. Dover, New York.
44. Zhao, X., J. Li, H. Li, and S. Li. 2012. Intrinsic and extrinsic defect relaxation behavior of ZnO ceramics. *Journal of Applied Physics* 111:124106.
45. Ellmer, K. 2001. Resistivity of polycrystalline zinc oxide films: current status and physical limit. *Journal of Physics D: Applied Physics* 34:3097–3108.
46. Kohan, A. F., G. Ceder, D. Morgan, and C. G. Van de Walle. 2000. First-principles study of native point defects in ZnO. *Physical Review B* 61:15019–15027.
47. Heo, Y. W., D. P. Norton, and S. J. Pearton. 2005. Origin of green luminescence in ZnO thin film grown by molecular-beam epitaxy. *Journal of Applied Physics* 98:073502.

7 Thin-Film Thermoelectrics

Materials, Devices, and Applications

Xizu Wang[a], *Ady Suwardi*[a], *Qiang Zhu*[a], *and Jianwei Xu*[a,b,c]
[a]Institute of Materials Research and Engineering (IMRE), Singapore
[b]National University of Singapore, Department of Chemistry, Singapore
[c]Institution of Sustainability for Chemical, Energy and Environment (ISCE2), Jurong Island, Singapore

CONTENTS

DOI: 10.1201/9781003141358-7

7.1 FUNDAMENTALS OF THIN-FILM THERMOELECTRIC MATERIALS

Machines all around the world ranging from airline, marine, factories, or even simple smart watches generate heat. More than two-third of energy utilized worldwide is dissipated as heat and released into the atmosphere, so it is important that the waste heat can be utilized to generate eco-friendly power for economic and environmental benefit. Thermoelectric (TE) materials have the ability to convert heat into electricity.[1,2] TE generators (TEG) are solid-state semiconductor devices that convert a temperature difference and heat flow into a useful direct current (DC) power source.[3,4] TEGs are essentially of solid-state, no movement, and no noise, making them ideal for power generation. With the help of Seebeck effect and Peltier effect, TE materials can generate useful electric (or electromagnetic) fields. In the presence of a temperature gradient, the Seebeck effect develops an electric potential.[5] The Peltier effect on the other hand can pass on heat energy against the temperature slope in which a current is driven concurrently against this potential. TE materials enable conversion of electricity into heat pump and vice versa.[5–7]

Overall, TE devices serve dual purposes as a power generator via the Seebeck effect or as a cooling device via the Peltier effect. The performance metrics of a TE material can be estimated by the figure of merit (ZT):

$$ZT = \frac{S^2\sigma T}{k_t} = \frac{S^2 T}{\rho k} = \frac{S^2 T}{\rho(k_L + k_e)} = \frac{PF}{k} \tag{7.1}$$

where S is the Seebeck coefficient, σ is the electrical conductivity, and κ is the thermal conductivity. S represents the voltage per unit temperature difference. In principle, to obtain a high ZT value, high S, high σ, and low κ are usually essential. Nevertheless, it is challenging to obtain above combinations due to the interdependencies between these three parameters via carrier concentration (n), which is expressed as

$$S = \frac{8\pi^2 k_B^2 \mathrm{T}}{3eh^2} m_S^* \left(\frac{\pi}{3n}\right)^{2/3} \tag{7.2}$$

$$\sigma = ne\mu \tag{7.3}$$

For most inorganic TE materials, n typically ranges from 10^{19} to 10^{21} cm^{-3}, whereas for organic materials, n is relatively hard to be quantified reliably and remains a challenge to be measured in the scientific communities especially for these materials with a low carrier mobility.[8] Apart from tuning the magnitude of n, strategies to improve S, σ, and κ synergistically have been the focus of TE material research for many years.[9–25] Furthermore, κ is also in part dependent on σ due to the electronic contribution to the total κ. To the same end, the relatively low κ (partly due to the low σ) in organic thermoelectrics has been a major advantage as compared to their inorganic counterparts.

Thermoelectrics found their first application as the radioisotope TEG in space by National Aeronautics and Space Administration in 1976.[2,23] Other uses including solid state cooling and self-powered temperature sensors have also been commercialized.[26–30] In recent years, TEGs have been touted for various applications, especially in flexible and wearable electronics such as watches.[30–41] In this domain, thin-film TEGs play an important role due to their tunable dimensions, and unique wearable characteristics that differ from traditional electronics. They are lightweight and can be easily mounted onto different kinds of substrates, which offer possibilities in developing TE devices with flexible, stretchable, bendable, and miniature requirements.[30,42–55]

7.1.1 Materials

TE materials can be broadly categorized into inorganic and organic thermoelectrics. In this section, the characteristics and performances of each of these classes of materials will be outlined. In addition, hybrid organic–inorganic materials and their potential will be summarized.

7.1.1.1 Inorganic Thin-Film Thermoelectrics

Bulk inorganic TE materials such as semiconductors, metals, and alloys have been widely studied to date. The following section discussed fabrication of some typical inorganic thin-film TE materials and their corresponding performance.

In the past two decades, Bi-Te-based bulk materials have leaped forward in terms of TE performance. In comparison to pristine Bi_2Te_3, Bi_2Te_3 doped with Ag nanoparticles demonstrated a significantly low κ and an improved power factor. Bulk Bi_2Te_3 with 2.0 vol% Ag nanoparticles dispersed throughout the matrix presented a *ZT* of 0.77, which is approximately threefold that of its pristine material at the measured temperature of 475 K.[56] Likewise, studies have been done to optimize doping of Bi-Te-based superlattices used for thin-film TE applications with success in improving *ZT*.[57] In a study conducted by Rama Venkatasubramanian *et al.*, *p*-type Bi_2Te_3/Sb_2Te_3 superlattices and *n*-type $Bi_2Te_3/Bi_2Te_{2.83}Se_{0.17}$ superlattices thin films were prepared, and demonstrated *ZT* values of ~2.4 and ~1.4 at 300 K, respectively.[57] *P*-type $Bi_{0.4}Te_{3.0}Sb_{1.6}$ and *n*-type $Bi_{2.0}Te_{2.7}Se_{0.3}$ thin films were fabricated by the flash evaporation method. This simple material preparation only requires three equipment components, a vacuum chamber with a particle holder, a tungsten heater, and a substrate holder. A power factor of 15.9 $\mu Wcm^{-1}K^{-2}$ for the *p*-type and 21.5 $\mu Wcm^{-1}K^{-2}$ for the *n*-type thin-film thermoelectrics was achieved.

In addition to Bi-Te thin films, recent studies on Zn- and Cu-based thin-film TE materials have also shown great promise. Additionally, relatively low cost of these alternatives makes them more attractive than expensive and scarce Bi-Te-based materials.[58]

Cu-based TE materials, in both bulk and thin film forms, have drawn tremendous attention in practical applications as well.[59] Bulk Cu_2Se showed comparable TE performance to the well-known Bi-Te thin film and superlattices, with a reported peak *ZT* value of 2.4 at 1000 K.[60] One contributing factor to this high *ZT* is its low lattice κ of Cu_2Se. However, so far, work on Cu_2Se thin films has shown to have inferior TE properties as compared to its bulk counterpart. For example, Cu_2Se powder was suspended in organic solvents and wet-deposited onto flexible substrates, giving a power factor of 0.62 $mWcm^{-1}K^{-2}$ at 684 K.[61]

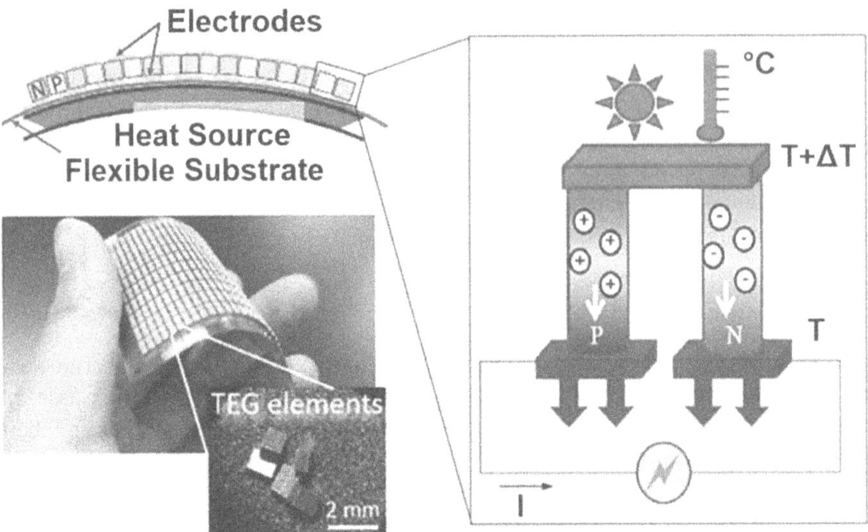

FIGURE 7.1 Illustration of a typical flexible TE module for power generation (Seebeck effect). (Reproduced with permission from reference [63] copyright 2019, WILEY-VCH GmbH & Co. KGaA, Weinheim and from reference [64]. Copyright 2018, Elsevier.)

Moreover, it is of significance to develop TEGs that can work efficiently within the range of room temperature to power wearable and portable electronic devices. Recent research has reported *p*-type CuI thin-films TE materials with high S and σ and low κ values.[62] In this work, the thin film was prepared by the reactive magnetron sputtering method onto a glass substrate at room temperature using a high-purity copper target, while iodine was injected by a needle valve, followed by annealing.[61] The undoped CuI thin film had a σ of 156 S cm^{-1}, while the iodine-doped thin film exhibited a higher σ of 283 S cm^{-1}. A recorded ZT value of CuI thin film is 0.21 at 300 K. These CuI thin-films with 200–300 nm thickness are capable of being assembled into transparent and flexible TE modules, carving a path for promising flexible TE device applications in the future.[62]

On the device level, the configuration of TEG modules does not differ much from that of TE-based TE cooler (TEC) modules, which function in reverse of a TEG. When a voltage is applied to a TEC, an electrical current is created, which induces the Peltier effect, and thus heat is moved from the cold side to the hot side. Fundamentally, as the voltage produced by each TE leg unit is very low even for TE materials with very high S value, multiple TE legs essentially need to be connected together to achieve a useful voltage range to power devices as depicted in Figure 7.1.

7.1.1.2 Organic Thin-Film Thermoelectrics

Other than inorganic TE materials, organic or semi-metallic polymers, for example, polyacetylene, polyaniline, polypyrrole, polythiophene, and poly(3,4-ethylenedioxythiophene)[65] are also potential contenders for TE device applications (Figure 7.2). These polymer materials inherently have the low κ due to their amorphous nature,

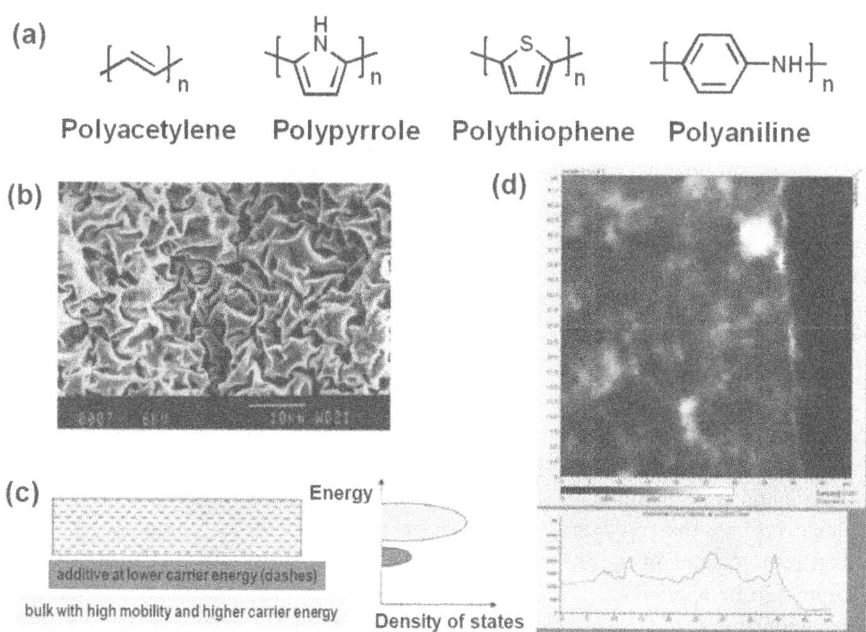

FIGURE 7.2 Conducting polymers and their composites used for thermoelectric materials. (Reproduced from reference[65] with permission; Copyright American Chemical Society 2010.) (a) Chemical structure of the polymers (b) SEM (scanning electron micrograph) morphology. (c) Energy levels of the bulk and additive, with the resultant density of states. (d) Atomic force microscopy morphology of the polymers.

giving them an advantage over conventional inorganic TE counterparts. However, most conducting polymers suffer from low-energy conversion efficiency due to their poor electrical transport properties. Various doping agents can be introduced to these conducting polymers to enhance their TE performance. Polyacetylene film doped with iodine has shown a high σ of 1×10^5 S cm^{-1}, a large improvement from the undoped polymer film with an σ of 3×10^3 S cm^{-1}.[66] The σ of polyacetylene film could be further enhanced by doping with $FeCl_3$, presenting a high σ of 3×10^5 S cm^{-1} measured at 220 K, but unfortunately polyacetylene is unstable at ambient atmosphere,[67] discouraging its application. Polyaniline is another important type of conducting polymers. Similar to other conducting polymers, its TE properties can be improved by various doping methods. Polyaniline was primarily prepared using chemical oxidative polymerization with acid doping, which exhibited a rather low σ of a few S cm^{-1}. Secondary doping was adopted to improve polyaniline microstructural ordering, which resulted in a greatly increased σ of 280 S cm^{-1} and a power factor of 11 μW m^{-1} K^{-2}, respectively.[68–70] Furthermore, a variety of polyaniline composites with different nanostructures including inorganic TE materials, carbon nanostructures such as carbon nanotubes and graphene have been studied, targeting at achieving a high σ or S.[71–73] In addition, high-quality free-standing polythiophene and poly(3-methylthiophene) nanofilms displayed fair TE performance with a ZT value of 0.03 at 250 K.

7.1.2 Key Parameters of Thin-Film Thermoelectric Materials Σ, *S*, and κ

The σ of a material is representative of the amount of charge flowing through it and also known as the reciprocal of resistivity, ρ. The simplest method to measure σ is to apply a DC through two electrical probes, and to measure the potential difference across them and then to calculate the electrical resistance (R) according to the Ohm's law ($I = V/R$). Subsequently, electrical resistivity ρ and σ (σ is the inverse of ρ) can be calculated using the dimensions of the wire ($\rho = RA/L$, where A is the cross-section area of the resistance and L is the length of the resistance.). However, using the two-probe method will obtain the cumulative resistance of all components of the circuit, comprising the contact resistance, probe resistance, and spreading resistance under the probes as well as the sample. Thus, it will be difficult to extract and analyze the resistance of the sample.

A popular alternative to measure σ is the four-probe method. This method employs two additional probes, specifically to detect the potential difference between two points using a high impedance voltmeter. As no current is applied to the pair of voltage probes, the resistance of the other components of the circuit is no longer taken into account in the measurement. For bulk samples, metal foils or soldered indium can be applied at both ends of the sample to ensure a uniform current flow throughout the material. Various four-probe configurations for σ measurement are illustrated in Figure 7.3. To measure σ of bulky TE samples, all four probes can be

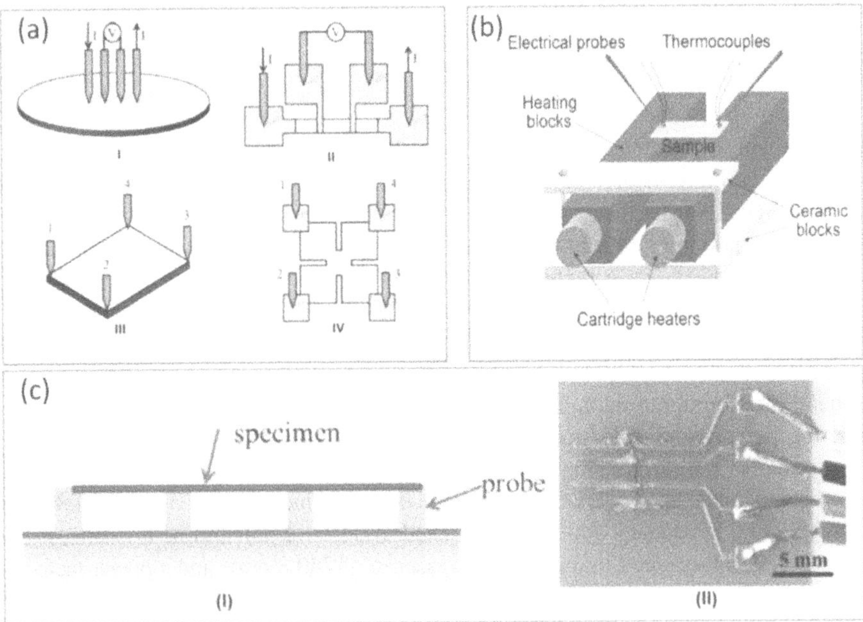

FIGURE 7.3 (a) Various four-probe configurations for σ measurement. (b) Schematic of an *S* measurement setup. (c) In-plane κ measurement of the poly(nickel-ethylenetetrathiolate) film. (Reproduced from reference[74] with permission; Copyright John Wiley & Sons, Inc. 2016.)

arranged collinearly, equidistant from each other (Figure 7.3(a-I)). For small samples, it is favorable to pattern the sample into a well-defined bar and deposit four Ohmic contacts at four points on the bar (Figure 7.3(a-II)), where the probes are situated. This can be achieved by electrodeposition of metal Ti or Au while keeping in mind to minimize the width of the bars to be comparatively thinner than the distance between them to reduce the uncertainty in length. The contact pads may have a large surface area to ensure good contact between the pads and the probes. This arrangement is also a favored choice when there is a need to measure the σ of anisotropic materials.

To study thin-films with an arbitrary shape, it is most appropriate to employ the van der Pauw (vdP) method (Figure 7.3(a-III) and 7.3(a-IV)). In this method, a specific perimeter on the sample is identified and the four probes are situated to enclose this selected area. Similar to the two-probe method, indium can be soldered to ensure good contact between the probes and the thin-film. Alternatively, gold contacts patterned through lithography can also be used as a substitute. Typically, square samples with a size of 1 cm × 1 cm are used. However, the vdP method is not sensitive to geometry, and the σ can still be accurately obtained from other-shaped samples.

The S of a TE material is an intrinsic property independent of the size or geometry of the sample. It is typically uniform with the exception of inhomogeneous materials that can have varying S values, depending on the position of the measurement. The sign of the S is indicative of the majority charge carrier of a TE material. It is negative for n-type and however positive for p-type materials. To measure the in-plane S of a material, a temperature gradient is imposed across a sample. The potential difference across it is then measured and divided by the difference in temperature. Experimentally, it is common to suspend the sample between two separated isothermal blocks such that each end of the sample is maintained at a different temperature. To measure the potential difference and temperature difference, electrical probes and thermocouples are used respectively at both ends of the sample as illustrated in Figure 7.3(b). Currently, there are several commercially available systems, for example, ULVAC ZEM-3, MMR SB-100, and Linesis LSR-3, for measuring S and σ. However, most of these are tailored to measure bulk samples in the millimetre-size range.

There are two popular methods in measuring the in-plane S of thin-films, the integral method and the differential method. For the integral method, one end of the sample is fixed at a known reference temperature, typically ambient temperature or 273 K. The other end is continuously heated up in the temperature range of interest, and a large range of thermal gradients data will be collected. The S is then estimated by taking the derivative of the V–T graph that can be expected to be nonlinear. It is important to note that the integral method assumes that thermal equilibrium is achieved immediately between the thermocouple–sample junctions, and the voltage and temperature measurements occur instantly as well. Comparatively, the differential method is more widely used for thin-film thermoelectrics. Small thermal gradients (ΔT) are applied along the length of the sample, in the temperature range where the Seebeck voltage is expected to have a linear relation to temperature. The potential difference is measured a few times at each ΔT. Similar to the integral method, the gradient of the V–T plot gives the S value at the defined temperatures. Measurements for thin-film TE materials specifically are difficult to obtain with a high accuracy.

To maximize precision, multiple readings are taken at each ΔT, perhaps up to a hundred to account for the spread of the data points. Variations in voltage (y-axis) suggest inconsistencies in the measuring apparatus. Meanwhile, spread along the temperature (x-axis) may be indicative of nonideal thermal resistances between the sample and the probes. Clustering of many points or a poor linear fit may suggest nonuniform heat flow, which is possibly attributed to the poor contact between the sample and probes. To further improve accuracy in measurement, it is good practice to include sufficient wait time after each temperature increment before voltage measurement to ensure that thermal equilibrium has been achieved. Moreover, an additional consideration is the calibration error of thermocouples, which can contribute to a 1–2 K variance in ΔT, with the total error for the in-plane S measurement amounting to a 10–15%.

Thermal conductivity κ describes the intrinsic capability of a material to transmit heat. The rate of heat gain or loss is dependent on the magnitude of the temperature gradient applied and thermal characteristics of a material such as thermal diffusivity. Similar to σ and S measurements, the experimental analysis of κ of thin films also requires additional considerations as the resulting manifest error can be as high as ~10%.[75] There have been some novel solutions to counter the experimental errors and offer fast measurements. One example is the "steady-state isothermal technique" which improves accuracy by acquiring readings under conditions where thermal losses and errors are inconsequential. Using this technique on a Peltier cooler has shown to decrease the error down to below 1%.[76] Likewise, for TEGs, experimental uncertainty was reduced to be lower than half the initial value. Other methods include "scanning hot-probe" and "lock-in transient Harman" which involve the treatment of the interfacial contact effects that encompass electrical contact resistance and thermal contact effects. To measure cross-plane and in-plane κ, the 3ω methods and time-domain thermoreflectance measurements are employed.[77] While additional precautions are taken to minimize the error in the measurements, these methods still have improvement room to be able to accurately measure κ of thin-films with rough surfaces. For the in-plane measurement, a self-heating 3ω method can be adopted in which a gold wire with comparable thickness to the film is used. This method (Figure 7.3(c)) employs four probes to measure the in-plane κ of the thin-film sample which functions as the heater to generate a change in temperature and its thermal response.[78]

7.2 THIN-FILM THERMOELECTRIC DEVICES

The Seebeck effect, the Peltier effect, and the Thomson effect pertaining to thermoelectrics were discovered by Thomas Seebeck in 1821, Jean Peltier in 1934, and William Thomson in 1851, respectively. TE devices are developed based on these three effects, which can be used for thermal power generation and cooling/heat flow control. TE devices can be divided into two types according to their applications, the power generator based on the Seebeck effect, and the cooling/heat flow control device based on both the Peltier effect and the Thomson effect. TEGs can be utilized to generate electrical power, while the cooling/heat flow control devices such as TEC have been potentially applied in human body, mobile phones, refrigerators, and chiller systems.

7.2.1 Traditional Thermoelectric Module

The first TEG using bulk TE materials was fabricated in 1948 in the Union of Soviet Socialist Republics (USSR).[79] This classical TEG module can convert heat to electricity, and actually it can control heat flow and build a temperature gradient as well. Based on the nature of TE materials, two types of TE materials, namely *p*-type and *n*-type, can be used for TE devices. Figure 7.4 shows the two charge carrier flows that are in line with the direction of heat flow from the heat side to the cold side. The electrons are major carriers in *n*-type TE materials and the holes are major carriers in *p*-type TE materials, both of which induce the current flow in an opposite direction in TE materials.

As illustrated in Figure 7.4, the TE material is the channel to connect the heat flow and the current flow. Therefore, there are two major device structures to fulfill two opposite functions: conversion of heat to electricity and the heat flow controlled by electricity. In general, the device that can directly convert heat to electricity is defined as a TEG module (left diagram in Figure 7.4(b)), while the device that can operate in a reverse way is named as a TEC module (right diagram in Figure 7.4(b)). These two TE devices correspond to the Seebeck and the Peltier effect of TE materials. A basic TE module consists of several pairs of TE legs terminated with electrode bridges to connect an *n*-type TE material with a negative charge carriers flow heat direction, and a *p*-type counterpart with a positive charge carriers reverse heat direction. Therefore, the electric current will flow in the circuit when there is a temperature difference between the cold end and the hot end of the TEG module. On the contrary, in the TEC device, the charge carriers of both types of TE materials are responsible for the heat transfer flow, in which the current flow affects the different heat flow in *p*-type and *n*-type TE materials. In *n*-type TE legs, electrons move from the electrode into the *n*-type material, they absorb heat energy from the environment to the material interface and then release the heat after moving to another side of legs. In the *p*-type TE leg, the holes replace the electrons to absorb and subsequently release the heat at both sides of legs. Therefore, the temperature difference is induced between the *n*-type and *p*-type TE legs since the current flow drives the heat flow from the top side to the bottom side. To put it simply,

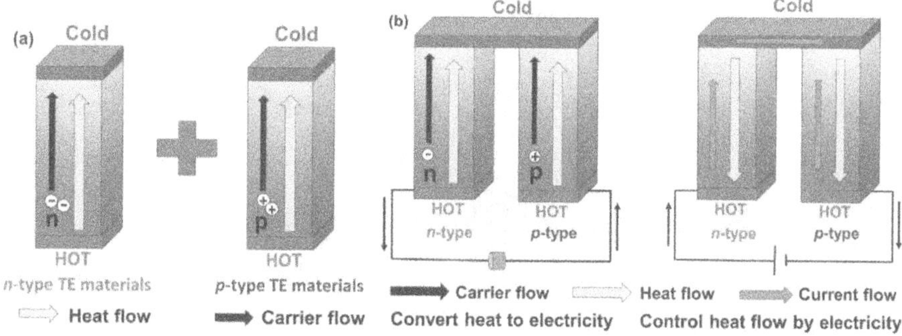

FIGURE 7.4 (a) A single *n*-type or *p*-type TE module. (b) Configurations of TEG (left) and TEC (right).

the functions of TEG and TEC can be regarded as the opposite processes in a same p–n pair legs structure. It is worth noting that there is a big difference between the TE device with other typical p–n junction electrical devices: the p-type and n-type legs are not physically contacted with each other in the TE devices.

In a TEG device, the voltage of one cell V_{cell} is proportional to the product of the sum of corresponding Seebeck coefficients S_n and S_p of n-type and p-type TE materials, and the temperature difference ΔT applied.

$$V_{cell} = \left(S_n + S_p\right)\Delta T \tag{7.4}$$

However, in light of the S of TE materials defined in the prior section, majority of the S values are less than 1 mV/K. Then the voltage of output of a p–n TEG cell is limited at the mV level, which would fail miserably to meet the demand of the current battery and electrical system requirements. Hence, apart from taking into consideration other factors, a usual attempt to increase the magnitude of output voltage V_{out} is to connect sufficient pairs of TEG cells,

$$V_{out} = n\ V_{cell} \tag{7.5}$$

For example, one TEG device with connecting 100 pairs of p–n TEG cells together can obtain V_{out} of 4 V with ΔT of 100 K assuming that both S_n and S_p are 200 μV/K. Moreover, this structure also suits for the TEC device for cooling which can pump heat from one side to another side.

Actually, being different from other p–n semiconductor electrical devices, only one-type TE legs can configurate a whole module of either TEG or TEC. According to the carrier and heat flow, one of the configurations of single n- or p-type TE legs connected in parallel electrically and thermally is shown in Figure 7.5(a). This configuration is impractical due to its limitation of low voltage output. In comparison, another configuration that connects single n- or p-type TE legs in series (Figure 7.5(b)) can drive out a high voltage and a low current assuming that sufficient TE legs are used. However, the internal connections between legs and electrodes lead to mediocre efficiency in heat transfer and complication in device fabrication. To sum up, current TEG and TEC devices use both n-type and p-type pairs legs to form a couple, but it is believed that these

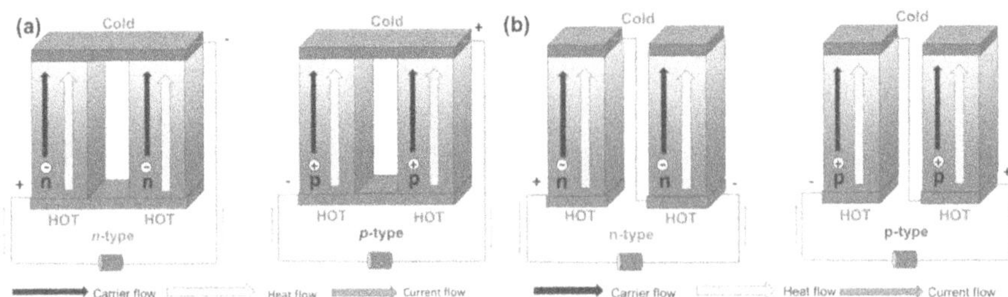

FIGURE 7.5 A TE module consisting of identical p- or n-type TE legs connected in parallel (a) and in series (b).

simple device configurations can be redesigned to a type of new generation TE device such as thin-film or in a two-dimensional (2D) scale. TEG devices are less efficient than other heat-electricity energy conversion technologies. In other words, less of the electricity is converted from thermal energy (heat) when the same amount of heat energy is input to TEGs. Therefore, most applications of TEGs usually focus on waste heat recovery where the heat is free since the efficiency may be not a big concern.

7.2.2 Thin-Film Thermoelectric Device

Traditionally, TE devices are fabricated in a bulk three-dimensional (3D) mode. Obviously, thin-film TE devices have a smaller structure space, which reduces the quantity of the materials required and improves the TE efficiency. We started with a rigid thin-film TE device followed by other three types of thin-film TE devices including flexible, stretchable, and transparent thin-film TE devices in this section.

7.2.2.1 Rigid Thin-Film Thermoelectric Device

Different from the vertical design of bulk TE modules, the most popular design of thin-film TE module used for waste heat harvesting is the planar structure. Since the TE module turns from a 3D structure to a 2D structure so that the planar structure helps to build up the temperature difference in the plane of substrate, whereas the temperature gradient can be generated along the thin-film TE materials between the hot side and the cold side, as shown in Figure 7.6.

The thin-film TE module is connected with the *p*- and *n*-type legs in series with electrode contacts so that the whole device is linked electrically in series and thermally in parallel, which is similar to the bulk module. In operation, the contacts at the hot end and the cold end are the main factor that determines the electricity power output of the TE device. Thus, this thin-film TE module demonstrates an in-plane heat harvesting model that is different from the out-plane model in bulk TE modules.

Some examples of rigid thin-film TE devices are shown in Figure 7.7. To build up sufficiently high-temperature difference between the two sides of the thin-film, *n*- and *p*-type thin-films were deposited onto the both sides of the insulating substrate.

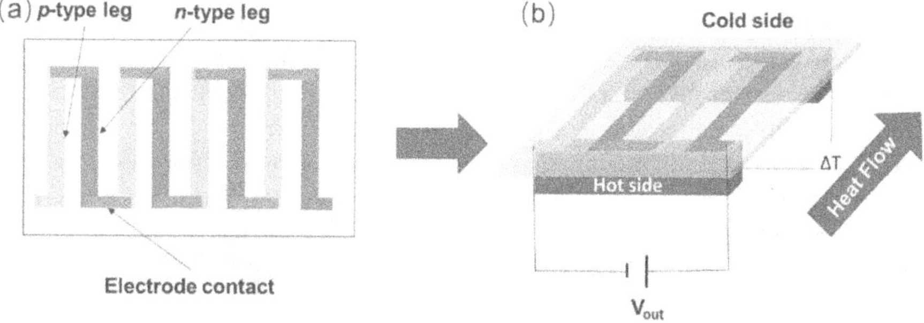

FIGURE 7.6 Configuration of a thin-film TE device. (a) Schematic of the in-plane planar p-type and n-type contacts. (b) Schematic showing the heat flow and current flow direction of the setup.

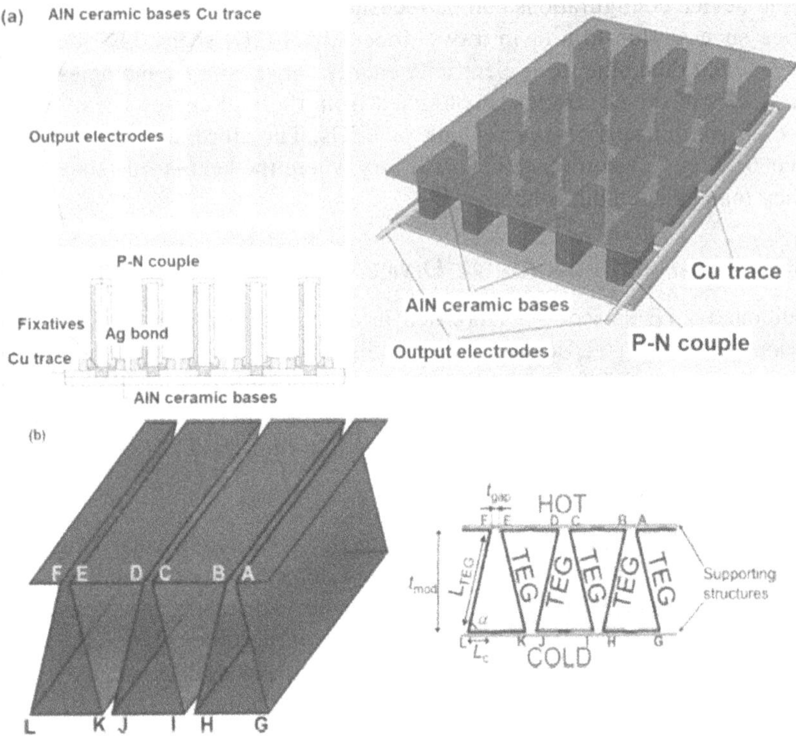

FIGURE 7.7 (a) Schematic illustration of the thin-film TEG.[80] (b) Design of a 3D TE module.[84] (Reproduced from reference[80] with permission; Copyright Springer 2014, from reference[84] with permission; Copyright Elsevier 2018.)

The *p–n* couple was formed through connection by depositing Cu thin-film onto the hot and cold sides of the glass substrate as contact electrodes and as output electrodes, respectively (Figure 7.7(a)).[80] The second TE thin-film module, called the vertical and multi-layered TE device,[81,82] is made of *p–n* thin-film couples arranged vertically between the heat source and the heat sink. One of the advantages of the thin-film TE devices is that it is easy to fabricate a micro- or nano-sized structure.[83] The micro-structure diagram of the thin-film thermoelectrics was developed with a silicon substrate coated with one insulating multilayer and the planar array of *p–n* thin-film couples connected in series. In the micro TE device, the TE *p–n* couple pair is Cu/constantan interconnected by Al contacts. Figure 7.7(b) shows one sketch and one left side view of the folded 2D thin-film TE device.[84] This thin-film TE configuration depicts a possible supporting structure with a thermally conductive layer, demonstrating that the 2D thin-film TE device harvests the out-plane heat flow from the top heat source to the bottom cold source.

Similar to a normal TEG, a thin-film TEG structure can be considered a thin-film TEC structure to build the temperature difference or pump heat from one side to another side in the plane. In addition, thin-film TEC can be used in microelectronics devices to function as micro-coolers or micro-refrigerators on chips because of

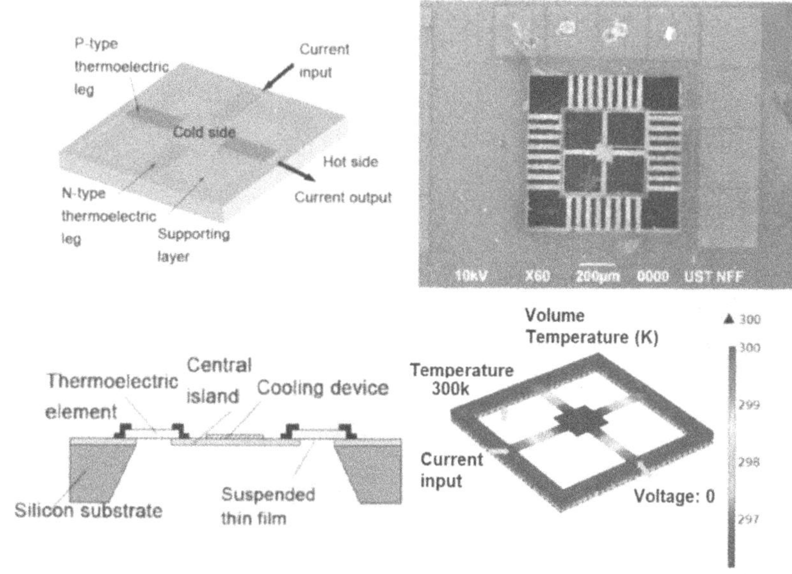

FIGURE 7.8 Conventional design of a TEC and schematic of a TE refrigerator (left); SEM picture and FEM result of TE microrefrigerator (right). (Reproduced from reference[85] with permission; Copyright Elsevier 2018.)

their ability of miniaturizing TE elements and excellent compatibility. For example, solid-state refrigeration for chip-scale electronics cooling was achieved by using thin-film TEC (thickness of around 100 μm) built with ultrathin nanostructured thin-film superlattice Bi_2Te_3-based materials (thickness between 5 and 8 μm range).[85] Figure 7.8 illustrates the structures of the micro-TEC integrated with a silicon chip on the top, in which an array of 7 × 7 *p–n* micro-couples are fabricated on the backside of the copper heat sink. Here, the copper heat sink is widely used in microprocessors cooling to enhance thermal diffusion and to protect the chips from damage. However, this kind of 3D TEC with top/bottom of contacts will impose restrictions on the efficiency of pumping out heat as the cooling area is only limited in the micro-level. Since *p–n* thin-film couples have a huge potential to increase the heat spreading area in the planar configuration, the planar thin-film TEC may be suitable for cooling in the microelectronic devices.[86]

7.2.2.2 Flexible Thin-Film Thermoelectric Device

As rapid development of the personal and portable electronic devices, researchers have paid much attention to developing a variety of flexible and wearable electronic devices including flexible TE devices.[87–89] Like other flexible electronic devices, thin-film TE couples can be mounted onto flexible substrates to create flexible TEG and TEC modules. For example, 32 pairs of *p*- and *n*-type thin-films were deposited on a flexible polyimide substrate to obtain a flexible thin-film TEG (Figure 7.9(a)).[87] This flexible TEG with an active device area of 50 cm^2 substantiates that the direct conversion human waste heat into electricity could be wrapped

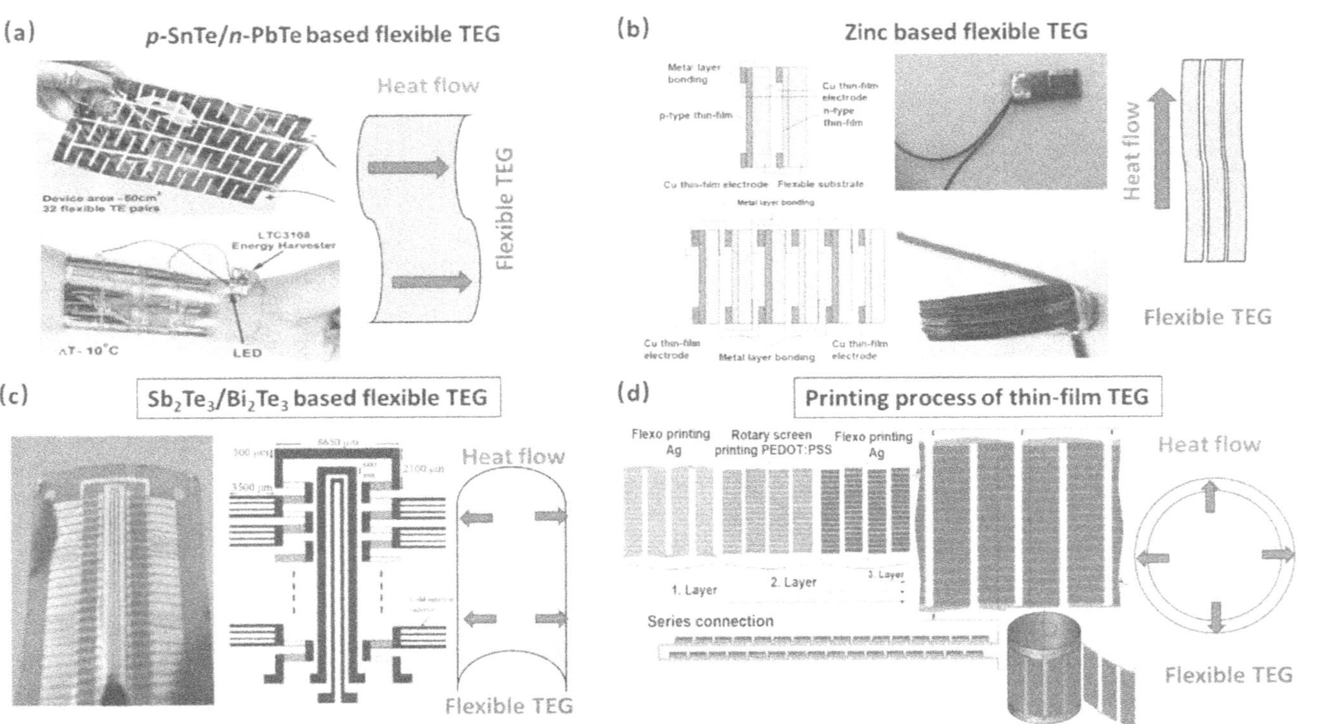

FIGURE 7.9 (a) Large-area flexible SnTe–PbTe TEG and a real-time demonstration of the TEG.[87] (b) Schematic illustration of the Zn-based thin-film TEG.[88] (c) Sb_2Te_3/Bi_2Te_3-based flexible TEG.[89] (d) Illustration of the printing process of TEGs.[90] Copyright Elsevier 2020 (a), 2015 (c), (b), Wiley 2013 (d). (Reproduced from references[87,89] with permission; Copyright Elsevier 2020 & 2015, from reference[88] with permission; Copyright AIP 2015, from reference[90] with permission; Copyright Wiley 2013.)

around the human wrist to power a small LED lamp. Another flexible thin-film TEG module was built on a Kapton-type polyimide substrate using aluminum-doped zinc oxide and Zn-Sb-based thin-film, forming a fence-like structure (Figure 7.9(b)). For a real TE application, the thin-film flexible TEG module was prepared with 10 identical *p–n* couples, which were connected in series to form a one-dimensional array.[88] While the heat flow is still in the in-plane of the thin-film, this design realizes the heat flow from bottom to top in a fence-like flexible thin-film TEG. In addition to this structure for the in-plane heat flow thin-film TEG module, another in-plane device can embed a thermometer to monitor the hot junction temperature as shown in Figure 7.9(c).[89] This flexible thin-film TEG not only demonstrates conversion of heat to electricity with in-plane heat flow, but also realizes conversion along the out-plane heat flow. Figure 7.9(d) shows a single *p*-type thin-film TE structure, in which each TE unit is serially connected to give a connection along the web.[90] The printed thin-film flexible TEG can roll up on some surfaces with multiple turns like cylinder.

7.2.2.3 Stretchable Thin-Film Thermoelectric Device

The recent advance of electronic mechanics and materials technology leads soft thermoelectrics to achieving a pathway to sophisticated embodiments, because their components can be made stretchable, compressible, twistable, bendable, and deformable into arbitrary shapes. The stretchable TE modules can be classified broadly into two types: intrinsically stretchable TEG (based on plastic materials) and extrinsically stretchable TEG (based on stretchable structure). An intrinsically stretchable thin-film TEG module (Figure 7.10) was obtained by printing 10 stretchable *p*-type conducting polymer composite legs on an elastic polydimethylsiloxane (PDMS) substrate in a lateral configuration.[91] At 40% tensile strain, the output voltage of this

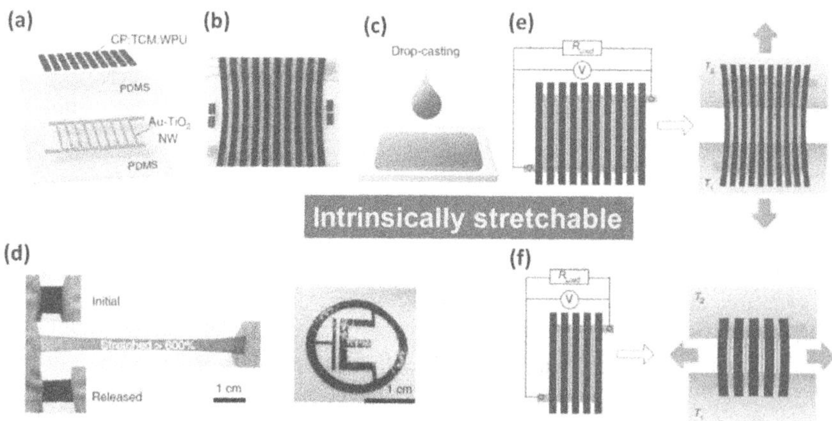

FIGURE 7.10 Intrinsically stretchable organic TE module.[91] (Reproduced with permission from reference.[91] Copyright Nature 2020.) (a-c): fabrication process of the stretchable thin films. (d) A free-standing composite film (CP:TCM:WPU = 15:25:85, w/w) can be stretched to over 600% and relax to its original shape with little hysteresis. (e) TE module stretched parallel with the thermodiffusion. (f)TE module stretched perpendicular to the thermodiffusion

TEG module was not affected, but the output power decreased with the increase in the internal resistance of the module, retaining 48% and 83% of the initial power under parallel and perpendicular strains, respectively.

The "origami" folding or corrugated architecture provides another feasible, stretchable solution for 2D thin-film TEG devices, and this design combines the advantages of large-area 2D printing fabrication and stretchable application. One of the folding thin-film TEG structures shown in Figure 7.11(a) is simplistic in design that it replaces the bottom electrode on a standard thin-film piezoelectric generator with continuous alternating *p*- and *n*-type thin-film legs. The flat diagram shows that by folding at the *p/n* junctions, a thermal gradient can be established across each element and a TEG is formed from a single film. In this configuration, the hybrid folding TE piezoelectric generator can harvest waste heat and mechanic vibration energy at the same time.[92] Figure 7.11(b) shows another folded stretchable TEG module with *p*- and *n*-TE elements printed in rows on a thin, flexible substrate and then folded into zigzag ribbons to form a stretchable TEG structure. At the elements of the series connection, the *n*-type and *p*-type legs overlap to ensure good electrical contact in the whole module.[93] Another stretchable thin-film TE module with elastics structure is a corrugated architecture in which thin-film TE couples are fabricated on a silicon elastomer substrate as depicted in Figure 7.11(c). The 10×10 TE *p-n* couple array is on a silicon matrix of 25 mm × 25 mm, and the *p–n* TE legs are connected thermally in parallel and electrically in series. The *p–n* unites are assembled with the "island-bridge" layout electrodes, embedded in the compliant and ultra-stretchable silicone elastomer. The excellent performance of stretchable TEG on both developable and non-developable hot surfaces is attributed to its outstanding stretchability to guarantee the surface attachment and the heat transfer from the waste heat.[94] Figure 7.11(d) illustrates a helical coil architecture that is usually used in a stretchable mechanical structure, such as spring. A mechanically guided assembly generates a 3D helical structure from a 2D serpentine through compressive buckling induced by relaxation of a stretched elastomer substrate to which the serpentines bond at selected locations. These TE thin-film 3D structures can be stretched in the in-plane direction by up to 60% for hundreds of cycles and can be vertically compressed up to 30%, with only minimal degradation in the electric properties.[95]

7.2.2.4 Transparent Thin-Film Thermoelectric Device

Besides flexible devices and stretchable thin-film TE modules, transparent thin-film TE modules are becoming more attractive for the applications in the area of smart and wearable devices.

To make transparent TE modules, a simple and direct method is to use a transparent thin-film TE material to directly deposit on a transparent substrate. As discussed previously, traditional TE devices are constructed by connecting *n*- and *p*-type TE pairs in series. For example, Figure 7.12(a) illustrates a transparent TE module, which consists of a single *p*-type TE leg of transparent γ-CuI thin-film on the polyethylene terephthalate (PET) substrate.[62] This transparent TE device exhibits interference effects because of its film thickness, and it has about 60–85% transmittance at the visible and near infrared regions (410–2,000 nm). With the use of transparent well-matched thin-film *p*- and *n*-type

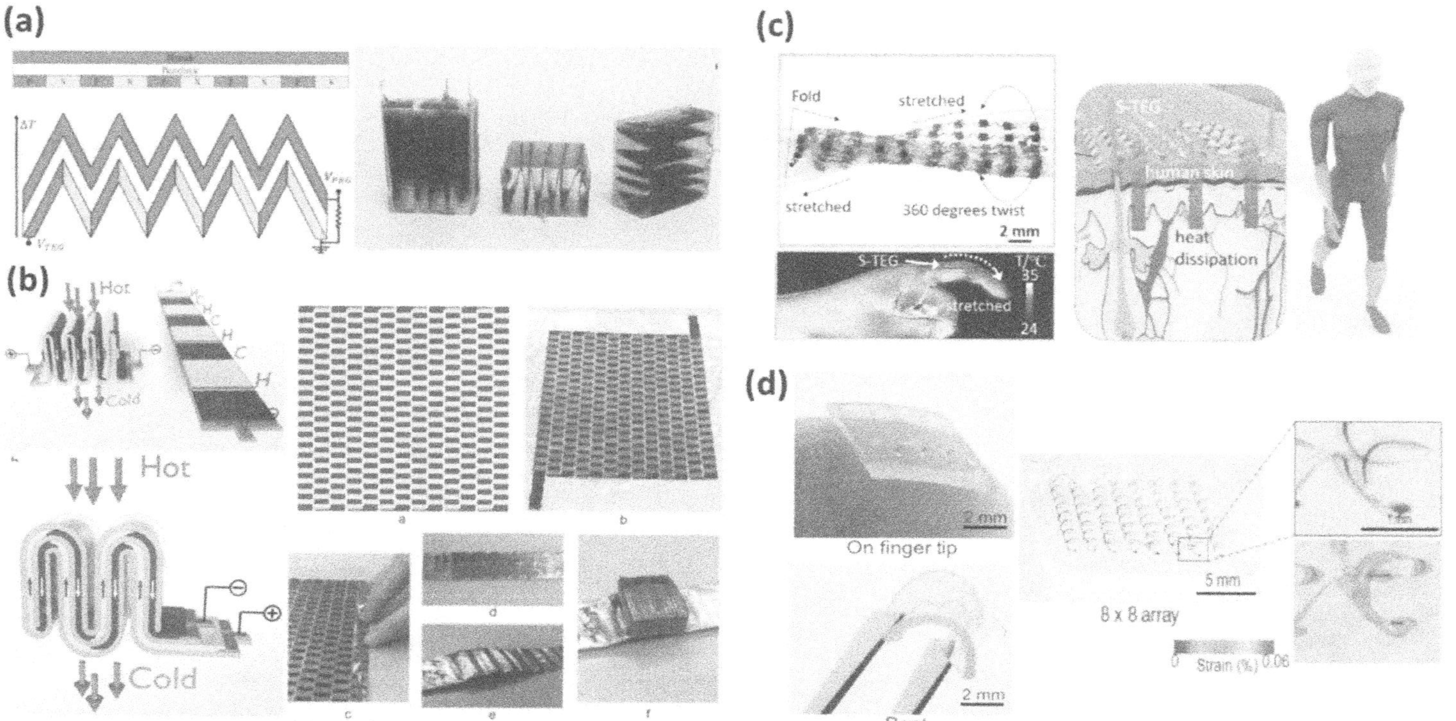

FIGURE 7.11 Illustrations of stretchable thin-film TE module with structure elastics. (Reproduced from reference[92] with permission; Copyright AIP 2016, from reference[93] with permission; Copyright Nature 2021, from reference[94] with permission; Copyright American Chemical Society 2020, from reference[95] with permission; Copyright American Association for the Advancement of Science 2018.)

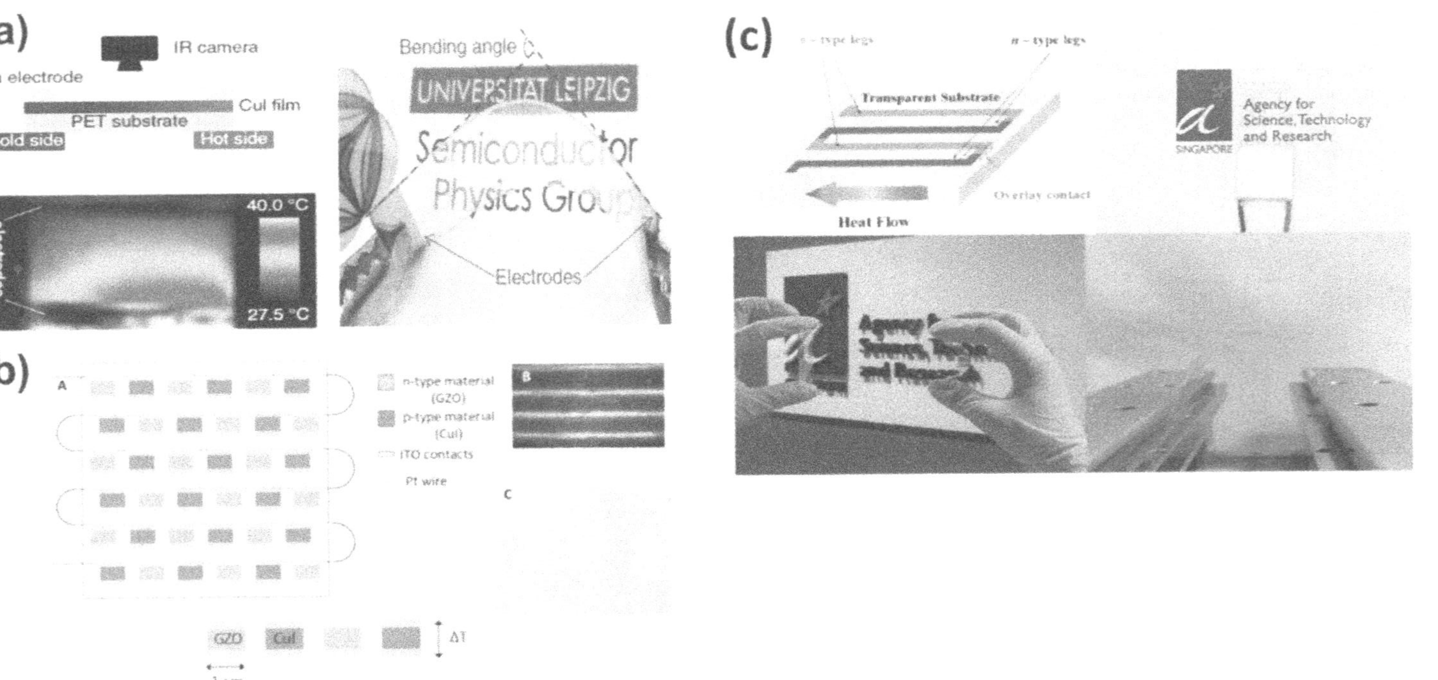

FIGURE 7.12 Transparent thin-film TEG devices.[54,62,96] (Copyright Royal Society of Chemistry 2019, from reference[54] with permission; Copyright Nature 2020; Reproduced from reference[62] with permission; Copyright Nature 2017, from reference[96] with permission.)

materials, it is possible to construct alternating *p–n* elements connected electrically in series and thermally in parallel for the transparent TEG module. As illustrated in Figure 7.12(b), a flexible transparent thin-film TEG module consisted of 17 pairs of transparent *p–n* thermocouples, gallium doped zinc oxide (GZO) as *n* -type and CuI as *p*-type legs.[96] The thermal mapping of transparency was measured by IR FLIRA310 thermal camera in the front side of the device. From these two TEG modules, it was worth noting that the major challenge of transparent TEG module is that all the components must be transparent. Figure 7.12(c) illustrates a new transparent module, comprising highly electrically conductive thin-film ITO as the *n*-type TE legs and the overlay contact bridges (functioning as electrodes), and doped PEDOT:PSS thin-film as the *p*-type TE leg.[54] The TE thin-films thermocouples on both glass and PET substrates were highly transparent with an optical transmittance of 84% over the visible and near infrared regions (350–1100 nm). Furthermore, the flexible transparent TE module was repeatedly bent up to 10,000 cycles with a bending angle up to 130°, and exhibited stable internal resistance (less than 3% decrease).

Various thin-film deposition methods such as thermal evaporation, flash evaporation, magnetic/reactive sputtering, atomic layer deposition, chemical vapor deposition, and molecular-beam epitaxy have been widely adopted to directly deposit inorganic TE materials on either rigid or flexible substrates acting as TE legs and electrodes. In contrast, the printed electronics technologies using standard graphic arts printing processes, such as screen printing, flexography, and inkjet printing, have been applied to fabricate flexible TE devices by using TE polymers or nanocomposites containing inorganic TE components on the substrates. Compared to the traditional bulk TE fabrication methods, these printing techniques are facile and scalable for the fabrication of large-scale and low-cost flexible thin-film TEG devices. In particular, using a roll-to-roll system will significantly reduce the manufacturing cost of large-area flexible thin-film TE devices in the future.

7.3 APPLICATIONS OF THIN-FILM THERMOELECTRIC DEVICES

Thin-film TE devices are mainly used for harvesting energy to power the pump heat that can be applied in any heat sources or environments with temperature difference. Although TEG and TEC have been made for at least six decades, flexible, stretchable, and transparent thin-film TEG and TEC are still in the infant stage mainly due to lack of suitable materials. It requires researchers to put more efforts to explore applications of thin-film TEG and TEC devices. This section will summarize some recent state-of-the-art applications of thin-film TEG and TEC devices. Based on electricity generation and cooling, the state-of-the-art applications of the thin-film TE devices broadly include on-chip cooling, wearable heat harvesting, and photovoltaic-thermal solar energy harvesting system.

In the past half a century, the rapid development of semiconductor devices along with the significant shrinkage in their feature size has resulted in a sharp enhancement in power density and junction temperature of the microelectronics devices. The heat power dissipated from the current microprocessor can be high over 30–50 W cm^{-2}, leading to great challenges in the thermal management of the modem electronic products. As can be seen in the prior session, advances of thin-film TE materials and modules have created great opportunities for using micro-sized TE devices in heat pumping applications, like on-chip cooling and heat-waste energy harvesting. Moreover, thin-film TEG and TEC are quite attractive for microelectronics cooling due to its all solid-state material and structure, high reliability, long lifetime, low voltage, and fast response. Simulation studies showed that on-chip heat energy harvesting can yield up to a power of 30 mW and a heat flux of 200 W cm^{-2} when thin-film TEC places on the die.[97] One of the on-chip thin-film TECs is based on a freestanding planar thin-film TE structure that helps address the heat dissipating surface cooling.[86] In this on-chip thin-film TE cooling module, the thermal and contact resistances are equally important for the device cooling efficiency. The structure of this thin-film micro-TEC was fabricated on a copper heat spreader and was integrated into a state-of-the-art silicon chip. The measured temperature of the localized high heat flux region was decreased and reached a minimum temperature when the TEC was powered by the DC drive current. The temperature of the localized heater region was 124.5°C in the absence of a TEC. In contrast, the temperature was dropped to 116.9°C when the TEC driven by a 3 A DC current was attached with the heat spreader (passive). When the on-chip TEC (active) was working, the temperature further went down to 109.4°C. In this on-chip cooler, the total temperature reduction by 14.9°C can be collectively achieved via passive and active cooling because of the use of the integrated thin-film micro-TE unit.

Wearable electronics and sensors are now widely integrated with smartphones, smart watches, and smart sensors for health monitoring, Internet of Things (IoT), etc. These devices are energy-consuming tools, in which a limited space is available for accommodating an energy storage unit to power them, hence requiring an additional power supply component, such as wireless or wired charging devices, lithium cells, or nickel–zinc batteries. It is interest and useful if wearable electronics and sensors are non-intermittently self-powered by using an energy-harvesting device, for example, TEG. Among various types of energy-harvesting techniques such as photovoltaic energy harvesting, piezoelectric energy harvesting, pyroelectric energy harvesting, wireless or electromagnetic energy harvesting, wind energy harvesting, and vibration energy harvesting, TEG is potential due to the fact that our human body is a heat source, from which TEG can harvest heat energy and convert it to electricity to power wearable electronics. Recent improvement in stretchable, flexible thin-film TEG devices has resulted in the development of a new generation of ultralight and wearable thin-film TEG, which can offer a noiseless, eco-friendly, and long lifetime form of direct conversion of our body heat energy to electricity suitable for driving personal electrical devices and sensors. For example, TEGs have been used for wearable electronics, such as flexible thin-film TEG wraps on wrist,[87–95] finger,[94] ankle,[95] textile,[98] T-shirt,[99] and fabric[100], some of which are shown in Figure 7.13.

Undoubtedly, such wearable thin-film TE devices will be more reliable, sustainable and allow for continuous battery-less operation for future applications. In addition to power wearable sensors and devices, some thin-film and fiber-based TE techniques show considerable potential for other applications, such as health, medical, aircraft, vehicle, and aerospace.

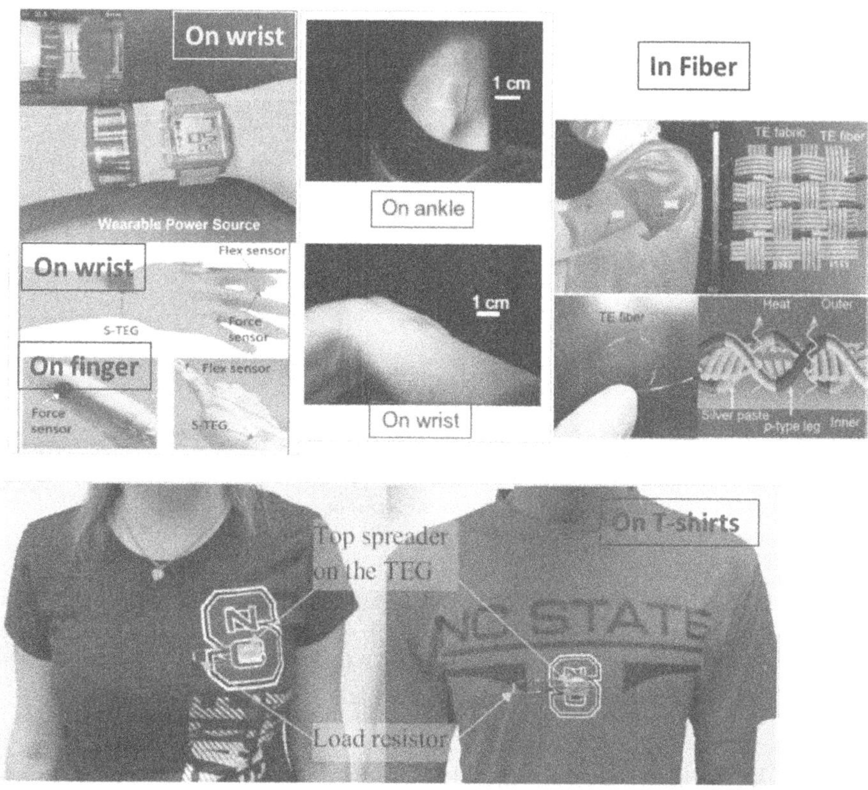

FIGURE 7.13 Thin-film TE modules for wearable electronics applications.[87–95,99,100] (Reproduced from references,[99,100] with permission; Copyright Elsevier 2020 & 2016, from references[94,100] with permission; Copyright American Chemical Society 2020, from reference[95] with permission; Copyright American Association for the Advancement of Science 2018.)

Solar power of the earth is the part of energy from the sun, and the primary forms of the solar energy are heat and light. As we know, the sunlight can be converted into electricity by the photovoltaic effect of the solar cells. The solar heat energy is transformed and harvested in multiple ways, such as for cooking, heating water, and generating electricity. Recently, it was reported that thin-film TEG could be integrated with solar panel systems. As shown in Figure 7.14(a–c), there is a multi-functional thin-film TE device for small-scale energy harvesting and self-powering light sensing, in which the Fresnel lens was incorporated into the thin-film TE device for energy conversion.[101]

Moreover, some researchers reported a hybrid photovoltaic-thermal solar energy harvesting system, which consists of the photovoltaic device for light energy harvesting and TE device for thermal energy harvesting. This is an interesting design for solar energy harvesting due to its improved power conversion efficiency over its monolithic counterparts. Then, a hybrid solar energy generator with TEG was composed of the thin-film TE module, photovoltaic solar module, hot mirror, and near infrared focusing

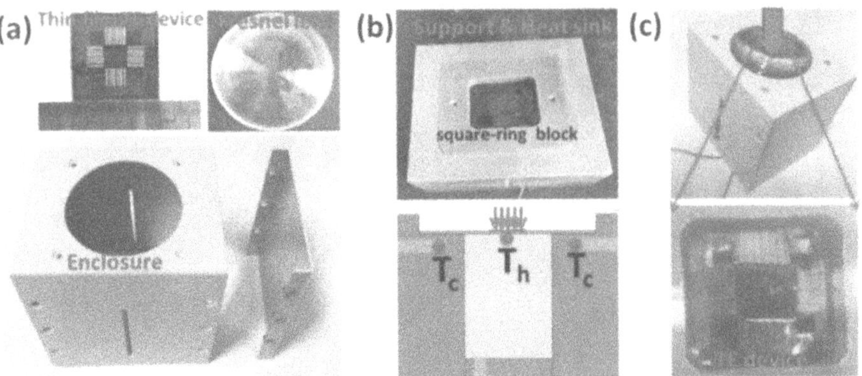

FIGURE 7.14 Integration of solar generator and sensor based on flexible TE device[101] (Reproduced from reference[101] with permission; Copyright Elsevier 2017.) (a) Setup of the placement of thin film device inside an enclosure. (b) The location of heat sink, with hot and cold thermocouple. (c) Overall location of thin film TE device within the enclosure.

cylindrical lens in system.[102] Compared to the open voltage of single photovoltaic solar cells, the energy conversion efficiency of this photovoltaic-thermal hybrid solar generator was increased by 1.3%, suggesting potential in increasing the energy conversion efficiency although this technology is still in its early stage.

7.4 CONCLUSIONS AND OUTLOOK

To sum up, this chapter briefly introduced the fundamentals of thin-film TE materials, thin-film TE modules including rigid, flexible stretchable, and transparent thin-film TE devices. The various applications of TE modules including on-chip cooling, wearable electronics well as photovoltaic-thermal solar energy harvesting system were summarized. For example, flexible thin-film TEG wraps on wrist, finger, ankle, textile, T-shirt, and fabric have been reported. Table 7.1 summarizes details of recently reported thin-film TE devices including active TE materials, device structure, substrate, process methods, power output or heat flux, and other device features. Apart from flexible thin-film TE-active materials, dielectrics as well as substrates must be stretchable and bendable to cater to wearable electronics uses in particular for e-skin, which requires key materials to meet the mechanical compliance with human body and skin. The energy conversion efficiency of existing thin-film TEGs is quite low and the manufacturing process is costly, prohibiting their wide applications. In addition to the low-energy conversion efficiency, long-term stability of TE devices and heat dissipation efficiency of heat sink need to be taken into consideration for further improvement. Thus, numerous efforts need to be devoted to the enhancement on power output of thin-film TEGs, device fabrication methods, long-term stability, etc. Moreover, it is of importance to develop more efficient TE-active materials and design various innovative TE systems to explore their applications in the areas of wearable electronics, mobile phones, 5G mobile networks, data center, etc., which provides additional exciting opportunities.

TABLE 7.1
A Summary of Reported Thin-Film TE Devices with Key Features

TE Function	Active TE Materials	Device Structure	Substrate	Process Method	Power Output or Heat Flux	Other Features	Ref
TEG	*n*-Bi_2Te_3 *p*-Sb_2Te_3	*p–n* thin-film in parallel	AIN ceramic	DC sputtering	19.13 μW (ΔT 85K)	Rigid	80
Cooling and TEG	*n*-Bi_2Te_3 *p*-Sb_2Te_3	*p–n* thin-film in series	Silicon die	CVD[b]	4.2 mW	Rigid	81
Mic TEG	*n*-CuNi *p*-Constantan	*p–n* thin- film in series	$SiO_2/Si_3N_4/SiO_2$	Evaporation	1.1 μ Wcm^{-2}	Rigid	82
Cooling	*n*- $Bi_2Te_3/Bi_2Te_{2.8}Se_{0.17}$ *p*- Bi_2Te_3/Sb_2Te_3	*p–p* thin- film in series	Cu	MOCVD [c]	42.7 Wcm^{-2}	Rigid	84
Micro-cooling	*n*-Bi_2Te_3 *p*-Sb_2Te_3	Free-standing planar	Si	CVD	42.7 Wcm^{-2}	Rigid	86
TEG	*n*-PbTe *p*-SnTe	*p–n* thin-film in series	Polyimide	thermal evaporation	8.4 m Wcm^{-2}	Wearable & flexible	85
TEG	*n*-AlZnO *p*- ZnSb	*p–n* thin-film in parallel	Polyimide	DC sputtering	246.3 μW (ΔT 180 K)	Flexible	87
TEG	*n*-Bi_2Te_3 *p*-Sb_2Te_3	*p–n* thin-film in parallel	Kapton polyimide	RF[d] sputtering	4.18 nW	Flexible	88
TEG	*p*- PEDOT:PSS	*p*-type thin-film in series	Substrate/ adhesive	Roll-to-roll processing	0.2 Wcm^{-2}	Flexible	89
TEG	*p*-poly(3,4-ethylenedioxythiophene)-polyurethane-ionic liquid composite	*p*-type thin-film in series	Free standing	Water-processable	25 nW (ΔT 30 K)	Flexible & stretchable	90

(Continued)

TABLE 7.1 *(Continued)*
A Summary of Reported Thin-Film TE Devices with Key Features

TE Function	Active TE Materials	Device Structure	Substrate	Process Method	Power Output or Heat Flux	Other Features	Ref
TEG	*n*-CNT *p*-Polyethyleneimine doping CNT	*p–n* thin-film in series	Piezoelectric	solution drop-casting	140 nW	Flexible & stretchable	91
TEG	n- TiS_2:Hexylamine-complex material *p*-PEDOT nanowires	*p–n* thin-film in series	PEN[a]	Printing	63.4 μW(ΔT 30 K)	Flexible & stretchable	92
TEG	*n*-Bi_2Te_3 *p*-Sb_2Te_3	*p–n* thin-film in series	Silicon elastomer	Assembled	0.15 m Wcm^{-2}	Flexible & stretchable	93
TEG	n-type Si p-type Si	*p–n* thin-film in series	elastomer	PECVD[e]	2 nW	Flexible & stretchable	94
TEG	*n*-ITO *p*- PEDOT	*p–n* thin-film in parallel	Glass & PET	RF sputtering & printing	22.4 Wcm^{-2}	Transparent & flexible	54

Notes: [a] PEN: poly(ethylene 2,6-naphthalate; CVD: [b] Chemical vapor deposition; [c] MOCVD: Metal organic chemical vapor deposition; [d] RF: radio frequency; [e] PECVD: plasma enhanced chemical vapor deposition.

ACKNOWLEDGMENT

We would like to acknowledge the A*STAR, SERC Thermoelectric Materials Program (grant numbers: 1527200019, 1527200024 & 1527200021) and Sustainable Hybrid Lighting System for Controlled Environment Agriculture Program (grant numbers: A19D9a0096).

REFERENCES

1. J.R. Sootsman, D.Y. Chung, M.G. Kanatzidis, New and old concepts in thermoelectric materials, *Angew. Chem. Int.* 2009, **48**, 8616–8639.
2. G.S. Nolas, D.T. Morelli, T.M. Tritt, Skutterudites: A phonon-glass-electron crystal approach to advanced thermoelectric energy conversion applications, *Annu. Rev. Mater. Sci.* 1999, **29**, 89–116.
3. D. M. Rowe, *Modules, systems, and applications in thermoelectrics*, CRC Press, 2012.
4. D. Kraemer, Q. Jie, K. McEnaney, F. Cao, W. Liu, L. A. Weinstein, J. Loomis, Z. Ren, G. Chen, Concentrating solar thermoelectric generators with a peak efficiency of 7.4%, *Nat. Energy*, 2016, **1**, 16153.
5. D. M. Rowe, *CRC handbook of thermoelectrics*, CRC Press, 1995.
6. H. J. Goldsmid, *Introduction to thermoelectricity*, Springer, 2010.
7. K. Behnia, *Fundamentals of thermoelectricity*, OUP Oxford, 2015.
8. S. D. Kang and G. J. Snyder, Charge-transport model for conducting polymers, *Nature Materials*, 2017, **16**, 252.
9. J. Yan, P. Gorai, B. Ortiz, S. Miller, S. A. Barnett, T. Mason, V. Stevanović and E. S. Toberer, Material descriptors for predicting thermoelectric performance, *Energy & Environmental Science*, 2015, **8**, 983–994.
10. Y. Takagiwa, Y. Pei, G. Pomrehn and G. J. Snyder, Dopants effect on the band structure of PbTe thermoelectric material, *Applied Physics Letters*, 2012, **101**, 092102.
11. Y. Pei, A. D. LaLonde, H. Wang and G. J. Snyder, Low effective mass leading to high thermoelectric performance, *Energy & Environmental Science*, 2012, **5**, 7963–7969.
12. E. S. Toberer, A. Zevalkink and G. J. Snyder, Phonon engineering through crystal chemistry, *Journal of Materials Chemistry*, 2011, **21**, 15843–15852.
13. G. J. Snyder and E. S. Toberer, Complex thermoelectric materials, *World Scientific*, 2011, pp. 101–110.
14. P. Pichanusakorn and P. Bandaru, Nanostructured thermoelectrics, *Materials Science and Engineering: R: Reports*, 2010, **67**, 19–63.
15. M. G. Kanatzidis, Nanostructured thermoelectrics: The new paradigm, *Chemistry of Materials*, 2009, **22**, 648–659.
16. K. N. Dinh, Y. Sun, Z. Pei, Z. Yuan, A. Suwardi, Q. Huang, X. Liao, Z. Wang, Y. Chen and Q. Yan, Electronic modulation of nickel disulfide toward efficient water electrolysis, *Small*, 2020, **16**, 1905885.
17. A. Suwardi, D. Bash, H. K. Ng, J. R. Gomez, D. M. Repaka, P. Kumar and K. Hippalgaonkar, Inertial effective mass as an effective descriptor for thermoelectrics via data-driven evaluation, *Journal of Materials Chemistry A*, 2019, **7**, 23762–23769.
18. A. Suwardi, J. Cao, L. Hu, F. Wei, J. Wu, Y. Zhao, S. H. Lim, L. Yang, X. Y. Tan and S. W. Chien, Tailoring the phase transition temperature to achieve high-performance cubic GeTe-based thermoelectrics, *Journal of Materials Chemistry A*, 2020, **8**, 18880–18890.
19. A. Suwardi, J. Cao, Y. Zhao, J. Wu, S. W. Chien, X. Y. Tan, L. Hu, X. Wang, W. Wang, D. Li, Y. Yin, W. -X. Zhou, D. V. M. Repaka, J. Chen, Y. Zheng, Q. Yan, G. Zhang, J. Xu, Achieving high thermoelectric quality factor toward high figure of merit in GeTe, *Materials Today Physics*, 2020, **14**, 100239.

20. A. Suwardi, L. Hu, X. Wang, X. Y. Tan, D. V. M. Repaka, L.-M. Wong, X. Ni, W. H. Liew, S. H. Lim and Q. Yan, Origin of high thermoelectric performance in earth-abundant phosphide–tetrahedrite, *ACS Applied Materials & Interfaces*, 2020, **12**, 9150–9157.
21. A. Suwardi, S. H. Lim, Y. Zheng, X. Wang, S. W. Chien, X. Y. Tan, Q. Zhu, L. M. N. Wong, J. Cao, W. Wang, Q. Yan, C. K. I. Tan and J. Xu, Effective enhancement of thermoelectric and mechanical properties of germanium telluride via rhenium-doping, *Journal of Materials Chemistry C*, 2020, **8**, 16940–16948.
22. A. Suwardi, B. Prasad, S. Lee, E.-M. Choi, P. Lu, W. Zhang, L. Li, M. Blamire, Q. Jia and H. Wang, Turning antiferromagnetic $Sm_{0.34}Sr_{0.66}MnO_3$ into a 140 K ferromagnet using a nanocomposite strain tuning approach, *Nanoscale*, 2016, **8**, 8083–8090.
23. L. P. Tan, T. Sun, S. Fan, L. Y. Ng, A. Suwardi, Q. Yan and H. H. Hng, Facile synthesis of Cu_7Te_4 nanorods and the enhanced thermoelectric properties of Cu_7Te_4–$Bi_{0.4}Sb_{1.6}Te_3$ nanocomposites, *Nano Energy*, 2013, **2**, 4–11.
24. F. Yang, J. Wu, A. Suwardi, Y. Zhao, B. Liang, J. Jiang, J. Xu, D. Chi, K. Hippalgaonkar and J. Lu, Gate-tunable polar optical phonon to piezoelectric scattering in few-layer Bi_2O_2Se for high-performance thermoelectrics, *Advanced Materials*, 2020, 33, 2004786.
25. Y. Zheng, H. Xie, Q. Zhang, A. Suwardi, X. Cheng, Y. Zhang, W. Shu, X. Wan, Z. Yang and Z. Liu, Unraveling the critical role of melt-spinning atmosphere in enhancing the thermoelectric performance of p-type $Bi_{0.52}Sb_{1.48}Te_3$ alloys, *ACS Applied Materials & Interfaces*, 2020, **12**, 36186–36195.
26. D. M. Rowe, *CRC handbook of thermoelectrics*, CRC Press, 2018.
27. L. Zhu, H. Tan and J. Yu., Analysis on optimal heat exchanger size of thermoelectric cooler for electronic cooling applications, *Energy Convers Manage*, 2013,**76**, 685–690.
28. M. Ma and J. Yu., An analysis on a two-stage cascade thermoelectric cooler for electronics cooling applications, *Int J Refrig* 2014, **38**, 352–357.
29. X. Lin, S. Mo, L. Jia, Z. Yang, Y. Chen and Z. Cheng, Experimental study and Taguchi analysis on LED cooling by thermoelectric cooler integrated with microchannel heat sink, *Applied Energy*, 2019, **242**, 232–238.
30. Z. Fan and J. Ouyang, PEDOT-based thermoelectrics, *Organic Thermoelectric Materials*, 2019, **24**, 117.
31. C. Wang, K. Sun, J. Fu, R. Chen, M. Li, Z. Zang, X. Liu, B. Li, H. Gong and J. Ouyang, Enhancement of conductivity and thermoelectric property of PEDOT:PSS via acid doping and single post-treatment for flexible power generator, *Advanced Sustainable Systems*, 2018, **2**, 1800085.
32. L. Hu, M. Li, K. Yang, Z. Xiong, B. Yang, M. Wang, X. Tang, Z. Zang, X. Liu and B. Li, PEDOT:PSS monolayers to enhance the hole extraction and stability of perovskite solar cells, *Journal of Materials Chemistry A*, 2018, **6**, 16583–16589.
33. E. Yildirim, G. Wu, X. Yong, T. L. Tan, Q. Zhu, J. Xu, J. Ouyang, J.-S. Wang and S.-W. Yang, A theoretical mechanistic study on electrical conductivity enhancement of DMSO treated PEDOT:PSS, *Journal of Materials Chemistry C*, 2018, **6**, 5122–5131.
34. S. Zhang, Z. Fan, X. Wang, Z. Zhang and J. Ouyang, Enhancement of the thermoelectric properties of PEDOT:PSS via one-step treatment with cosolvents or their solutions of organic salts, *Journal of Materials Chemistry A*, 2018, **6**, 7080–7087.
35. F. Wu, P. Li, K. Sun, Y. Zhou, W. Chen, J. Fu, M. Li, S. Lu, D. Wei and X. Tang, Conductivity enhancement of PEDOT: PSS via addition of chloroplatinic acid and its mechanism, *Advanced Electronic Materials*, 2017, **3**, 1700047.
36. Y. Xia, J. Fang, P. Li, B. Zhang, H. Yao, J. Chen, J. Ding and J. Ouyang, Solution-processed Highly superparamagnetic and conductive PEDOT:PSS/Fe_3O_4 nanocomposite films with high transparency and high mechanical flexibility, *ACS Applied Materials & Interfaces*, 2017, **9**, 19001–19010.

37. Z. Fan, P. Li, D. Du and J. Ouyang, Significantly enhanced thermoelectric properties of PEDOT: PSS films through sequential post-treatments with common acids and bases, *Advanced Energy Materials*, 2017, **7**, 1602116.
38. Z. Fan, D. Du, H. Yao and J. Ouyang, Higher PEDOT molecular weight giving rise to higher thermoelectric property of PEDOT: PSS: a comparative study of clevios P and clevios PH1000, *ACS Applied Materials & Interfaces*, 2017, **9**, 11732–11738.
39. Z. Fan, D. Du, Z. Yu, P. Li, Y. Xia and J. Ouyang, Significant enhancement in the thermoelectric properties of PEDOT: PSS films through a treatment with organic solutions of inorganic salts, *ACS Applied Materials & Interfaces*, 2016, **8**, 23204–23211.
40. C. X. Guo, K. Sun, J. Ouyang and X. Lu, Layered V_2O_5/PEDOT nanowires and ultrathin nanobelts fabricated with a silk reelinglike process, *Chemistry of Materials*, 2015, **27**, 5813–5819.
41. Y. Xia and J. Ouyang, Highly conductive PEDOT: PSS films prepared through a treatment with geminal diols or amphiphilic fluoro compounds, *Organic Electronics*, 2012, **13**, 1785–1792.
42. T. A. Yemata, Y. Zheng, A. K. K. Kyaw, X. Wang, J. Song, W. S. Chin and J. Xu, First-principles study of the adsorption behaviors of Li atoms and LiF on the CFx (x= 1.0, 0.9, 0.8, 0.5,~ 0.0) surface, *RSC Advances*, 2020, **10**, 1786–1792.
43. T. A. Yemata, A. K. K. Kyaw, Y. Zheng, X. Wang, Q. Zhu, W. S. Chin and J. Xu, Enhanced thermoelectric performance of poly(3,4-ethylenedioxythiophene):poly(4-styrenesulfonate) (PEDOT:PSS) with long-term humidity stability via sequential treatment with trifluoroacetic acid, *Polymer International*, 2019, **69**, 84–92.
44. E. Yildirim, Q. Zhu, G. Wu, T. L. Tan, J. Xu and S.-W. Yang, Self-organization of PEDOT: PSS induced by green and water-soluble organic molecules, *The Journal of Physical Chemistry C*, 2019, **123**, 9745–9755.
45. T. A. Yemata, Y. Zheng, A. K. K. Kyaw, X. Wang, J. Song, W. S. Chin and J. Xu, Continuous flow synthesis of high valuable N-heterocycles via catalytic conversion of levulinic acid, *Frontiers in Chemistry*, 2019, **7**, 870.
46. Q. Zhu, E. Yildirim, X. Wang, X. Y. D. S. Soo, Y. Zheng, T. L. Tan, G. Wu, J. Xu and S.-W. Yang, Improved alignment of PEDOT:PSS induced by in-situ crystallization of "green" dimethylsulfone molecules to enhance the polymer thermoelectric performance, *Frontiers in Chemistry*, 2019, **7**, 783.
47. A. K. K. Kyaw, T. A. Yemata, X. Wang, S. L. Lim, W. S. Chin, K. Hippalgaonkar and J. Xu, Enhanced thermoelectric performance of PEDOT: PSS films by sequential post-treatment with formamide, *Macromolecular Materials and Engineering*, 2018, **303**, 1700429.
48. X. Wang, A. K. K. Kyaw, C. Yin, F. Wang, Q. Zhu, T. Tang, P. I. Yee and J. Xu, Enhancement of thermoelectric performance of PEDOT:PSS films by post-treatment with a superacid, *RSC Advances*, 2018, **8**, 18334–18340.
49. H. Ling, G. Ding, D. Mandler, P. S. Lee, J. Xu and X. Lu, Facile preparation of aqueous suspensions of WO_3/sulfonated PEDOT hybrid nanoparticles for electrochromic applications, *Chemical Communications*, 2016, **52**, 9379–9382.
50. G. Ding, C. M. Cho, C. Chen, D. Zhou, X. Wang, A. Y. X. Tan, J. Xu and X. Lu, Black-to-transmissive electrochromism of azulene-based donor–acceptor copolymers complemented by poly(4-styrene sulfonic acid)-doped poly(3,4-ethylenedioxythiophene), *Organic Electronics*, 2013, **14**, 2748–2755.
51. J. Song, D. Jańczewski, Y. Ma, L. van Ingen, C. E. Sim, Q. Goh, J. Xu and G. J. Vancso, Electrochemically controlled release of molecular guests from redox responsive polymeric multilayers and devices, *European Polymer Journal*, 2013, **49**, 2477–2484.
52. T. Ye, X. Wang, X. Li, A. Q. Yan, S. Ramakrishna and J. Xu, Ultra-high Seebeck coefficient and low thermal conductivity of a centimeter-sized perovskite single crystal acquired by a modified fast growth method, *Journal of Materials Chemistry C*, 2017, **5**, 1255–1260.

53. J. Recatala-Gomez, A. Suwardi, I. Nandhakumar, A. Abutaha and K. Hippalgaonkar, Toward accelerated thermoelectric materials and process discovery, *ACS Applied Energy Materials*, 2020, **3**, 2240–2257.
54. X. Wang, A. Suwardi, S. L. Lim, F. Wei and J. Xu, Transparent flexible thin-film *p–n* junction thermoelectric module, *npj Flexible Electronics*, 2020, **4**, 1–9.
55. X. Wang, A. Suwardi, Y. Zheng, H. Zhou, S. W. Chien and J. Xu, Enhanced thermoelectric performance of nanocrystalline indium tin oxide pellets by modulating the density and nanoporosity via spark plasma sintering, *ACS Applied Nano Materials*, 2020, **3**, 10156–10165.
56. Q. Zhang, X. Ai, L. Wang, Y. Chang, W. Luo, W. Jiang and L. Chen, Improved thermoelectric performance of silver nanoparticles-dispersed Bi_2Te_3 composites deriving from hierarchical two-phased heterostructure, *Advanced Functional Materials*, 2015, **25**, 966–976.
57. R. Venkatasubramanian, E. Siivola, T. Colpitts and B. O'quinn, Thin-film thermoelectric devices with high room-temperature figures of merit, *Nature*, 2001, **413**, 597–602.
58. P. Fan, Y.-z. Li, Z.-h. Zheng, Q.-y. Lin, J.-t. Luo, G.-x. Liang, M.-q. Zhang and M.-c. Chen, Thermoelectric properties optimization of Al-doped ZnO thin films prepared by reactive sputtering Zn–Al alloy target, *Applied Surface Science*, 2013, **284**, 145–149.
59. X. Q. Chen, Z. Li and S. X. Dou, Ambient facile synthesis of gram-scale copper selenide nanostructures from commercial copper and selenium powder, *ACS Applied Materials & Interfaces*, 2015, **7**, 13295–13302.
60. R. Nunna, P. Qiu, M. Yin, H. Chen, R. Hanus, Q. Song, T. Zhang, M.-Y. Chou, M. T. Agne and J. He, Ultrahigh thermoelectric performance in Cu_2Se-based hybrid materials with highly dispersed molecular CNTs, *Energy & Environmental Science*, 2017, **10**, 1928–1935.
61. Z. Lin, C. Hollar, J. S. Kang, A. Yin, Y. Wang, H. Y. Shiu, Y. Huang, Y. Hu, Y. Zhang and X. Duan, A solution processable high-performance thermoelectric copper selenide thin film, *Advanced Materials*, 2017, **29**, 1606662.
62. C. Yang, D. Souchay, M. Kneiß, M. Bogner, H. Wei, M. Lorenz, O. Oeckler, G. Benstetter, Y. Q. Fu and M. Grundmann, Transparent flexible thermoelectric material based on non-toxic earth-abundant p-type copper iodide thin film, *Nature Communications*, 2017, **8**, 1–7.
63. T. Sugahara, Y. Ekubaru, N. V. Nong, N. Kagami, K. Ohata, L. T. Hung, M. Okajima, S. Nambu and K. Suganuma, Fabrication with semiconductor packaging technologies and characterization of a large-scale flexible thermoelectric module, *Advanced Materials Technologies*, 2019, **4**, 1800556.
64. Y. Du, J. Xu, B. Paul and P. Eklund, Flexible thermoelectric materials and devices, *Applied Materials Today*, 2018, **12**, 366–388.
65. J. Sun, M.-L. Yeh, B. J. Jung, B. Zhang, J. Feser, A. Majumdar and H. E. Katz, Simultaneous increase in Seebeck coefficient and conductivity in a doped poly(alkylthiophene) blend with defined density of states, *Macromolecules*, 2010, **43**, 2897–2903.
66. R. Zuzok, A. Kaiser, W. Pukacki and S. Roth, Thermoelectric power and conductivity of iodine-doped "new"polyacetylene, *The Journal of Chemical Physics*, 1991, **95**, 1270–1275.
67. Y. Park, C. Yoon, B. Na, H. Shirakawa and K. Akagi, Metallic properties of transition metal halides doped polyacetylene: The soliton liquid state, *Synthetic Metals*, 1991, **41**, 27–32.
68. A. G. MacDiarmid and A. J. Epstein, Secondary doping in polyaniline, *Synthetic Metals*, 1995, 69, 85–92.
69. Y. Min, Y. Xia, A. G. MacDiarmid and A. J. Epstein, Vapor phase "secondary doping" of polyaniline, *Synthetic. Metal*, 1995, **69**, 159–160.

70. Q. Yao, Q. Wang, L. Wang, Y. Wang, J. Sun, H. Zeng, Z. Jin, X. Huang and L. Chen, The synergic regulation of conductivity and Seebeck coefficient in pure polyaniline by chemically changing the ordered degree of molecular chains, *Journal Materials Chemistry A*, 2014, **2**, 2634–2640.
71. B. Zhang, J. Sun, H. E. Katz, F. Fang, R. L. Opila, Promising thermoelectric properties of commercial PEDOT:PSS materials and their Bi_2Te_3 powder composites, *ACS Applied Materials Interfaces*, 2010, **2**, 3170–3178.
72. C. Yu, K. Choi, L. Yin and J. C. Grunlan, Light-weight flexible carbon nanotube based organic composites with large thermoelectric power factors, *ACS Nano*, 2011, **5**, 7885–7892.
73. P. Zong, R. Hanus, M. Dylla, Y. Tang, J. Liao, Q. Zhang, G. J. Snyder and L. Chen, Realizing a thermoelectric conversion efficiency of 12% in bismuth telluride/skutterudite segmented modules through full-parameter optimization and energy-loss minimized integration, *Energy Environmental Science*, 2017, **10**, 183–191.
74. Y. Sun, L. Qiu, L. Tang, H. Geng, H. Wang, F. Zhang, D. Huang, W. Xu, P. Yue and Y. S. Guan, Flexible n-type high-performance thermoelectric thin films of poly (nickel-ethylenetetrathiolate) prepared by an electrochemical method, *Advanced Materials*, 2016, **28**, 3351–3358.
75. H. Wang, W. D. Porter, H. Böttner, J. König, L. Chen, S. Bai, T. M. Tritt, A. Mayolet, J. Senawiratne and C. Smith, Transport properties of bulk thermoelectrics: an international round-robin study, part II: thermal diffusivity, specific heat, and thermal conductivity, *Journal of Electronic Materials*, 2013, **42**, 1073–1084.
76. P. J. Taylor, J. R. Maddux and P. N. Uppal, Measurement of thermal conductivity using steady-state isothermal conditions and validation by comparison with thermoelectric device performance, *Journal of Electronic Materials*, 2012, **41**, 2307–2312.
77. G.-H. Kim, L. Shao, K. Zhang and K. P. Pipe, Engineered doping of organic semiconductors for enhanced thermoelectric efficiency, *Nature Materials*, 2013, **12**, 719–723.
78. L. Lu, W. Yi and D. Zhang, 3ω method for specific heat and thermal conductivity measurements, *Review of Scientific Instruments*, 2001, **72**, 2996–3003.
79. M. V. Vedernikov and E. K. Iordanishvili, A. F. Ioffe and origin of modern semiconductor thermoelectric energy conversion, *A. F. 17th Int. Conf. on Thermoelectrics* 1998, **1**, 37–42. https://www.tib.eu/en/search/id/BLCP%3ACN027865466/A-F-Ioffe-and-Origin-of-Modern-Semiconductor-Thermoelectric/
80. P. Fan, Z. Zheng, Z. -K. Cai, T. Chen, P. Liu, X. Cai, D. Zhang, G. Liang and J. Luo, The high performance of a thin film thermoelectric generator with heat flow running parallel to film surface, *Applied Physics Letters*, 2013, **102**, 033904.
81. N. Jaziri, A. Boughamoura, J. Müller, B. Mezghani, F. Tounsi and M. Ismail, A comprehensive review of thermoelectric generators: Technologies and common applications, *Energy Reports*, 2020, **6**, 264–287.
82. M. P. Markowski Multilayer thick-film thermoelectric microgenerator based on LTCC technology, *Multilayer Microelectronics International*, 2016, **33**, 155–161.
83. S. Pelegrini, A. Adami, C. Collini, P. Conci, C. I. L. de Araújo, V. Guarnieri, S. Güths, A. A. Pasa and L. Lorenzelli, Development and characterization of a micro-thermoelectric generator with plated copper/constantan thermocouples, *Microsystem Technologies*, 2014, **20**, 585–592.
84. K. Tappura, A numerical study on the design trade-offs of a thin-film thermoelectric generator for large-area applications, *Renewable Energy*, 2018, **120**, 78–87.
85. Y. Su, J. Lu and B. Huang, Free-standing planar thin-film thermoelectric microrefrigerators and the effects of thermal and electrical contact resistances, *International Journal of Heat and Mass Transfer*, 2018, **117**, 436–446.

86. I. Chowdhury, R. Prasher, K. Lofgreen, G. Chrysler, S. Narasimhan, R. Mahajan, D. Koester, R. Alley and R. Venkatasubramanian, On-chip cooling by superlattice-based thin-film thermoelectrics, *Nature Nanotechnology*, 2009, **4**, 235–238.
87. V. Karthikeyan, J. U. Surjadi, J. C. K. Wong, V. Kannan, K.-H. Lam, X. Chen, Y. Lu and V. A. L. Roy, Wearable and flexible thin film thermoelectric module for multi-scale energy harvesting, *Journal of Power Sources*, 2020, **455**, 227983.
88. P. Fan, Z. Zheng, Y. Li, Q. Lin, J. Luo, G. Liang, X. Cai, D. Zhang and F. Ye, Low-cost flexible thin film thermoelectric generator on zinc based thermoelectric materials, *Applied Physics Letters* 2015, **106**, 073901.
89. L. Francioso, C. de Pascali, I. Farella, C. Martucci, P. Cretì, P. Siciliano and A. Perrone, Flexible thermoelectric generator for ambient assisted living wearable biometric sensors, *Journal of Power Sources*, 2011, **196**, 3239–3243.
90. R. R. Søndergaard, M. Hösel, N. Espinosa, M. Jørgensen and F. C. Krebs, Practical evaluation of organic polymer thermoelectrics by large-area R2R processing on flexible substrates, *Energy Science & Engineering*, 2013, **1**, 81–88.
91. N. Kim, S. Lienemann, I. Petsagkourakis, D. A. Mengistie, S. Kee, T. Ederth, V. Gueskine, P. Leclère, R. Lazzaroni, X. Crispin and K. Tybrandt, Elastic conducting polymer composites in thermoelectric modules, *Nature Communications*, 2020, **11**, 1424.
92. D. S. Montgomery, C. A. Hewitt and D. L. Carroll, Hybrid thermoelectric piezoelectric generator, *Applied Physics Letters*, 2016, **108**, 263901.
93. A. G. Rösch, A. Gall, S. Aslan, M. Hecht, L. Franke, M. M. Mallick, L. Penth, D. Bahro, D. Friderich and U. Lemmer, Direct roll transfer printed silicon nanoribbon arrays based high-performance flexible electronics, *npj Flexible Electronics*, 2021, **5**, 1.
94. Y. Yang, H. Hu, Z. Chen, Z. Wang, L. Jiang, G. Lu, X. Li, R. Chen, J. Jin, H. Kang, H. Chen, S. Lin, S. Xiao, H. Zhao, R. Xiong, J. Shi, Q. Zhou, S. Xu and Y. Chen, Stretchable nanolayered thermoelectric energy harvester on complex and dynamic surfaces, *Nano Letters*, 2020, **20**, 4445–4453.
95. K. Nan, S. D. Kang, K. Li, K. J. Yu, F. Zhu, J. Wang, A. C. Dunn, C. Zhou, Z. Xie, M. T. Agne, H. Wang, H. Luan, Y. Zhang, Y. Huang, G. J. Snyder and J. A. Rogers, Compliant and stretchable thermoelectric coils for energy harvesting in miniature flexible devices, *Science Advances*, 2018, **4**, eaau5849.
96. J. C. logoa, B. M. Morais Faustino, A. Marquesa, C. Bianchia, T. Koskinen, T. Juntunen, I. Tittonen and I. Ferreiraa, Compliant and stretchable thermoelectric coils for energy harvesting in miniature flexible devices, *RSC Adv.*, 2019, **9**, 35384–35391.
97. S. H. Choday, M. S. Lundstrom and K. Roy, Prospects of thin-film thermoelectric devices for hot-spot cooling and on-chip energy harvesting, *IEEE Transactions on Components, Packaging and Manufacturing Technology*, 2013, **3**, 2059–2067.
98. S. J. Kim, J. H. We and B. J. Cho, A wearable thermoelectric generator fabricated on a glass fabric, *Energy & Environmental Science*, 2014, **7**, 1959–1965.
99. M. Hyland, H. Hunter, J. Liu, E. Veety and D. Vashaee, Wearable thermoelectric generators for human body heat harvesting, *Applied Energy*, 2016, **182**, 518–524.
100. H. Xu, Y. Guo, B. Wu, C. Hou, Q. Zhang, Y. Li and H. Wang, Highly integrable thermoelectric fiber, *ACS Applied Materials & Interfaces*, 2020, **12**, 33297–33304.
101. W. Zhu, Y. Deng and L. Cao, Light-concentrated solar generator and sensor based on flexible thin-film thermoelectric device, *Nano Energy*, 2017, **34**, 463–471.
102. M. Mizoshiri, M. Mikami and K. Ozaki, Thermal–photovoltaic hybrid solar generator using thin-film thermoelectric modules, *Japanese Journal of Applied Physics*, 2012, **51**, 06FL07.

8 Smart Piezoelectric Films for Sensors and Actuators Applications

Hua-Feng Pang[a,b] *and Yong-Qing Fu*[b]
[a]Xi'an University of Science and Technology, Xi'an, People's Republic of China
[b]Northumbria University, Newcastle upon Tyne, UK

CONTENTS

8.1 INTRODUCTION

Piezoelectricity is an intrinsic electromechanical transduction property of a material. Self-reorientation of dipole moments of the material caused by a mechanical force such as compression, twisting, or expansion will result in the variations in the electric polarization of the molecules and the production of an electrical potential; therefore, the applied strain is then converted into electrical charges on the surface of the material. Conversely, an electrical potential applied across the piezoelectric material can induce a change in the dipole moments and result in a significant mechanical

DOI: 10.1201/9781003141358-8

deformation of the material's structure. After the discovery of the piezoelectric effect in quartz and Rochelle salts by Jacques and Pierre Cuire, the exploration of the piezoelectricity in various materials without center of symmetry has seen a rapid and explosive growth due to their great potential applications in sensors, actuators, sonar, and ultrasonic devices[1–3]. Later on, bulk materials including quartz (SiO_2), lithium niobate ($LiNbO_3$), potassium niobate ($KNbO_3$), lithium tantalate ($LiTaO_3$), and langasite ($La_3Ga_5SiO_{14}$) dominate the fields of applications for modern telecommunication serving as piezoelectric single crystals in the frequency-stabilized oscillators and surface acoustic wave (SAW) devices in television filters and analog signal correlators[4]. Moreover, significant advances on improving the electromechanical coefficients and material properties of the piezoelectric crystals have been achieved through reducing the defects and precisely controlling the stoichiometric ratio during their growth processes[5,6]. Intensive and fundamental studies of the primary piezoelectric materials with wurtzite and perovskite crystal structures such as zinc oxide (ZnO), aluminum nitride (AlN), and lead zirconate titanate (PZT) have significantly expanded the research areas toward acoustofluidics, quantum acoustics, and energy-harvesting applications[7–9].

Compared with piezoelectric crystals and synthetic piezoelectric materials which are normally rigid, expensive, and fragile, piezoelectric thin films have the advantages of low-cost, easy integration with microelectronics and other technologies, microminiaturization, and applicability to various substrates. Moreover, piezoelectric thin films provide integration of multiple functions onto one substrate such as silicon, glass, metal, or polymer. Piezoelectric films can also be applied onto the flexible substrate resulting in the rapid development of wearable sensing devices, energy-harvesting devices, and tactile transducers[10–12]. Furthermore, high-quality piezoelectric films can be deposited using advanced thin-film growth techniques, providing film-based applications with low-cost and mass production. Therefore, piezoelectric films have become key materials of the new-generation functional devices.

Sensors and actuators have played important roles in modern society. Sensors are the devices or instruments that change the physical/chemical/electrical/mechanical measurands into visible or readable measurement values. The input and output quantities obtained from the two ports in various forms (e.g., electrical or optical signals) normally exist in a definitely functional relationship due to the different sensing mechanisms of the sensing materials. The sensors are normally utilized for monitoring devices, widely applied in industrial environment and daily living, including the natural sensors in living organisms (e.g., fingers, nose, and eye) and the artificial sensors (e.g., gas sensor, physical sensor, and biochemical sensor)[13,14]. A remarkable progress of sensor technologies has been made due to the rapid developments on the techniques of processing and manufacturing at the micro or nano scale[13]. Therefore, the integrated and multifunctional sensing system could gain a significantly higher selectivity and sensitivity different from those of the macroscopic counterparts based on the smart designs of the microsensors and nanosensors. For instance, an acoustic signal with a frequency at a scale of GHz can be excited from the film bulk acoustic resonator (FBAR), which presents an excellent sensitivity to the variation of mass load[15]. Furthermore, advances on micro/nanosensors and wireless sensor networks show that the novel structures, latest manufacturing methods, and optimized algorithms of the signal processing could be introduced to enhance the performance of the sensing system[16]. Up to date, the integrated sensors have become indispensable parts that exist widely in the equipment of medical diagnosis, chemical testing and biological recognition, smart wearing, intelligent manufacturing, and environment monitoring system.

As mentioned above, piezoelectric actuators can change electrical quantities into mechanical quantities. The actuator system often has an electrical input and a mechanical output, normally exhibiting an excellent performance such as a simple structure, a quick response, compactness, a low power consumption, low noise, and strong vibration, which is not simply influenced by the electromagnetic interference in the environments with a strong magnetic field[17]. The piezoelectric actuators have both a resonant and nonresonant type according to the employed vibration states. The resonant-type piezoelectric actuators generally involve in the ultrasonic motor, ultrasonic pump, or acoustic wave devices operated at their resonant frequencies, whereas the non-resonant-type actuators generate the mechanical motions at the non-resonant frequency by the piezoelectric deformation and static friction force[17].

This chapter is mainly focused on the fundamentals of piezoelectricity thin films, acoustic wave-based sensor, and actuator applications. Some other applications of the piezoelectric films, such as acoustofluidics, quantum acoustics, and nanogenerator, are not discussed here, and these can be obtained from the other references[18–20].

8.2 SMART PIEZOELECTRIC FILMS

8.2.1 Fundamentals of Piezoelectricity Thin Films

Good understanding of the fundamental and critical characteristics of piezoelectric films can provide a breakthrough point to design and develop various acoustic wave-based sensors and actuators with high performance. Piezoelectricity of thin films is normally dependent on the film quality including microstructure, crystallinity, orientation, thickness, and surface roughness. Therefore, the film-quality control of piezoelectric films is a key factor to push the applications toward high performance and smart integration of novel configurations. Furthermore, among thin-film piezoelectric materials, wurtzite ZnO and AlN thin films have attracted much attention and have often been chosen for the fabrication of acoustic wave-based sensors and actuators[21].

8.2.1.1 ZnO Films

ZnO as a semiconductor possesses a wide-bandgap (3.437 eV at 2 K) and a large exciton binding energy (60 meV). ZnO thin film with thicknesses ranging from a few nanometers (monolayer) to tens of micrometers can be grown on different substrates. Lots of studies on ZnO thin films are mainly devoted to their properties, growth mechanism, preparation technologies, and device applications. Crystalline and mechanical properties, including the texture, orientation, microstructure, morphology, stress, adhesion, substrate, and defects, are critical for piezoelectric functions and sensing performance. However, the growth conditions of the films have significant influence on their growth dynamics. To acquire high-quality ZnO thin films, the growth parameters need to be optimized, and this optimization work needs to be performed before the application of ZnO thin films to fabricate high-performance ZnO-based devices[2,22]. The developments of the deposition techniques using techniques such as physical vapor deposition (PVD), chemical vapor deposition (CVD), and wet chemical method have realized a good control of the crystallinity, piezoelectric, and electrical properties of ZnO thin films. Moreover, large-scale production (in size of a few inches) of ZnO thin films with high quality generally requires a reproducible, stable, and robust process by optimizing the growth conditions. The texture, substrates, interlayer, and stress control are thought as the key factors for growing ZnO thin films with high quality using different deposition techniques.

In a highly textured ZnO thin film as shown in Figure 8.1(a), the preferred orientation is commonly along the [0002] direction of the hexagonal wurtzite structure, and the cross-sectional microstructure is generally columnar, rod-like, or nanofiber-like, whereas nontextured polycrystalline ZnO thin films consist of particle-like fine grains in microstructure. Furthermore, the columnar structures can be grown at an inclined angle toward the sputtering source using the oblique angle deposition or glancing angle deposition via tilting the substrates at a large angle in the direction of the incident flux[23,24]. The inclined columnar structures can be further modified into various shapes of nanopillars, zigzag, or nanoscale helical structures, which allow dual acoustic wave mode or multiple acoustic modes to be generated in the acoustic devices[25–27]. Hence, the film textures and microstructures have important influences on the hardness, elastics, strain and piezoelectricity. Furthermore, good control of the film texture is vital to achieve the high performance of the ZnO thin film-based devices.

Substrates and buffer layer (or inter-layer) are also one key factor for obtaining the ZnO thin films with high quality. Various substrates such as silicon, glass, fused silica, sapphire, diamond, metal foils, MgO, and flexible polymer have been used for the deposition of ZnO thin films[28,29]. Various microstructures, morphologies, and strains in the ZnO thin films originate from the lattice and thermal mismatches

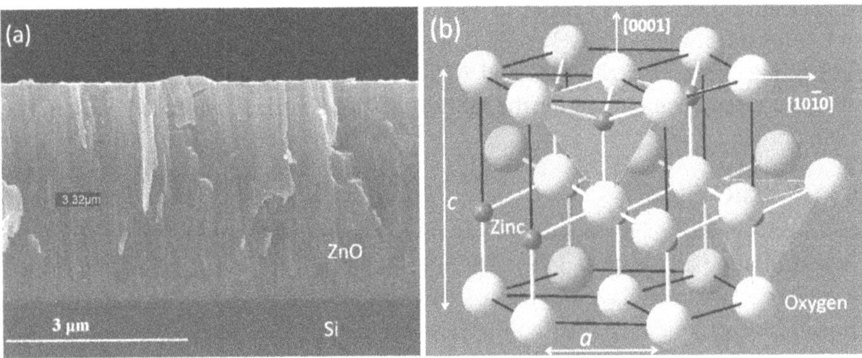

FIGURE 8.1 (a) The typical columar microstructure of the ZnO film with a preferred orientation, (b) wurtzite ZnO crystal structure with tetrahedral coordination [2]. (Redrawn with permission from figure 8.2 in [Ref. 2].)

between the substrate and the deposited film. The *c*-axis-oriented ZnO thin films generally grow much faster on special crystallographic surfaces than those on amorphous substrates due to the lowest surface energy density (0.99 eV·Å^{-2}) of the crystal plane [0001][29]. Furthermore, The oriented growth of the ZnO films could be directed by a buffer layer of ZnO that plays a role as a seed layer[30,31]. Inter-layers of silicon carbide (SiC), GaN, nanocrystalline diamond (NCD), or diamond-like carbon (DLC) on Si substrates have been used to improve the film texture and crystallinity[32–34].

Low film stress and good adhesion of the ZnO thin films to substrate are very beneficial to successfully fabricate the ZnO film-based devices in the MEMS process. Stresses in the ZnO thin films usually originate from the lattice mismatch and the difference of the thermal expansion coefficients between the ZnO films and substrates, which is sensitively changed and dependent on different growth conditions using various growth techniques. Quantitative analysis of the film stress can be performed using Stoney formula[35]. Take magnetron sputtering, for instance, the compressive stresses often arise through bombardment of the growing film with energetic ions and atoms controlled by the radio-frequency (RF) power, chamber pressure, and deposition rate. Extremely large stress in the ZnO thin films usually generates a poor film adhesion to substrate, resulting in early adhesion failure. This is one of the major obstacles in fabricating ZnO-based devices with a high performance. Therefore, reducing the stress in ZnO thin film is very helpful for improving the film adhesion and realizing the ZnO film-based devices with high performance. Recently, some strategies were adopted for this purpose, such as introducing a buffer layer or interlayer, using free-standing substrate or self-standing without substrates, annealing treatment, as well as the optimization of the growth conditions[36,37]. A new deposition technique, high-target utilization sputtering (HiTUS), has also been developed to significantly reduce the film stress[30]. Helicon-wave-excited-plasma sputtering was also reported to prepare ZnO thin films that could form a smooth surface morphology with 0.26-nm monolayer atomic steps[38]. In addition, good stoichiometric ratio, low defects and dislocations, and surface roughness of the ZnO thin films also contribute to the good crystalline properties. However, it is complex to improve the piezoelectric, optical, and electrical properties of ZnO thin films.

As an important characteristic of hexagonal wurtzite ZnO, piezoelectricity originates from the polarity that is composed of tetrahedral coordination as shown in Figure 8.1(b). The direction of the polarity along the *c*-axis from cation to anion results in the primary polar plane (0001) with the lowest energy. The hexagonal ZnO has two polar surfaces in its wurtzite structure including one surface that is terminated with zinc ions (Zn^+) in the (0001) plane and the other that is terminated with oxygen ions (O^-) in the ($000\bar{1}$) plane. When the external mechanical stress is applied and lattice distortion of the wurtzite ZnO is induced, the centers of Zn cation and O anion are displaced in the non-centrosymmetric structure and local dipole moments are formed. Accordingly, piezoelectricity along the [0001]-direction appears due to the macroscopic polarization in the ZnO crystal. The spontaneous polarization of bulk ZnO is estimated to be −0.057 C·m^{-2}, whereas it can be modified as 0.6 μC·m^{-2} for ZnO thin films[39]. A large electromechanical coupling of k^2 ranging from 1% to 5.2% can be obtained due to the large piezoelectric tensor of the tetrahedrally bonded ZnO in the II–VI compounds with a wurtzite structure[22,40].

In view of practical applications, in order to achieve a good piezoelectricity, the ZnO thin film is required to possess a strong texture, low defects, an accurate stoichiometric ratio of Zn atoms to O atoms, a smooth surface with a low roughness, and an appropriate thickness. Various technologies and methods have been developed to obtain high-quality piezoelectric ZnO materials[41–44]. In order to further increase the piezoelectric constants, different transition-metal atoms (e.g., Fe, V, Cr) were reported to be doped into ZnO[41,42]. The piezoelectric properties can also be tailored by controlling the Mg composition in Mg-doped ZnO[43,44]. Furthermore, reducing the size of the ZnO materials to the nanoscale could enhance the piezoelectricity. Recent advance on the theoretical computations using the first-principles method has shown that the effective piezoelectric constant of ZnO nanowire is much larger than that of bulk ZnO material due to their abundant free grain boundaries[45]. Giant size effects of piezoelectricity in the ZnO nanowire were also reported, and the piezoelectric coefficient of 50.4 $C{\cdot}m^{-2}$ can be obtained when the diameter of the nanowires was reduced to 0.6 nm calculated using the density functional theory[46].

8.2.1.2 AlN Films

AlN belongs to group III-V-nitride semiconductors, which possesses excellent properties such as a wide band-gap (6.2 eV), high piezoelectricity, high surface acoustic wave velocity (11050 m/s), high thermal stability (melting point of 2100 ºC), high thermal conductivity, and low electron affinity[47,48]. The piezoelectric coefficient (d_{33}) of AlN is reported to be 6.4 $pC{\cdot}N^{-1}$, and the piezoelectric coefficient (d_{33}) of ZnO is reported to be 12.3 $pC{\cdot}N^{-1}$ as listed in Table 8.1[2,49]. Although the piezoelectric coupling coefficient of AlN is slightly lower than that of ZnO, the phase velocity of acoustic waves in AlN is much larger than that of ZnO. It suggests that the AlN-based acoustic devices could be designed for high-frequency sensing applications[50]. The good chemical inertness of AlN allows the AlN-based acoustic wave devices to be operated with good sensitivity and performance in harsh environments[51].

TABLE 8.1
Comparison of Various Piezoelectric Materials[2,49]

Material	ZnO	AlN	GaN	PZT	128º cut $LiNbO_3$	36º YX-cut $LiTaO_3$	ST-cut Quartz
Piezoelectric constant d_{33} (pC/N)	12.3	4.5, 6.4	4.5	289–380, 117	12	12	2.3 (d_{11})
Effective coupling coefficient, k^2 (%)	1.5–1.7	3.1–8	20–35	8.8–16	2–11.3	0.66–0.77	2.9
Acoustic velocity by transverse (m/s)	6336 (2650)	11050 (6090)	4500 (2200)	5960 (3310)	3970	3230–3295	2600
Dielectric constant	8.66	8.5–10	380	4.3	85 (29)	54 (43)	6–8

Hexagonal AlN single crystal has a distorted tetrahedron structure where each Al atom is surrounded by four N atoms. Three of the four Al-N bonds are equivalent, and the remaining one is unique that has the *c*-axis direction or the (0002) orientation. Therefore, the surface energy of the (0002) face is lower than those of the other faces. The high piezoelectric constant exists in the *c*-axis direction; and AlN films with *c*-axis orientation have shown much better piezoelectricity. Furthermore, AlN thin films generally have been grown with a columnar microstructure in the (0002) crystallographic direction perpendicular to the substrate surface as shown in Figure 8.2(a). Various substrates such as Si, SiO_2, SiC, sapphire, diamond, and polymers have been used to deposit AlN films[52,53]. The highly *c*-axis-oriented AlN films are very suitable for the fabrication of thin-film acoustic wave devices. For instance, layered structures of piezoelectric AlN thin films deposited on high-velocity substrates (i.e. sapphire, diamond) have been reported, providing a high velocity of 6.700 m/s[54,55]. Figure 8.2(b) shows the typical microstructure of the AlN film on the ultra-nanocrystalline diamond interlayer. Furthermore, scandium doping of AlN thin films on SiC or Si substrates was used to enhance the electromechanical coupling coefficient of the SAW devices up to 3.8% to 4.5%[56,57]. Recently, considering the change of the ionic radius of the dopant, single element or dual elements such as Y, Cr, Ta, Er, Mg-Zr, and Mg-Hf have been doped into AlN films to significantly improve the values of d_{33}[2].

Deposition of the AlN films could be realized using various techniques such as radio-frequency (RF)[58] and direct current (DC) sputtering[59], pulsed laser deposition (PLD)[60], metal organic chemical vapor deposition (MOCVD)[61], and molecular beam epitaxy (MBE)[62]. Optimization of the deposition parameters makes it possible to modify the piezoelectricity of AlN films and enhance the performance of subsequent acoustic devices. All the above methods can be used to grow the *c*-axis-preferred films; however, the growth mechanism is significantly different among these techniques. Taking the polarity of AlN films on sapphire substrates for instance, the N-face is preferentially grown using MBE, while the Al-face is normally observed in film prepared using MOCVD[63].

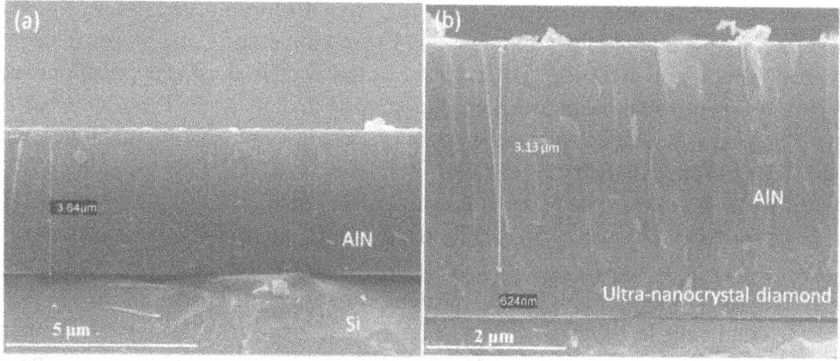

FIGURE 8.2 The highly *c*-axis-oriented AlN films deposited on (a) Si, and (b) ultra-nanocrystal diamond.

Among the different techniques, magnetron sputtering operated using DC or RF sources or reactive metal target is a typical and prevalent choice to deposit thin films and coatings. This technique can be used to deposit the thin films of most piezoelectric materials, such as ZnO, AlN, and PZT. It exhibits the advantages of structural simplicity, good reproducibility, low cost, better adhesion, low operating temperature, and compatibility with microelectronics and MEMS processing. In the basic sputtering process, the energetic particles are generated in a glow discharge plasma that bombard on or beneath the target surface. The target atoms or ions are produced in the bombardment process and transported toward the substrate surface. Subsequently, the sputtered particles or atoms with large energy and momentum collide with the substrate surface and are adsorbed and diffused to form the islands of the nanocrystals, and thin films are developed through further condensation and recrystallization. The modified Thornton model presents the correlations among the deposition parameters (e.g., chamber pressure, substrate temperature, sputter power, bias voltage, and deposition rate), microstructure, and the film growth[64]. The generation of different defects and their influences on the properties of the piezoelectric film such as crystal orientation, texture, grain size, and intrinsic stress have been studied[65,66]. In a HITUS system, a side chamber has been developed to generate the high-density plasma that can be launched into the main chamber and further steered onto the target using an electromagnet [67,68]. Thus, The surface of the growing film is not influenced by the plasma bombardment, which is very helpful for the reduction of the ion-induced damage and good control of the film roughness and stress.

8.2.1.3 PZT Films

As for the perovskite piezoelectric materials, the piezoelectric material of barium titanate ($BaTiO_3$) ceramic has been synthesized for the first time in the early 1940s that exhibits good ferroelectric and piezoelectric properties[69]. Moreover, it possesses a relatively high dielectric constant and a high electromechanical coupling coefficient with good thermal and mechanical stability[70]. However, the ceramic material is inert and their dipoles are normally oriented randomly; therefore, a poling process that applies a strong electrical field at a temperature slightly lower than its Curie point is often used to realize the permanent polarization of the ceramic material. PZT ceramic with its piezoelectric capability was discovered in 1954[71]. A higher electromechanical coupling coefficient and Curie temperature and higher dielectric constant were found in PZT, which has been widely used in sensing applications such as pressure sensors and cantilevers for bio-sensing and energy harvesting[72,73]. PZT thin films have attracted remarkable attention because they can be integrated with MEMS and complementary metal-oxide-semiconductor (CMOS) technology. PZT thin film can be easily grown on silicon and other substrates using a simple sol-gel method, which allows it to be utilized in the actuator and transducer applications [74].

PZT belongs to ABO_3 family of perovskite mineral, where A and B are metal ions. The total charge of the A and B metal ions with different radii is +6. A set of BO_6 octahedra is arranged in the simple cubic pattern, where the A atoms is located in spaces between the octahedra. $PbZr_{1-x}Ti_xO_3$ is a solid solution of $PbTiO_3$ and $PbZrO_3$ compounds, whose properties depend on the fraction x of $PbTiO_3$ and temperature according to a $PbTiO_3$-$PbZrO_3$ phase diagram[75]. A "morphotropic phase boundary" (MPB) could be observed in the phase diagram, which divides the ferroelectric region into a tetragonal Ti-rich region with a space group symmetry P4 mm and a rhombohedral Zr-rich region with symmetries R3m at high temperature and R3c at low-temperature phases, respectively. The superior ferroelectric and piezoelectric properties in this material have been found at $x \approx 0.47$ near the MPB, and the maximum value of the d_{33} existed at the MPB[76]. A revised phase diagram indicates that a narrow region of monoclinic phases with Cm and Cc symmetries at compositions corresponds to the MPB, where the monoclinic phase is responsible for the high piezoelectric response of PZT [76]. However, the existence of the monoclinic phase and the phase composition at MPB is still debatable[77]. Recent studies showed that the MPB appeared approximately at $PbZr_{0.52}Ti_{0.48}O_3$, where both the dielectric permittivity and piezoelectric coefficient exist at their maximum values[78].

Although PZT thin films exhibit superior piezoelectric properties, they are considered environmentally hazardous and toxic materials[79], unsuitable for many biological or biomedical applications. Lead-free piezoelectric materials including potassium niobate ($KNbO_3$) and bismuth titanate (e.g., $Bi_4Ti_3O_{12}$) families were used as alternative materials in the biomedical and environmental applications[80].

PZT films can be deposited using reactive sputtering, RF sputtering, and sol-gel techniques. In the sputtering process, the metal target with mixed materials of Pb, Zr, and Ti is used to reactively deposit the films and the ceramic target of $PbZr_{1-x}Ti_xO_3$ is utilized for RF sputtering. The key problems of the sputter deposition of PZT films are how to accurately control the Pb and O content within the films and how to maintain an appropriate substrate temperature during deposition[81]. In the deposition process, excess PbO in the ceramic target is required for compensating the lead loss due to the high volatility of PbO above 500°C, and the stoichiometry of the reactively sputtered film is normally modified by changing the chamber pressure[82]. Different substrate temperatures result in various film textures and the variation of active-site number of the perovskite phase nucleation[83].

8.2.2 Different Acoustic Modes of the Thin Films

The constitutive equations for piezoelectricity show the relations among the electrical displacement D_m, stress vector T_{ij}, electric field vector E_n, and strain vector S_{kl} in a strain–charge formula of a piezoelectric material, which have the following relationship:

$$D_m = \varepsilon_{mn}^S E_n + e_{mkl} S_{kl} \tag{8.1}$$

$$T_{ij} = -e_{ijn} E_n + c_{ijkl}^E S_{kl} \tag{8.2}$$

where ε_{mn}^S, e_{mkl}, and c_{ijkl}^E are the electric permittivity (F·m^{-1}) for constant strain, the piezoelectric stress constant (C·m^{-2}), and elastic stiffness constant (N·m^{-2}), respectively[65]. It reveals that 18 possible modes could couple the components of electrical and mechanical fields within piezoelectric materials. All of them are collated as four specific modes of longitudinal, transverse, longitudinal shear, and transverse shear mode. The longitudinal mode is generated when the normal stress is accompanied by the variation of the electric polarization in the same direction; and the transverse mode is induced by the change of electric polarization that occurs perpendicular to the applied mechanical load. Longitudinal shear mode is produced when the applied shear stress results in the change of the polarization perpendicular to the plane in which the piezoelectric material is sheared. When the electric polarization changes in the plane, the transverse shear mode in the piezoelectric material occurs[84].

Lamb wave can be excited across a thin substrate when the substrate thickness is smaller than or comparable to the wavelength of the wave. It is a superposition of both longitudinal and transverse components and constrained by the elastic properties of the substrate boundaries. Typical Lamb waves that propagate in thin plates or membranes have two modes at low frequencies, namely, the zero-order antisymmetric mode (A_0) the zero-order symmetrical Lamb wave mode (S_0)[2]. The A_0 mode named as flexural plate wave is highly dispersive in the regime of low frequency[85]. The S_0 mode is an extensional mode because the thin plate expands in the direction of wave propagation and contracts in the thickness direction. Both the A_0 and S_0 modes will converge toward the Rayleigh mode at the higher frequency. The energy of the S_0 mode dissipated into the liquid is small, which is suitable for sensing in the liquid environment.

Rayleigh mode has both longitudinal and vertical shear components that are coupled at the surface of the piezoelectric material. The surface particles of the material move into an elliptical trajectory and the particle oscillation is rapidly decayed with the depth into the substrate. The harmonic mode of Rayleigh wave in the layered structure is named as Sezawa mode. Sezawa waves are excited in the layered structure when the phase velocity of the SAWs in the substrate is larger than that of the overlaid piezoelectric film, which propagate around the interfaces with reflections and refractions[86]. Sezawa waves show an intrinsic feature of the guided wave. Moreover, the ratio of the piezoelectric film thickness to the wavelength of the SAW device has an important influence on the generation of the Sezawa waves.

The shear horizontal (SH) SAW is induced when the transverse shear distortion causes the displacement of SAW parallel to the substrate and the in-plane acoustic propagation is confined around the top surface and the interface between the piezoelectric film and substrate. SH-SAW has a minimal damping and weak attenuation inside the liquid that could enhance the efficient acquisition of the mechanical and electrical change of the liquids[87]. Lots of studies showed that a high sensitivity of the SAW device with SH-SAW was realized in the liquid environments[88,89].

Love mode SAW is generated by depositing a thin wave-guide layer (such as SiO_2, ZnO, or polymers) with microns or sub-micron thickness on the surface of the SH-SAW device, which results in the shear-horizontal acoustic wave propagating in a guiding layer on the top of the piezoelectric substrate. The velocity of the Love mode SAW is very sensitive to the mass change and photoconductivity variation, which is well suited for applications in liquid biosensing, gas sensing, and photodetection[90,91].

In principle, the above SAW modes can be excited by applying an alternating electric field on a piezoelectric substrate using a pair of interdigital transducers (IDTs). The profile, number of fingers, spacing, and aperture (overlap space) of the IDTs normally determine the wave propagation path and mode of the SAW.

8.3 DEVELOPMENT OF SMART-PIEZOELECTRIC-FILM-BASED DEVICES

8.3.1 Different Acoustic Wave Devices Based on the Piezoelectric Film

The fundamental properties of the piezoelectric films determine the design and fabrication of the piezoelectric devices. Many of novel configurations have been developed in the SAW devices, bulk wave devices, actuators, energy-harvesting devices, and sensors (i.e., biosensor, UV sensor, humidity sensor, and temperature sensor). The quartz crystal microbalance (QCM) with a transverse shear mode wave is the earliest well-known BAW device that is composed of an AT-cut quartz crystal plate sandwiched between two electrodes, as shown in Figure 8.3(a). The QCM device has been extensively used in various liquid-based sensing applications[92]. Film bulk acoustic wave resonator (FBAR) was first demonstrated in 1980, consisting of a piezoelectric film (i.e., ZnO, AlN, or PZT) sandwiched between two thin electrodes[93]. There are three typical structures that are designed for the back electrodes of FBARs including back trench, air-gap type, and Bragg reflector type with stacks of alternate layers of low and high acoustic impedance, as shown in Figures 8.3(b)–8.3(d). FBARs with high frequencies (typically GHz level) have attracted much attention on high precision sensor applications due to their small size, large quality (*Q*) factor, high sensitivity, good linearity and reliability, and low cost[94]. The growth of the *c*-axis-inclined wurtzite ZnO and AlN films deposited using an oblique angle method can be used to generate a dual mode of FBARs including the longitudinal mode and shear mode[95,96]. The thickness effect for FBARs shows that a thinner piezoelectric film results in a higher resonant frequency, which is a dispersive effect of the resonant frequency with the film thickness.

The SAW devices include filters, resonators, and delay lines (see Figure 8.4), which have a pair of different interdigital transducers (IDTs) consisting of the straight fingers, curved fingers, or circle fingers patterned on the piezoelectric films.

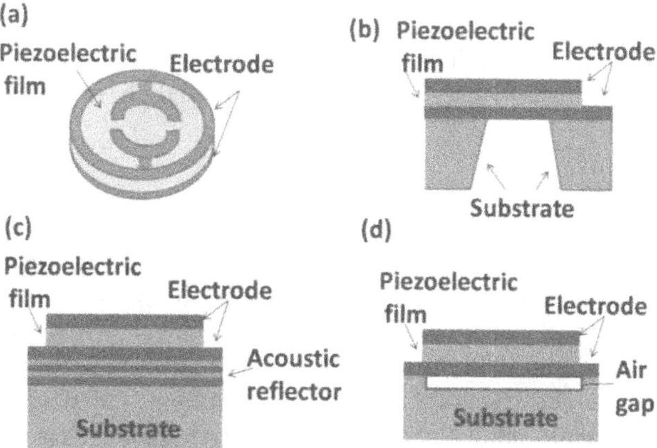

FIGURE 8.3 The configurations of the BAW devices (a) QCM, (b) back trench FBAR, (c) SMR, (d) air-gap FBAR [49]. (Redrawn with permission from table 8.1 in [Ref. 49].)

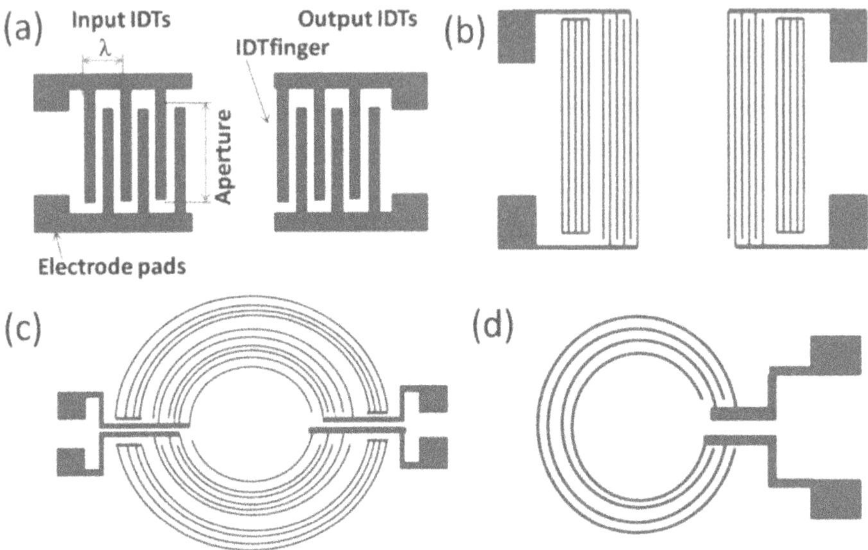

FIGURE 8.4 Different designs of the IDTs for the SAW devices: (a) slanted IDTs of the delay line structure, (b) IDTs with reflector, (c) semi-circular IDTs, and (d) circular IDTs.

IDTs developed by White and Voltmer in 1965 generally possess two comb-shaped metallic electrodes with inter-crossing periodic fingers[97]. The acoustic waves are generated by applying the electric signals on the input IDT, and subsequently the received electric signals of the output IDT can be obtained through the piezoelectric effect. Different configurations of the IDTs with focused, curved, circular/annular, or randomly shaped patterns in the SAW devices allow the wave generation, acoustic energy, and propagation direction to be more efficient and flexible by controlling the parameters of electrode materials, shape/dimensions, positions, substrate anisotropic properties, aperture, and number of reflective electrodes[2,98]. The acoustic properties of the devices are often determined by the resonant frequencies (f_0) of the SAW device, which is used to evaluate the performance for the ZnO SAW device. The resonant frequency is related to the wavelength (λ) and phase velocity (v) of SAW device, e.g.,

$$v = \lambda f_0 \tag{8.3}$$

where λ corresponds to the spatial period of the fingers in IDT. When the thickness of the piezoelectric substrate is fixed, the ultrahigh-frequency at a range of about 4–23.5 GHz can be obtained by combining the UV-based nanoimprint lithography with lift-off processes[99]. Furthermore, some factors such as the thickness, IDT parameters, substrates, triple-transit effect, metallization ratio effect, SAW reflections and diffractions, and bulk wave generations should be thoroughly considered to improve the generation efficiency of acoustic waves and spurious signal suppressions and to reduce signal distortion and insertion loss for a high-performance SAW device.

The present piezoelectric actuators have been widely applied in the field of ultrasonic transducers, micropumps, micromotors, positioning systems with sub-nanometer accuracy and microswitches for high current using the inverse piezoelectric effect of PZT, AlN, or ZnO piezoelectric films[100–102]. An ultrasonic transducer consists of the top electrode, piezoelectric film, back electrode, and the matching layer on backing materials. The performance of ultrasonic transducers is mainly determined by the piezoelectricity of the films, acoustic wave modes, electrode size, effects of backing and matching, and array configuration[103,104]. The piezoelectric pumps can convert electrical energy into mechanical energy to drive the fluid moving into a channel or small pool, and it is integrated with the chambers, valves, and vibrators to overcome the leakage, wear, and fatigue damage and to reduce the energy loss between different components in the traditional pumps[100]. One of the common micromotors is the piezoelectric ultrasonic motor (PUSM), which is a resonant piezoelectric motor that can reach much higher speeds even exceeding one meter per second[101]. The principle of the PUSM is that the friction force coupling between the runner and the stator drives the runner by the vibration of the elliptical motion parallel to the moving direction of the runner. The traveling wave ultrasonic motors using PZT thin films were developed at the mm scale exhibiting low power and high torque[105]. The stepping piezoelectric actuator with large working stroke is a common actuator that can achieve nanoscale positioning resolution for the positioning systems. Different types of the stepping piezoelectric actuators have been designed and developed such as inchworm type, friction-inertia type, and parasitic type. The first one mimics the motion process of the real inchworms consisting of one driving unit and two clamping units. The second one realizes the stepping motion of the slider/rotor by tuning the friction and inertia forces between two mass blocks. The last one adopts the parasitic motion of compliant mechanisms that an input force can generate two motions in different directions[106].

8.3.2 Development Trend of Smart-Piezoelectric-Film-Based Devices

Different research fields such as sensing, actuating, communication, mesoscopic system, collective excitations, quantum acoustics, and single quantum system are incorporated into various acoustic wave devices, which stimulate the development toward the highly diverse, interdisciplinary, and continuously expanding fields. Engineering and development in this direction are focused on enhancing the performance of the piezoelectric-film-based devices including the reproducibility, stability, sensitivity, and efficiency. Furthermore, the significant reduction of dimensions and manufacturing cost for the materials and devices also attracts attention to obtain the highly integrated devices with multiple functionalities.

Flexible and novel designs of the IDT configurations, layered films on different substrates and the direction of the acoustic propagations offer great versatility of SAW devices, which provides good opportunities for the fabrication and development of the SAW devices including delay lines, filters, resonators, pulse compressors, and convolvers. Moreover, many films with good piezoelectricity have been explored to improve the performance of the SAW devices by using doped piezoelectric materials.

Fundamental studies of the piezoelectric films further allow different acoustic modes to be excited in one substrate, which dictates the requirements of functional diversity in a lab-on-chip system. Meanwhile, appropriately arrangement of different acoustic units in the two-dimensional or three-dimensional space will improve the structural compatibility and integration to develop the new-generation SAW-integrated devices. The interactions of surface acoustic phonons with electrons and photons have induced lots of new phenomena that attract much attention for physics, engineering science, and applications[9,20]. For example, hybrid acoustic devices were developed which combine the SAW technique with the superconductivity technique, optoelectronic technique, quantum technique, and metastructures of photonic or phononic crystals[9,20]. In addition, wireless and flexile SAW devices have also been made to realize the wearable, portable, and smart functions, where the SAW devices can be worked at low power consumption and controlled for monitoring and detecting the physical, chemical, and biological variations[107,108].

FBARs research is focused on the electrode designs, orientation control of the piezoelectric films, and matching circuits for the output signals. The lattice match of the hexagonal piezoelectric films (i.e., ZnO and AlN) with the face-centered cubic or hexagonal electrode materials (i.e., Au, Al, Mo, Pt, Ti, or Ru) promises a better c-axis orientation[2]. The conducting and transparent oxides (i.e., AZO and ITO), carbon nanotubes, and graphene have been reported in the structures of FBARs, which could remarkably improve the Q-factor and sensitivity for biosensing[109]. The c-axis-inclined wurtzite films have been used to excite both the pure thickness longitudinal and shear modes in the FBARs, which can achieve a good mass sensitivity for biosensing due to the lower damping of the shear wave in the liquid[110].

As for the piezoelectric actuators, the mechanical outputs including the output force, travel, and velocity are still required to be improved. The hybrid working modes of direct drive, inertial drive, and inchworm drive are proposed for the new design and development of non-resonant piezoelectric actuators to realize the high resolution, long travel range, large output force, and high speed[17]. New friction materials and friction-resistant technologies with a better anti-friction performance are still necessary for the piezoelectric micromotor to make them withstand wear and tear with a long working life. The piezoelectric actuators also face the challenges of the miniaturization from centimeter level to micrometer level and the enhancement of the positioning accuracy for ultra-precision positioning and manipulating applications[111].

8.4 PIEZOELECTRIC FILMS FOR ACOUSTIC WAVE SENSOR

Acoustic wave sensor can sensitively convert mechanical, optical, electrical, and chemical changes (i.e., temperature ΔT, pressure ΔP, mass load Δm, mechanical constant Δc, conductivity $\Delta\sigma$, dielectric constant $\Delta\varepsilon$, the viscosity $\Delta\eta$, and the density $\Delta\rho$) or chemical stimulations (i.e., biochemical concentration) into acoustic signals of the frequency and phase for quantitative measurement and analysis. The fundamental sensing principle using acoustic waves is that the changes in the propagation velocity of the waves can be measured from the changes in the resonant frequency, phase angle, or occasionally amplitude of reflection or transmission signals. The perturbations in the velocity of the SAWs are induced by the physical or chemical changes, which can be described using the following formula[2,91]:

$$\frac{\Delta V}{V} = \frac{1}{V}\left(\frac{\partial V}{\partial m}\Delta m + \frac{\partial V}{\partial \sigma}\Delta\sigma + \frac{\partial V}{\partial \varepsilon}\Delta\varepsilon + \frac{\partial V}{\partial c}\Delta c + \frac{\partial V}{\partial \eta}\Delta\eta + \frac{\partial V}{\partial T}\Delta T + \frac{\partial V}{\partial P}\Delta P + \frac{\partial V}{\partial \rho}\Delta\rho\right) \quad (8.4)$$

where the variations in phase velocity of the acoustic wave are dominated by the intrinsic factors (i.e., density, elasticity, viscosity, conductivity, permittivity, changes in carrier concentration and mobility) and extrinsic factors such as mass loading, temperature, deformation, pressure, strain, stress, humidity, pH values, ultraviolet (UV) and infra-red (IR) sources, externally applied electric/magnetic fields and charge injection. A change in acoustic velocity results in a variation of resonant frequency due to the relation between the acoustic phase velocity (V) and the resonant frequency. The performance of the sensors is generally estimated from the parameters such as sensitivity, detect limit, selectivity, response time, recovery time, stability, and reproducibility. The different sensors have been developed to detect the changes in the temperature, moisture, strain, pressure, acceleration, viscosity, pH levels, magnetic and radiation fields, and gas detection.

8.4.1 SAW Sensors

For the operating principle of the SAW sensors, the characteristic parameters of the SAWs including the frequency, phase, and magnitude are modulated on the path of the SAW propagation by the sensing materials that respond to the adsorbed gas molecules, UV light, biological molecules in liquid, or temperature variation. The performance of the SAW sensors can be estimated by the phase changes, reduction of the insertion loss, or frequency variations of the output SAW signals.

For the SAW gas sensors, a sensing layer was coated on the area between the two IDTs. The interactions of the SAWs with the adsorbed gas molecules in the sensing layer or receptor layer will cause the changes in the characteristic parameters of the SAWs on the SAW propagating path. Subsequently, the variations of the attenuation and velocity of the SAWs are observed. The perturbation of the SAWs is remarkably dependent on the number of the adsorbed molecules and a large concentration of the gas molecules results in strong modulation of the SAWs. The response of the sensing layer to the gas molecules is intrinsically related to the variations of mass loading due to the coupling of the recognition layer with the analyte[112].

The dual-device configuration matched with the oscillation circuits is often used to improve the sensitivity and to avoid the interference of the environment (e.g., humidity and temperature). Furthermore, the high performance of the SAW gas sensor requires a good compatibility among the sensor design and fabrication, the highly sensitive sensing layer, and the integration of the sensor element with the detection system and appropriate analysis methods of the sensing data. When the electrical and viscoelastic effects are weak and negligible, there is a simple relationship between the resonant frequency and the film properties which reads as follows:

$$\Delta f = \frac{(k_1 + k_2)mf_0^2}{A} \quad (8.5)$$

where Δf, m, A, and f_0 are the frequency shift, the area of the sensing layer, the mass of the gas molecules absorbed on the sensing layer, and the center frequency of the SAW sensor, respectively; k_1 and k_2 are piezoelectric material constants[113,114]. The SAW gas sensors possess the advantages of real-time detection, low power consumption, high sensitivity, stability in harsh environments[115].

Accurately controlling and monitoring of the hazardous gases including ammonium (NH_3), hydrogen (H_2), hydrogen sulfide (H_2S), nitric dioxide (NO_2), carbon monoxide (CO), and volatile organic compounds are critical for the public security, environmental protection, and health care of the human. NH_3 is highly toxic, volatile chemical gas that is normally used in chemical industries, food processing, and medical diagnosis. It can be sensitively detected by the sensing materials such as ZnO films, composite films of Co_3O_4/SiO_2, TiO_2/SiO_2, ZnO/SiO_2, graphene oxide(GO), and glutamic acid hydrochloride (GAH) using ST-cut quartz surface acoustic wave devices at room temperature (RT) as listed in Table 8.2[116–121]. The water absorptions

TABLE 8.2
Sensing Performance of the SAW Gas Sensors for Different Hazardous Gases

Hazardous Gases	Sensing Materials	Detection Limit (ppm)	Frequency Shift (kHz)	Working Temperature(°C)	Response/ Recovery Times (s)
NH_3	ZnO films[116] and nanowire and nanowire	10	0.11	RT	50/34
	Co_3O_4/SiO_2[117]	1	3.5	RT	35/100
	SiO_2/TiO_2[118]	1	2	RT	80/60
	ZnO /SiO_2[119]	10	1.132	RT	~60/60
	GO[120]	0.5	0.62	RT	~240/450
	GAH[121]	5.02	0.11	RT	–/–
	Polyaniline[122]	20.45	-	RT	<150/-
H_2	Pd doped SnO_2[124]	100	-	175	1/312
	ZnO nanorods[125]	500	~150	265	28/36
	Au–WO_3[126]	600	187	270	240/480
	Pd doped graphene[127]	2500	15	RT	1/9
	Pt/ZnO[128] nanoparticle	10000	55	RT	60/120
H_2S	CuO film[130]	0.5	1.2	25	–/–
	SnO_2 film[131]	68.5	112.232	120	–/–
	SnO_2/CuO composite[132]	20	230	160	55/45
	Cu NP[133] -SWCNT	5	~50	175	7/9
NO_2	InO_x nanofilms[134]	0.51	~70	246	180/360
	ZnO nanofilms[135]	0.4	6	RT	-/-
	$Pb(NO_3)_2$-treated PbS CQDs[136]	10	9.8	RT	45/58

on dangling Si bonds of the colloidal silica in the composite film contribute to the significant rise of the film conductivity, which causes a negative frequency shift of the SAW sensor[117,118]. The polyaniline coated on a $LiNbO_3$ SH-SAW device was reported to achieve a detection limit of 20.45 ppm[122].

Hydrogen gas exhibits high energy density, highly explosive and dangerous at a concentration larger than 4%, which requires real-time and strictly monitoring and detecting the concentration of hydrogen gas at trace levels in the production, storage, and transportation of hydrogen gas[123]. The SAW gas sensors operated using the platinum (Pd)-doped SnO_2 film, ZnO nanorods, and Au modified WO_3 film exhibit a good sensitivity for H_2 gas[124–126]. However, these sensors require high operating temperatures that limit the wide applications in various extreme environments. The sensing coatings of the Pd-doped graphene films or Pd-modified ZnO nanoparticles achieve a fast response at RT[127,128]. The ball-type structure of the SAW sensor was adopted and a Pt-coated ZnO film was sputtered as the sensing film, which can trace a concentration of 20 ppm for H_2 gas without apparent influences from the water moisture[129].

Toxic H_2S gas is colorless, which can be generated from the decomposition of organic matter, human and animal's waste, food processing, cooking stove, craft paper mills, and oil refineries[130]. Hazardous H_2S gas is harmful to the eyes, skin, mucous membranes, and nervous systems of human beings even at low concentrations. CuO film has been coated on the SAW sensor for monitoring the H_2S gas, which can detect a low concentration of 0.5 ppm at room temperature. The response/recovery time has been obviously improved using SnO_2 film, SnO_2/CuO composite, and single-wall carbon nanotube decorated with copper nanoparticles (Cu NP-SWCNT) at a relatively high temperature (i.e., 175 °C)[131–133]. The research for the SAW NO_2 sensors is also focused on the exploration of highly sensitive materials such as the nanofilms of InO_x, ZnO, and $Pb(NO_3)_2$-treated PbS colloidal quantum dots (CQDs), which could exhibit better detect limits and the faster responses and recovery times[134–136]. In addition, an electrospun ZnO nanostructured thin film on quartz Rayleigh SAW device exhibited good responses to the volatile organic compounds including acetone, trichloroethylene, chloroform, ethanol, *n*-propanol, and methanol vapor, which can be designed for the sensor array for detecting different volatile organic compounds in a complex environment[137].

Moreover, a comprehensive integrated control strategy for the improvement of the performance of the SAW gas sensors includes exploring new sensing materials, designing a sensor array, and adopting various analytical methods (i.e., cluster analysis, factorial analysis, and regression analysis)[138].

The SAW UV sensor is critical for the applications of environmental light monitoring, space research, high-temperature flame detection, health care, and optical communication, and it can detect the UV radiation with a wavelength ranging from 400 to 10 nm[139]. In a typical SAW UV sensor, a UV-sensing coating is deposited on the SAW propagation path. Its main operating principle is originated from the acoustic–electric interaction that the attenuation and velocity variations with the conductivity are induced by the moving electric field under the mechanical deformation. The free electron–hole pairs will be formed through exciting the electrons into the conduction band and interact with the propagating SAWs by exposing the sensing layer using UV light. Thus, the insertion loss and frequency of the SAW device will be changed with the power density of the UV light. When the thickness of the sensing film is smaller than the acoustic wavelength, the velocity shift (Δv) and the attenuation ($\Delta\Gamma$) and the change in the phase angle ($\Delta\phi$) could be evaluated using the following equations[91,140]:

$$\frac{\Delta v}{v_0} = -\frac{k^2}{2}\frac{1}{1+(v_0 C_s/\sigma_{sh})^2} \tag{8.6}$$

$$\Delta\Gamma = \frac{k^2}{2}\frac{\sigma_{sh}/C_s}{1+(\sigma_{sh}/v_0 C_s)^2} \tag{8.7}$$

$$\Delta\phi = \frac{2\pi L}{\lambda}\frac{\Delta v}{v_0} \tag{8.8}$$

where v_0, k^2, C_s, and σ_{sh} are the SAW velocity on free surface, coupling coefficient, capacitance per unit length of the surface, and the sheet conductivity of the piezoelectric film, respectively; λ and L are the SAW wavelength and the acoustic path length, respectively. The mass-loading effect induced by adsorbing and desorbing the oxygen molecules also contributes to the frequency shift[141]. The photocapacitive effect is proposed to explain the mechanism that the phase change of the SAW signals has been influenced by the UV-induced changes of the IDT capacitance[142].

The layered structures of the SAW UV sensors such as ZnO/$LiNbO_3$, ZnO/Quartz, ZnO/3C-SiC/Si, ZnO/Si have been used to detect UV radiations using Rayleigh SAWs and Sezawa wave as listed in Table 8.3[143–148]. The different performances indicate that the variations of the velocity and amplitude for the SAW signal are remarkably modulated by the physical changes of sensing layers, piezoelectric substrate, and ambient temperature. A low-level intensity of 450 nW·cm^{-2} for the UV radiation at 365 nm was detected by depositing a 71-nm-thick ZnO thin film on the $LiNbO_3$ SAW filter with a working frequency of 36.3 MHz[144]. Furthermore, nanocrystalline ZnO films including nanoparticles, nanorods, or nanowires were coated on the Rayleigh SAW devices to improve the photoconductivity due to large specific surface area and strong electrochemical activity[145–148]. A bilayer of Ag nanowire (NW) and ZnO has also been used to obtain a fast response of the $LiNbO_3$ SAW UV sensor due to the rapid oxygen adsorption on the ZnO surface[149]. The lowest UV irradiation of 0.04 μW/cm^{-2} at 365 nm was detected using Ag-doped ZnO nanoparticles (NPs) on the $LiNbO_3$ SAW device[150].

The Sezawa wave has also been utilized to detect the low-intensity UV light, which exhibits a better performance than that using Rayleigh SAWs[140,151,152]. Most of the UV sensing materials are ZnO or metal-doped ZnO because their large direct bandgap and high exciton binding energy are very suitable for UV photon-assisted transition of the electrons and the stable formation of the excitons in the sensing layer. For instance, a Mg:ZnO layer was sandwiched into the ZnO/Mg:ZnO/ZnO/Si structure, and the fabricated ZnO SAW UV sensor achieved the power density limit of 810 μW·cm^{-2} for the Sezawa wave[140]. Besides ZnO, TiO_2 nanorods have also been

TABLE 8.3
Performance of the ZnO SAW UV Sensors with Different Acoustic Modes

Sensing Layer	Piezoelectric Substrate	Resonant Mode	Resonant Frequency (MHz)	UV Density Limit (μW cm^{-2})	Frequency Shift (kHz)
ZnO film[143]	$LiNbO_3$	Rayleigh	37	10	170
ZnO film[144]	$LiNbO_3$	Rayleigh	36.3	0.45	28
ZnO film[145]	$LiNbO_3$	Rayleigh	145	500	40
ZnO particle[146]	$LiNbO_3$	Rayleigh	64	691	–
ZnO film[147]	Quartz	Rayleigh	–	19,000	45
ZnO film[148]	3C-SiC/Si	Rayleigh	122.25	600	150
MoS_2 nanosheets[142]	ZnO/Si	Rayleigh	1020	1466	3500
Ag NW/ZnO[149]	$LiNbO_3$	Rayleigh	242.25	3916	68
Ag doped ZnO NPs[150]	$LiNbO_3$	Rayleigh	156	0.05	–
ZnO/Mg:ZnO[140]	ZnO/r-Al_2O_3	Sezawa	711.3	810	1360
ZnO film[151]	ZnO/Si	Sezawa	842.8	551	1017
ZnO nanorods[152]	ZnO/Si	Sezawa	271.83	600	25
ZnO nanorods[154]	quartz	Love	117	–	–
ZnO film[91]	$LiTaO_3$	Love	41.5	350	150

hydrothermally grown as a sensing film on the SAW UV sensor using the 128° Y-cut $LiNbO_3$ and ST-cut quartz substrates[153]. In addition, the amplifier and oscillator circuit were designed with the ZnO SAW UV oscillator to enhance the sensitivity[144].

Love mode SAW devices have a better sensitivity to the variations of the mass load and photoconductivity, which has been widely applied for liquid biosensing, gas sensing, and UV detection. ZnO nanorods grown on the ZnO/ST-cut (42°45′) quartz SAW device showed an enhanced UV sensitivity[154]. It indicates that the crystalline structure and defect properties of ZnO film are very important to the photoconductivity, which allows the sensitivity of the Love mode SAW UV sensor to be improved by controlling the film growth conditions. Compared with a low k^2 value of 0.11% and a low dielectric constant of 4.5 for ST-cut quartz, the 36° Y-cut $LiTaO_3$ possesses a larger k^2 value of 4.7% and higher dielectric constant of 47[154,155]. Love mode SAW UV sensor using the sputtered ZnO film on 36° Y-cut $LiTaO_3$ substrate has been developed, which can detect a low power density of 350 μW·cm^{-2} under a 254-nm illumination and a corresponding frequency shift of 150 kHz[91].

For the SAW biosensor, the sensing layer is functionalized with the biomolecules. When the biochemical recognition components biochemically react with the analyte, the physical, chemical, and/or biochemical variations occur due to the mass loading and viscosity changes. The traveling SAWs are modulated by the above variations, which can be monitored by the changes in velocity, attenuation, and resonant frequency of the acoustic waves. Moreover, the micro-fluidic channels or reservoir are fabricated on the top of the sensor, which allows the biological liquids to flow on the surface of the sensing layer[156]. Therefore, the SAW biosensors could combine the

SAW and microfluidic techniques with the biological technique to form an integrated analytical micro-system, which can exhibit selective and quantitative responses to the trace amounts of biological samples.

SAW biosensors generally have three parts: SAW device functionalized with bio-specific layer, the electronic circuitry that can read and amplify the output signal, and the microfluidic units integrated with the biosensor chip[156–158]. The biochemical recognition is a bioreceptor that can interact with the target analyte, which determines the sensing functionality of the SAW biosensor. The SAW biosensor with a label-free detection shows a high selectivity because the bioreceptor selectively interacts with the target analyte against the various chemical and biological components[156]. Up to date, the involved biochemical reactions in the SAW biosensors include enzymes, antibody/antigen, nucleic acids/DNA, and cellular structures/cells[159]. The performance of SAW biosensors is summarized in Table 8.4.

SH-SAW and Love mode wave propagate into the liquid with very weak dissipation of coupling acoustic energy, which are optimal candidates for biosensing application. Love mode wave confines the shear-horizontal wave into the guiding layer which is sensitively influenced by the changes of the sensing layer on the propagation

TABLE 8.4
Performance of SAW Biosensors in Different Resonant Modes

Sensing Layer	Layered Piezoelectric Structure	Resonant Mode	Resonant Frequency (MHz)	Target Analyte	Sensitivity
Au film[160]	ST-cut quartz	Love	89	Rat Ig G	950 $cm^2 \cdot g^{-1}$
Gluteraldehyde[161]	ZnO/SiO_2	Love	747.7	Interleukin-6	4.456 $\mu m^2 \cdot pg^{-1}$
Glucose oxidase[162]	Mn-ZnO/SiO_2	Love	433	Glucose	7.184 $MHz \cdot mM^{-1}$
Streptavidin[163]	Au/ZnO	Love	3219	Mammoglobin	12.495 $pg \cdot Hz^{-1}$
C_{60}-hemoglobin C_{60}-myoglobin[164]	LiTaO3	SH-SAW	145	Anti-hemoglobin Anti-myoglobin	1.27 $kHz \cdot (\mu g/mL)^{-1}$
Anti-CEA antibody/Au film[165]	SiO_2/ ST-cut quartz	Love	120	Carcinoembryonic antigen	37 $pg \cdot mL^{-1}$
Anti-CEA antibody/Au nanoparticles[166]	SiO_2/ ST-cut quartz	Love	160	Carcinoembryonic antigen	1 $ng \cdot mL^{-1}$
Anti-CEA antibody[167]	Au/ST-cut quartz	Love	123	Carcinoembryonic antigen	0.37 $ng \cdot mL^{-1}$
Protein A[168]	AlN/PEN	Lamb	500	Escherichia coli	6.54×10^5 $CFU \cdot mL^{-1}$ 6.54×10^5 CFU/mL
Anti-DNP antibody[169]	Au/ZnO	Sezawa	179	Anti-DNP antigen	0.2 $kHz \cdot (ng/mL)^{-1}$
Human Ig-E antibody[170]	Cr/Au/ZnO /Si_3N_4	Sezawa	200/1497	Human IgE	4.44×10^6 $cm^2 \cdot g^{-1}$

path, where the guiding layers are generally coated with ZnO, SiO_2, polymer, or Au films[160–170].

A sensing layer of Au film on the Love mode SAW device with a ZnO/ST-cut quartz structure was used to detect rat immunoglobulin G (Ig G), which achieves a high sensitivity of 950 $cm^2 \cdot g^{-1}$[160]. Gluteraldehyde film was functionalized on the Love mode SAW biosensor with a layered structure of ZnO/SiO_2/Si, which showed a successful detection of IL-6 protein with a low level of 4.456 $\mu m^2 \cdot pg^{-1}$[161]. Glucose biosensor has been developed using the SAW device with a Mn-ZnO/SiO_2/Si structure, which showed a good response to blood sugar levels with a sensitivity of 7.184 MHz· mM^{-1}[162]. A novel design of the ZnO SAW biosensor integrated with MEMS and COMS processes were demonstrated, which could detect mammaglobin using the streptavidin sensing layer, resulting in a sensitivity of 8.704 $pg \cdot Hz^{-1}$[163]. The C_{60}-hemoglobin composite was coated as the sensing layer on the surface of the $LiTaO_3$ SAW device to monitor the anti-hemoglobin with the sensitivity of 12.495 $pg \cdot Hz^{-1}$ [164]. The carcinoembryonic antigen (CEA) is an indicator of disease recurrence, which could be tested using the Love mode SAW biosensors with the limit of the concentration low to 0.37 $ng \cdot mL^{-1}$[165–167].

Recently, Lamb wave in the AlN/polyethylene naphtholate (PEN) SAW device was exploited for detecting the *Escherichia coli*, which exhibited a sensitivity of 6.54×10^5 colony-forming units (CFU) per milliliter[168]. A 6-(2,4-dinitrophenyl) aminohexanoic acid (DNP) antigen has been tested using Sezawa wave exciting from the ZnO SAW device, where the anti-DNP antibody was adsorbed on the Au film and the sensitivity of 0.2 $kHz \cdot (ng/mL)^{-1}$ was obtained [169]. Moreover, Sezawa mode wave in the ZnO/Si_3N_4 SAW biosensor for detecting human immunoglobulin E (IgE) showed a sensitivity of 4.44×10^6 $cm^2 \cdot g^{-1}$, which is higher than that of Rayleigh SAW[170].

8.4.2 FBAR Sensors

The FBAR sensors are new and one of the popular acoustic sensors, which possess significant advantages over the SAW sensors due to their high quality factor, small size, and large operating frequency in the GHz regime. The study of SMR sensors showed a high sensitivity of 500 $Hz \cdot cm^2 \cdot ng^{-1}$ greatly higher than that of 0.057 $Hz \cdot cm^2 \cdot ng^{-1}$ for the commercial QCM[171]. As for the sensing principle of the FBAR sensors, the resonant frequency responds quickly to the mass change of the sensing layer that has been coated on the working area and that interacts with the gaseous or biochemical molecules. The qualitative analysis of the frequency shift based on the Sauerbrey relationship allows the FBAR sensors to be widely applied for biosensor, gas sensor, UV sensor, and humidity sensor, exhibiting high sensitivity, low hysteresis, label free, good selectivity, and excellent compatibility with the integrated circuit[172]. Table 8.5 lists the performance of selected FBAR biosensors with different resonant modes.

In the FBAR biosensors, the immobilized antibodies have been adsorbed on the top electrode or the back membrane. The target antigens were captured by the antibodies through a specific molecular recognition in a liquid environment. The channel or cavity structures were introduced to confine the bio-liquid flow over the functionalized area, which could keep the Q-factor from significantly damping[173].

TABLE 8.5
Performance of FBAR Biosensors in Different Resonant Modes

Sensing Layer	Piezoelectric Structure	Resonant Mode	Resonant Frequency(GHz)	Target Analyte	Sensitivity
Protein A[174]	ZnO SMR	longitudinal	3.94	Human IgG	8970 Hz·cm²·ng⁻¹
Odorant binding protein[175]	ZnO FBAR	longitudinal	1.5	DEET	2 kHz·cm²·ng⁻¹
Biofunctional linkers[176]	$Mg_xZn_{1-x}O$ FBAR	longitudinal	1.5628	DNA	103 Hz·cm²·ng⁻¹
Avidin[177]	ZnO FBAR	Shear	0.79	Anti-avidin	585 Hz·cm²·ng⁻¹
CNTs[178]	AlN FBAR	longitudinal	2	Binding of BSA	2.4 kHz·cm²·ng⁻¹
Bare surface[179]	ZnO FBAR	longitudinal	1.55	Blood proteins	1358 Hz·cm²·ng⁻¹
Streptavidin[110]	ZnO SMR	shear	1.1	Rabbit IgG	4.95 kHz·cm²·ng⁻¹
Monoclonal antibodies[180]	ZnO SMR	shear	2.26	Cardiac troponins I	20 pg·mL⁻¹
Aptamer[181]	AlN FBAR	longitudinal	2	CEA proteins	2284 Hz·cm²·ng⁻¹
Streptavidin[182]	AlN FBAR	shear	1.5	biotinated BSA	2.9 Hz·cm²·ng⁻¹
Anti-CEA aptamer[183]	AlN SMR	shear	1.2	CEA antigen	2045 Hz·cm²·ng⁻¹
IgE antibody[184]	AlN FBAR	shear	1.175	Human IgE	1.425×10^5 cm²·g⁻¹
Streptavidin[185]	AlN SMR	shear	1.3	biotinylated receptors	1800 kHz·cm²·pg⁻¹
Human IgE antibody[186]	AlN SMR	longitudinal	2.22	Goat anti-human IgG antigen	3.15 kHz·cm²·pg⁻¹

Up to now, great efforts have been made to improve the sensitivity, enhance the integration with microfluidic functions, and expand the boundary of the biological detections[174–186]. The highest sensitivity of the ZnO solidly mounted resonator (SMR) consisting of piezoelectric stack and Bragg acoustic reflector is 8970 Hz·cm²·ng⁻¹ for detecting the human IgG using longitudinal mode at 3.94 GHz[174]. The odorant biosensors have been developed using ZnO FBARs with a resonant frequency of 1.5 GHz that was functionalized by a layer of N-diethyl-meta-toluamide (DEET)[175]. A ZnO-nanotip film functionalized on FBAR using the piezoelectric magnesium zinc oxide ($Mg_x Zn_{1-x}O$) film was exploited to enhance the mass sensitivity for selectively immobilizing DNA[176]. Carbon nanotubes (CNTs) were coated as a biosensing layer on the AlN FBARs, which could obtain a good sensitivity of 2.4 kHz·cm²·ng⁻¹ for the detection of bovine serum albumin (BSA)[178].

The shear mode of FBARs exhibits a lower dissipation of the acoustic energy in a liquid environment than that of the longitudinal mode, which produces the shear-mode FBAR biosensors or dual mode FBAR biosensors using the inclined wurtzite films. A ZnO piezoelectric film with an inclined angle of 16° was used to generate the shear mode for the FBAR biosensor with a sensitivity up to 585 $Hz \cdot cm^2 \cdot ng^{-1}$ in the measurement of the avidin/anti-avidin model[177]. The AlN SMRs with a shear mode have also been applied on the biosensing of CEA and biotinylated receptors[183,185]. The competitive adsorption and exchange behavior among the proteins of albumin (Alb), IgG, and fibrinogen (Fib) were observed, where a mass detect limit was for 1.35 $ng \cdot cm^{-2}$ [187].

A ZnO SMR gas sensor was reported to monitor the nerve gas using the poly(vinylidene fluoride) film, which achieved a sensitivity of 718 $kHz \cdot ppm^{-1}$ and showed a good linear correlation between the frequency shifts and the concentrations[188]. The hydrogen gas sensor using ZnO SMR with a Pd thin film as the sensing layer could reach a detection limit of 0.05% at room temperature operated at 2.39 GHz[189]. The piezoelectric layer in the ZnO FBARs with a resonant frequency of 1.14 GHz could act as a sensitive layer by designing microscale through-hole arrays in the top electrode, which showed a sensitivity of 21.2 kHz/1% for the humidity variation[190]. A novel FBAR design based on the polyimide film was employed for humidity sensing; and finite element method (FEM) simulation proved that the sensitivity reached 67.3 kHz/% RH[191]. The ZnO FBARs UV sensors have been reported with a sensitivity reaching 9.8 kHz for the 365-nm UV radiation with an intensity of 600 $\mu W \cdot cm^{-2}$, which showed that the resonant frequency shifts were related to the temperature-dependent Young modulus of the ZnO film[192]. Further studies indicated that the UV sensitivity was dependent on the temperature and relative humidity because the adsorbed oxygen was influenced by the temperature, water molecular, and gas molecular[193]. The FBARs with thickness field excitation and lateral field excitation were used to detect the IR light with the detection limits of 0.7 and 2 $\mu W \cdot mm^{-2}$, respectively[193].

8.5 PIEZOELECTRIC ACTUATORS WITH SMART STRUCTURES

The piezoelectric film is the most common and important one of smart materials, which can be fabricated for different piezoelectric actuators with the advantages of quick response, wide bandwidth, and easy implementation[194]. The smart structures based on the piezoelectric films generally undergo a macroscopic change from its mechanical property in a controlled manner due to an externally electric stimulus. They are capable of alternating the mechanical states (i.e., position and velocity) or mechanical stiffness, which have been integrated into various piezoelectric actuators. Moreover, they can extend the control functions into an intelligent, efficient, and adaptable way for various engineering applications such as biomimetic flapping wings and structural health monitoring[195–197].

Combined with control algorithms, smart structures using the distributed actuators are arranged in a three-dimensional geometric space, which exhibit complex control capability[196–199]. An iteratively calibrated incremental method is used to optimize control voltages that can be designed for static shape control of structures with nonlinear piezoelectric actuators[196]. Piezoelectric fiber composite actuators have been designed for the flapping wing, which can improve the aerodynamic properties of the flapping wing[197]. Furthermore, piezoelectric T-beam actuators were introduced into the biomimetic four-winged flapping system device, which provided a high movement amplification and achieved a maximum thrust of 1.34 mN at 25.5 Hz[198]. Two piezoelectric bending actuators were employed to develop the micromechanical flying insect, which produced a high lift force of 1.4 mN at the wing beat frequency of 275 Hz with a feedback control[199]. A simple flapping wing rotor composed of piezoelectric actuators, shaft, and flapping wings could improve the lift force up to 5.35 mN at the beat frequency of 125 Hz[200]. The lightweight piezocomposite actuator can be used to actuate a flapping-wing system for wing rotation, wing corrugation, and wing clap, which showed a flapping angle of 92° at the optimum frequency of 17 Hz[201,202]. When the new and high technologies including the piezoelectric film deposition, three-dimensional printing, and additive manufacturing are combined with the structural design and control, the development of flapping-wing system will be further accelerated toward a miniaturized, lightweight, and smart system that can flap with high amplitudes at high frequencies.

8.6 SUMMARY AND FUTURE OUTLOOK

The fundamentals and advances of the piezoelectric films and their applications were reviewed in the chapter. Different acoustic wave modes have been summarized that could be generated in the piezoelectric devices with various configurations. The acoustic wave devices including BAW devices, SAW devices, and piezoelectric actuators were presented, as well as the development trends of the future studies on the performance improvement and novel configurations. The latest progress of the acoustic wave sensor with different new techniques was highlighted. The applications of the piezoelectric actuators on smart structures were discussed.

Although the smart piezoelectric films have been successfully applied to the acoustic wave sensors and actuators with high performance, great challenges have been posed by the rapid changes of the human requirements from the extensive development with low attention on the environmental and healthcare issues to the intensive development on a pollution-free, friendly, and safe environment. Based on the characteristics of acoustic waves, technical progress, and the challenges, the future trends of the piezoelectric films on the sensor and actuator applications are listed as follows:

1. The film piezoelectricity can be significantly improved by applying new material designs and advanced techniques of the film deposition. The studies on the doping strategies, optimization, and hybrid composite of the typical piezoelectric films can gain the giant piezoelectric coefficient. They will fulfill the urgent requirement of the acoustic wave devices with high performance, which is the basis of the acoustic wave-based sensors with a high sensitivity. The appropriate engineering of the inclined orientation in the wurtzite piezoelectric films allows modulating the piezoelectricity of different-mode acoustic waves for the multifunctional device applications.
2. The novel design and development of the acoustic wave devices can be integrated with microelectronics, superconductivity, optoelectronics, quantum technique, and phononic crystals toward new generation of the hybrid

acoustic devices. These will be used to monitor multiple parameters of the environment, to detect the variations of quantum states, and to digitally control and sense the bioliquid or nanoparticles in the microsystems.

3. The sensing mechanisms of the various new sensing films on the acoustic devices are still undergoing exploration under the multiple perturbations of the complicated environment. The sensitivities on the changes of the mass, gases, UV, humidity, and bioreactions are closely related to the material properties, device performance, test methods, and system construction. The sensors and actuators with low cost, microminiaturization, high throughput, trace level, strong stability, and long life are concentrated with the development of MEMS techniques in the future.

ACKNOWLEDGMENTS

This work was financially supported by the National Natural Science Foundation of China (11504291, 12074309) and the UK Engineering and Physical Sciences Research Council (EPSRC) under grant EP/P018998/1, Newton Mobility Grant (IE161019) through Royal Society, and International Exchange Grant (IEC/NSFC/201078) through Royal Society and the National Natural Science Foundation of China.

REFERENCES

1. J. Curie, P. Curie, Development by pressure of polar electricity in hemihedral crystals with inclined faces, Bulletin de la Societe de Minerologique de France, 3, 90–93 (1880).
2. Y.Q. Fu, J.K. Luo, X. Du, A.J. Flewitt, Y. Li, A. Walton, W.I. Milne, Recent developments on ZnO films for acoustic wave based bio-sensing and microfluidic applications: a review, Sens. Actuat. B., 143, 606–619 (2010)
3. R. Manenti, A. F. Kockum, A. Patterson, T. Behrle, J. Rahamim, G. Tancredi, F. Nori, P. J. Leek, Circuit quantum acoustodynamics with surface acoustic waves, Nat. Commun. 8, 975 (2017).
4. E. K. Akdogan, M. Allahverdi, A. Safari, Piezoelectric composites for sensor and actuator applications, IEEE Trans. Ultrason. Ferroelectr. Frequency Control 52(5), 746–775 (2005).
5. R. Herdier, M. Detalle, D. Jenkins, C. Soyer, D. Remiensa, Piezoelectric thin films for MEMS applications—A comparative study of PZT, 0.7 PMN–0.3 PT and 0.9 PMN–0.1 PT thin films grown on Si by RF magnetron sputtering, Sens. Actuators A, 148, 122–128 (2008).
6. R. Zhu, Z. Wang, Z. Cheng, X. Guo, T. Zhang, Z. Cai, H. Kimura, T. Matsumoto, N. Shibata, Y. Ikuhara, Composition gradient (1-x) Ba $(Zr_{0.2}Ti_{0.8})$ O_{3-x} $(Ba_{0.7}Ca_{0.3})$ TiO_3 film with improved dielectric, piezoelectric and ferroelectric temperature stability, Ceram. Int. 46(12), 20284–20290 (2020).
7. A. Khan, Z. Abas, H. S. Kim, I.-K. Oh, Piezoelectric thin films: an integrated review of transducers and energy harvesting, Smart Mater. Struct. 25, 053002 (2016).
8. L. Y. Yeo, J. R. Friend, Surface acoustic wave microfluidics, Annu. Rev. Fluid Mech., 46, 379–406 (2014).
9. M. J. A. Schuetz, E. M. Kessler, G. Giedke, L. M. K. Vandersypen, M. D. Lukin, J. I. Cirac, Universal quantum transducers based on surface acoustic waves, Phys. Rev. X 5, 031031 (2015).

10. H. Zhou, Y. Zhang, Y. Qiu, H. Wu, W. Qin, Y. Liao, Q. Yu, H. Cheng, Stretchable piezoelectric energy harvesters and self-powered sensors for wearable and implantable devices, Biosens. Bioelectron.,168, 112569 (2020).
11. V. M. Mastronardi, L. Ceseracciu, F. Guido, F. Rizzi, A. Athanassiou, M. De Vittorio, et al. Low stiffness tactile transducers based on AlN thin film and polyimide, Appl. Phys. Lett., 106, 162901 (2015).
12. Smecca E, Maita F, Pellegrino G, Vinciguerra V, Magna LL, Mirabella S, et al. AlN texturing and piezoelectricity on flexible substrates for sensor applications, Appl. Phys. Lett., 106, 232903 (2015).
13. K. D. Wise, Integrated sensors, MEMS, and microsystems: Reflections on a fantastic voyage, Sens. Actuators A 136, 39–50 (2007).
14. A. Wei, L. Pan, W. Huang, Recent progress in the ZnO nanostructure-based sensors, Mater. Sci. Eng. B 176, 1409–1421 (2011).
15. Z. Yan, X. Y. Zhou, G. K. H. Pang, T. Zhang, W. L. Liu, J. G. Cheng, Z. T Song, S. L. Feng, L. H. Lai, J. Z. Chen, Y. Wang, ZnO-based film bulk acoustic resonator for high sensitivity biosensor applications, Appl. Phys. Lett., 90, 143503 (2007).
16. I. Voiculescu, A. N. Nordin, Acoustic wave based MEMS devices for biosensing applications, Biosens. Bioelectron., 33, 1–9 (2012).
17. L. Wang, W. Chen, J. Liu, J. Deng, Y. Liu, A review of recent studies on non-resonant piezoelectric actuators, Mech. Syst. Signal Pr., 133, 106254 (2019).
18. R. Friend, L. Y. Yeo, Microscale acoustofluidics: Microfluidics driven via acoustics and ultrasonics, Rev. Modern Phys., 83, 647–704 (2011).
19. D. Hu, M. Yao, Y. Fan, C. Ma, M. Fan, M. Liu, Strategies to achieve high performance piezoelectric nanogenerators, Nano Energy 55, 288–304 (2019).
20. D. A. Golter, T. Oo, M. Amezcua, I. Lekavicius, K. A. Stewart, H. Wang, Coupling a surface acoustic wave to an electron spin in diamond via a dark state, Phys. Rev. X 6, 041060 (2016).
21. P. Muralt, Recent progress in materials issues for piezoelectric MEMS, J. Am. Ceram. Soc., 91, 1385–1396 (2008).
22. H. F. Pang, L. Garcia-Gancedo, Y. Q. Fu, S. Porro, Y. W. Gu, J. K. Luo, X. T. Zu, F. Placido, J. I. B. Wilson, A. J. Flewitt, W. I. Milne, Characterization of the surface acoustic wave devices based on Zn O/nanocrystalline diamond structures, Phys. Stat. Sol. (a) 210, 1575–1583 (2013).
23. K. Robbie, J. Sit, C. M. J. Brett, Advanced techniques for glancing angle deposition, J. Vac. Sci. Technol. B, 16, 1115–1122 (1998).
24. L.-C. Chen, C.-H. Tien, X. Liu, B. Xu, Zigzag and helical AlN layer prepared by glancing angle deposition and its application as a buffer layer in a GaN-based light-emitting diode, J. Nanomater., 2012, 409123 (2012).
25. T. Yanagitani, M. Suzuki, Significant shear mode softening in a c-axis tilt nanostructured hexagonal thin film induced by a self-shadowing effect, Sci. Mater., 69, 724–727 (2013).
26. J. K. Kwan, J. C. Sit, Acoustic wave liquid sensors enhanced with glancing angle-deposited thin films, Sens. Actuators B, 181, 715–719 (2013).
27. D. Toledano, R.E. Galindo, M. Yuste, J. M. Albella, O. Sánchez, Compositional and structural properties of nanostructured ZnO thin films grown by oblique angle reactive sputtering deposition: effect on thc refractive index, J. Phys. D: Appl. Phys. 46, 045306 (2013).
28. Y. Yoshino, K. Inoue, M. Takeuchi, T. Makino, Y. Katayama, T. Hata, Effect of substrate surface morphology and interface microstructure in ZnO thin films formed on various substrates, Vacuum 59(2-3), 403–410 (2000).
29. M. Novotný, J. Čížek, R. Kužel, J. Bulíř, J. Lančok, J. Connolly, E. McCarthy, S. Krishnamurthy, J.-P. Mosnier, W. Anwand, G. Brauer, Structural characterization of

ZnO thin films grown on various substrates by pulsed laser deposition, J. Phys. D: Appl. Phys. 45, 225101 (2012).

30. P. Misra, L. M. Kukreja, Buffer-assisted low temperature growth of high crystalline quality ZnO films using pulsed laser deposition, Thin Solid Films 485, 42–46 (2005).
31. V. Khranovskyy, R. Minikayev, S. Trushkin, G. Lashkarev, V. Lazorenko, U. Grossner, W. Paszkowicz, A. Suchocki, B. G. Svensson, R. Yakimova, Improvement of ZnO thin film properties by application of ZnO buffer layers, J. Cryst. Growth 308, 93–98 (2007).
32. D.-T. Phan, H.-C. Suh, G.-S. Chung, Surface acoustic wave characteristics of ZnO films grown on a polycrystalline 3C-SiC buffer layer, Microelectr. Eng., 88, 105–108 (2011).
33. J. H. Kim, E.-M. Kim, D. Andeen, D. Thomson, S. P. DenBaars, F. F. Lange, Growth of heteroepitaxial ZnO thin films on GaN-buffered Al_2O_3 (0001) substrates by low-temperature hydrothermal yynthesis at 90 °C, Adv. Funct. Mater., 17, 463–471 (2007).
34. H. Nakahata, S. Fujii, K. Higaki, A. Hachigo, H. Kitabayashi, S. Shikata, N. Fujimori, Diamond-based surface acoustic wave devices, Semicond. Sci. Technol., 18, S96–S104 (2003).
35. L. Garcia-Gancedo, J. Pedros, Z. Zhu, A. J. Flewitt, W. I. Milne, J. K. Luo, C. J. B. Ford, Room-temperature remote-plasma sputtering of c-axis oriented zinc oxide thin films, J. Appl. Phys., 112, 014907 (2012).
36. A. Gulino, F. Lupo, M. E. Fragalà, Substrate-free, self-standing ZnO thin films, J. Phys. Chem. C 112(36), 13869–13872 (2008).
37. Z. Xu, Z. Li, Design and fabrication of ZnO-based SAW sensor using low power homo-buffer layer for enhanced humidity sensing, IEEE Sensors J. 21(6), 7428–7433(2021)
38. H. Amaike, K. Hazu, Y. Sawai, S. F. Chichibu, Helicon-wave-excited-plasma sputtering as an expandable epitaxy method for planar semiconductor thin films, Appl. Phys. Expr., 2, 105503 (2009).
39. Ü. Özgür, Y. I. Alivov, C. Liu, A. Teke, M. A. Reshchikov, S. Doğan, V. Avrutin, S.-J. Cho, H. Morkoç, A comprehensive review of ZnO materials and devices, J. Appl. Phys., 98, 041301 (2005).
40. A. D. Corso, M. Posternak, R. Resta, A. Balderschi, Ab initio study of piezoelectricity and spontaneous polarization in ZnO, Phys. Rev. B 50, 10715 (1994).
41. J. T. Luo, F. Pan, P. Fan, F. Zeng, D. P. Zhang, Z. H. Zheng, G. X. Liang, Cost-effective and high frequency surface acoustic wave filters on ZnO:Fe/Si for low-loss and wide-band application, Appl. Phys. Lett., 101, 172909 (2012).
42. G. Chen, J. J. Peng, C. Song, F. Zeng, F. Pan, Interplay between chemical state, electric properties, and ferromagnetism in Fe-doped ZnO films, J. Appl. Phys. 113, 104503 (2013).
43. N. W. Emanetoglu, S. Muthukumar, P. Wu, R. Wittstruck, Y. Chen, Y. Lu, $Mg_xZn_{1-x}O$: a new piezoelectric material, IEEE Trans. Ultrason. Ferroelect. Freq. Contr., 50, 537–543 (2003).
44. Y. Chen, N. W. Emanetoglu, G. Saraf, P. Wu, Y. Lu, Analysis of SAW properties in $ZnO/Al_xGa_{1-x}N/c-Al_2O_3$ structures, IEEE Trans. Ultrason., Ferroelect. Freq. Contr., 52, 1161–1169 (2005).
45. H. J. Xiang, J. Yang, J. G. Hou, Q. Zhu, Piezoelectricity in ZnO nanowires: a first-principles study, Appl. Phys. Lett., 89, 223111 (2006).
46. R. Agrawal, H. D. Espinosa, Giant piezoelectric size effects in zinc oxide and gallium nitride nanowires. A first principles investigation, Nano Lett., 11, 786–790 (2011).
47. A.V. Singha, S. Chandra, A.K. Srivastava, B.R. Chakroborty, G. Sehgal, M.K. Dalai, G. Bose, Structural and optical properties of RF magnetron sputtered aluminum nitride films without external substrate heating, Appl. Surf. Sci. 257, 9568–9573 (2011).
48. R. Supruangnet, W. Sailuam, W. Busayaporn, C. Wattanawikkam, A. Jiamprasertboon, A. Ruangvittayanon, W. Sangsai, A. Pirasampansiri, S. Limpijumnong, R. Yimnirun,

A. Bootchanont, Effects of N_2-content on formation behavior in AlN thin films studied by NEXAFS: Theory and experiment, J. Alloys Comp., 844, 156128 (2020).

49. Y. Q. Fu, J. K. Luo, N. T. Nguyen, A.J. Walton, A.J. Flewitt, X.T Zu, Y. Li, G. McHale, A. Matthews, E. Iborra, H. Du, W.I. Milne, Advances in piezoelectric thin films for acoustic biosensors, acoustofluidics and lab-on-chip applications, Prog. Mater Sci., 89, 31–91 (2017).
50. Y. Yu, T.-L. Ren, L.-T. Liu, High quality silicon-based AlN thin films for MEMS application, Integr. Ferroelectr. Int. J. 69, 367–374 (2005).
51. G. Wingquist, J. Bjurstrom, L. Liljeholm, L. Liljeholm,V. Yantchev, J. Katardjiev, Shear mode AlN thin film electro-acoustic resonant sensor operation in viscous media, Sens. Actuators B, 123, 466–473 (2007).
52. G. F. Iriarte, J. Bjurstrom, J. Westlinder, F. Engelmark, I. V. Katardjiev, Synthesis of c-axis-oriented AlN thin films on high-conducting layers: Al, Mo, Ti, TiN, and Ni, IEEE Trans. Ultrason. Ferroelectr. Freq. Control, 52(7), 1170–1174 (2005).
53. G. F. Iriarte, J. G. Rodriguez, F. Calle, Synthesis of c-axis oriented AlN thin films on different substrates: A review, Mater. Res. Bull., 45(9),1039–1045 (2010).
54. L. Zhang, F. Xu, M. Wang, Y. Sun, N. Xie, T. Wang, B. Dong, Z. Qin, X. Wang, B. Shen, High-quality AlN epitaxy on sapphire substrates with sputtered buffer layers, Superlatt. Microstruct., 105, 34–38 (2017).
55. A. Benetti, D. Cannata, F. Di Pietrantonio, E. Verona, A. Generosi, B. Paci, V.R. Albertini, Growth and characterization of piezoelectric AlN thin films for diamond-based surface acoustic wave devices, Thin Solid Films 497(1-2), 304–308 (2006).
56. K. Y. Hashimoto, S. Sato, A. Teshigahara, T. Nakamura, K. Kano, High-performance surface acoustic wave resonators in the 1 to 3 GHz range using a ScAlN/6H-SiC structure, IEEE Trans. Ultrason. Ferroelectr. Freq. Control 60(3), 637–642 (2013).
57. W. Wang, P. M. Mayrhofer, X. He, M. Gillinger, Z. Ye, X. Wang, A. Bittner, U. Schmid, J. K. Luo, High performance AlScN thin film based surface acoustic wave devices with large electromechanical coupling coefficient, Appl. Phys. Lett., 105, 133502 (2014).
58. A. Sanz-Hervás, L. Vergara, J. Olivares, E. Iborra, Y. Morilla, J. García-López, M. Clement, J. Sangrador, M. A. Respaldiza, Comparative study of c-axis AlN films sputtered on metallic surfaces, Diamond Relat. Mater. 14, 1198–1202 (2005).
59. M. Dubois, P. Muralt, Stress and piezoelectric properties of aluminum nitride thin films deposited onto metal electrodes by pulsed direct current reactive sputtering, J. Appl. Phys. 89, 6389–6395 (2001).
60. W. Wang, W. Yang, Z. Liu, H. Wang, L. Wen, G. Li, Interfacial reaction control and its mechanism of AlN epitaxial films grown on Si (111) substrates by pulsed laser deposition, Sci. Rep., 5, 11480 (2015).
61. C. P. Huang, C. H. Wang, C. P. Liu, and K. Y. Lai, High-quality AlN grown with a single substrate temperature below 1200 °C, Sci. Rep. 7, 7135 (2017).
62. M. Nemoz, R. Dagher, S. Matta, A. Michon, P. Vennégués, J. Brault, Dislocation densities reduction in MBE-grown AlN thin films by high-temperature annealing, J. Cryst. Growth, 461, 10–15 (2017).
63. E. S. Hellman, The polarity of GaN: a critical review, MRS Internet J. Nitride Semicond. Res. 3, e11 (1998).
64. O. Kluth, G. Schope, J. Hüpkes, C. Agashe, J. Müller, B. Rech, Modified Thornton model for magnetron sputtered zinc oxide: film structure and etching behaviour, Thin Solid Films 442, 80–85 (2003).
65. J. Zhou, H.-F. Pang, L. Garcia-Gancedo, E. Iborra, M. Clement, M. De Miguel-Ramos, et al., Discrete microfluidics based on aluminum nitride surface acoustic wave devices, Microfluid. Nanofluid., 18, 537–548 (2015).

66. D. Zhang, P. Fan, X. Cai, J. Huang, L. Ru, Z. Zheng, et al., Properties of ZnO thin films deposited by DC reactive magnetron sputtering under different plasma power, Appl. Phys. A, 97, 437–441 (2009).
67. S. Calnan, H. M. Upadhyaya, M. J. Thwaites, A. N. Tiwari, Properties of indium tin oxide films deposited using high target utilisation sputtering, Thin Solid Films 515(15), 6045–6050 (2007).
68. F. M. Li, C. B. Bayer, S. Hofmann, S. P. Speakman, C. Ducati, W. I. Milne, A. J. Flewitt, High-density remote plasma sputtering of high-dielectric-constant amorphous hafnium oxide films, Phys. Status Solidi (b) 250, 957–967 (2013).
69. M. Vijatović, J. Bobić, B. Stojanović, History and challenges of barium titanate: Part I, Sci. Sintering, 40(2), 155–165 (2008).
70. M. Vijatović, J. Bobić, B. Stojanović, History and challenges of barium titanate: Part II, Sci. Sintering, 40(3), 235–244 (2008).
71. X. Xing, X. Zhu, J. Li, Structure of $Pb(Zr,Ti)O_3$(PZT) for power ultrasonic transducer, J. Wuhan Univ. Techn.-Mater. Sci. Ed., 33, 884–887 (2018).
72. G. H. Haertling, Ferroelectric ceramics: history and technology, J. Am. Ceram. Soc., 82, 797–818 (1999).
73. M. Siddiqui, J. J. Mohamed, Z. A. Ahmad, Structural, piezoelectric, and dielectric properties of PZT-based ceramics without excess lead oxide, J. Austr. Ceram. Soc., 56, 371–377 (2020).
74. N. Setter, D. Damjanovic, L. Eng, G. Fox, S. Gevorgian, S. Hong, et al., Ferroelectric thin films: Review of materials, properties, and applications, J. Appl. Phys., 100, 051606 (2006).
75. N. Izyumskaya, Y.-I. Alivov; S.-J. Cho, H. Morkoç, H. Lee,Y.-S. Kang, Processing, structure, properties, and applications of PZT thin films, Crit. Rev. Sol. Stat. Sci., 32(3-4), 111–202 (2007).
76. B. Jaffe, R. S. Roth, S. Marzullo, Piezoelectric properties of lead zirconate-lead titanate solid-solution ceramics, J. Appl. Phys., 25(6), 809–810 (1954)
77. D. I. Woodward, J. Knudsen, I. M. Reaney, Review of crystal and domain structures in the $PbZr_xTi_{1-x}O_3$ solid solution, Phys. Rev. B. 72, 104110 (2005).
78. D. Damjanovic, Ferroelectric, dielectric and piezoelectric properties of ferroelectric thin films and ceramics, Rep. Prog. Phys. 61 (9), 1267–1324 (1998).
79. A. M. Roji M, J. G, and A. B. Raj T, A retrospect on the role of piezoelectric nanogenerators in the development of the green world, RSC Adv., 7, 33642–33670 (2017).
80. D. Damjanovic, N. Klein, J. I. N. Li, V. Porokhonskyy, What can be expected from lead-free piezoelectric materials? Funct. Mater. Lett., 03, 5–13 (2010)
81. W. R. Ali, M. Prasad, Piezoelectric MEMS based acoustic sensors: A review, Sens. Actuators A 301, 111756 (2020).
82. C.K. Kwok, S.B. Desu, Formation kinetics of $PbZr_xTi_{1-x}O_3$ thin films, J. Mater.Res. 9 (7), 1728–1733 (1994).
83. A. Bose, M. Sreemany, S. Bysakh, Role of TiO_2 seed layer thickness on the nanostructure evolution and phase transformation behavior of sputtered PZT thin films during post-deposition air-annealing, J. Am. Ceram. Soc. 94(11), 4066–4077 (2011).
84. S. J. Rupitsch, Piezoelectric sensors and sctuators: fundamentals and applications, Springer-Verlag GmbH Germany, part of Springer Nature, 2019, p. 57
85. P. Muralt, N. Ledermann, J. Paborowski, A. Barzegar, S. Gentil, B. Belgacem, S. Petitgrand, A. Bosseboeuf, N. Setter, Piezoelectric micromachined ultrasonic transducers based on PZT thin films, IEEE Trans. Ultrason., Ferroelectr., Freq. Control 52(12), 2276–2288 (2005).
86. F. Hadj-Larbi, R Serhane, Sezawa SAW devices: Review of numerical-experimental studies and recent applications, Sens. Actuators A 292, 169–197 (2019).

87. F. Martina, M. I. Newton, G. McHale, K. A. Melzak, E. Gizeli, Pulse mode shear horizontal-surface acoustic wave (SH-SAW) system for liquid based sensing applications, Biosens. Bioelectron., 19(6), 627–632 (2004).
88. T. Kogai, H. Yatsuda, J. Kondoh, Rayleigh SAW-Assisted SH-SAW Immunosensor on X-Cut 148-Y $LiTaO_3$, IEEE Trans. Ultrason. Ferroelectr. Freq. Control, 4(9), 1375–1381 (2017).
89. K. Takayanagi, J. Kondoh, Improvement of estimation method for physical properties of liquid using shear horizontal surface acoustic wave sensor response, Japan. J. Appl. Phys., 57(7S1), 07LD02 (2018).
90. N. Moll, E. Pascal, D.H. Dinh, J.-P. Pillot, B. Bennetau, D. Rebiere, D. Moynet, Y. Mas, D. Mossalayi, J. Pistre, C. Dejous, A Love wave immunosensor for whole E. coli bacteria detection using an innovative two-step immobilisation approach, Biosens. Bioelectron., 22, 2145–2150 (2007).
91. H.-F. Pang, Y.-Q. Fu, Z.-J. Li, Y.-F. Li, F. Placido, A. Walton, X.-T. Zu, Love mode surface acoustic wave ultraviolet sensor using ZnO films deposited on 36° Y-cut $LiTaO_3$, Sens. Actuator A, 193, 87–94 (2013).
92. J. Kankare, Sauerbrey equation of quartz crystal microbalance in liquid medium, Langmuir, 18, 7092–7094 (2002).
93. T. W. Grudkowski, J. F. Black, T. M. Reeder, D. E. Cullen, R. A. Wagner, Piezoelectric materials parameters for piezoelectric thin films in GHz applications, Appl. Phys. Lett., 37, 993–995 (1980).
94. Y. Zhang, J. Luo, A. J. Flewitt, Z. Cai, X. Zhao, Film bulk acoustic resonators (FBARs) as biosensors: A review, Biosens. Bioelectron., 116, 1–15 (2018).
95. C.-J. Chung, Y.-C. Chen, C.-C. Cheng, K.-S. Kao, An improvement of tilted AlN for shear and longitudinal acoustic wave, Appl. Phys. A, 94(2):307–313 (2009).
96. X. Chen, Y. Yang, H. L. Cai, C. J. Zhou, T. L.R en, M. A. Mohammad, A multiple resonant mode film bulk acoustic resonator based on silicon-on-insulator structures, Chin. Phys. Lett., 31(12), 124302 (2014).
97. R. White, F. Voltmer, Direct piezoelectric coupling to surface elastic waves, Appl. Phys. Let., 7, 314–316 (1965).
98. S. Lehtonen, V.P. Plessky, C.S. Hartmann, M. Salomaa, Extraction of the SAW attenuation parameter in periodic reflecting gratings, IEEE Trans. Ultra. Ferroelectr. Freq. Contr., 51, 1697–1703 (2004).
99. S. Büyükköse, B. Vratzov, J. van der Veen, P. V. Santos, W. R. van der Wiel, Ultrahigh-frequency surface acoustic wave generation for acoustic charge transport in silicon, Appl. Phys. Lett., 102, 013112 (2013).
100. H. Li, J. Liu, K. Li, Y. Liu, A review of recent studies on piezoelectric pumps and their applications, Mech. Syst. Signal Pr., 151, 107393 (2021).
101. D. Balma, A. Lamberti, S.L. Marasso, D. Perrone, M. Quaglio, G. Canavese, S. Bianco, M. Cocuzza, Piezoelectrically actuated MEMS microswitches for high current applications, Microelectron. Eng., 88, 2208–2210 (2011).
102. X. Tian, Y. Liu, J. Deng, L. Wang, W. Chen, A review on piezoelectric ultrasonic motors for the past decade: Classification, operating principle, performance, and future work perspectives, Sens. Actuators A, 306, 111971 (2020).
103. Q. Zhou, S. Lau, D. Wu, K. K. Shung, Piezoelectric films for high frequency ultrasonic transducers in biomedical applications, Prog. Mater. Sci., 56, 139–174 (2011).
104. H. F. Pang, Y. Q. Fu, R. Hou, K. J. Kirk, D. Hutson, X. T. Zu, F. Placido, Annealing effect on the generation of dual mode acoustic waves in inclined ZnO films, Ultrasonics 53, 1264–1269 (2013).
105. G. L. Smith, R. Q. Rudy, R. G. Polcawich, D. L. Devoe, Integrated thin-film piezoelectric traveling wave ultrasonic motors, Sens. Actuator A, 188, 305–311 (2012).
106. J. Li, H. Huang, T. Morita, Stepping piezoelectric actuators with large working stroke for nano-positioning systems: A review, Sens. Actuators A, 292, 39–51 (2019).

107. R. Tao, W.B. Wang, J.T. Luo, S. Hasan, H. Torun, P. Canyelles-Pericas, J. Zhou, W.P. Xuan, M. Cooke, D. Gibson, Q. Wu, W.P. Ng, J.K. Luo, Y.Q. Fu, Thin film flexible/bendable acoustic wave devices: Evolution, hybridization and decoupling of multiple acoustic wave modes, Surf. Coatings Technol., 357, 587–594 (2019).
108. D. C. Malocha, M. Gallagher, B. Fisher, J. Humphries, D. Gallagher, N. Kozlovski, A passive wireless multi-sensor SAW technology device and system perspectives, Sensors, 13(5), 5897–5922 (2013).
109. A. Muller, D. Neculoiu, D. Vasilache, G. Konstantinidis, K. Grenier, D. Dubuc, L. Bary, R. Plana, E. Flahaut, High performance thin film bulk acoustic resonator covered with carbon nanotubes, Appl. Phys. Lett., 89, 143122 (2006).
110. G. Rughoobur, M. DeMiguel-Ramos, J.-M. Escolano, E. Iborra, A. J. Flewitt, Gravimetric sensors operating at 1.1 GHz based on inclined c-axis ZnO grown on textured Al electrodes, Sci. Rep., 7, 1367 (2017).
111. S. Wang, W. Rong, L. Wang, H. Xie, L. Sun, J. K. Mills, A survey of piezoelectric actuators with long working stroke in recent years: Classifications, principles, connections and distinctions, Mech. Syst. Signal Pr., 123, 591–605 (2019).
112. A. Afzal, N. Iqbal, A. Mujahid, R. Schirhagl, Advanced vapor recognition materials for selective and fast responsive surface acoustic wave sensors: A review, Analytica Chimica Acta 787, 36–49 (2013).
113. M. Penza, E. Milella, V.I. Anisimkin, Gas sensing properties of Langmuir-Blodgett polypyrrole film investigated by surface acoustic waves, IEEE Trans. Ultrason. Ferroelectr. Freq. Control. 45, 1125–1132 (1998).
114. H. Wohltjen, Mechanism of operation and design considerations for surface acoustic wave device vapour sensors, Sens. Actuators B 5, 307–325 (1984).
115. D. Kwak, Y. Lei, R. Maric, Ammonia gas sensors: A comprehensive review, Talanta, 204, 713–730 (2019).
116. W. Li, Y. Guo, Y. Tang, X. Zu, J. Ma, L. Wang, Y. Q. Fu, Room-temperature ammonia sensor based on ZnO nanorods deposited on ST-cut quartz surface acoustic wave devices, Sensors 17, 1142 (2017).
117. Y.-L. Tang, Z.-J. Li, J.-Y. Ma, H.-Q. Su, Y.-J. Guo, L.u Wang, B. Du, J.-J. Chen, W. Zhou, Q.-K. Yu, X.-T. Zu, Highly sensitive room-temperature surface acoustic wave (SAW) ammonia sensors based on Co_3O_4/SiO_2 composite films, J. Hazard. Mater., 280, 127–133 (2014).
118. Y. Tang, D. Ao, W. Li, X. Zu, S. Li, Y. Q. Fu, NH_3 sensing property and mechanisms of quartz surface acoustic wave sensors deposited with SiO_2, TiO_2, and SiO_2-TiO_2 composite films, Sens. Actuators B, 254, 1165–1173 (2018).
119. S.-Y. Wang, J.-Y. Ma, Z.-J. Li, H. Q. Su, N. R. Alkurd, W.-L. Zhou, L. Wang, B. Du, Y.-L. Tang, D.-Y. Ao, S.-C. Zhang, Q.K. Yu, X.-T. Zu, Surface acoustic wave ammonia sensor based on ZnO/SiO_2 composite film, J. Hazard. Mater., 285, 368–374 (2015).
120. Q. B. Tang, Y. J. Guo, Y. L. Tang, G. D. Long, J. L. Wang, D. J. Li, X. T. Zu, J. Y. Ma, L. Wang, H. Torun, Y. Q. Fu, Highly sensitive and selective Love mode surface acoustic wave ammonia sensor based on graphene oxides operated at room temperature, J. Mater. Sci., 54, 11925–11935 (2019).
121. Y. Pan, N. Mu, B. Liu, B. Cao, W. Wang, L. Yang, A novel surface acoustic wave sensor array based on wireless communication network, Sensors 18, 2977 (2018).
122. C.-Y. Shen, S.-Y. Liou, Surface acoustic wave gas monitor for ppm ammonia detection, Sens. Actuators B 131, 673–679 (2008).
123. P. S. Chauhan, S. Bhattacharya, Hydrogen gas sensing methods, materials, and approach to achieve parts per billion level detection: A review, Int. J. Hydrogen Energ., 44(47), 26076–26099 (2019).

124. L. Yang, C. Yin, Z. Zhang, J. Zhou, H. Xu, The investigation of hydrogen gas sensing properties of SAW gas sensor based on palladium surface modified SnO_2 thin film, Mater. Sci. Semicond Process 60, 16e28(2017).
125. C. Tasaltin, M. A. Ebeoglu, Z. Z. Ozturk, Acoustoelectric effect on the responses of saw sensors coated with electrospun ZnO nanostructured thin film, Sensors 12, 12006 (2012).
126. S.J. Ippolito, S. Kandasamy, K. Kalantar-Zadeh, W. Wlodarski, Layered SAW hydrogen sensor with modified tungsten trioxide selective layer, Sens. Actuators B 108, 553–557 (2005).
127. N. H. Ha, N. H. Nam, D. D. Dung, N. H. Phuong, P. D. Thach, H. S. Hong, Hydrogen gas sensing using palladium-graphene nanocomposite material based on surface acoustic wave, J. Nanomater., 2017,1e6 (2017).
128. D.-T. Phan, G.-S. Chung, Surface acoustic wave hydrogen sensors based on ZnO nanoparticles incorporated with a Pt catalyst, Sens. Actuators B 161, 341–348 (2012).
129. H. Nagai, S. Kawai, O. Ito, T. Oizmi, T. Tsuji, N. Takeda, K. Yamanaka, Possibility for sub-ppm hydrogen detection with the ball SAW sensor, The 20th International Congress on Acoustics, Sydney, Australia, 23–27 August 2010.
130. D. Li, X. Zu, D. Ao, Q. Tang, Y.Q. Fu, Y. Guo, K. Bilawal, M. B. Faheem, L. Li, S. Li, Y. Tang, High humidity enhanced surface acoustic wave (SAW) H_2S sensors based on sol–gel CuO films, Sens. Actuators B, 294, 55–61 (2019).
131. W. Luo, Q. Fu, D. Zhou, J. Deng, H. Liu, G. Yan, A surface acoustic wave H_2S gas sensor employing nanocrystalline SnO_2 thin film, Sens. Actuators B 176, 746–752 (2013).
132. X. Wang, W. Wang, H. Li, C. Fu, Y. Ke, S. He, Development of a SnO_2/CuO-coated surface acoustic wave-based H_2S sensor with switch-like response and recovery, Sens. Actuators B 169, 10–16 (2012).
133. M. Asad, M. H. Sheikhi, Surface acoustic wave based H_2S gas sensors incorporating sensitive layers of single wall carbon nanotubes decorated with Cu nanoparticles, Sens. Actuators B 198, 134–141 (2014).
134. S.J. Ippolito, S. Kandasamy, K. Kalantar-zadeh, W. Wlodarski, K. Galatsis, G. Kiriakidis, N. Katsarakis, M. Suchea, Highly sensitive layered ZnO/$LiNbO_3$ SAW device with InO_x selective layer for NO_2 and H_2 gas sensing, Sens. Actuators B 111–112, 207–212 (2005).
135. L. Rana, R. Gupta, M. Tomar, V. Gupta, ZnO/ST-Quartz SAW resonator: An efficient NO_2 gas sensor, Sens. Actuators B, 252, 840–845 (2017).
136. M. Li, H. Kan, S. Chen, X. Feng, H. Li, C. Li, C. Fu, A. Quan, H. Sun, J. Luo, X. Liu, W. Wang, H. Liu, Q. Wei, Y. Fu, Colloidal quantum dot-based surface acoustic wave sensors for NO_2-sensing behavior, Sens. Actuators B, 287, 241–249 (2019).
137. C. Tasaltin, M. A. Ebeoglu, Z. Z. Ozturk, Acoustoelectric effect on the responses of saw sensors coated with electrospun ZnO nanostructured thin film, Sensors 12, 12006 (2012).
138. D. L. García-González, R. Aparicio, Sensors: From biosensors to the electronic nose, Grasas y Aceites, 53(1), 96–114 (2002).
139. Y. Zhang Y. Cai, J. Zhou, Y. Xie, Q. Xu Y. Zou, S. Guo, H. Xu, C. Sun, S. Liu, Surface acoustic wave-based ultraviolet photodetectors: a review, Sci. Bull., 65, 587–600 (2020).
140. N. W. Emanetoglu, J. Zhu, Y. Chen, J. Zhong, Y. Chen, Y. Lu, Surface acoustic wave ultraviolet photodetectors using epitaxial ZnO multilayers grown on r-plane sapphire, Appl. Phys. Lett., 85, 3702–3074 (2004).
141. P. Zhou, C. Chen, X. Wang, B. Hu, H. San, 2-Dimentional photoconductive MoS_2 nanosheets using in surface acoustic wave resonators for ultraviolet light sensing, Sens. Actuators A, 271, 389–97 (2018).

142. V. S. Chivukula, D. Ciplys, R. Rimeika, M. S. Shur, J. Yang, R. Gaska, Highly sensitive radio-frequency UV sensor based on photocapacitive effect in GaN, IEEE Sens. J., 10, 883–887 (2010).
143. P. Sharma, K. Sreenivas, Highly sensitive ultraviolet detector based on ZnO/$LiNbO_3$ hybrid surface acoustic wave filter, Appl. Phys. Lett., 83, 3617–3619 (2003).
144. S. Kumar, P. Sharma, K. Sreenivas, Low-intensity ultraviolet light detector using a surface acoustic wave oscillator based on ZnO/$LiNbO_3$ bilayer structure, Semicond. Sci. Technol., *20*, L27–L30 (2005).
145. D.-T. Phan, G.-S. Chung, Characteristics of SAW UV sensors based on a ZnO/Si structure using third harmonic mode, Curr. Appl. Phys., 12, 210–213 (2012).
146. W.-S. Wang, T.-T. Wu, T.-H. Chou, Y.-Y. Chen, A ZnO nanorod-based SAW oscillator system for ultraviolet detection, Nanotechnology 20, 135503 (2009).
147. S. Kumar, G.-H. Kim, K. Sreenivas, R. P. Tandon, ZnO based surface acoustic wave ultraviolet photo sensor, J. Electroceram. 22, 198–202 (2009).
148. T. Phan, G.-S. Chung, Fabrication and characteristics of a surface acoustic wave UV sensor based on ZnO thin films grown on a polycrystalline 3C–SiC buffer layer, Curr. Appl. Phys. 12, 521–524 (2012).
149. B.-W. Jeon, Y. H. Kwak, B.-K. Ju, K. Kim, Ultraviolet sensor with fast response characteristics based on an AgNW/ZnO bi-layer, Sens. Actuators A, 311, 112044 (2020).
150. C. Fu, K.J. Lee, K. Lee, S.S. Yang, Low-intensity ultraviolet detection using a surface acoustic-wave sensor with a Ag-doped ZnO nanoparticle film, Smart Mater. Struct. 24 (1), 015010 (2014).
151. C.-L. Wei, Y.-C. Chen, C.-C. Cheng, K.-S. Kao, D.-L. Cheng, P.-S. Cheng, Highly sensitive ultraviolet detector using a ZnO/Si layered SAW oscillator, Thin Solid Films 518, 3059–3062 (2010).
152. Y J Guo, C Zhao, X S Zhou, Y Li, X T Zu, D Gibson,Y Q Fu, Ultraviolet sensing based on nanostructured ZnO/Si surface acoustic wave devices, Smart Mater. Struct. 24, 125015 (2015).
153. W. Water, C.-W. Wen, Application of TiO_2 thin film with nanorods to surface acoustic wave type ultraviolet photo detection, J. Electroceram., 36(1-4), 94–101 (2016).
154. W. Water, R.-Y. Jhao, L.-W. Ji, T.-H. Fang S.-E. Chen, Love wave ultraviolet photodetector using ZnO nanorods synthesized on 90°-rotated ST-cut (42° 45′) quartz Sens. Actuators A 161, 6–11 (2010).
155. S. Y. Chu, W. Water, J. T. Liaw, An investigation of the dependence of ZnO film on the sensitivity of Love mode sensor in ZnO/quartz structure, Ultrasonics 41(2), 133–139 (2003).
156. I. Voiculescu, A. N. Nordin, Acoustic wave based MEMS devices for biosensing applications, Biosens. Bioelectron., 33, 1–9 (2012).
157. R.-C. Chang, S.-Y. Chu, C.-S. Hong, Y.-T. Chuang, A study of Love wave devices in ZnO/Quartz and ZnO/$LiTaO_3$ structures, Thin Solid Films 498, 146–151 (2006).
158. K. Länge, B. E. Rapp, M. Rapp, Surface acoustic wave biosensors: a review, Anal. Bioanal. Chem., 391, 1509–1519 (2008).
159. M.-I. Rocha-Gaso, C. March-Iborra, Á. Montoya-Baides, A. Arnau-Vives, Surface generated acoustic wave biosensors for the detection of pathogens: A review, Sensors 9, 5740 (2009).
160. K. Kalantar-Zadeh, W. Wlodarski, Y. Y. Chen, B. N. Fry, K. Galatsis, Novel Love mode surface acoustic wave based immunosensors, Sens. Actuators B 91, 143–147 (2003).
161. S. Krishnamoorthy, A. A. Iliadis, T. Bei, G. P. Chrousos, An interleukin-6 ZnO/SiO_2/Si surface acoustic wave biosensor, Biosens. Bioelectron., 24, 313–318 (2008).
162. J. Luo, P. Luo, M. Xie, K. Du, B. Zhao, F. Pan, P. Fan F. Zeng, D. Zhang, Z. Zheng, G. Liang, A new type of glucose biosensor based on surface acoustic wave resonator using Mn-doped ZnO multilayer structure, Biosens. Bioelectron., 49, 512–518 (2013).

163. O. Tigli, L. Bivona, P. Berg, M. E. Zaghloul, Fabrication and characterization of a surface-acoustic-wave biosensor in CMOS technology for cancer biomarker detection, IEEE Trans. Biomed. Circuits Syst., 4, 62–73 (2010).
164. H.-W. Chang, J.-S. Shih, Surface acoustic wave immunosensors based on immobilized C_{60}-proteins, Sens. Actuators B 121, 522–529 (2007).
165. S. Li, Y. Wan, Y. Sua, C. Fan, V. R. Bhethanabotla, Gold nanoparticle-based low limit of detection Love wave biosensor for carcinoembryonic antigens, Biosens. Bioelectron., 95, 48–54 (2017).
166. X. Zhang, Y. Zou, C. An, K. Ying, X. Chen, P. Wang, Sensitive detection of carcinoembryonic antigen in exhaled breath condensate using surface acoustic wave immunosensor, Sens. Actuators B 217, 100–106 (2015).
167. P. J. Jandas, J. Luo, A. Quan, C. Qiu, Y. Q. Fu, Highly selective and label-free Love-mode surface acoustic wave biosensor for carcinoembryonic antigen detection using a self-assembled monolayer bioreceptor, Appl. Surf. Sci., 518, 146061 (2020).
168. L. Lamanna, F. Rizzi, V. R. Bhethanabotla, M. De Vittorio, Conformable surface acoustic wave biosensor for E-coli fabricated on PEN plastic film, Biosens. Bioelectron., 163, 112164 (2020).
169. D.-S. Lee, J. H. Lee, J. Luo, Y. Fu, W. I. Milne, S. Maeng, M. Y. Jung, S. H. Park, H. C. Yoon, A surface acoustic wave-based immunosensing device using a nanocrystalline ZnO film on Si, J. Nanosci. Nanotechnol. 9, 7181–7185 (2009)
170. Y. C. Chen, W. T. Chang, C. C. Cheng, J. Y. Shen, K. S. Kao, Development of human IgE biosensor using Sezawa-mode SAW devices, Curr. Appl. Phys., 14, 608–613 (2014).
171. M. L. Johnston, I. Kymissis, K. L. Shepard, FBAR-CMOS oscillator array for mass-sensing applications, IEEE Sensors J., 10, 1042–1047 (2010).
172. L. Mai, D.-H. Kim, M. Yim, G. Yoon, A feasibility study of ZnO-based FBAR devices for an ultra-mass-sensitive sensor application, Microw. Opt. Technol. Lett., *42*, 505–507 (2004).
173. S. S. Pottigari, J. W. Kwon, Vacuum-gapped film bulk acoustic resonator for low-loss mass sensing in liquid, IEEE International Conference on Solid-State Sensors, Actuators and Microsystems, Denver, 21–25 June, 2009, 156–159.
174. C. Klingshirn, J. Fallert, H. Zhou, H. Kalt, Comment on "Excitonic ultraviolet lasing in ZnO-based light emitting devices", Appl. Phys. Lett., 91, 126101 (2007).
175. X. Zhao, G. M. Ashley, L. Garcia-Gancedo, H. Jin, J. Luo, A. J. Flewitt, J. R. Lu, Protein functionalized ZnO thin film bulk acoustic resonator as an odorant biosensor, Sensors and Actuators B, 163(1), 242–246 (2012).
176. Y. Chen, P. I. Reyes, Z. Duan, G. Saraf, R. Wittstruck, Y. Lu, O. Taratula, E. Galoppini, Multifunctional ZnO-based thin-film bulk acoustic resonator for biosensors, J. Electron. Mater., 38 (8), 1605–1611 (2009).
177. J. Weber, W. Albers, J. Tuppurainen, M. Link, R. Gabl, W. Wersing, and M. Schreiter, Shear mode FBARs as highly sensitive liquid biosensors, Sens. Actuators, A 128, 84–88 (2006).
178. G. Rughoobur, H. Sugime, M. DeMiguel-Ramos, T. Mirea, S. Zheng, J. Robterson, E. Iborra, A.J. Flewitt, Carbon nanotube isolation layer enhancing in-liquid quality-factors of thin film bulk acoustic wave resonators for gravimetric sensing, Sens. Actuat B, 261, 398–407 (2018).
179. W. Xu, X. Zhang, S. Choi, J. Chae, A high-quality-factor film bulk acoustic resonator in liquid for biosensing applications, J. Microelectromech. Syst., 20, 213–220 (2011)
180. J. Liu, D. Chen, P. Wang, G. Song, X. Zhang, Z. Li, Y. Wang, J. Wan, J. Yang, A micro-fabricated thickness shear mode electroacoustic resonator for the label-free detection of cardiac troponin in serum, Talanta, 215, 120890 (2020).

181. D. Zheng, J. Xiong, P. Guo, Y. Li, S. Wang, H. Gu, Detection of a carcinoembryonic antigen using aptamer-modified film bulk acoustic resonators. Mater. Res. Bull., 59, 411–415(2014)
182. G. Wingqvist, H. Anderson, C. Lennartsson, T. Weissbach, V. Yantchev, A. Lloyd Spetz, Carbon nanotube isolation layer enhancing in-liquid quality-factors of thin film bulk acoustic wave resonators for gravimetric sensing, Biosens. Bioelectron., 24(11), 3387–3390 (2009).
183. D. Zheng, J. Xiong, P. Guo, S. F. Wang, H. S. Gu, AlN-based film buck acoustic resonator operated in shear mode for detection of carcinoembryonic antigens, RSC Adv., 6, 4908–4913 (2016).
184. Y. C. Chen, W. C. Shih, W. T. Chang, C. H. Yang, K. S. Kao, C. C. Cheng, Biosensor for human IgE detection using shear-mode FBAR devices, Nanoscale Res. Lett., 10, 69 (2015).
185. M. DeMiguel-Ramos, B. Díaz-Durán, J.M. Escolano, M. Barba, T. Mirea, J. Olivares, M. Clement, E. Iborra, Gravimetric biosensor based on a 1.3 GHz AlN shear-mode solidly mounted resonator, Sens. Actuators B 239, 1282–1288 (2017).
186. C. Han, X. Wang, Q. Zhao, L. Teng, S. Zhang, H. Lv, J. Liu, H. Ma, Y. Wang, Solidly mounted resonator sensor for biomolecule detections, RSC Adv., 9, 21323–21328 (2019).
187. S. Choi, M. Goryll, L. Y. M. Sin, P. K. Wong, J. Chae, Microfluidic-based biosensors toward point-of-care detection of nucleic acids and proteins, Microfluid. Nanofluidics, 10, 231–247 (2011/12).
188. D. Chen, J. Wang, D. Li, Y. Liu, H. Song, Q. Li, A poly (vinylidene fluoride)-coated ZnO film bulk acoustic resonator for nerve gas detection, J. Micromech. Microeng., 21, 085017 (2011).
189. D. Chen, J. Wang, Q. Liu, Y. Xu, D. Li, Y. Liu, Highly sensitive ZnO thin film bulk acoustic resonator for hydrogen detection, J. Micromech. Microeng., 21, 115018 (2011).
190. M. Zhang, L. Du, Z. Fang, Z. Zhao, Micro through-hole array in top electrode of film bulk acoustic resonator for sensitivity improving as humidity sensor, Procedia Eng. 120, 663–666 (2015).
191. J. Liu, Z. Zhao, Z. Fang, Z. Liu, Y. Zhu, L. Du, High-performance FBAR humidity sensor based on the PI film as the multifunctional layer, Sens. Actuators B 308, 127694 (2020).
192. Z. Wang, X. Qiu, S. J. Chen, W. Pang, H. Zhang, J. Shi, H. Yu, ZnO based film bulk acoustic resonator as infrared sensor, Thin Solid Films 519, 6144–6147 (2011).
193. X. Qiu, R. Tang, J. Zhu, J. Oiler, C. Yu, Z. Wang, H. Yu, The effects of temperature, relative humidity and reducing gases on the ultraviolet response of ZnO based film bulk acoustic-wave resonator, Sens. Actuators B 151, 360–364 (2011).
194. J. W. Sohn, S.-B. Choi, H. S. Kim, Vibration control of smart hull structure with optimally placed piezoelectric composite actuators, Int. J. Mech. Sci., 53, 647–659 (2011).
195. J. Qiu, H. Ji, K. Zhu, Semi-active vibration control using piezoelectric actuators in smart structures, Front. Mech. Eng. China, 4(3), 242–251 (2009).
196. D. Sun, L. Tong, D. Wang, An incremental algorithm for static shape control of smart structures with nonlinear piezoelectric actuators, Int. J. Solids Struct., 41(9–10), 2277–2292 (2004).
197. X. Shen, Y. Liu, J. Zhang, Study of piezoelectric fiber composite actuators applied in the flapping wing, 18th IEEE International Symposium on Applications of Ferroelectrics Xian, China, 23–27 August 2009,1-4.
198. J. C. Duran, J. A. Escareno, G. Etcheverry, M. Rakotondrabe, Getting started with peas-based flapping-wing mechanisms for micro aerial systems, Actuators 5(2), 14 (2016).

199. E. Steltz, S. Avadhanuka, R. Fearing, High lift force with 275 Hz wing beat in MFI, Proceedings of the International Conference on Intelligent Robots and Systems, San Diego, CA, USA, 29 Oct.–2 Nov., 2007, 3987–3992.
200. S. Guo, D. Li, J. Wu, Theoretical and experimental study of a piezoelectric flapping wing rotor for micro aerial vehicle, Aerosp. Sci. Technol., 23, 429–438 (2011).
201. Q. Nguyen, M. Syaifuddin; H. Park, D. Byun, N. Goo, K. Yoon, Characteristics of an insect-mimicking flapping system actuated by a unimorph piezoceramic actuator, J. Intell. Mater. Syst. Struct., 19, 1185–1193 (2008).
202. Q. Truong, Q. Nguyen, H. Park, D. Byun, N. Goo, Modification of a four-bar linkage system for a higher optimal flapping frequency, J. Intell. Mater. Syst. Struct., 22, 59–66 (2011).

9 Nanomaterials for Lithium(-ion) Batteries

Yi Li, Sam Zhang, and Maowen Xu
Southwest University, Chongqing,
People's Republic of China

CONTENTS

9.1 INTRODUCTION

9.1.1 History of Lithium-ion Batteries

In 1973, M. Stanley Whittingham found the layered TiS_2 could realize reversible storage of Li^+; based on this, a prototype of a rechargeable lithium metal battery was constructed [1]. Later in 1988, the first commercial rechargeable lithium battery was launched by Moli Energy Co., adopting lithium metal as the anode and layered MoS_2 as the cathode. However, the growth of lithium dendrites during cycling caused the short circuit of batteries, leading to a fire or even an explosion. As a result, security incidents terminated the research and development of lithium batteries.

In 1980, Michel Armand proposed whether it is possible to construct a new type of secondary lithium battery system with an intercalated lithium storage mechanism

DOI: 10.1201/9781003141358-9

for both cathode and anode. This system can be regarded as the reversible shuttle of Li^+ between the cathode and the anode during the charge/discharge process; thus, it is named a "rocking chair battery" [2]. Lithium-ion batteries began to brew in the scientific community, although the term "lithium-ion battery" didn't exist at the time.

During the same year, John B. Goodenough put forward metal oxides containing lithium to replace metal sulfides used as cathodes of lithium batteries, with a higher operating voltage and stability. Later, he uncovered layered $LiCoO_2$ as a promising cathode material [3]. The discovery of this crucial material provides the ideal cathode material for building a rocking chair lithium-ion battery.

Moreover, searching for an intercalated anode material for the reversible electrochemical storage of lithium ions with low voltage became an urgent problem to be solved. Scientists focused on layered graphite first, while the propylene carbonate (PC), widely adopted as a solvent for electrolyte at that time, would irreversibly co-insert with Li^+ to break the structure of graphite. The incompatibility of graphite with PC gives rise to the misconception that it is impossible to use graphite as an anode material. Finally, in 1983 after a long search journey, Akira Yoshino screened petroleum coke out as anode material, coupling with $LiCoO_2$ cathode to build the first lithium-ion battery prototype [4]. After a few years of investigation by their team, the first commercial LIB was launched by Sony Co., constituting $LiCoO_2$ as the cathode, petroleum coke as the anode, and $LiPF_6$ dissolved in PC as the electrolyte. The lithium-ion battery was discovered and continues to flourish until today [5].

The 2019 Nobel Prize in Chemistry has been awarded to John B. Goodenough, M. Stanley Whittingham, and Akira Yoshino for their remarkable contribution to lithium-ion batteries. The role of M. Stanley Whittingham laid on discovering an extremely energy-rich material TiS_2, which was used to create an innovative cathode for a lithium battery. Meanwhile, John B. Goodenough came up with the first Li-contained cathode material $LiCoO_2$ to realize the rocking chair-type lithium-ion battery concept. At the same time, Akira Yoshino utilized petroleum coke as a carbon material that intercalated lithium ions as anode material, coupled with $LiCoO_2$ to create the first commercially viable lithium-ion battery in 1985.

9.1.2 Working Principle

The typical LIBs are mainly composed of four parts: cathode electrode, an anode electrode, electrolyte, and separator. Besides, it also contains other components such as current collectors, binder, conductive additive, and packaging materials. Generally, the compounds with lithium intercalated are applied as cathode, such as $LiCoO_2$, while anode materials are compounds with potential close to lithium, like graphite, carbon fiber, and metal oxide. The electrolyte used in batteries is an organic solvent with lithium salts. Take an example of cell fabrication by $LiCoO_2$ as a cathode and graphite as an anode (Figure 9.1) [6]. During the charging process, Li^+ extracts from the lattice of $LiCoO_2$, combines with electrolyte molecule and passes through the membrane to insert into the interlayer of graphite, along with Co^{3+} turns into Co^{4+} and electrons pass toward the anode along external circuit to ensure charge balance. In the discharge process, the reverse process is conducted that Li^+ extracts and reinserts into the lattice of $LiCoO_2$ and oxidizes Co^{3+} into Co^{4+}, while electrons

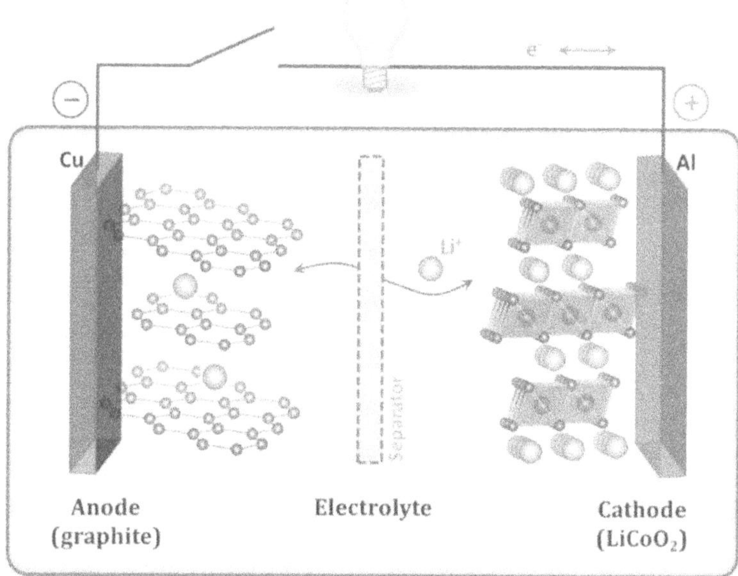

FIGURE 9.1 Schematic illustration of the first Li-ion battery ($LiCoO_2$/Li^+ electrolyte/graphite). (Reproduced with permission [6]. Copyright 2009, American Chemical Society.)

transfer to cathode along outer circuit. Li^+ shuttles back and forth between the cathode and the anode; hence lithium-ion batteries are called rocking chair batteries (RCB). Typically, the reaction of lithium-ion battery is a type of reversible chemical reaction.

9.2 CATHODE MATERIALS

The key component in modern LIBs is a cathode, determining the overall performance of the cell, which is Li^+ donor. Usually, a cathode material should meet the requirements given below [7]: (1) It should offer a high potential to ensure a high battery output voltage and maintain a stable voltage platform during the operation. (2) It should provide enough Li^+ for intercalation/extraction and be consumed for SEI firm forming on an anode-electrode surface during the first cycle. This hinges on the number of available lithium sites and the accessibility of multiple valence states for transition metal in the insertion host. (3) The structure of cathode materials should be stable enough and possesses high reversibility to guarantee a good cycling performance. (4) It should have a high diffusion coefficient of Li^+, a stable electrode interface, and a high power density, which enable LIBs with high rate capability to meet the demand of high power supply. (5) It should have a high electronic and ionic conductivity. (6) It should be rich in resources, inexpensive, thermally and chemically stable.

The most common commercial cathode materials mainly include layered oxides with 2D Li-ion diffusion channels (for instance, $LiCoO_2$), spinel oxide with 3D

TABLE 9.1
Some Parameters of Cathode Materials

Formula	**$LiCoO_2$**	**$LiMn_2O_4$**	**$LiFePO_4$**	**$LiNi_xCo_yMn_{1-x-y}O_2$**
Crystal pattern	Layered	Spinel	Olivine	Layered
Theoretical capacity (mA h g^{-1})	274	148	170	273–285
Actual capacity (mA h g^{-1})	135–150	100–120	130–140	155–220
Voltage platform (V)	3.7	3.8	3.4	3.6
Cyclicity (times)	500–1,000	500–2,000	2,000–6,000	800–2,000
Tap density (g cm^{-3})	2.8–3.0	2.2–2.4	0.80–1.10	2.6–2.8
Compacted density (g cm^{-3})	3.6–4.2	>3.0	2.20–2.30	>3.4
Cost (dollar kg^{-1})	40–47	14–23	23–31	24–25
Reserves (metal)	Low	Rich	Rich	-
Applications	3C products	Electric tools, bicycles, vehicles, large-scale electricity storage	Electric vehicles, large-scale electricity storage	Electric tools, bicycles, vehicles, large-scale electricity storage

Source: Ref. [9] Copyright 1948, with permission from IOP science, Ref. [10] Copyright 1985, with permission from Elsevier, Ref. [11] Copyright 1994, with permission from Elsevier, Ref. [12] Copyright 2002, with permission from Elsevier.

Li-ion diffusion channels (represented by $LiMn_2O_4$) polyanion with 1D Li-ion diffusion channels (represented by $LiFePO_4$), and ternary $LiNi_xCo_yMn_{1-x-y}O_2$ and $LiNi_xCo_yAl_{1-x-y}O_2$ [8]. Some basic parameters are listed in Table 9.1 [9–12].

9.2.1 Layered Oxides

Since proposed by John B. Goodenough in 1980, $LiCoO_2$ received enough attention from scientists and LIBs manufacturers and was then widely applied as the cathode material of the first commercial LIBs. $LiCoO_2$ is a representatively layered structure, as shown in Figure 9.2, where Li^+ and Co^{3+} occupy the alternate (111) planes to present a layered sequence of –O–Li–O–Co–O– along the *c*-axis. Li-ions diffuse along 2D channels in the crystalline structure of $LiCoO_2$. For $LiCoO_2$, there are two kinds of synthetic phase at low temperature and high temperature, respectively. When synthesized at high temperature above 800°C [13], $LiCoO_2$ is the O_3 layered structure with a good ordering of the Li^+ and Co^{3+} on the alternate (111) planes of the rock salt lattice, during the significant charge and size distinction between the Li^+ and Co^{3+} results in this good ordering. Consequently, this O_3-layered structure endows $LiCoO_2$ on many merits. First, the Li^+ and Co^{3+} ions order well in alternate layers, assuring excellent structural stability of cathode material. Second, the direct Co–Co interaction in this structure leads to high electronic conductivity. Third, the interconnected Li^+ sites facilitate good

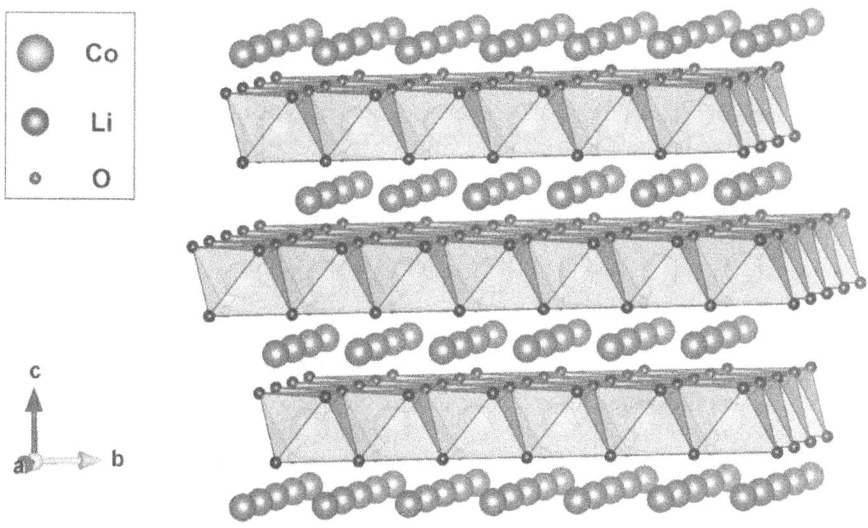

FIGURE 9.2 Crystal structure of layered $LiCoO_2$.

lithium-ion mobility and conductivity. When synthesized at a low temperature of about 400°C [14], $LiCoO_2$ is the O_2-layered structure with a disordering of the Li^+ and Co^{3+} ions, leading to the formation of a lithiated spinel-like phase, which demonstrates poor electrochemical performance.

No matter in an O_3 or O_2 layered structure, the rearrangement of Co and O arrays will occur with the constant change of Li content during cycling, leading to the emergence of new phases. Typically, when adopting the O_3-layered structure, $LiCoO_2$ undergoes three types of phase change during the extraction of Li^+. Take $Li_{1-x}CoO_2$ ($0 < x < 1$) to represent the phases emerging during the extraction of Li^+. When the extraction ratio of Li^+ is up to 7%–25% (x = 0.07–0.25), the first phase change occurs with an enlarger c-axis by 2% while the length of Co–Co is significantly reduced [15]. As a result, energy band dispersion, valence band, and conduction band overlap, and the material converses from semiconductor to metal conductor. When x = 0.25–0.5, the structure of $Li_{1-x}CoO_2$ remains unchanged, as well as the conductivity. The second and third phase changes occur when the extraction ratio of Li^+ is around 50%, Li^+ transits between disorder state and order state, followed by the transition from hexagonal to monoclinic phase. In Lu's work in 2012 [16], the O_2 structure has been directly observed in $Li_{1-x}CoO_2$ for the first time with ABF-STEM technology. In their opinion, the O_3 structure was converting to O_2 structure when $0.07 \leq x \leq 0.25$, O_2 structure was converting to O_1 structure when $0.25 \leq x \leq 0.43$, and when $0.43 \leq x \leq 0.52$ O_2 structure completed the transition to O_1 structure.

$LiCoO_2$, with a theoretical specific capacity of 274 mA h g^{-1}, can offer a capacity less than 150 mA h g^{-1} for practical LIBs merely, for the reason that when over 50% of Li-ions extract from the lattice, structure collapse occurs, leading to a sudden cycle stability deterioration. To limit the extraction of Li^+, the operational voltage for $LiCoO_2$ is usually limited to 4.35 V (vs Li^+/Li). The voltage platform of $LiCoO_2$

ranges from 3.6 V to 4.2 V, and the capacity retention rate reaches over 80% even after 500 cycles.

To get more Li^+ extract out of the crystal structure reversibly, heteroatomic doping and surface coating are widely employed to modify $LiCoO_2$. Up to now, a series of nanostructured oxides such as TiO_2, ZrO_2, SiO_2, and Al_2O_3 have been applied as coating materials for $LiCoO_2$, which effectively increase the reversible capacity up to 200 mA h g^{-1}, corresponding to a reversible extraction of 70% lithium [17]. Cho et al. utilized several metal oxides to coat $LiCoO_2$ [18], resulting in improved cycle stability at a high voltage window (2.75–4.4 V), and a higher reversible capacity of 170 mA h g^{-1} was obtained. From their view, the metal oxides coating materials would react with $LiCoO_2$ to generate an ultrathin film of solid solution $LiCo_{1-x}M_xO_2$(M = Zr, Al, Ti, B), which have been adequate to inhibit lattice expansion/shrinkage along the *c*-axis during the charge/discharge process. On the other hand, heteroatomic doping is an effective method to ameliorate the electrochemical stability of $LiCoO_2$ at high voltage. Li's group [19] reported $LiCoO_2$ doped with a trace amount of Ti, Mg, Al. Interestingly, they found that Mg and Al tend to insert in the inner lattice of $LiCoO_2$ to suppress dramatic phase transition occurring at high voltage operation. In comparison, Ti usually remains at the surface of $LiCoO_2$ to adjust the charge distribution of surrounding oxygen atoms in the state of delithiation and further reduce its oxidation activity.

The rapid development of electric vehicles induces a sharply increased demand for cathode materials. At the same time, the market put forward a new requirement of energy density at a higher level; while the conventional $LiCoO_2$ system has been developed and mature, its capacity has reached the limit. Ternary materials derived from layered $LiCoO_2$ captured researchers' attention to ease this dilemma. A complete replacement of Co by Ni/Mn/Li or any other transition metals gives rise to a new layered $LiNiO_2$ and $LiMnO_2$. However, partial replacement of Co by Ni and Mn produces a ternary material, i.e., $LiNi_xCo_yMn_{1-x-y}O_2$. Ternary material has a variety of combinations, as long as the average charge of the M site is +3. Ternary material $LiNi_xCo_yMn_{1-x-y}O_2$ has the same layered structure (R-3m space group) as $LiCoO_2$, in which Mn maintains a valence of +4 and has no electrochemical activity but contributes to stabilizing the layered structure. At the same time, Ni occupies a valence of +2 to exhibit electrochemical activity, as for Co, which occupies a valence of +3 and exhibits electrochemical activity as well, and more importantly, it improves the electronic conductivity of the material and inhibits the mutual occupition of Li/Ni.[20–22]. Notably, rate performance can be enhanced by appropriate Co doping. $LiNi_xCo_yMn_{1-x-y}O_2$ ternary material has a low cost compared with $LiCoO_2$ and has been widely used in commercial LIBs in the field of electric tools, bicycles, vehicles, and large-scale energy storage. At present, the typical ratio of Ni, Co, Mn in ternary materials is 4:2:4, 3:3:3, 5:2:3, 2:6:2, 8:1:1, which are called NCM424, NCM333, NCM523, NCM262, NCM811. In addition, in the lattice of $LiNiO_2$, Ni can be substituted by Al/Co. Typically, the reversible capacity of $LiNi_{0.80}Co_{0.15}Al_{0.05}O_2$ material is more than 180 mA h g^{-1}. The composite doping of Co and Al can promote the oxidation of Ni^{2+}, reduce the content of Ni^{2+} at 3a position, and inhibit the irreversible phase transition from H_2 to H_3 during charge and discharge, improving the cycling stability of the material.

When Co is practically replaced by Li, the material is categorized as Li-rich cathode materials. $Li[Li_{1/3}Mn_{2/3}]O_2$ (commonly known as Li_2MnO_3) has a layered structure similar to $LiCoO_2$, with one-third of the transition metal planes occupied by Li^+ ions. This type of layered Li-rich cathode materials exhibits a capacity of ~250 mA h g^{-1}, yet the charge–discharge mechanism of lithium-rich cathode materials involves many aspects to be investigated, such as the source of high capacity of the material, the platform at 4.5V in the first cycle, the maximum irreversible capacity in the first cycle, and the gradual reduction of the discharge platform during the cycle.

9.2.2 Spinel Oxides

Spinel $LiMn_2O_4$ is another valuable cathode material that has been commercialized, as Mn is inexpensive and environmentally friendly compared to Co and Ni, which was put forward by Thackeray et al. [23] in 1983. $LiMn_2O_4$ has good stability and excellent electronic and Li-ion conductivity. Significantly, $LiMn_2O_4$ cathode material effectively reduces the safety concern of LIBs for its high decomposition temperature and much lower oxidability than $LiCoO_2$. Even if there is a short circuit or overcharging, the danger of combustion and explosion can also be avoided. Among all the merits, the high working voltage at a platform of 3.8 V is the most remarkable one and laid the status of $LiMn_2O_4$ among the commercial LIBs cathode materials.

Figure 9.3 illustrates the lithium ions transport via three-dimensional channels in the lattice of $LiMn_2O_4$. When charged, $LiMn_2O_4$ would undergo two-step extraction of lithium ions from the $LiMn_2O_4$ framework [24]. Concretely, lithium ions extraction occurs around 4 V in the first step, corresponding to extraction from the 8a tetrahedral sites. At this time, $LiMn_2O_4$ maintains a good cubic symmetry of spinel structure and

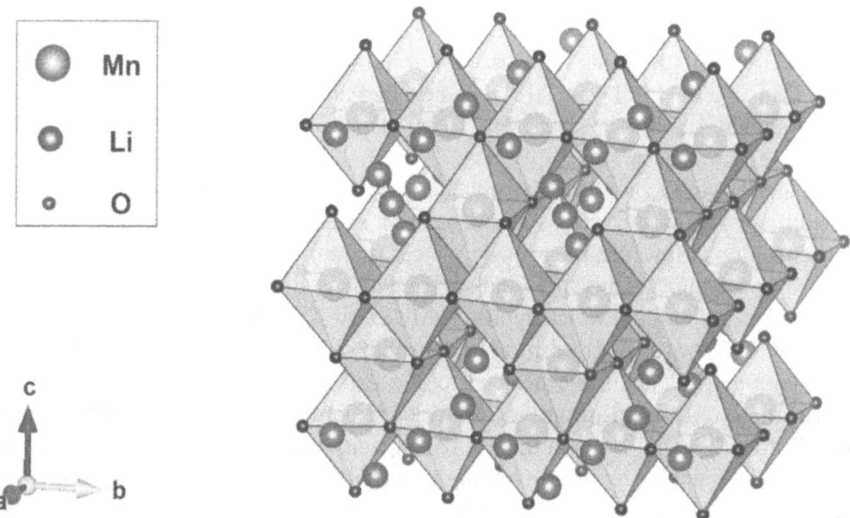

FIGURE 9.3 Crystal structure of spinel $LiMn_2O_4$.

a good cycling performance. Lithium ions extraction occurs around 3 V in the second step, corresponding to extraction from the 16c octahedral sites. A two-phase structure transition from the cubic spinel $LiMn_2O_4$ to the tetragonal lithiated spinel $Li_2Mn_2O_4$ occurs with the valence state of Mn reduced from +3.5 to 3.0. The Jahn–Teller distortion associated with this transition resulted in a 6.5% increment in unit cell volume with surface cracking and drastic capacity fading in the 3-V region.

Ultimately, the primary concern of $LiMn_2O_4$ is the severe capacity fade, even in the 4-V region. This fade originates from several aspects involving Jahn–Teller distortion and a surface passivation layer ($Li_2Mn_2O_4$) forming after cycling. The incompatibility between the surface-distorted tetragonal crystal system and the cubic crystal system inside the particles severely deteriorates the structural integrity affecting Li^+ diffusion and the conductivity between the particles, resulting in capacity loss [25]. The main factor for capacity fade is transition metal dissolution (Mn dissolution), stemming from the reaction of electrolyte salt $LiPF_6$ with trace amounts of water in the electrolyte. The parasitic side reactions produce HF to further induce a disproportionation of Mn^{3+} into Mn^{4+} and Mn^{2+}, while Mn^{2+} would dissolve in electrolyte [26]. Another factor for capacity fade is the decomposition of the electrolyte. When operating at a high voltage, electrolyte tends to decompose, and Li_2CO_3 film is formed on the surface of the material, increasing the polarization of the cell, resulting in capacity attenuation.

To suppress the severe capacity fade of $LiMn_2O_4$, scientists have explored several strategies. Reducing the surface area by adjusting particle size to control the contact with electrolytes is effective. Particularly, replacing Mn with other metals (such as Li, Cu, Mg, Al, Ti, Cr, Ni, Co) proves the most effective. Banov et al. [27] conducted a small quantity of cobalt ($x = 0.1$) doping in the spinel structure of $LiMn_{2-x}Co_xO_2$ that restrained the variations of crystal lattice during cycling, leading to improved cycling stability of LIBs. Modifying the surface of the material also worked to suppress capacity fading and polarization. In the work of Huang's group [28], the surface of spinel $LiMn_2O_4$ was modified with $LiAlO_2$ by a sol-gel method. Al^{3+} diffused into the lattice to form a surface solid solution $LiAl_xMn_{2-x}O_4$ on $LiMn_2O_4$. In this layer, Al^{3+} replaced the site of Mn^{3+} to prevent the cubic-to-tetragonal transition restricting from disproportionation of Mn^{3+}, therefore improving its structural stability.

Spinel $LiMn_2O_4$, with a low cost and no pollution, is suitable for high-power, cost-effective power battery systems such as electric vehicles and energy storage power stations. However, the primary concerns are the electrochemical cycle performance under high temperature and severe capacity decay. Europe's largest energy storage battery facility, which applies $LiMn_2O_4$ to store electricity, will be launched in Leighton Bazard, Bedfordshire, in the south of England, to supply on electricity consumption peak and meet the needs of the grid.

$LiMn_{1.5}Ni_{0.5}O_4$ is an attractive cathode candidate with a higher operating voltage than $LiMn_2O_4$ at 4.7 V and a theoretical capacity of 146 mA h g^{-1}. $LiMn_{1.5}Ni_{0.5}O_4$ has a cubic spinel structure involving two spatial structures [29] with one is the space group of Fd3m, called disordered structure owing to the random occupation of Ni/Mn. The other is the space group of P4332, an ordered structure with Ni atom occupying 4a position and Mn atom, 12d site. In $LiMn_{1.5}Ni_{0.5}O_4$, Mn maintains the +4 oxidation state and undergoes no redox reaction during cycling, avoiding the Jahn–Teller distortions of Mn^{3+} and further dissolution of Mn^{2+}. At the same time, Ni possesses +2 valence and acts as electrochemically active metal ions in the structure [30]. During high-temperature synthesis, the formation of NiO impurity is inevitable, and the disordering between Mn^{4+} and Ni^{2+} leads to inferior performance compared to the disordered phase. This issue can be controlled by cationic doping with Co and Zn. One major concern with the spinel $LiMn_{1.5}Ni_{0.5}O_4$ cathode is its chemical stability with electrolyte at the high voltage of 4.7 V. In this regard, surface modification with oxides such as ZnO, Al_2O_3, $AlPO_4$, and Bi_2O_3 has been employed by Jun Liu and Arumugam Manthiram [31]. As shown in Figure 9.4 [31], these modification layers not only act as a protective shell between the active cathode material surface and the electrolyte but also facilitate with the fast Li^+ diffusion and electron-transfer channel, resulting in enhanced cycle life and rate performance.

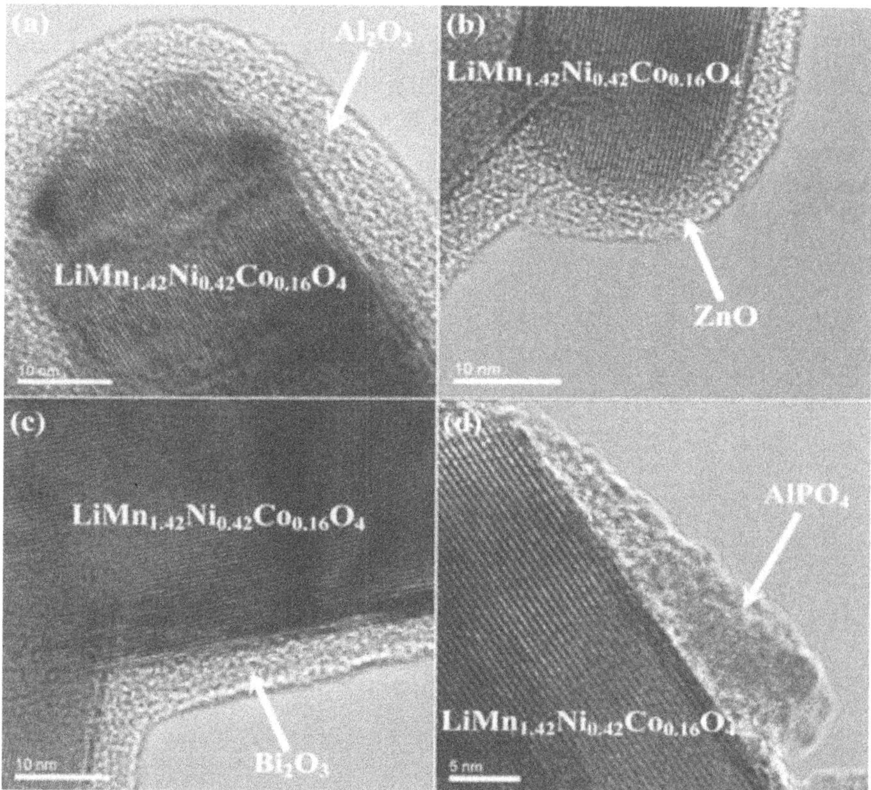

FIGURE 9.4 High-resolution TEM images of 2 wt % (a) Al_2O_3, (b) ZnO, (c) Bi_2O_3, and (d) $AlPO_4$-coated $LiMn_{1.42}Ni_{0.42}Co_{0.16}O_4$. (Reproduced with permission [31]. Copyright 2009, American Chemical Society.)

9.2.3 Polyanion Compound

$LiFePO_4$ was put forward as a cathode material by Goodenough's group in 1997, crystallizing in the olivine structure, belonging to the space group of Pnma. In the crystalline structure of $LiFePO_4$, as shown in Figure 9.5, O atoms are arranged in a slightly distorted hexagonal close-packed with P occupying the 4c site of the O tetrahedron, forming a PO_4 tetrahedron. Fe and Li occupy the 4c and 4a positions of the O octahedron, respectively, creating FeO_6 and LiO_6 octahedrons [32]. Li^+ forms a co-edge straight-chain at position 4a and is parallel to the *b*-axis so that Li^+ can be freely extracted during charging and discharging. The strong P–O covalent bond in PO_4 crystal makes the structure have thermodynamically and kinetically stable. The extraction of Li^+ will not cause rapid contraction or expansion of the material volume. During the charging process, the $LiFePO_4$ phase transforms into the $FePO_4$ phase, and the discharge process is the opposite. The mechanism of phase transformation is discussed as multispecies models, such as core/shell, shrinking shell, domino-cascade, phase transformation wave, and so on [33].

$LiFePO_4$ has high specific capacity, good safety, low cost, and less pollution to the environment. In the lattice, electrons are transported through the transition metal layer. However, the O in the PO_4 tetrahedron separates the FeO_6 octahedrons, and each FeO_6 octahedron is only connected by a common vertex, which makes it impossible to form a continuous transition metal layer network in the unit cell. Therefore, the electronic conductivity of $LiFePO_4$ is as low as 10^{-9} to 10^{-10} S cm^{-1} at room temperature [34]. The volume change in unit cell is limited by the PO_4 tetrahedron structure with strong covalent bonds between the octahedrons, and the transmission of Li^+ is hindered, resulting in a low ion diffusivity of $LiFePO_4$. These factors lead to

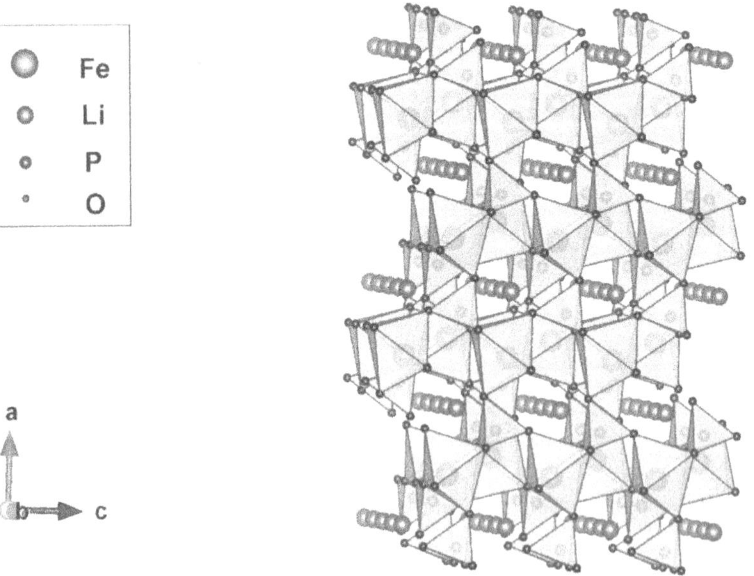

FIGURE 9.5 Crystal structure of polyanion $LiFePO_4$.

poor rate performance of $LiFePO_4$. Another drawback of $LiFePO_4$ is the poor performance at low temperature. The poor performance might be related to the increased electrolyte viscosity, slow Li^+ diffusion rate, easy precipitation of lithium on anode at low temperatures, and lithium reaction with electrolyte resulting in thickness increasing of SEI film [35].

Strategies to enhance the performance of $LiFePO_4$ were brought forward continuously, mainly including surface coating, element doping, and the addition of Li-rich materials. By coating the surface of the material with electron/ion conductive materials, the performance of $LiFePO_4$ can be improved. In Liu's work [36], the surface of $LiFePO_4$ is coated by complete and uniform carbon layer with good consistency, controllable thickness, and close contact with $LiFePO_4$, which well inhibits the overgrowth and agglomeration of $LiFePO_4$ particles. The doping of elements can increase the lattice defects of $LiFePO_4$, which is beneficial to improve the diffusion rate of Li^+ and the internal conductivity of the particles. Heteroatom doping strategies mainly include doping at Li-site, Fe-site, O-site and co-doping at two sites. Chung et al. [37] reported that Li sites are doped with elements such as Ti, Al, and Mg. The doping of multivalent positive ions will produce defects. The concentration of Fe^{3+}/Fe^{2+} increases, and the mixed-valence states promote the formation of P-type semiconductors. Thereby, the electronic conductivity of $LiFePO_4$ significantly increased. Zhang et al. [38] conducted research on Mo-doped $LiFe_{1-x}Mo_xPO_4/C$ materials. The results showed that the doping of Mo did not change the olivine structure of $LiFePO_4$, but formed Li^+ vacancies, which helped improve the Li^+ diffusion rate. The addition of Li-rich materials (Li_2O, LiF, Li_3N, and Li_2S, etc.) to $LiFePO_4$ is effective as the decomposition of these additives releases excessive lithium during the charging process to compensate for the irreversible lithium loss caused by the formation of SEI film on the anode electrode. For example, Zhan et al. [39] designed a core-shell Li_2S/KB/PVP nanocomposite to coat $LiFePO_4$ (Figure 9.6), which can

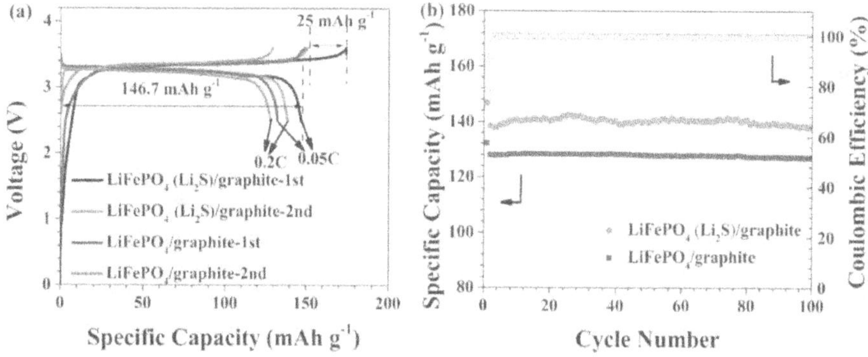

FIGURE 9.6 Comparison of the electrochemical performance for the $LiFePO_4$ (Li_2S)/graphite and $LiFePO_4$/graphite full cells. (a) The 1st and 2nd cycle voltage profiles for the $LiFePO_4$ (Li_2S)/graphite and $LiFePO_4$/graphite full cells. (b) The cycle performance of the $LiFePO_4$ (Li_2S)/graphite and $LiFePO_4$/graphite full cells. The rate for the initial cycle was 0.05C, and that for the following cycles was 0.2C. (Reproduced with permission [39]. Copyright 2017, Elsevier.)

be fully released during the first charge and offered enough Li^+ to compensate for the loss of active lithium and enhance the specific energy of lithium-ion batteries. As a result, the $LiFePO_4$ (Li_2S) electrode had higher specific energy than without Li_2S, and there was almost no capacity or Coulombic efficiency deterioration.

Fe is the key transition metal element of $LiFePO_4$, making it a promising material in terms of cost and environmental issues. The cycle life of $LiFePO_4$ reaches far more than 2000 cycles, fast charge and discharge lifespan can also reach more than 1,000 times. With a longer lifespan, higher stability, safer and more environmentally friendly and affordable, and better performance, batteries applying $LiFePO_4$ as cathode materials have been widely used in electric vehicles, large-scale energy storage, backup power supply, and so on.

Since the redox voltage of Mn^{2+}/Mn^{3+}, Co^{2+}/Co^{3+}, and Ni^{2+}/Ni^{3+} is higher than that of Fe^{2+}/Fe^{3+}, replacing the transition-metal ion Fe^{2+} by Mn^{2+}, Co^{2+}, and Ni^{2+} increases the redox potential significantly, and the corresponding energy density of $LiMnPO_4$/$LiCoPO_4$/$LiNiPO_4$ is also higher than that of $LiFePO_4$. Significantly, the voltage platform of $LiNiPO_4$ is 5.2 V, which almost exceeds the electrochemical window of all electrolytes. $LiMnPO_4$ has a voltage platform of 4.0 V, and $LiCoPO_4$ has a voltage of 4.8 V [40, 41]. Compared to $LiFePO_4$, the wider band gap and lower electronic conductivity make $LiMnPO_4$ outputs low practical capacity even at low current density. For $LiCoPO_4$, the cycle life is short because its high operating voltage would cause the decomposition of the commonly used electrolyte. The drawbacks of these cathode materials are their extremely low electronic and ionic conductivity. At the same time, reducing material particle size and carbon-coating strategies are considered effective methods to improve the electrochemical performance of $LiMPO_4$ materials. In recent years, the mixed transition metal system has attracted widespread attention. Among them, the $LiFe_{1-x}Mn_xPO_4$ exhibits higher energy density and improved redox kinetics due to the increased electronic conductivity. Other multielement phosphate compounds like $LiFe_{1/4}Mn_{1/4}Co_{1/4}N_{i1/4}PO_4$ and $LiFe_{1/3}Mn_{1/3}Co_{1/3}PO_4$ also present good electrochemical performance and have similar enhanced activity [42].

$Li_3V_2(PO_4)_3$ cathode material is the other cathode material applied in small batches in the industry. The monoclinic $Li_3V_2(PO_4)_3$ belongs to the P21/n space group and has a higher Li^+ diffusion coefficient and higher energy density than the rhombic system. $Li_3V_2(PO_4)_3$ has a similar discharge platform and specific energy as $LiCoO_2$ but much better thermal stability and safety [43]. However, an inferior electronic conductivity limits its rate performance, while carbon coating and element doping are effective methods to enhance the performance of $Li_3V_2(PO_4)_3$. Liao et al. [44] employed vanadium metal-organic framework precursor as both carbon sources and vanadium sources, obtained carbon-coated $Li_3V_2(PO_4)_3$ nanocomposites, and improved electrochemical stability and enhanced superior rate performance at various current densities from 0.1 C to 10 C, as seen in Figure 9.7. Dai et al. [45] prepared $Li_3(V_{1-x}Mg_x)_2(PO_4)_3$ ($x = 0.04$, 0.07, 0.10, and 0.13) with the same monoclinic structure as $Li_3V_2(PO_4)_3$ and the expansion of c-axis increased with the dose of Mg^{2+}. After Mg doping in raw materials, the particle size became smaller, and charge transfer resistance was lower. As a result, $Li_3(V_{0.9}Mg_{0.1})_2(PO_4)_3$ exhibited a discharge-specific capacity of 107 mA h g^{-1} at a current density of 20 C.

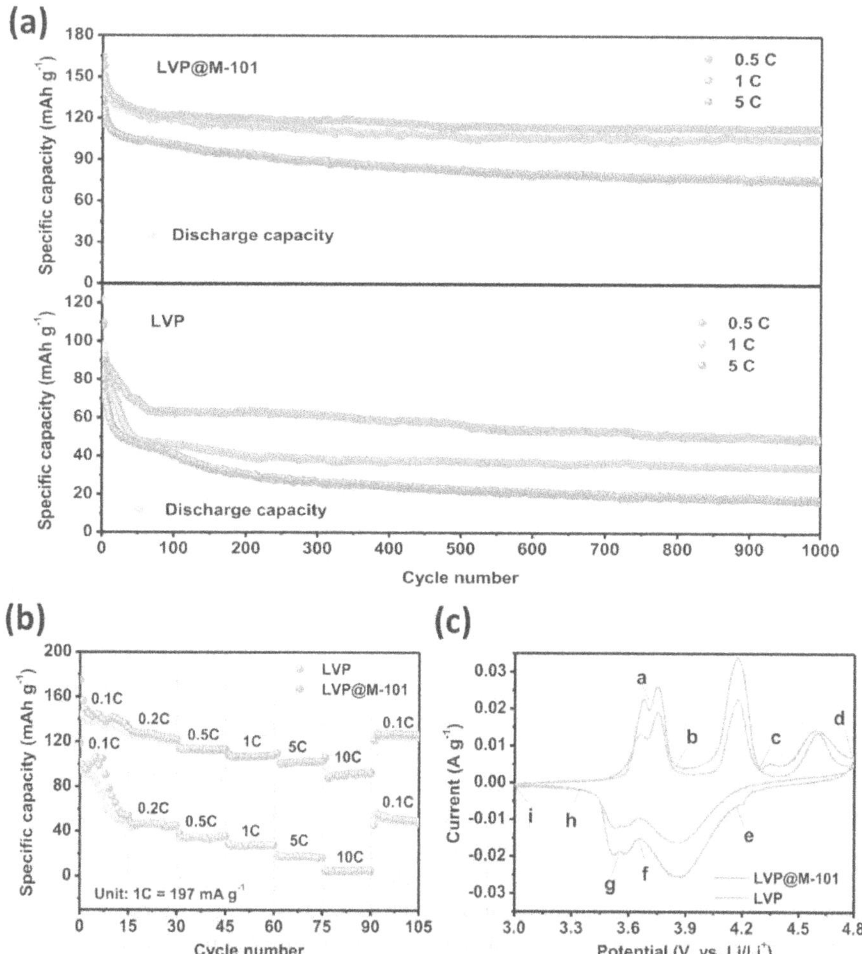

FIGURE 9.7 (a) Cycle performance of LVP@M-101 and LVP electrodes at a current rate of 0.5 C, 1 C, and 5 C. (b) Rate capability of LVP@M-101 and LVP electrode at different current densities from 0.1 C to 10 C in the range of 3.0–4.8 V versus Li/Li^+. (c) Cyclic voltammograms of LVP@M-101 at a scanning rate of 0.1 mV s^{-1}. (Reproduced with permission [44]. Copyright 2018, Elsevier.)

9.3 ANODE MATERIALS

Pursuing high energy density is the eternal theme of LIBs research and development. One of the effective methods of enhancing the energy density of LIBs is to look for high-capacity anode materials. Studies on anode materials date back to the 1980s. In 1983, Akira Yoshino screened petroleum coke as anode material to build the first-generation LIBs [4]. After 1993, commercial LIBs began to use stable graphite as anode material. For the low working potential, stable cyclic ability, and environmentally friendly, graphite is widely adopted as anode material in the commercial LIBs.

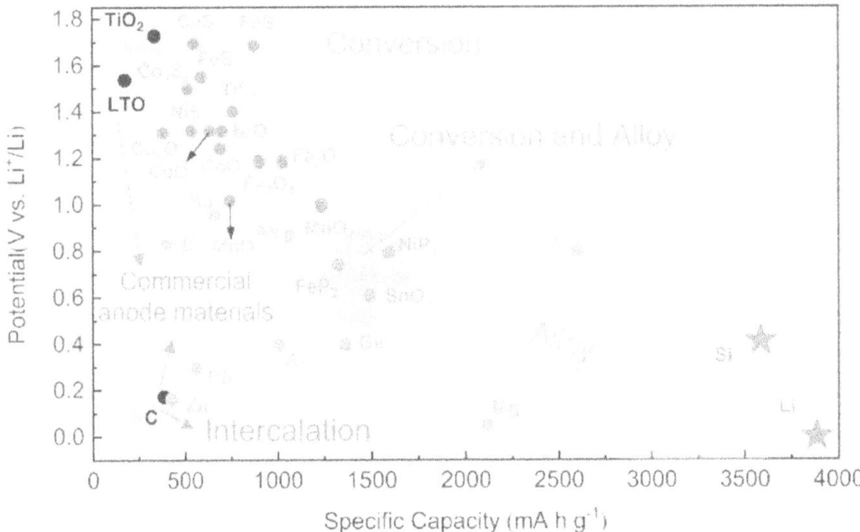

FIGURE 9.8 Schematic illustration of active anode materials for the next-generation lithium batteries. (Reproduced with permission [48]. Copyright 2020, Royal Society of Chemistry.)

However, the theoretical specific capacity of graphite is as low as 372 mA h g^{-1}, and the power density stands at a moderate level [46]. Thus, searching for alternative anode materials with high capacity and improved power densities is necessary.

Anode materials should meet the requirements given below [47]: (1) it should have a low Li^+ (de)intercalation potential to ensure the high output voltage of LIBs. (2) It should have a high irreversible capacity to guarantee the high energy density of LIBs. (3) It should be chemically and thermally stable to assure good cycle life. (4) It should be easy to form a stable SEI film on the surface of anode material to stabilize electrochemical reaction. (4) It should be low in price, environmentally friendly, and easy to mass-produce. Anode materials that have been investigated are numerous in variety, as shown in Figure 9.8 [48].

Anode materials could be classified into three main groups (Figure 9.9): intercalation anodes including carbon-based materials and $Li_4Ti_5O_{12}$; alloy anodes such as Si, Sb, Sn; conversion anodes referring to transition metal oxides, metal sulfides, phosphides, and nitrides [49].

Among the numerous anode materials that mainly have been commercialized are graphite and $Li_4Ti_5O_{12}$. Besides, hard carbon, soft carbon, and silicon/carbon anode materials have been produced and applied in small batches.

9.3.1 Carbon-Based Anode Materials

Graphite is the most crucial LIBs anode material, occupying a market share as high as over 90%. Mining natural graphite or graphitization of petroleum coke at high temperatures of 2500–3000°C can produce graphite. Graphite is a semimetal with high conductivity for the conjugated π-band electron network within the graphene

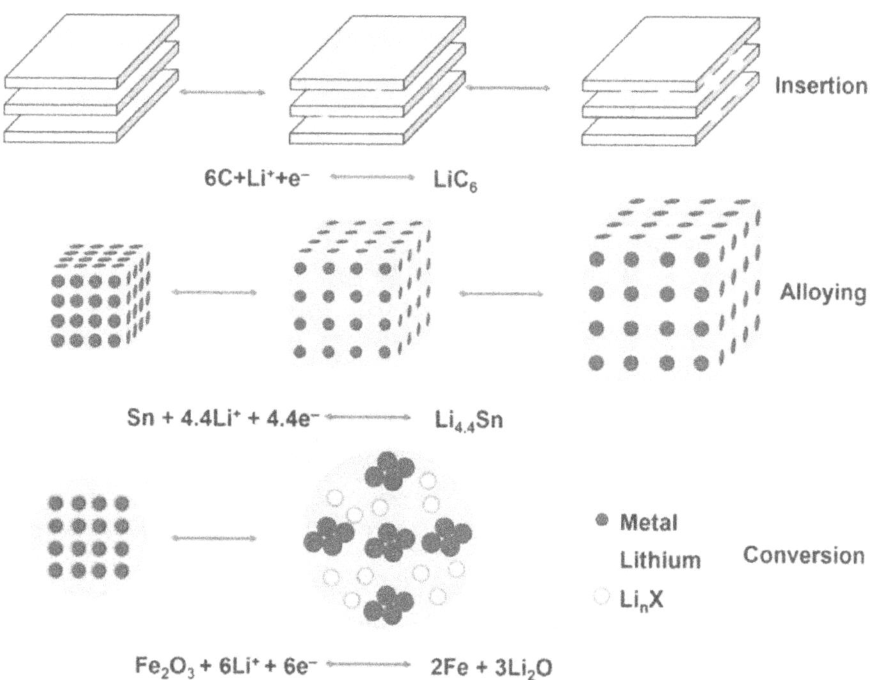

FIGURE 9.9 Schematic illustration of three different types of anodes based on the lithium storage mechanism. (Reproduced with permission [49]. Copyright 2021, Royal Society of Chemistry.)

plane. Since weak van-der-Waals bonding forces held adjacent graphene layers, covalent and ionic compounds are easy to intercalate in the layers (e.g., ionic graphite intercalation compounds, GICs for short). For example, the lithium occupies the sites between two adjacent graphene planes, and within a plane, each lithium reversibly inserts/extracts to form a stoichiometric LiC_6 [50], in which a Li–Li distance is 0.430 nm, and deliver a theoretical specific capacity of 372 mA h g^{-1}, as shown in Figure 9.10(a). LiC_2 with a super high volumetric lithium concentration can be generated under certain conditions (Figure 9.10(b)) [51]. During electrochemical intercalation, graphite layers undergo expansion in the *c*-axis perpendicular to the (002) hexagonal plane. Besenhard et al. [52] thought the electrochemical reactions of Li^+ ions occur not only in graphitic domains but also in cavities simultaneously (Figure 9.10(c)). The discharge/charge curves of graphite show a slope region upper 0.3 V and a plateau region at 0.05–0.3 V along with main capacity contribution (Figure 9.10(d)) [52]. The main drawbacks of graphite anode material are the low redox voltage platform of lithium storage and the growth of lithium dendrite during the repetitive charging and discharging process. After the lithium dendrite penetrates the membrane, it leads to the internal short circuit of the battery, and the cell releases a lot of heat to decompose the electrolyte rapidly, causing safety problems. According to the degree of crystallinity, natural graphite can be divided into flake graphite and amorphous graphite (or microcrystalline graphite).

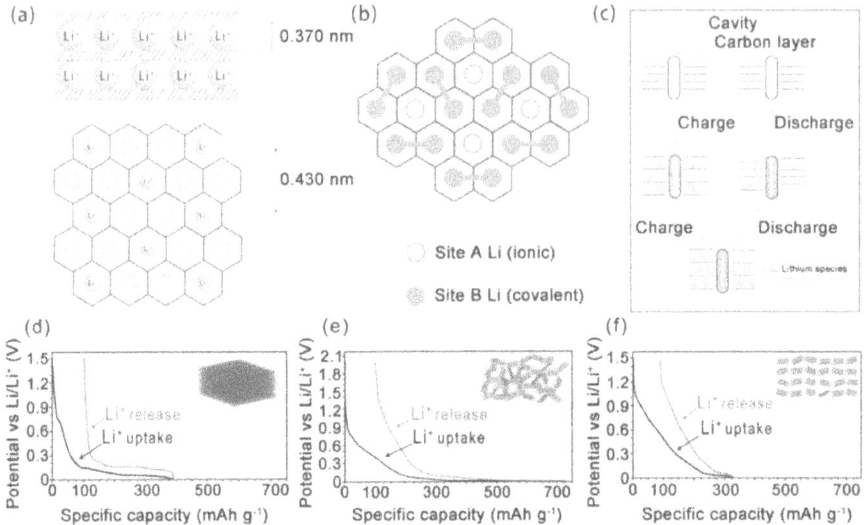

FIGURE 9.10 (a) The structure of LiC_6. (b) Schematic illustration for the coexistence of two types of Li sites in poly(p-phenylene)-based carbons. (c) Model for the Li^+ ions storage mechanism involving the cavities. Constant current discharge/charge curves of (d) graphite, (e) hard carbons, and (f) soft carbons. ((a), (b), (c) reproduced with permission [50]. Copyright 2021, Wiley; (d), (e), (f) reproduced with permission [52]. Copyright 1998, Wiley.)

The graphitization degree of flake graphite exceeds 98%, showing anisotropy. When flake graphite is applied as an anode of LIBs, the Coulombic efficiency is 90% to 93% for the first cycle. And it could deliver performance close to the theoretical specific capacity at a low current density. However, as the weak van-der-Waals force bonds adjacent graphene layers, solvents easily enter the layers and lead to graphene layers stripping, resulting in unsatisfactory cycling performance of the battery. Coating, pore-creating, and recombination methods are mainly used to improve the electrochemical performance of flake graphite. Jian et al. [53] modified natural flake graphite by mild oxidation and carbon coating, and the crystal structure, morphology, and electrochemical performance were investigated. Oxidation treatment can increase the layer spacing of natural flake graphite, while pyrolytic carbon coating treatment can form a compact amorphous carbon film on the surface of flake graphite. As a result, the first Coulombic efficiency of graphite anode material was improved, the electrochemical polarization was reduced, and the diffusion coefficient of lithium ion in the electrode is increased by nearly 34%. As for amorphous graphite, the graphitization degree is usually less than 93%, showing isotropy. Usually, the amorphous graphite contains a small amount of impurities such as Fe, S, P, N, Mo, or others. There are many crystal defects in the structure of amorphous graphite, and the reversible capacity is generally less than 300 mA h g^{-1} [54]. Reducing the particle size, recombination and surface coating are effective methods to enhance the reversible capacity of amorphous graphite. The reversible lithium storage capacity of amorphous graphite is

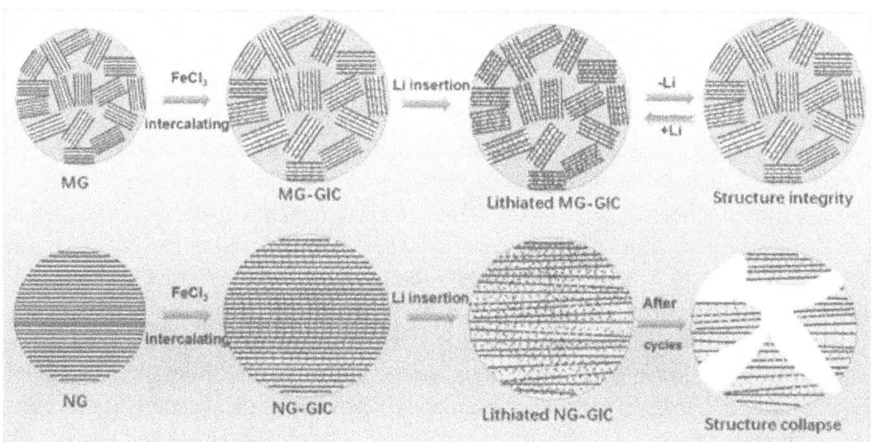

FIGURE 9.11 Schematic illustration for the structure change of GICs with MG and NG as the hosts during the preparation process and Li^+ insertion/extraction. (Reproduced with permission [54]. Copyright 2019, Wiley.)

significantly affected by particle size since materials in small particle sizes usually have a high specific surface area, which can achieve rapid lithium-ion insertion and high reversible specific capacity. Sun et al. [54] prepared $FeCl_3$-intercalated graphite intercalation compounds to improve the performance of amorphous graphite. $FeCl_3$-intercalated MG achieves the integration of isotropic orientation structure, amorphous carbon ingredients between graphite grains, and crumpled graphite layers, thus effectively buffering structure expansion and suppressing the dissolution of soluble FeCl3 guest and LiCl discharge products, as shown in Figure 9.11. As a result, the anode shows a high volumetric capacity of 859 mA h cm^{-3} and stable cycle performance for lithium-ion storage.

Artificial graphite is mainly made by high-temperature treating petroleum coke, asphalt, needle coke, polymer fiber, or others. MCMB (mesophase carbon micro-beads) is a type of important artificial graphite. In 1973, Yamada et al. [55] prepared micron-level spherical carbon material from mesophase asphalt and named it MCMB, which aroused great interest of carbon material researchers and further carried out in-depth research. In 1993, Osaka Gas Company successfully commercialized MCMB as the anode of LIBs. The diameter of MCMB is usually 1–100 μm, while commercial MCMB size is generally between 5 and 40 μm. The MCMB has a smooth surface, and high compaction density, whose surface is hydrophobic with high reaction activity; introducing hydrophilic groups on the surface can improve its conductivity. The excellent electrochemical performance of MCMB is mainly because the outer surface of MCMB particles is almost graphite-structured edge, and it is easy to form stable SEI film, which is conducive to insertion/extraction of lithium. However, repeated insertion of Li^+ at the edge of MCMB easily leads to carbon layer stripping and deformation, leading to fast capacity attenuation. Surface coating effectively suppresses the stripping phenomenon. Mehrdad G. et al. [56] sputtered TiO_2 film on the surface of bare artificial graphite through the physical vapor deposition (PVD), increasing the initial discharge capacity of artificial graphite around 9% and greatly improving the rate performance.

No matter for natural graphite or artificial graphite, element doping effectively enhances lithium storage property [57]. Introducing nonmetallic elements in the carbon crystal, for example, nitrogen atoms that are very close in size and less electronegative significantly changes the electronic properties of the graphene layer and increases the number of active sites in the material, enhancing electrochemical performance. Besides, some positive metal ions can be introduced to the interlayer and surface of graphite, which increases the binding force between graphite layers and improves the reversibility of lithium storage.

Compared to well-ordered graphite, disordered carbons, including hard carbon and soft carbon, are very important anode materials produced in small batches. The crystallinity of hard/soft carbon is lower, and the structure is less orderly than graphite, as shown in Figure 9.12 [58]. In the structure of hard carbon and soft carbon, the decrease in the stacking number of graphene layers leads to a substantial loss of the 2D band in Raman spectra. The increase in average interlayer spacing from graphite to soft carbon then to hard carbon shows up as the (002) peak moved to lower degrees in XRD curves. Hard carbon cannot be graphitized, usually made from thermal cracking of thermosetting precursors (e.g., sugars, phenolic resins, biomass), and has good lithium storage property and better safety than graphite. Sony Co. first employed hard carbon obtained by pyrolysis of polyfurfuryl alcohol (PFA) as the anode material in 1991 [59]. In the structure of hard carbon, there exist a large number of open or closed pores, edges, defects, and microporous surfaces. Based on the mechanisms recognized by most researchers, called as "intercalation-filling mechanism" [60], the lithium storage mechanisms of hard carbon are discussed as three types (Figure 9.13): the filling of Li^+ ions in nanopores, the absorption of Li^+ ions at defect/edge/surface active sites, and the intercalation of Li^+ ions into graphitic layers. Hard carbon has a relatively high capacity (1.5–0 V vs. Li/Li^+, 200–600 mA h g^{-1}), and the voltage platform curve is composed of two parts: the one part is a slope with a voltage range of 1.0–0.1 V and a capacity of about 150–250 mA h g^{-1}

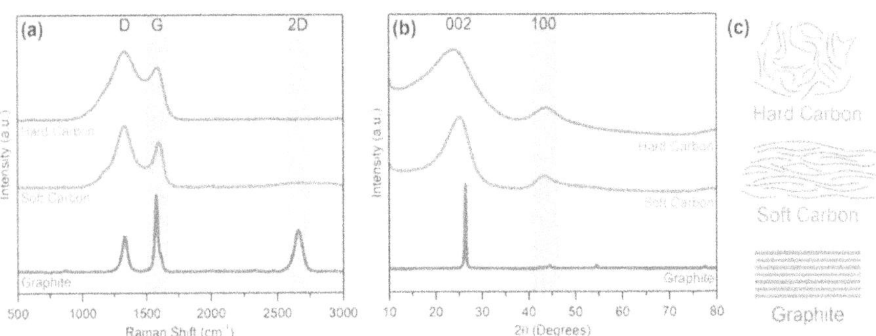

FIGURE 9.12 Schematic illustrations of (a) Raman spectra, (b) XRD patterns, and (c) nanostructures of graphite, soft carbons, and hard carbons. (Reproduced with permission [58]. Copyright 2019, Wiley.)

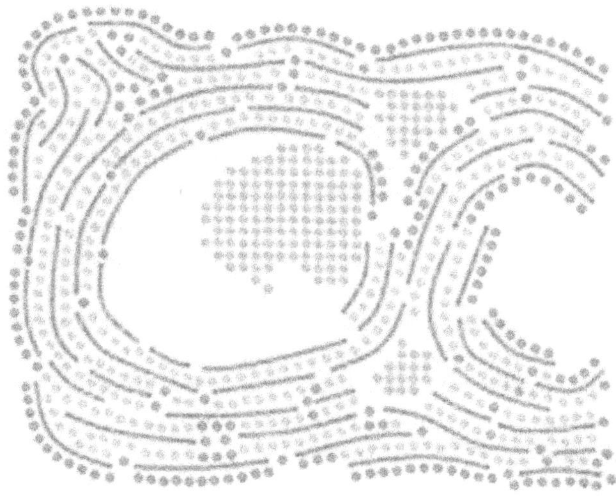

FIGURE 9.13 Schematics of Li^+ ions storage mechanism in hard carbons. (Reproduced with permission [60]. Copyright 2020, Wiley.)

corresponding to the adsorption of Li^+ ions on defects or edge sites. The other part is the platform corresponding to the electrochemical intercalation of Li^+ ions into the graphitic layers, representing a capacity of 100–400 mA h g^{-1} (Figure 9.10(e)) [52]. Although advantages hard carbon has, like excellent rate capacity, long cycling ability, and good low-temperature performances, yet low efficiency at the first cycle, potential lag, poor rate performance, and other shortcomings affect the application of hard carbon. The optimal strategy mainly focuses on the selection of different carbon sources, improvement of preparation technics, recombination, and coating. The nanotexture of precursors and synthesis conditions have a great influence on microstructures and lithium storage properties of hard carbons. The capacity of hard carbon reaches 400 mA h g^{-1} by using PFA-C as raw material and 580 mA h g^{-1} by using polyphenol as raw material. Fey et al. [61] obtained hard carbon anode material by pyrolysis of rice husk, which attained a reversible capacity as high as 1055 mA h g^{-1}.

Soft carbons are graphitizable at high temperatures (>2500°C), by pyrolysis of hydrogen-containing thermoplastic precursors (e.g., hydrocarbons, polyvinyl chloride, pitch, benzene). The first-generation commercial LIBs introduced by Sony Co. in 1991 applied petroleum coke, a soft carbon, as anode material. When the pyrolysis temperature is lower than 1000°C, the precursors decompose to release small gas molecules, such as H_2O, CH_4, and CO_2. If these gas molecules are in a semifluid state, soft carbon can be produced at elevated temperatures. Differing from hard carbons, when heated above 2000°C, the graphene layers in soft carbons rotate into the graphite stacking arrangement, evolving progressively toward the graphite structure when treated at high temperature up to 3000°C [62]. The interlayer spacings of the (002) hexagonal plane are 3.4–3.6 Å for soft carbon, located between graphite (3.35 Å) and hard carbons (>3.7 Å) [63]. Soft carbon has strong adaptability to the electrolyte, good resistance to overcharge, good cycling performance, and low cost. Moreover, soft carbons show higher diffusion coefficients of lithium than hard carbons, enabling better rate performance. However, soft carbons show no obvious voltage plateau as hard carbons or graphite, whose charge/discharge curves are much steeper than that of hard carbons and present a high average voltage of 1.0 V versus Li/Li^+, thus limiting the energy density of cells, as seen in Figure 9.10(f) [52]. Promotion strategies focus on element doping, coating, and recombination, same as hard carbon and graphite.

9.3.2 $Li_4Ti_5O_{12}$

$Li_4Ti_5O_{12}$, as an anode material with great concern has been commercialized in a small branch. In 1956, Jonker et al. [64] first put forward $Li_4Ti_5O_{12}$ with a spinel structure. In 1994, Ferg et al. [65] studied $Li_4Ti_5O_{12}$ as anode material for LIBs, later Ohzuku et al. [66] systematically investigated and emphasized its zero-strain characteristics. The earliest industrialization report can be traced back to Co. Toshiba's power battery applied $Li_4Ti_5O_{12}$ anode in 2008, with a nominal voltage of 2.4 V and an energy density of 67.2 Wh kg^{-1}. The crystal structure is shown as Figure 9.14.

$Li_4Ti_5O_{12}$ is a semiconductor (bandwidth of 2 eV) with an electronic conductivity of 10^{-9} S m^{-1} at room temperature. $Li_4Ti_5O_{12}$ has a spinel structure depicted as

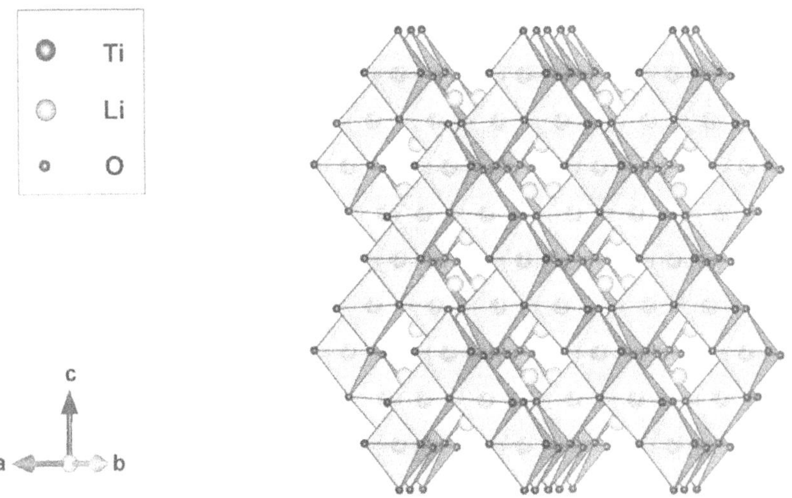

FIGURE 9.14 Structure of $Li_4Ti_5O_{12}$.

$Li[Li_{1/3}Ti_{5/3}]O_4$ with a cubic space group of Fd3m. In the lattice of $Li_4Ti_5O_{12}$, each crystal cell contained eight $Li[Li_{1/3}Ti_{5/3}]O_4$ units, which provides three-dimensional pass channels for Li^+ diffusion at coplanar 8a tetrahedral and 16c octahedral position [67]. During lithiation at a voltage of 1.55 V (vs. Li/Li^+), one lithium moves to the 16c sites along with 60% of Ti^{4+} reduced to Ti^{3+} delivering a theoretical specific capacity of 175 mA h g^{-1} [68]. It is generally believed that electrolyte decomposition occurs below 1.2 V. In comparison, the lithiation of the material is around 1.55 V, so there might be no SEI film growth in this potential range [69]. Additionally, the crystal structure changes from spinel to rock salt structure after Li^+ is inserted in the lattice. Such a process exhibits excellent reversibility of lithium insertion/extraction with little volume change (~0.2%), as seen in Figure 9.15(a) [66, 70]. Thus, $Li_4Ti_5O_{12}$ is an extremely rare zero-strain material for lithium insertion that is capable of achieving an extremely long cycle lifespan for its structural stability.

The lithiation voltage of 1.55 V (vs. Li/Li^+) attaches $Li_4Ti_5O_{12}$ with both advantages and disadvantages. Compared to graphite used primarily on commercial LIBs, since the SEI film is easily decomposed and the heat is continuously accumulated due to the exothermal reaction between lithiated graphite and electrolyte, a thermal runaway associated with fire or explosion of the battery would eventually be triggered [71]. The property of no SEI film growth during lithiation endows $Li_4Ti_5O_{12}$ with better safety. For disadvantages, as anode material of LIBs, the lithiation voltage of $Li_4Ti_5O_{12}$ at 1.55 V is much higher than graphite, reducing the power density of cell when coupled with cathode material. Additionally, $Li_4Ti_5O_{12}$ is a poor

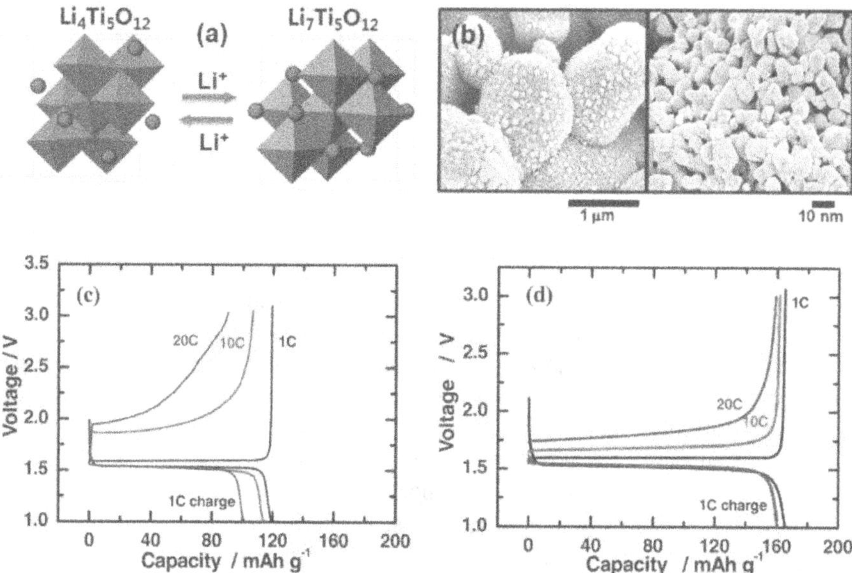

FIGURE 9.15 (a) Structure of $Li_4Ti_5O_{12}$ and $Li_7Ti_5O_{12}$. (b) Scanning electron microscopy under low and high magnification of MSNP-LTO. (c, d) Charge and discharge curves of micron-size LTO and MSNP-LTO. (Reproduced with permission [70]. Copyright 2010, Wiley.)

electronic conductor (~10^{-13} S cm^{-1}) and interfacial reactivity between lithiation product $Li_7Ti_5O_{12}$ and the electrolyte resulting in the undesirable gas release, causing capacity attenuation, life shortening, and safety decline of LIBs.

To enhance the electronic conductivity of $Li_4Ti_5O_{12}$, surface treatment and downsizing to nanoscale are effective. During the past few decades, a variety of nanostructured $Li_4Ti_5O_{12}$ have been prepared and demonstrated improved electrochemical performance. However, these nano-materials usually have a high surface area and low packing density leading to low volumetric energy density. Amine et al. [70] prepared $Li_4Ti_5O_{12}$ shaped with micron-size (~0.5–2 μm) secondary particles composed of nanometer-size (<10 nm) primary particles (micron secondary nano primary-$Li_4Ti_5O_{12}$, MSNP-LTO for short), as shown in Figure 9.15(b). The nanoporous structure allows the electrolyte to infiltrate inside the particles, and the nanoprimary particles allow for fast lithium diffusion for the short lithium pathway. As a result, better rate capability can be achieved at 20 C (Figure 9.15(c) and (d)). Surface modification on the $Li_4Ti_5O_{12}$ anode appears to be more promising in terms of avoiding the gas release. For example, Li et al. [72] modified $Li_4Ti_5O_{12}$ by AlF_3 to suppress the gas generation. The results indicate that most of the Al^{3+} and F^- remain on the surface of the $Li_4Ti_5O_{12}$ particles to form an AlF_3 coating layer, improving high-rate charge/discharge performance and suppressing the gassing behavior of $Li_4Ti_5O_{12}$ anode.

9.3.3 Silicon-Based Anode

Some elements are accessible to alloy with Li, such as Si, Ge, Sn, and P. The alloying mechanism confirms the general reaction of $xLi^+ + Xe^- + M \rightarrow Li_xM$. Compared with intercalation anode materials, the formation of lithium alloy compounds could accept more Li, thus achieving higher theoretical capacity. For example, when pure Si material is entirely lithiation, the ultra-high specific capacity can be obtained by a simple calculation. Further, the change of each crystal plane of silicon obtained by theoretical calculation in the process of alloying also reflects more lithium ions received by Si-Li alloy reaction [73], as seen in Figure 9.16. In fact, Si has a theoretical specific capacity of 4200 mA h g^{-1} [74], which is about 11 times graphite or 24 times $Li_4T_{i5}O_{12}$, categorizing its great potential as novel super high-capacity anode materials for the new generation of batteries. Furthermore, delithiation potential of Si is 0.45 V, which is low enough to maximize the discharge voltage for cells when coupled with high potential cathode materials. Finally, unlike the metal lithium anode with high theoretical capacity, which is easy to grow dendrites on the surface, silicon anode is relatively safe and more suitable for practical applications.

Two challenging issues that silicon anode material faces are volume expansion and unstable SEI film [75]. During lithiation, Si anode material has a large volume change as high as 320%, which leads to falling off of the active material from a conductive network, and results in particle cracked, which seriously affects the cycling performance of the Si-base anode. At the same time, due to the ever-changing volume, the SEI layer constantly varies, which in turn leads to the constant consumption of Li^+. This process manifests itself as a lower Coulombic efficiency in measured electrochemical performance.

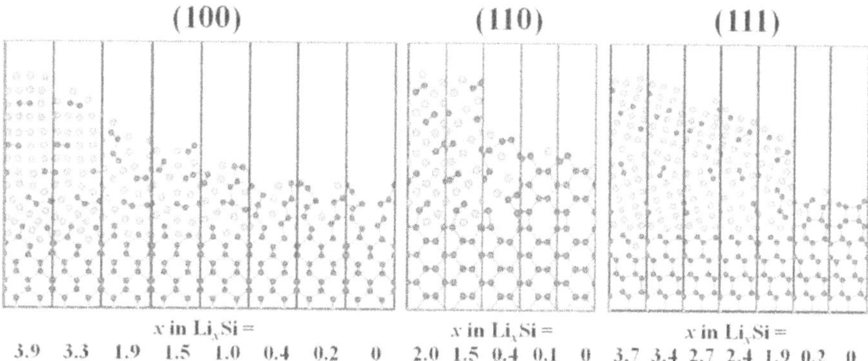

FIGURE 9.16 Configurations of the lithiated Si (100), (110), and (111) surfaces during different stages of delithiation. (Reproduced with permission [73]. Copyright 2012, American Chemical Society.)

Some of the methods are proposed to relieve the above issues. 1) Reducing the particle size is helpful to ease volume change. Huang et al. [76] found that the particle diameter of silicon nanoparticles plays an essential role in the cyclic stability during the initial alloying process. According to this study, 150 nm is the critical size for the fracture behavior of silicon nanoparticles (Figure 9.17). Below this value, the fracture will not occur, while fracture will occur above this value. 2) Porous materials provide buffer space for volume expansion through their reserved space. Zhu et al. [77] prepared silicon oxide materials with different oxygen content by controlling the ratio of water and silicon through a simple wet ball mill method, then transformed SiO_x into Si and SiO_2 through the disproportionation reaction under heating condition, and finally obtained porous silicon by etching SiO_2 with HF solution. The obtained material delivered a high capacity even at a high mass

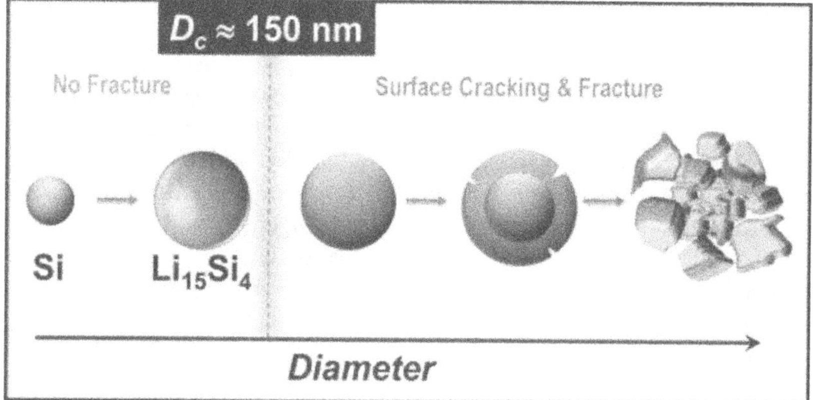

FIGURE 9.17 Critical numerical study on alloying fracture of silicon nanoparticles, fracture of silicon nanoparticles will not occur if the critical size is below150 nm, and will occur if is above. (Reproduced with permission [76]. Copyright 2012, American Chemical Society.)

loading of 1.98 mg cm^{-2}. 3) Decomposition is an effective method because it can concentrate the advantages of several materials together and significantly improve the electrochemical performance. Carbon materials possess good conductivity and small volume expansion. Decomposition of carbon and silicon not only improves the electrical conductivity of Si, but also alleviates the volume expansion and reduces the formation of unstable SEI film. Guo's group developed an amorphous carbon-coated Si nanoparticle by colloid reaction [78]. After treatment, its surface is connected with a long alkyl chain so as to obtain better dispersion, which is conducive to the uniform coating of resorcinol-formaldehyde resin. The material performs well at high mass loading of 4.5 mg cm^{-2} and can maintain a capacity of about 600 mA h g^{-1} when cycled at 100 times. 4) Binder directly affects the properties of Si anode materials, especially those containing carboxyl groups and conductive properties. The polyvinylidene fluoride binder (PVDF) of traditional LIBs is not suitable for silicon anode materials, because PVDF is mainly connected to silicon materials through van-der-Waals force, and the weak interaction cannot stabilize the silicon-based anode materials well. As a binder for LIBs anode material, the carboxyl group on the polymer chain of such binder can be esterified with hydroxyl groups on the surface of nano-Si material, thus increasing the bonding performance and stabilizing the electrode structure. Choi et al. [79] designed a supramolecular network binder of polyacrylic acid (PAA)-polyrotaxane. The ring structure of polyrotaxane allows it to slide on the chain of polyacrylic acid when crosslinking with polyacrylic acid, which makes the polymer structure highly elastic and can accommodate drastic volume changes of the electrode. 5) Development of silicon oxide anode materials includes reducing the size of silicon to a certain extent, which can effectively reduce the volume expansion, increasing the cost of its production. While silicon oxide material (SiO_x, $1 \geq x \leq 2$) does not require such an extremely small particle size to achieve relatively stable cyclic performance, it captured strong attention recently. Different from the lithium storage mechanism of silicon-based compounds, the mechanism of SiO_x is more complex, which involves the generation of several by-products. Still, the existence of these by-products greatly alleviates the volume expansion effect. Moreover, silicon oxide still has the advantage of silicon because it can be converted into simple silicon. However, silicon oxide materials still face challenges like the consumption of lithium in the system during the first cycle, resulting in a low initial Coulombic efficiency, high resistivity of SiO_x further limits its development, and SiO_x also has a certain volume expansion issue. Mai et al. [80] mixed carbon precursor with silicon precursor and obtained SiO_x/C microspheres with controllable carbon content by monodisperse sol-gel method (Figure 9.18). The strong binding strength of SiO_x to carbon greatly reduced the resistivity of SiO_x complex and inhibited its volume expansion, resulting in capacity retention of 91% after 400 cycles.

9.3.4 Other Anode Materials

Conversion anode materials include mainly transition metal components, taking M_aX_b as representation, M on behalf of transition metal, and X represents for O, N, P, S, Se, Te, and so on. They have a higher capacity but lower cost of raw materials compared to silicon-based anode materials. In 2000, Tarascon's group [81] reported

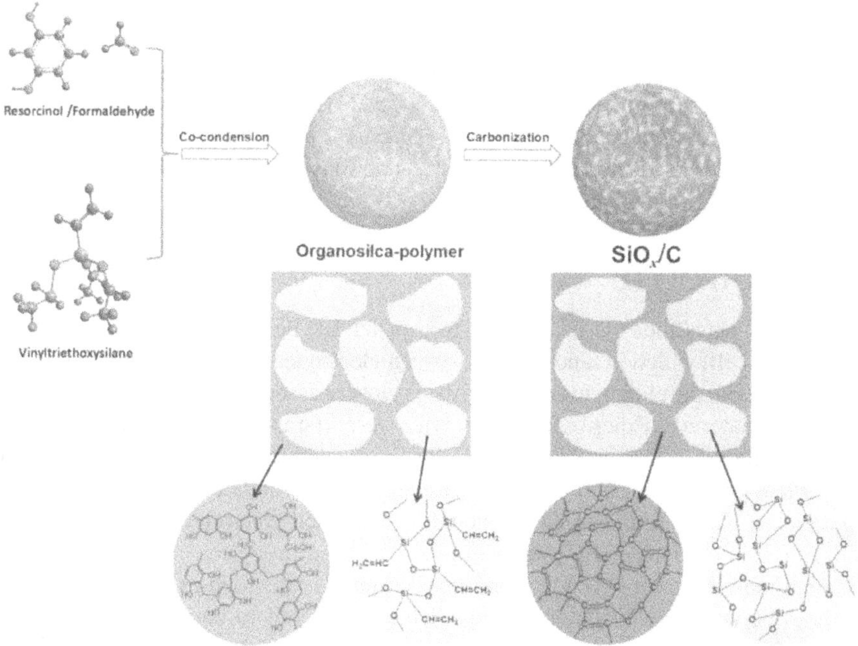

FIGURE 9.18 Schematic illustration for the preparation of SiO_x/C microspheres. (Reproduced with permission [80]. Copyright 2018, Elsevier.)

transition-metal oxides (Co, Ni, Cu, or Fe) with electrochemical capacities of 700 mA h g^{-1} and 100% capacity retention for 100 cycles. They found that differing from the classical Li insertion/desertion or Li-alloying processes, this type of mechanism involves the formation and decomposition of Li_2O, accompanying the reduction and oxidation of metal nanoparticles (in the range 1–5 nanometers), respectively. Most of these oxides crystallize in a rock-salt structure that does not contain any available empty sites for Li-ions, and also that none of the 3d metals considered forms alloys with Li. As a result, they proposed the lithium storage mechanism can be formulated as a displacive redox reaction: $M_xO_y + 2yLi = xM + yLi_2O$. Although bulk phase Li_2O is neither an electronic conductor nor an ionic conductor and cannot participate in electrochemical reaction at room temperature, nanoscale complexes are formed after Li is inserted into transition metal oxides. The particle size of transition metal M and Li_2O is less than 5 nm, which is advantageous from dynamic consideration. It is the main reason for the high electrochemical activity of Li_2O at room temperature. Up to now, conversion anode materials have not been applied practically for several issues they faced; researchers should devote more time and energy to push the pace of its application.

Li metal anode has a very high specific capacity of 3860 mA h g^{-1} [82], while unfortunately, Li dendrite growth and low Coulombic efficiency during operation have immensely impeded its application. These challenges date from two aspects, mechanical and chemical properties of Li metal. For mechanical properties, during operation, the uncontrolled deposition of Li creates significantly volumetric change, growth of dendrites, and cracks in electrodes and the SEI. For chemical properties, Li is so active to react with almost all the chemical species. Recent studies focus on the development of host materials, interfacial design, electrolyte additives, and solid electrolyte approaches to ease the problems it faces. Although some progress has been achieved, deep fundamental investigations should be made to capacitate applications of Li metal anode.

9.4 ELECTROLYTE

Electrolyte plays the role of medium to transfer Li^+ between the cathode and the anode in a cell, which is sandwiched between electrodes and has a close interaction with both. The development of electrolyte is closely associated with the evolution of electrode materials. Ether-based electrolytes were first used to intercalation cathode TiS_2 invented by Whittingham in 1972 for the moderate operating potential of the cell, while it is unstable above 4.0 V [1]. Later, to accommodate high-voltage cathode materials $LiCoO_2$ found by Goodenough, ester-based electrolytes replaced ethers [3]. Ethylene carbonate (EC) was less preferred than propylene carbonate (PC) for its high melting point (~37°C); therefore, PC mainly was used as a solvent of electrolytes between the 1950s and 1990s. While PC trends to co-insert with Li^+ and further exfoliate layers of graphite, it cannot match with this anode. After which, in the 1980s, Yoshino and colleagues at Asahi Kasei Co. filtered out petroleum coke as novel anode materials to match PC. Based on that, Sony Co. launched the first-generation LIBs [4], in which electrolytes were based on PC. Later in the 1990s, Fujimoto and colleagues screened out EC that does not exfoliate graphitic layer [83]. They mixed EC with linear carbonate to attempt high melting point and high viscosity of EC. As a result, they invented LIBs electrolytes as $LiPF_6$ dissolved in a mixture of EC and a linear carbonate selected from dimethyl carbonate (DMC), diethylene carbonate (DEC), or ethyl methyl carbonate (EMC).

Generally, electrolytes should meet the following criteria [84]: (1) It should be an excellent ionic conductor and electronic insulators to facilitate the transfer of lithium ions and minimize self-discharge level. (2) It should also be hard to decompose and be shelved for a long time without side effects and be inert to other cell components. (3) It has good film-forming characteristics on the surface of positive and negative electrodes and can form stable solid electrolyte interphase with low impedance during the charging and discharging processes. (4) It should have a wide electrochemical window to enable the electrode surface to be passivated, thus letting the cell operate over a wide voltage range. (5) It should be environmentally friendly, low cost, and inflammable. Electrolytes mainly include nonaqueous electrolytes and solid electrolytes.

9.4.1 Non-aqueous Electrolyte

At present, commercial Li-ion mainly employs non-aqueous organic electrolyte that contains non-aqueous organic solvent and lithium salt. As a single solvent cannot meet multiple electrolyte requirements of electrolytes, the electrolyte is mainly a mixture of several organic solvents with different properties. However, a mixture of salts is rarely used for advantages or improvements are not readily demonstrated.

The properties of the LIBs electrolyte are closely related to the properties of the solvent. Generally speaking, the choice of solvent should meet the following basic requirements [85]: (1) It should have a higher dielectric constant (ε), so that it has a high ability to dissolve enough lithium salts. (2) It should have a low viscosity (η), so Li^+ in the electrolyte is easy to migrate. (3) The organic solvent must be inert to each component in the battery. (4) The organic solvent or its mixture must have a low melting point (T_m) and a high boiling point (T_b), in other words, a relatively wide liquid range, so that the battery has a relatively wide operating temperature range. (5) Organic solvents must be safe (high flash point T_f), harmless, and low cost.

Since the birth of nonaqueous electrolytes, researchers have conducted extensive research on various polar solvents, most of which belong to the following two categories: organic esters and ethers. Table 9.2 lists the most commonly used solvents in these families and their physical properties.

Common ester solvents mainly include cyclic PC and EC and linear carbonates. Propylene carbonate (PC) was the first to be studied for its wide liquid range, high dielectric constant, stability to lithium, and possesses a stable SEI film, thereby preventing further reaction between the electrolyte and anode material. However, the disadvantages are the severe capacity degradation mainly caused by the reaction with newly formed Li, co-intercalation with Li^+ during the charging process results in exfoliation and decomposition of the graphite, and uneven deposition of Li^+ will lead to the formation of lithium dendrites. As a result, PC is difficult to be used as a single solvent in lithium batteries and lithium-ion batteries. While, on the other hand, PC has the potential to be adopted by low-temperature electrolyte systems.

Ethylene carbonate (EC) has a slightly higher dielectric constant than PC. In 1964 EC was used as a co-solvent to increase the ionic conductivity of the electrolyte. In the 1970s, Scrosati et al. found that after adding lithium salt and a small amount of PC, the melting point of the electrolyte would be greatly reduced [86]. Then in the 1990s, Fujimoto et al. mixed EC with other linear carbonate esters to ease its high melting point and high viscosity and achieve almost all the theoretical capacity of LiC_6 [83].

TABLE 9.2
The Physical Properties of Some Organic Esters and Ethers for Electrolyte in Lithium Battery

Solvent	ε (25°C)	η/cP (25°C)	T_m (°C)	T_b (°C)	T_f (°C)
PC	64.92	2.53	–48.8	242	132
EC	89.78	1.9(40°C)	36.4	248	160
DMC	3.107	0.59(20°C)	4.6	91	18
DME	7.2	0.46	–58	84	0
THF	7.4	0.46	–109	66	–17

Source: Xu K. Reproduced with permission [83]. Copyright 2009, Royal Society of Chemistry.

Later in November 1991, they defined modern LIBs electrolytes in their patent (mentioned above as $LiPF_6$ dissolved in a mixture of EC and a linear carbonate). The major drawback of EC is the high melting point; thus, it is inappropriate for electrolytes systems applied at low temperatures.

The research on linear carbonates started mainly from the research on dimethyl carbonate (DMC) by Tarascon and Guyomard in 1994 [87, 88], who found that linear carbonates can mix with EC at any ratio. The mixed electrolytes profit from both merits of each individual solvent: high anodic stability of EC on cathode surfaces, high solubility of EC toward lithium salts, and low viscosity of DMC to promote ion transport. Researchers and LIBs manufacturers also explored other linear carbonates, including DEC, ethyl methyl carbonate (EMC), and propyl methyl carbonate (PMC). Nowadays, the electrolyte solvents used in commercial lithium-ion cells are almost based on the mixture of EC with these linear carbonates.

Ether-based electrolytes with low viscosity and high ionic conductivity, and better lithium morphology make them a potential candidate for LIBs. The first intercalated cathode (TiS_2) developed by Whittingham in 1972 used ether-based electrolyte. [1]. However, some problems hinder its further application: the poor capacity retention rate, growth of lithium dendrites in the long cycle, and instability at potentials above 4.0 V. Nowadays, although ether-based electrolyte seems to be left behind by lithium-ion battery, it is favored by alkali metal–chalcogen battery like Li-S battery and Li-Se battery.

Lithium salt is mainly the supplier of lithium ions in electrolytes, and its anions are the main factors that determine the physical and chemical properties of electrolytes. In the 1970s, researchers began to pay attention to the morphology of lithium dendrites on the surface of metallic Li anode and have found that lithium salts also play a vital role.

Lithium salt needs to meet the following basic requirements: (1) It has a relatively high solubility in appropriate solvents and is easy to dissociate so as to ensure that the electrolyte has relatively high electrical conductivity. (2) It has relatively high thermal and electrochemical stability and does not undergo any electrochemical reactions with organic solvents, electrode materials, or other battery components. (3) Lithium salt anion must be nontoxic, harmless, and environmentally friendly. (4) The production cost is low, and it is easy to prepare and purify.

At present, the most studied lithium salts are mainly compounds based on complex anions composed of a simple anion restabilized by a Lewis acid agent. Table 9.3 lists some of the salts with basic physical properties.

Lithium perchlorate ($LiClO_4$) is known as moderately soluble in many nonaqueous organic solvents with high conductivity and high anodic stability. It is rarely less hygroscopic and stable to ambient moisture, making it a popular electrolyte solute in laboratory application. Since LiF is not generated, the impedance of the SEI film formed by $LiClO_4$-based electrolytes is lower than that of $LiPF_6$ or $LiBF_4$-based electrolytes. Nonetheless, $LiClO_4$ is a strong oxidant for the high oxidation state of chlorine (VII), readily reacts with organic species at a high temperature or a high current, together with its insecurity in transportation, makes it inappropriate for industry [89].

Lithium hexafluorophosphate ($LiPF_6$) is currently the most widely used lithium salt in commercial LIBs. The success of $LiPF_6$ does not only lie in its single

TABLE 9.3
The Physical Properties of Some Lithium Salt for Electrolyte in Lithium Battery

Lithium Salt	M.W$_t$	Al-corrosion	H_2O-sensitivity	σ/mS·cm^{-1}(1M in EC/DMC, 20°C)
$LiPF_6$	151.91	N	Y	10.00
$LiBF_6$	93.74	N	Y	4.50
$LiClO_4$	106.40	N	N	9.00
$LiAsF_6$	195.85	N	Y	11.10 (25°C)
$LiCF_3SO_3$	156.01	Y	Y	1.70 (in PC,25°C)
LiTFSI	287.08	Y	Y	6.18
LiFSI	187.07	Y	Y	10.40 (25°C)

Source: Xu K. Reproduced with permission [83]. Copyright 2009, Royal Society of Chemistry.

outstanding performance in all aspects, but it also owns the majority of the fundamental and mutually exclusive properties that are required for durable cells. It exhibits characteristics such as moderate ion migration number, dissociation constant, and aluminum foil passivation ability. Despite all its virtues, it is unstable when heating; even worse, it would decompose into LiF(s) and PF_5(g) at room temperature [90]. PF_5 is a strong Lewis acid, which can easily attack lone pair electrons on oxygen atoms in organic solvents, resulting in cationic polymerization of organic solvents and ether bond cracking. Yet, the P-F bond is labile toward hydrolysis by even trace amounts of moisture in nonaqueous solvents, producing HF and LiF. The presence of LiF will increase the interface resistance and affect the cycle life of LIBs.

Although owning multiple advantages like less toxicity than $LiAsF_6$ and higher safety than $LiClO_4$, the negligible dissociation constant of lithium tetrafluoroborate ($LiBF_4$) endows it with moderate ionic conductivity. However, the most amazing property of $LiBF_4$ should be the better performance at both elevated temperatures and low temperatures [91].

Armand et al. took the lead to suggest employing Lithium Bis(trifluoromethanesulfonyl)imide (LiIm) in solid polymer electrolytes based on oligomeric or macromolecular ethers. Meanwhile, it was proved to be satisfactory salts in liquid electrolytes matching with carbonaceous anode materials [92]. LiIm was commercialized by 3M Corporation in the 1990s to defeat many salts and be the victor of commercial lithium battery for its advantages in conductivity, safety, thermally stability, and nontoxicity. Besides, LiIm-based electrolytes have a better morphology of lithium than other salt-based electrolytes. Nevertheless, LiIm causes severe Al corrosion to hinder its further application.

9.4.2 Ionic Liquid Electrolyte

The ionic liquid is composed of discrete and free-moving cations and anions in the liquid state at temperatures below 100°C. Usually, cations include quaternary ammonium, and anions prefer either organic anions or inorganic anions like DCA^-, $TFSI^-$, PF_6^-. Ionic liquid exhibits unique properties, such as nonvolatile, long liquid range, high thermal stability, high conductivity, and a broad electrochemical window, making it an interesting solvent for electrolytes [93]. The high viscosity, poor compatibility with the cathode and anode materials, and high price restrict its large-scale application.

Ionic liquid electrolytes are expected to completely solve the safety problem of LIBs based on reducing the viscosity and improving the migration rate of Li^+.

9.4.3 Solid Electrolyte

Since E. Warburg found some solid compounds were practically pure ionic conductors in the 1990s, the study of solid electrolytes was born [94]. Featured with absolute safety, no leakages of toxic organic solvents, low flammability, mechanical and thermal stability, easy processability and low self-discharge, solid electrolyte seem to overcome most of drawbacks of traditional liquid electrolyte. However, poor ionic conductivity and high impedance between solid interfaces are the main limitations. Solid electrolytes are classified into the all-solid-state electrolyte (SSE) and quasi-solid-state electrolyte (QSSE).

The ionic migration of SSE proceeds in the form of a solid state. Completely removing all liquid components, SSE dramatically enhances the safety of LIBs but reduces the ionic conductivity compared with a liquid counterpart. Inorganic solid electrolyte (ISE), solid polymer electrolyte (SPE), and composite polymer electrolyte (CPE) are three types of SSE. ISE is constituted by an inorganic material in the crystalline or glassy state that conducts ions by diffusion through the lattice. They are high in ionic conductivity, modulus, and transfer number compared to other SSEs. While ISE is brittle, it possesses low compatibility and stability to the electrode. SPE are well-ordered high polymers that serve as both solvents to dissolve lithium salts and mechanical matrices to support the process ability that conducts ions through the polymer chains. They are easy to process, have high elasticity and plasticity, mechanical stability, low-temperature sensitivity and have the good dissolution ability of Li salts. Low ionic conductivity restricts their rate capability yet. CPE is obtained when introducing particles as fillers inside the polymer solution. The particles can be Al_2O_3, TiO_2, SiO_2, and so on.

QSSE is composed of a liquid electrolyte and a solid matrix, in which liquid electrolyte acts as a pathway to ensure ion conduction. In contrast, the solid matrix adds mechanical stability to the material as a whole [95]. There are serial categories of QSSEs, GPEs (gel polymer electrolytes) is the most common type featured with conducting ions through the interaction with the substitutional groups of the polymer chains. The solvent of GPEs increases the ionic conductivity and softens the electrolyte to ensure better interfacial contact, while the matrix of GPEs contains the active ions. Thus, the electrolyte contains both mechanical properties of solids and the high transport properties of liquids.

9.5 SUMMARY

After decades of investigation and development, LIBs are now ubiquitous in daily life, whose diverse applications include digital products, electric vehicles, power tools, military equipment, and large-scale energy storage. For commercial LIBs devices, cathode materials mainly include $LiCoO_2$, $LiMn_2O_4$, and $LiFePO_4$; anode materials are graphite, hard/soft carbon, MCMB, and Si/C alloy, while electrolyte consists of $LiPF_6$ in the mixture of EC and other linear carbonates. The constant upgrading of these commercial products leads to continuous market demand growth toward higher energy density LIBs. Therefore, improving energy density while also maintaining lifetime and safety is the eternal theme of LIBs research.

REFERENCES

1. Whittingham, M. S., Electrical energy storage and intercalation chemistry, *Science*, 1976, *192* (4244), 1126–1127.
2. Armand, M. B., Intercalation Electrodes. In *Materials for advanced batteries*, Murphy, D. W., Broadhead, J., Steele, B. C. H., Eds. Springer US: Boston, MA, 1980, 145–161.
3. Mizushima, K., Jones, P. C., Wiseman, P. J., et al. $LixCoO_2$ ($0 < x < 1$): A new cathode material for batteries of high energy density, *Mater. Res. Bull.* 1980, *15* (6), 783–789.
4. Yoshino A., The birth of the lithium-ion battery, *Angew. Chem. Int. Ed.*, 2012, *51* (24), 5798–800.
5. Winter M., Barnett B., Xu K., Before Li ion batteries, *Chem. Rev.*, 2018, *118* (23), 11433–11456.
6. Goodenough J. B., Park K.-S., The Li-Ion rechargeable battery: a perspective. *J. Am. Chem. Soc.*, 2013, *135* (4), 1167–1176.
7. Whittingham M. S., Lithium batteries and cathode materials., *Chem. Rev.*, 2004, *104* (10), 4271–4302.
8. Ma C., Lv Y., Li H., Fundamental scientific aspects of lithium batteries (VII)–Positive electrode materials, *Energy Storage Sci. Technol.*, 2014, *3* (1), 53–65.
9. Thomas M. G. S. R., Bruce P. G., Goodenough J. B., AC impedance analysis of polycrystalline insertion electrodes: application to $Li_{1-x}CoO_2$, *J. Electrochem. Soc.* 1985, *132* (7), 1521–1528.
10. Thomas M. G. S. R., Bruce P. G., Goodenough, J. B., Lithium mobility in the layered oxide $Li_{1-x}CoO_2$, *Solid State Ionics*, 1985, *17* (1), 13–19.
11. Barker J., Pynenburg R., Koksbang R., Determination of thermodynamic, kinetic and interfacial properties for the $Li//Li_xMn_2O_4$ system by electrochemical techniques, *J. Power Sources*, 1994, *52* (2), 185–192.
12. Prosini P. P., Lisi M., Zane D., et al. Determination of the chemical diffusion coefficient of lithium in $LiFePO_4$, *Solid State Ionics*, 2002, *148* (1), 45–51.
13. Delmas C., Fouassier C., Hagenmuller P., Structural classification and properties of the layered oxides, *Physica B+C*, 1980, *99* (1), 81–85.
14. Mendiboure A., Delmas C., Hagenmuller P., New layered structure obtained by electrochemical deintercalation of the metastable $LiCoO_2$ (O_2) variety, *Mater. Res. Bull.*, 1984, *19* (10), 1383–1392.
15. Goodenough J. B., Mizushima K., Takeda T., Solid-solution oxides for storage-battery electrodes, *Jpn. J. Appl. Phys.*, 1980, *19* (S3), 305.
16. Shao-Horn Y., Croguennec L., Delmas C., et al. Atomic resolution of lithium ions in $LiCoO_2$, *Nat. Mater.*, 2003, *2* (7), 464–467.
17. Kim J., Noh M., Cho J., et al. Controlled nanoparticle metal phosphates (metal = Al, Fe, Ce, and Sr) coatings on $LiCoO_2$ cathode materials, *J. Electrochem. Soc.*, 2005, *152* (6), A1142.
18. Cho J., Kim Y. J., Park B., $LiCoO_2$ cathode material that does not show a phase transition from hexagonal to monoclinic phase, *J. Electrochem. Soc.*, 2001, *148* (10), A1110.
19. Zhang J.-N., Li Q., Ouyang C., et al. Trace doping of multiple elements enables stable battery cycling of $LiCoO_2$ at 4.6 V, *Nat. Energy*, 2019, *4* (7), 594–603.

20. MacNeil D. D., Lu Z., Dahn J. R., Structure and electrochemistry of $Li[Ni_xCo_{1-2x}Mn_x]O_2$ ($0 \leq x \leq 1/2$), *J. Electrochem. Soc.*, 2002, *149* (10), A1332.
21. Tsutomu O., Yoshinari M., Layered lithium insertion material of $LiCo_{1/3}Ni_{1/3}Mn_{1/3}O_2$ for Lithium-ion batteries, *Chem. Lett.*, 2001, *30* (7), 642–643.
22. Sun Y., Ouyang C., Wang Z., et al. Effect of Co content on rate performance of $LiMn_{0.5-x}Co_{2x}Ni_{0.5-x}O_2$ cathode materials for lithium-ion batteries, *J. Electrochem. Soc.*, 2004, *151* (4), A504.
23. Thackeray M. M., David W. I. F., Bruce P. G., et al. Lithium insertion into manganese spinels, *Mater. Res. Bull.*, 1983, *18* (4), 461–472.
24. Hosono E., Kudo T., Honma I., et al. Synthesis of Single Crystalline Spinel $LiMn_2O_4$ nanowires for a lithium ion battery with high power density, *Nano Lett.*, 2009, *9* (3), 1045–1051.
25. Eriksson T., Gustafsson T., Thomas J. O., Surface structure of $LiMn_2O_4$ electrodes, *Electrochem. Solid-State Lett.*, 2002, *5* (2), A35.
26. Park S. B., Shin H. C., Lee W.-G., et al. Improvement of capacity fading resistance of $LiMn_2O_4$ by amphoteric oxides, *J. Power Sources*, 2008, *180* (1), 597–601.
27. Scrosati B., Recent advances in lithium ion battery materials, *Electrochim. Acta*, 2000, *45* (15), 2461–2466.
28. Sun Y., Wang Z., Chen L., et al. Improved electrochemical performances of surface-modified spinel $LiMn_2O_4$ for long cycle life lithium-ion batteries, *J. Electrochem. Soc.*, 2003, *150* (10), A1294.
29. Kim J. H., Myung S. T., Yoon C. S., et al. Comparative study of $LiNi0.5Mn1.5O_{4-\delta}$ and $LiNi_{0.5}Mn_{1.5}O_4$ cathodes having two crystallographic structures: Fd3m and P4332, *Chem. Mater.*, 2004, *16* (5), 906–914.
30. Wang L., Li H., Huang X., et al. A comparative study of Fd-3m and P4332 "$LiNi_{0.5}Mn_{1.5}O_4$", *Solid State Ionics*, 2011, *193* (1), 32–38.
31. Liu J., Manthiram A., Understanding the improvement in the electrochemical properties of surface modified 5V $LiMn_{1.42}Ni_{0.42}Co_{0.16}O_4$ spinel cathodes in lithium-ion cells, *Chem. Mater.*, 2009, *21* (8), 1695–1707.
32. Tian L., Yu H., Zhang W., et al. The star material of lithium ion batteries $LiFePO_4$: Basic properties, optimized modification and future prospects, *Mater. Rep.*, 2019, 33 (21), 3561–3579.
33. Gao J., Lv Y., Li H., Fundamental scientific aspects of lithium batteries (III)—Phase transition and phase diagram, *Energy Stor. Sci. Technol.*, 2013, *2*(3), 250–266.
34. Nitta N., Wu F., Lee J. T., et al. Li-ion battery materials: present and future, *Mater. Today*, 2015, *18* (5), 252–264.
35. Ji Y., Zhang Y., Wang C.-Y., Li-ion cell operation at low temperatures, *J. Electrochem. Soc.*, 2013, *160* (4), A636–A649.
36. Liu Y., Wang J., Liu J., et al. Origin of phase inhomogeneity in lithium iron phosphate during carbon coating, *Nano Energy*, 2018, *45*, 52–60.
37. Chung S.-Y., Bloking J. T., Chiang Y.-M., Electronically conductive phospho-olivines as lithium storage electrodes, *Nat. Mater.*, 2002, *1* (2), 123–128.
38. Zhang Y., Shao Z., Zhang Y., Preparation of Mo-doping $LiFePO_4$/C by carbon reduction method, *Mater. Manuf. Processes*, 2021, *36* (4), 419–425.
39. Zhan Y., Yu H., Ben L., et al. Using Li_2S to compensate for the loss of active lithium in Li-ion batteries, *Electrochim. Acta*, 2017, *255*, 212–219.
40. Oh S.-M., Yoon C.-S., Scrosati B., et al. High-performance carbon-$LiMnPO_4$ nanocomposite cathode for lithium batteries, *Adv. Funct. Mater.*, 2010, *20* (19), 3260–3265.
41. Wang F., Yang J., NuLi Y., et al. Highly promoted electrochemical performance of 5V $LiCoPO_4$ cathode material by addition of vanadium, *J. Power Sources*, 2010, *195* (19), 6884–6887.

42. Wang X. J., Yu X. Q., Li H., et al. Li-storage in $LiFe_{1/4}Mn_{1/4}Co_{1/4}Ni_{1/4}PO_4$ solid solution, *Electrochem. Commun.*, 2008, *10* (9), 1347–1350.
43. Huang H., Yin S.-C., Kerr T., et al. Nanostructured Composites: a high capacity, fast rate $Li_3V_2(PO_4)_3$/carbon cathode for rechargeable lithium batteries, *Adv. Mater.*, 2002, *14* (21), 1525–1528.
44. Liao Y., Li C., Lou X., et al. Carbon-coated $Li_3V_2(PO_4)_3$ derived from metal-organic framework as cathode for lithium-ion batteries with high stability, *Electrochim. Acta*, 2018, *271*, 608–616.
45. Dai C., Chen Z., Jin H., et al. Synthesis and performance of $Li_3(V_{1-x}Mg_x)_2(PO_4)_3$ cathode materials, *J. Power Sources*, 2010, *195* (17), 5775–5779.
46. Long J. W., Dunn B., Rolison D. R., et al. Three-dimensional battery architectures, *Chem. Rev.*, 2004, *104* (10), 4463–4492.
47. Luo F., Chu G., Huang J., et al. Fundamental scientific aspects of lithium batteries (VIII)—-anode electrode materials, *Energy Stor. Sci. Technol.*, 2014, *3* (2), 146–163.
48. Wu F., Maier J., Yu Y., Guidelines and trends for next-generation rechargeable lithium and lithium-ion batteries, *Chem. Soc. Rev.*, 2020, *49*, 1569–1614.
49. Liu J., Xie D., Shi W., Cheng P., Coordination compounds in lithium storage and lithium-ion transport, *Chem. Soc. Rev.*, 2020, *49*, 1624–1642.
50. Xie L., Tang C., Bi Z., et al. Hard carbon anodes for next-generation Li-ion batteries: review and perspective, *Adv. Energy Mater.*, 2021, *11* (38), 2101650.
51. Sato K., Noguchi M., Demachi A., et al. A mechanism of lithium storage in disordered carbons, *Science*, 1994, *264* (5158), 556–558.
52. Winter M., Besenhard J. O., Spahr M. E., et al. Insertion electrode materials for rechargeable lithium batteries, *Adv. Mater.*, 1998, *10*(10),725–63.
53. Jian Z, Liu H, Kuang J, et al. Natural flake graphite modified by mild oxidation and carbon coating treatment as anode material for lithium ion batteries, *Procedia Eng.*, 2012, *27*, 55–62.
54. Sun Y, Han F, Zhang C, et al. $FeCl_3$ intercalated microcrystalline graphite enables high volumetric capacity and good cycle stability for lithium-ion batteries, *Energy Technol.*, 2019, *7*(4), 1801091.
55. Honda H., Yamada Y., Meso-carbon microbeads., *J. Japan Petrol Inst.*, 1973, *16*, 392–397.
56. Gholami M., Zarei-jelyani M., Babaiee M., et al. Physical vapor deposition of TiO_2 nanoparticles on artificial graphite: an excellent anode for high rate and long cycle life lithium-ion batteries, *Ionics*, 2020, *26*(9), 4391–9.
57. Ou J., Yang L., Zhang Z., et al. Honeysuckle-derived hierarchical porous nitrogen, sulfur, dual-doped carbon for ultra-high rate lithium ion battery anodes, *J. Power Sources*, 2016, *333*, 193–202.
58. R. A. Adams, A. Varma, V. G. Pol, Carbon anodes for non-aqueous alkali metal-ion batteries and their thermal safety aspects, *Adv. Energy Mater.*, 2019, *9*, 1900550.
59. Huang J., Peng Z., Xiao Y., et al. Hierarchical nanosheets/walls structured carbon-coated porous vanadium nitride anodes enable wide-voltage-window aqueous asymmetric supercapacitors with high energy density, *Adv. Sci.*, 2019, *6*(16), 1900550.
60. Zhao L. F., Hu Z., Lai W. H., et al. Hard carbon anodes: fundamental understanding and commercial perspectives for Na-ion batteries beyond Li-ion and K-ion counterparts, *Adv. Energy Mater.*, 2021, 11, 2002704.
61. Fey G. T. K., Chen C. L., High-capacity carbons for lithium-ion batteries prepared from rice husk, *J. Power Sources*, 2001, *97*, 47–51.
62. Saurel D., Orayech B., Xiao B., et al. From charge storage mechanism to performance: a roadmap toward high specific energy sodium-ion batteries through carbon anode optimization, *Adv. Energy Mater.*, 2018, *8*(17), 1703268.

63. Emmerich F G., Evolution with heat treatment of crystallinity in carbons, *carbon*, 1995, *33*(12), 1709–15.
64. Jonker G. H., Magnetic compounds with perovskite structure IV conducting and non-conducting compounds[C]//Madrid: Proceedings 3rd Symposium on Reactivity of Solids, 1956, 707–722.
65. Ferg E., Gummow R. J., De Kock A., et al. Spinel anodes for lithium-ion batteries, *J. Electrochem. Soc.*, 1994, *141*(11), L147–L150.
66. Ohzuku T., Ueda A., Yamamoto N., Zero-strain insertion material of $Li[Li_{1/3}Ti_{5/3}]O_4$ for rechargeable lithium cells, *J. Electrochem. Soc.*, 1995, *142*(5), 1431–1435.
67. Robertson A. D., Trevino L., Tukamoto H., et al. New inorganic spinel oxides for use as negative electrode materials in future lithium-ion batteries, *J. Power Sources*, 1999, (*81-82*), 352–357.
68. Peramunage D., Abraham K. M., Preparation of Micron-sized $Li_4Ti_5O_{12}$ and its electrochemistry in polyacrylonitrile electrolyte-based lithium cells, *J. Electrochem. Soc.*, 1998, *145*(8), 2609–2615.
69. Julien C. M., Massot M., Zaghib K., Structural studies of $Li_{4/3}Me_{5/3}O_4$ (Me=Ti, Mn) electrode materials: local structure and electrochemical aspects, *J. Power Sources*, 2004, *136*(1), 72–79.
70. Amine K., Belharouak I., Chen Z., et al. Nanostructured anode material for high-power battery system in electric vehicles, *Adv. Mater.*, 2010, *22*(28), 3052–3057.
71. Lu J., Chen Z., Pan F., et al., High-performance anode materials for rechargeable lithium-ion batteries, *Electrochem. Energy Rev.*, 2018, *1*, 35–53.
72. Li W., Li X., Chen M., et al. AlF_3 modification to suppress the gas generation of $Li_4Ti_5O_{12}$ anode battery, *Electrochim. Acta*, 2014, *139*, 104–110.
73. Chan M. K. Y., Wolverton C., Greeley J. P., First principles simulations of the electrochemical lithiation and delithiation of faceted crystalline silicon. *J. Am. Chem. Soc.*, 2012, *134*(35),14362–14374.
74. Huggins, R. A., Lithium alloy negative electrodes, *J. Power Sources*, 1999, *81*, 13–19.
75. Lee S. J., Kim H. J., Hwang T. H., et al. Delicate structural control of Si-SiO_x-C composite via high-speed spray for Li-ion battery anodes. *Nano Lett.*, 2017, *17*, 1870–1876.
76. Liu X. H., Zhong L., Huang S., et al. Size-dependent fracture of silicon nanoparticles during lithiation, *ACS Nano*, 2012, *69*, 1522–1531.
77. Zong L., Jin Y., Liu C., et al. Precise perforation and scalable production of Si particles from low-grade sources for high-performance lithium ion anodes, *Nano Lett.*, 2016, *16*, 7210–7215.
78. Su H., Barragan A. A., Geng L., et al. Colloidal synthesis of silicon-carbon composite material for lithium-ion batteries, *Angew. Chem.*, 2017, *129*(36), 10920–10925.
79. Choi S., Kwon T. W., Coskun A., et al. Highly elastic binders integrating polyrotaxanes for silicon microparticle anodes in lithium ion batteries, *Science*, 2017, *357*(6348):279.
80. Liu Z, Guan D, Qiang Y, et al. Monodisperse and homogeneous SiOx/C microspheres: a promising high-capacity and durable anode material for lithium-ion batteries. *Energy Stor. Mater.*, 2018, *13*: 112–118.
81. Poizot P., Laurelle S., Grugeon S., et al. Nano-sized transition-metal oxides as negative-electrode materials for lithium-ion batteries, *Nature*, 2000, *407*(6803), 496–499.
82. Xu W., Wang J., Ding F., et al. Lithium metal anodes for rechargeable batteries, *Energy Environ. Sci.*, 2014, *7*, 513–537.
83. Fujimoto M., et al. Lithium secondary batteries. *Japanese Patent 3*, 229, 635 (1991).
84. Xu K., Nonaqueous liquid electrolytes for lithium-based rechargeable batteries., *Chem. Rev.*, 2004, *104*(10), 4303–4317.
85. Jean M., Chausse A., Messina R., Analysis of the passivating layer and the electrolyte in the system:petroleum coke/solution of $LiCF_3SO_3$ in mixed organic carbonates, *Electrochim. Acta*, 1998, *43*(12–13), 1795–1802.

86. Pistoia G., Derossi M., Scrosati B., Study of behavior of ethylene carbonate as a non-aqueous battery solvent, *J. Electrochem. Soc.*, 1970, *117*, 500–502.
87. Guyomard D., Tarascon J. M., Rechargeable $Li_{1+x}Mn_2O_4$/carbon cells with a new electrolyte-composition-potentiostatic studies and application to practical cells, *J. Electrochem. Soc.*, 1993, *140*, 3071–3081.
88. Tarascon J. M., Guyomard D., New electrolyte compositions stable over the 0 V to 5 V voltage range and compatible with the $Li_{1+x}Mn_2O_4$ carbon Li-ion cells. *Solid State Ionics*, 1994, *69*(3-4),293–305.
89. Aurbach D., Zaban, A., Schecheter, A., et al. The study of electrolyte solutions based on ethylene and diethyl carbonates for rechargeable Li batteries: I. Li metal anodes, *J. Electrochem. Soc.* 1995, *142*, 2873.
90. Sloop S. E., Pugh J. K., Wang S., et al. Chemical reactivity of PF_5 and $LiPF_6$ in ethylene carbonate/dimethyl carbonate solutions. *Electrochem. Solid State Lett.*, 2001, *4* (4), A42–A44.
91. Zhang S. S., Xu K., Jow T. R., Study of $LiBF_4$ as an electrolyte salt for a Li-ion battery, *J. Electrochem. Soc.*, 2002, *149*, A586.
92. Dahn J. R., Sacken U. V., Juzkow M. W., et al. Rechargeable $LiNiO_2$/Carbon cells. *J. Electrochem. Soc.*, 1991, *138*(8), 2207–2211.
93. Welton T., Room-temperature ionic liquids, solvents for synthesis and catalysis. *Chem. Rev.*, 1999, *99*(8), 2071–2083.
94. Warburg, E. and Tegetmeyer, F. (1888) Wiedemann. *Ann. Phys.*, *32*, 455.
95. Stephan A. M., Review on gel polymer electrolytes for lithium batteries, *Eur. Polym. J.*, 2006, *42*(1):21–42.

10 Applications of Thin Films in Metallic Implants

[a]Katayoon Kalantari, [a]Bahram Saleh, and [b]Thomas J. Webster

[a]Northeastern University, Boston, Massachusetts
[b]Hebei University of Technology, Tianjin, China; UFPI, Teresina, Brazil; and School of Engineering, Saveetha University, Chennai, India

CONTENTS

DOI: 10.1201/9781003141358-10

10.1 INTRODUCTION

Thin films are known as materials with specific characteristics such as a high ratio of surface area to volume and thicknesses in the range of nanometers (nm) to micrometers (μm); their behavior is different from the bulk materials [1]. In many fundamental technologies, thin films play important roles. They can be used on different surfaces considering their corrosion properties, conductivity, wettability, and reactivity characteristics. Various techniques have been applied to deposit thin films onto surfaces. Their structure is categorized from polycrystalline to amorphous based on fabrication conditions and also the type of material. Last, but not the least, their fabrication techniques, components, and processing conditions should not be toxic nor adversely affect normal functions of cells in the human body [2].

This chapter describes the fundamentals of thin-film processing for metallic implants, including thin-film fabrication and surface modification.

10.2 THIN-FILM COATING TECHNOLOGIES FOR METALLIC IMPLANTS

10.2.1 Sol-Gel Technique

Thin-film coatings constitute one of the most effective surface modification techniques for improving metallic implant performance. This technique is generally used for the fabrication of oxide materials [2, 3]. Sol-gel processes are frequently used to create thin-film coatings. The "sol" and "gel" in sol-gel means the formation of a colloidal suspension and sol conversion to viscous gels or solid materials, respectively. In this technique, under low temperature, the solution (sol) undergoes a sol–gel transition in which the gel becomes a moderately porous and rigid mass [4]. Metal alkoxides with organic ligands, such as silicon tetra-ethylorthosilicate (TEOS), have been used as the colloidal solution precursors. They can attach to the atoms of metals making them readily reactable with water [5]. The sol-gel method has several advantages, for example it is simple and can fully cover complex structures as well as it can maintain a high degree of homogeneity. Thin films fabricated by this technique are categorized into two groups: organic–inorganic hybrid thin films and inorganic oxides. The former has been developed to solve the limitation of the latter, specifically brittleness. Moreover, by using an organic component into the inorganic network, a relatively high temperature treatment is not required after deposition. This process has been known as an ecofriendly technique in comparison with traditional methods [6]. It is well documented that in metallic implants, the corrosion resistance improved in organic–inorganic hybrid thin films in comparison with inorganic thin films. Nevertheless, the inorganic thin-film adhesion to metallic substrates is better than that of organic–inorganic thin films [7].

10.2.2 Spin-Coating Technique

This technique has been used widely for the deposition of uniform thin films on flat substrates and it has been performed using different steps: deposition, spin-up/off, and evaporation (Figure 10.1).

The material deposition is the first step which occurs on the rotating platform and then at the second step, spin up and spin off happen sequentially while the evaporation occurs throughout the process. Centrifugal force is used for the distribution of an applied liquid on the platform. This step is followed by later drying. The high speed of spinning leads to a thinner layer and uniform solvent evaporation. Components that are highly volatile can be removed from the substrate due to the evaporation or simply drying, while less volatile components of the solution remain on the substrate surface. The rotation speed and viscosity of the coating liquid effect the thickness of the deposited layer [8]. This technique has been used for coatings on synthetic grafts, intraluminal stents, and stent coverings. To fabricate different therapeutic compositions, a polymeric matrix can be mixed with several therapeutic agents [2]. The substrate size is the main disadvantage of this method. With size increments and

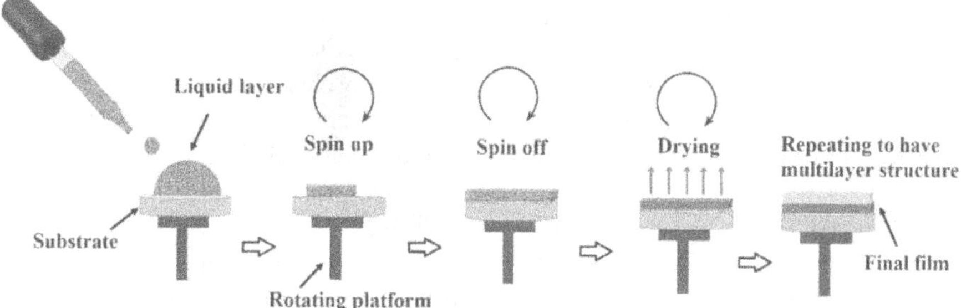

FIGURE 10.1 Schematic of the spin-coating technique.

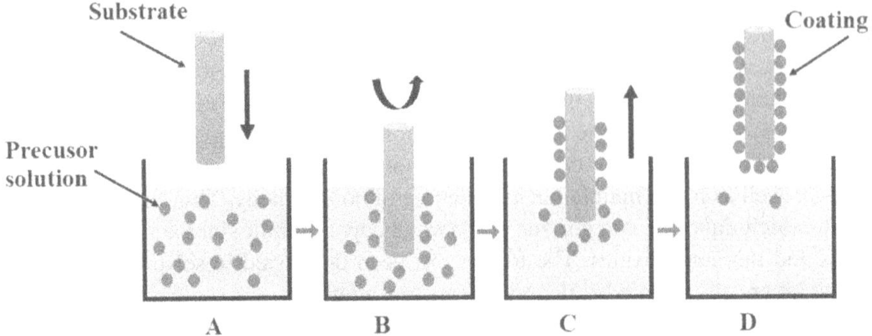

FIGURE 10.2 Graphical illustration of the dip-coating technique, redrawn from [10].

difficulty in film thinning, rapid spinning becomes difficult. Moreover, only 2–5% of the material is able to dispense onto the substrate and the rest is flung off during the process; thus, sustainability when using this technique is very low [5, 9].

10.2.3 Dip-Coating Technique

In the dip-coating technique (Figure 10.2), the substrate is immersed in a solution containing hydrolysable metal compounds (or readily formed particles). Then, in a controlled atmosphere containing water vapor and at a particular rate, the substrate is withdrawn from the liquid. After substrate removal, a homogeneous liquid film is formed on the surface. Finally, after the elimination of volatile solvents and the chemical reactions that occur at room temperature, a thin film results on the surface of the substrate. The viscosity of the liquid and withdrawal speed can be used to control the coating thickness. This method can be utilized to coat the cylinder, flat plane, and complex geometries with a large surface. This technique is a reproducible method which is low-cost and simple. The dip coating technique has applications in both the laboratory and industry due to the low-cost equipment and raw materials used, its easy steps and the development of products with good quality [8, 10].

10.2.4 Electrodeposition

Electrodeposition (ED) is a useful and cost-effective technique for metal, ceramic, and alloy materials. This method provides several advantages, such as controlling the chemical composition and thickness of the coatings by properly changing the deposition potential/current. There are more options in this technique to create high-quality coatings due to the use of the pulsed current mode instead of the conventional direct current mode [11].

10.2.4.1 Electrolytic Deposition (ELD) Coating

ELD is known as an electrochemical method used to make a uniform thickness coating of dense metals on conductive substrates. Deposition and substrate materials are selected as the anode and cathode while located inside an electrochemical unit cell. As shown in Figure 10.3, a potential difference is imposed between the

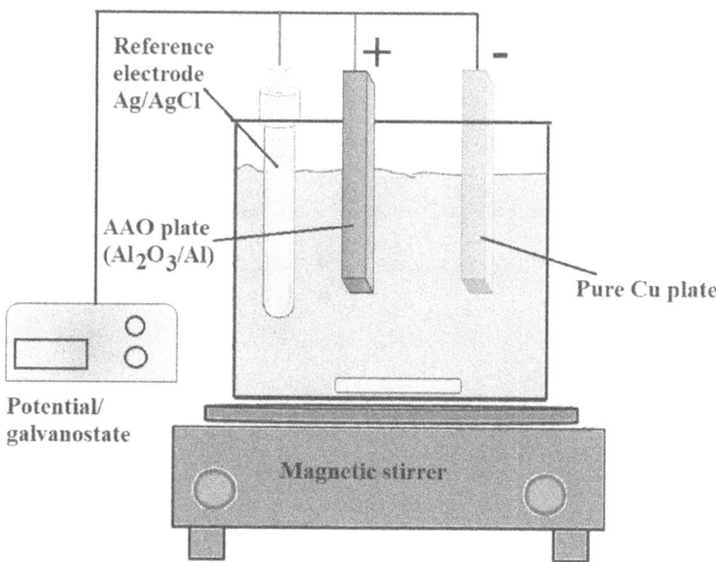

FIGURE 10.3 Schematic setup for the electrodeposition of copper metal particles over aluminum oxide, redrawn from [12].

cathode and anode poles which leads to movement of metallic ions toward the working electrolyte and then the substrate. During the process of coating, the concentration of electrolyte metallic ions remains constant [12]. This technique has been used in different fields including biomedical, optical, and as solid-oxide fuel cells [13, 14].

10.2.4.2 Electrophoretic Deposition (EPD) Coating

EPD, as a colloidal process, includes the movement of charged molecules or particles which are suspended in a solution and deposition of them occurs on an electrode with opposite charge to make films or monoliths under an applied electric field in a unit cell [15]. Different functional coatings can be made using this method to improve biocompatibility, magnetic susceptibility, and conductivity. Solid particles of polymers, ceramics, and metals with particle sizes lower than 30 μm can be easily deposited by EPD [16, 17]. This technique possesses several benefits such as a short time of deposition at room temperature, simplicity, and final homogeneous coating [15]. As shown in Figure 10.4, a typical EPD cell has two electrodes (cathode and anode), a power supply, and a suspension. Electrophoresis and deposition are two main steps of this process. In the first step, charged particles that are dispersed in an aqueous solution or solvent move toward the working electrode. In the deposition step, a homogeneous and rigid deposited particle is formed on the electrode. Charged particles deposit on the electrode with opposite charge. Cathodic deposition and anodic deposition are known as the coagulation of particles with positive charge on the cathode (negative electrode), whereas negatively charged particles deposit on the positive electrode (the anode) [18].

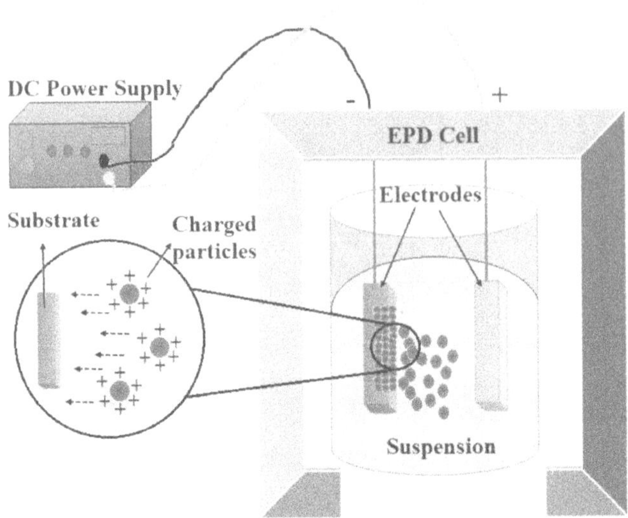

FIGURE 10.4 Schematic illustration of an EPD cell, redrawn from [15].

10.2.5 Physical Vapor Deposition (PVD) Process

Physical vapor deposition (PVD) is a term that is generally utilized to quantify the various deposition processes for thin films. In these methods, a condensed vaporized solid material is deposited on the top of a solid material surface under a partial vacuum environment [19]. Generally, these processes are used for the deposition of thin films with thickness values in the range of nanometers to thousands of nanometers. Moreover, for the deposition of a multilayer coating, high thickness, or graded composition deposits, hybrids with other deposition methods and free-standing structures can be accomplished [20].

10.2.5.1 Electron Beam Evaporation

This evaporation type is another physical deposition technique where an intensive electron beam is normally accelerated with voltages from about 5 to 20 kV, and is created by a filament and steered through both electric and magnetic fields to hit the target and vaporize it under vacuum condition as shown in Figure 10.5 [21]. By using an electron beam with high energy, the target materials placed in a crucible are bombarded [22]. Depending on the materials of the target, the crucible can be made from copper, graphite, tungsten, boron nitride, and nickel [23]. By using this method, most pure metals can be vaporized including metals with high melting points. Moreover, in comparison with other PVD methods, the highest deposition rate (as high as 50 $\mu m\ s^{-1}$) can be achieved using this technique.

10.2.5.2 Sputtering Deposition

Basically, in the sputtering technique, as shown in Figure 10.6, a glow-discharge plasma is located in front of the target, generates energetic ions that bombard the

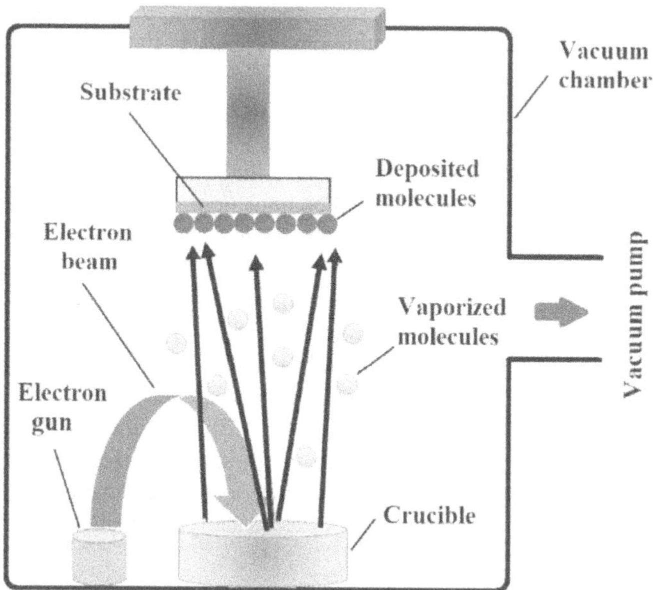

FIGURE 10.5 Schematic of an electron-beam evaporation system.

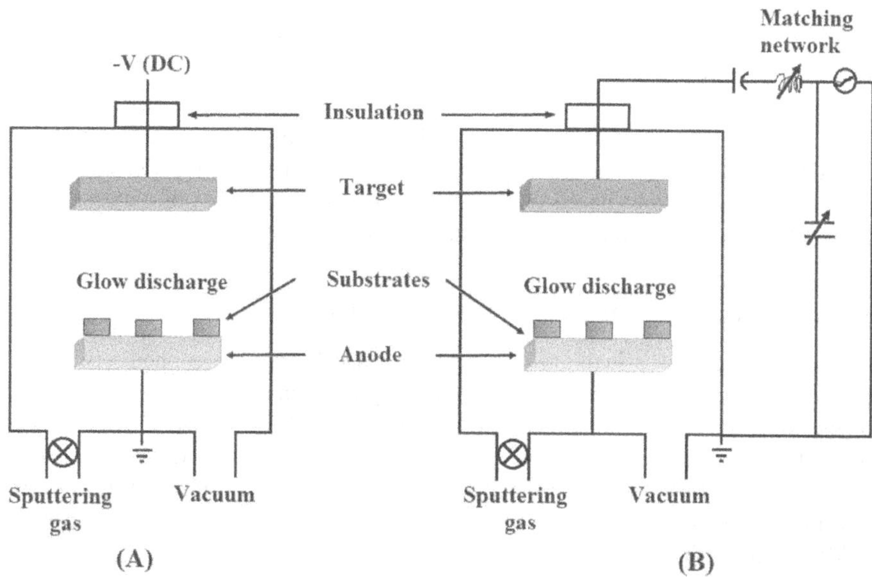

FIGURE 10.6 Schematic diagram of the principles of (A) direct current (DC) and (B) radio-frequency sputtering systems.

cathode made of the target material including an alloy, element, compound, or a mixture of them. The process of bombardment leads to the sputtering or removal of target atoms by transferring the momentum from the bombarding energetic gas ions accelerated in an electric field. Then, these sputtered atoms condense on the substrate as a thin film [24].

The sputtering deposition technique can be utilized to deposit materials and alloys as films and maintain the target material composition. This is possible because the material is removed from the target in a layer-by-layer structure, which is one of the main benefits of the process [25]. Nevertheless, this technique is often suitable for the manufacturing of small-scale specimens and is not generally suitable for large-scale structural applications. Moreover, the final products consistently contain contamination and residual porosity [26]. Materials with a high melting point can be formed easily by this technique. Moreover, the sputtering sources are compatible with oxygen as a reactive gas. Radio frequency (RF) and direct current (DC) are two main types of the sputtering technique. The first one utilizes RF power for most of the dielectric materials. The radio frequency sputtering depends on DC power, which is usually used with a target made of electrically conductive materials [27].

10.2.6 Chemical Vapor Deposition (CVD) Coating

CVD is known as another type of vapor deposition technique. In this technique, the substrate, called a wafer, is exposed to a set of volatile material precursors where a chemical reaction creates a deposition layer on the surface of the material [12, 28]. In chemical vapor deposition, a wide range of materials of different shapes and structures (polycrystalline, single crystal, and amorphous) can be used including carbon in the form of graphene, diamond, fluorocarbons, nanotubes/fibers/nanofibers, nitrides, titanium, carbides, and tungsten. This method has been known as an effective process for improving implantable medical devices. A CVD schematic is presented in Figure 10.7. The vaporized materials for CVD are pumped from the right side and the temperature is kept high enough using heaters to facilitate the chemical reaction between the substrate and vaporized materials. There are several CVD classifications based on pressure conditions, properties of materials, and plasma types used in vaporizing materials [29, 30].

10.2.7 Micro-Arc Oxidation (MAO) Coating

As shown in Figure 10.8, in this process, a high voltage difference is applied between the cathode and the anode which leads to micro-arc generation as plasma channels. Based on the micro-arc intensity in the substrate-hitting process, they are melted in a portion of the surface. Simultaneously, plasma channels release their pressure which supports the deposition of coating materials in the working electrolyte on the substrate surface. An oxidation reaction happens using the electrolyte oxygen, making a deposition layer of oxides on the surface of the substrate. The MAO technique is known as a flexible coating process regarding the composition of the coating layers with titanium, magnesium, aluminum, and their alloys [32]. The main characteristics of a layer treated by MAO are high corrosion resistance due to the formation of a uniform oxide layer [33].

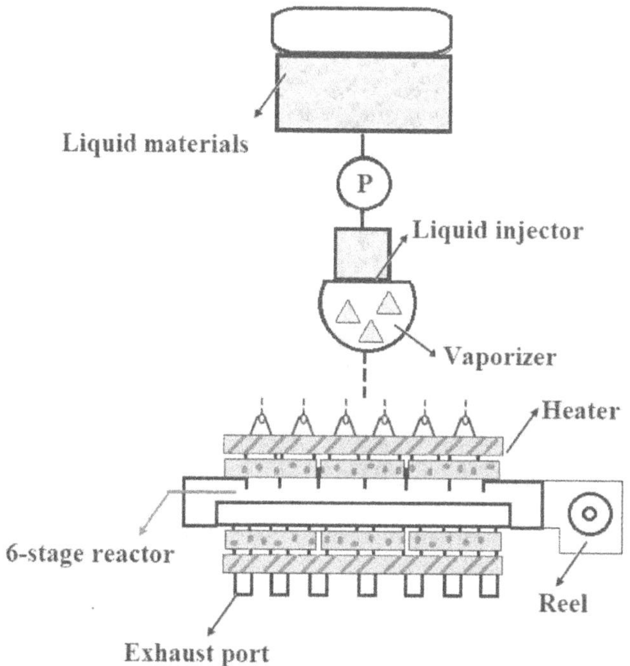

FIGURE 10.7 Schematic diagram of a six-stage CVD system, redrawn from [31].

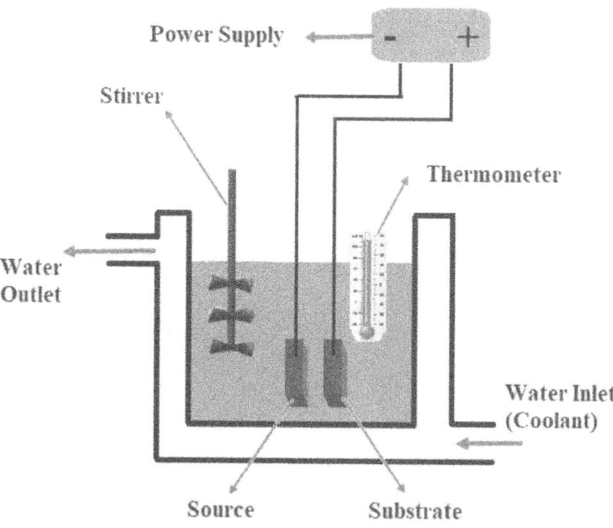

FIGURE 10.8 Schematic view of the micro-arc oxidation (MAO) process, redrawn from [12].

The coating surface made by MAO shows high adherence and hardness properties while it has several scales of porosity throughout its structure. Process time, voltage, electrolyte and current type, pulsate current, and current density are all factors that have an effect on the quality of the coating. There are limitations in substrates that are mainly valve metals like aluminum, niobium, magnesium, zirconium, tantalum, and titanium [34].

10.2.8 Plasma Spray Coating

Plasma spraying is one of the thermal spraying techniques that uses a heat source with high energy to melt and to accelerate fine particles onto a prepared surface [35]. Upon impact, these molten particles cool down and solidify rapidly by heat transfer to the underlying substrate and by accumulation, they create a coating consisting of lamellae.

In this technique, a plasma torch (also known as a plasma gun) is used to create a plasma as the high temperature source [36]. As shown in Figure 10.9, the plasma gun has a tungsten thoriated cathode, which is cone-shaped and a cylindrical copper anode.

There is a flow of gases forming a plasma (like Ar, He, H_2, and N_2) via the annular space between the two electrodes and an arc is started by high-frequency discharge. The stream of gas which flows between the two electrodes stretches the arc, so that in its course from one electrode to another, the arc loops out of the torch nozzle as a plasma flame. The temperatures in a plasma flame are usually 10000–15000°C. So, in principle, almost any feedstock material (metal or ceramic) including refractory metals or oxides and in different forms can be melted and deposited to form a coating by plasma spraying. The plasma jet heats and accelerates the melted materials and they are flattened on the target surface (metal substrate) to create a coating via successive impingement [36].

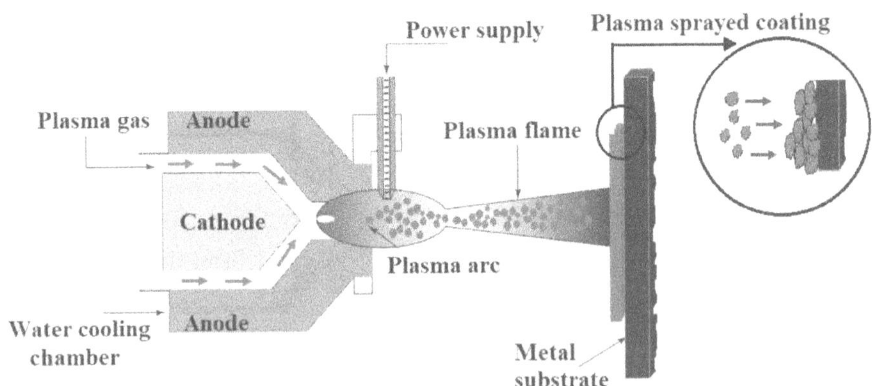

FIGURE 10.9 Schematic diagram of the plasma-spraying technique for forming coatings, redrawn from [36].

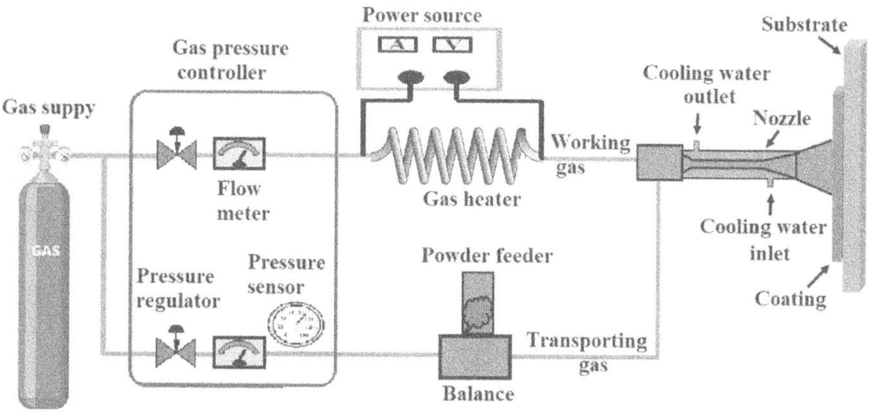

FIGURE 10.10 Schematic diagram of a typical cold spray system, redrawn from [38].

10.2.9 Cold Spray Coating

In cold spray processes as a solid-state, a supersonic particulate jet uses as molten droplet jet to create coatings, which avoids/minimizes many deleterious high-temperature reactions [37]. Figure 10.10 presents a schematic diagram of a typical cold spray process.

The cold spray system is generally composed of four sections. The first part includes the gas source which supplies gas and a pressure controller. This unit provides and adjusts the working gas to a preset value of pressure. Preheating of the gas unit is known as the second unit which includes a heater and a power source and heats the working gas to a specific temperature value. The powder supply part is the next unit that feeds the cold spray powder to the gun at a preset rate. The last part is a cold spray gun unit, which heats and accelerates the particles to a high speed with a de Laval nozzle. The cold spray systems can be categorized into two main groups as the high and low pressure [38]. The cold spray beam in this technique has a very narrow footprint which leads to a beam with a high density particle, thus, as a result there is better control on the coating shape without requiring masking. Moreover, the coatings are free of oxides and other inclusions [37].

10.2.10 Pulsed Laser Deposition (PLD)

Pulsed laser deposition (PLD) is one of the most used high-temperature methods developed for thin-film growth of different materials. PLD is a physical vapor deposition technique in which a high-power pulsed laser is directed toward a rotating target material with the aim of the target to be vaporized, that is, to create a plasma plume which should be deposited on a substrate (Figure 10.11). The plasma plume is a combination of ions, atoms, and molecules. This technique is commonly utilized for oxide deposition while the oxygen is applied as the background gas [39]. Several hundred Celsius degrees should be applied to substrates for heating them up causing atomic evaporation and sufficient atomic diffusion. In addition, for the prevention

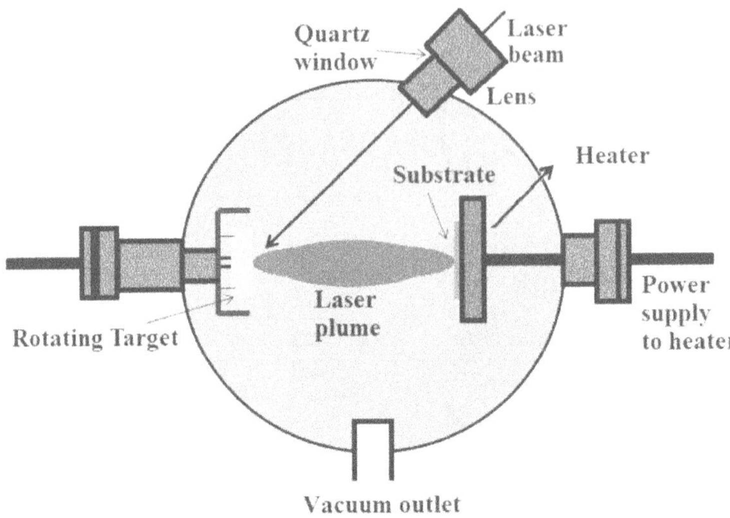

FIGURE 10.11 A schematic diagram of the pulsed laser deposition setup.

of too many species scattering in the plume as well as to decrease the number of impurities in the film, a high vacuum should be applied [40]. The PLD technique has the advantages of the high adhesion, absence of contamination and pores, and stoichiometry transfer of the target composition [41].

10.2.11 Biomimetic Coating Process

The biomimetic coating technique is one of the most useful methods developed for the surface coating of orthopedic implants. This technique can be utilized to deposit calcium phosphate (CaP) layers on medical devices for improving their osteoconductivity and osseointegration (direct bonding with bone) capabilities. This technique involves bone-like crystal nucleation and growth on a pretreated substrate by immersing it in a CaP supersaturated solution and heating to a temperature (37°C) and pH (7.4) similar to physiological conditions. By using biomimetic coatings, osteogenic agents can be released slowly and can be sustainable at the implantation site [42]. A biomimetic coating technique can be utilized in different types of materials, such as bio-ceramics, metals, and organic polymers. However, by using the original biomimetic technique, successful coatings of the materials can be achieved only on those surfaces with active chemical groups, which serve as nucleation sites for mineralization [43].

In the above discussion, widely used methods for surface treatments and depositing a protective layer's deposition were introduced in detail. Nevertheless, only a few of these methods are sufficiently reliable to be utilized for biomedical applications such as electrophoretic coatings, plasma spray, laser deposition, biomimetic deposition, and wet methods, such as sol-gel-based spin- and dip- or spray-coating.

10.2.12 Liquid–Liquid Interface Reaction Technique

Nanomaterials are synthesized by applying different chemical and electrochemical methods. In liquid–liquid interface synthesis, the properties of the interfaces depend on the nature of the species, their concentrations, the polarity of immiscible solvents, and temperature. Among different applied methods, reduction of the metallic precursor is better due to our good experience. This facile method is developed for the synthesis of monometallic nanostructures, bi, tri, and tetrametallic alloys and metal-organic frameworks. Toluene–water interface assembly was used for the preparation of monometallic nanomaterials, multimetallic nanoalloys, and metal-organic frameworks. In this assembly, organometallic precursors such as [$PtCl_2$(cod)], [$PdCl_2$(cod)] and also some other precursors such as [Ni$(acac)_2$], [Zn$(acac)_2$], [Fe$(acac)_3$], etc. are dissolved in toluene by using ultrasonic waves and contacted with water (as a second phase). Very good solubility in the organic solvents such as toluene and also easy loss of (cod) ligand are the main reasons for choosing (cod) complexes as precursors. The reaction is initiated by the dropwise addition of an aqueous solution of Na_BH_4 that leads to the formation of metallic thin films after 24 h [44–51].

10.3 METALLIC IMPLANTS

Metallic materials play significant roles in the human body as implants and metallic alloys are most commonly utilized in dental implants and bone joint replacements. Most applications of metallic implants are in orthopedic surgery because they show higher fatigue resistance and tensile strength in comparison to polymers and ceramics [52]. As early as 200 A.D., they have been widely exploited in clinical practice when an iron-based material was implanted in human bone [53]. The most commonly used metallic materials in implants are inert metals including stainless steels (SS), titanium (Ti) alloys, platinum (Pt), and cobalt (Co)-chromium (Cr) alloys [54], while biodegradable metals, such as zinc (Zn), magnesium (Mg), and iron (Fe), have been introduced recently as a new generation of biomaterials for temporary applications [55, 56]. Nevertheless, the bioactivities of these materials (inert and biodegradable) are somewhat limited, and thus some functionalities are needed when utilized in specific clinical practice [57]. Moreover, because of the nature of metallic interaction with various media, corrosion may occur and degradation happens after implementation of these materials inside the human body. The important aspects of material suitability for bio-applications are to have controlled biodegradability, higher corrosion resistance, and lower toxicity due to released metallic ions [58]. Implant/bone bonding is one of the largest issues today caused by differences between the chemical composition of bone and metallic implants [59]. Discussed thin-film coating techniques have been suggested by researchers to overcome these issues. After exposure to harsh environments, biocompatibility and corrosion resistance are the major compatibility indicators of metallic implants. The environment of the human body or physiological fluids includes a corrosive saline medium that causes remarkable corrosion of metallic implants if there are no protective oxide layers or other covers. Implants degrade because of corrosion, leading to a reduction in their mechanical stability and premature failure [60]. Moreover, due to the corrosion reactions inside the human body, some highly toxic byproducts like nickel (Ni), cobalt (Co), and vanadium (V) release which can be harmful for living organs. On the other hand, biocompatibility should be considered as an essential functional specification of implants. The tissues around biocompatible materials show a positive response in connection with implant surface properties, if, the surface of implants should provide a suitable environment for cell growth and sustainability. Biocompatibility depends on the surface quality of materials, as well as the chemical composition and microstructure of the substrate [61, 62]. For instance, in temporary implants, there are some requirements for surface modification, like preventing infections and promoting bone formation. For example, when Ti alloys are implanted into the human body, a calcium phosphate layer is formed, resulting in more bone formation. By the removal of these alloys from bone, bone re-fracture can happen [63]. Depending on the anatomical location, another requirement is the adhesion to soft tissue. Inflectional diseases and inflammation as well can be the result of non-fully adherence soft tissue which may lead to implant failure. Therefore, improving implant adhesion to soft tissue by using a suitable surface modification method is needed. Increasing implant wear resistance and prevention of biofilm formation are additional required factors [64].

10.3.1 Stainless Steel (SS)

316L and 316LVM stainless steel have been widely applied as materials for the creation of cardiovascular valves/stents, orthopedic prosthesis, craniofacial and dentistry implants/devices used in biomedicine because of their malleability and resistance to corrosion and fatigue [65]. The first use of SS in the 1920s was as a bone implant but since then, the applications of SS in medical devices have significantly expanded. SS can be used as a temporary implant (to be later removed) in such applications as screws, plates, pins, and networks/threads used in fixing fractures and as a permanent implant such as artificial joints [66].

316L SS contains some ions such as Fe, Cr, Ni, and Mo. There is low content of carbon as well that prevents corrosion in the human body. Most implants made of 316 L SS have pitting and crevice corrosion problems, but coating their surfaces with ceramic materials helps prevent these issues. 316LVM is another type of SS which is more resistant against corrosion, and has been utilized in numerous biomedical fields [67]. The Young's modulus of SS316 L is 200 GPa which is more than that of bone and will lead to detachment because of the stress shielding phenomenon. Stainless steel has a low cost and it is easy to manufacture; thus, it is a suitable candidate for utilization as a nonpermanent device in fracture fixation and for short-term implant applications [68].

As mentioned before, the biological environment of the human body is extremely harsh on SS due to the lack of inherent bio-functional properties like bioactivity, blood compatibility, and osteo-conductivity [69]. The unmodified SS surface is relatively hydrophobic which adsorbs proteins and can lead to biofilm formation and corrosion [70, 71]. Several bio-functionalization and coating techniques have been used to achieve the above-mentioned properties on SS without sacrificing bulk properties [72].

10.3.2 Magnesium (Mg)

Magnesium (Mg), as a lightweight metal, is naturally found in bone tissue and is vital to human metabolism as a cofactor for various enzymes. This metal is considered to be one of the most important elements in the human body, and the toxicity level of its ions is within the tolerable limit [73]. For the first time in the 1930s, Mg alloys were introduced as implant materials for orthopedic and trauma surgery. Mg and its alloys have attracted considerable attention as a temporary implant because of their outstanding biocompatibility with the physiology of the human body [74]. Moreover, the Mg ions (Mg^{2+}) are not toxic to cells and are known to help heal tissues. These ions can be expelled via urination [75, 76]. Despite the excellent properties of Mg, the main disadvantage of using Mg alloys as implants is their remarkably high rate of corrosion in the physiological environment (pH 7.4–7.6) which leads to a loss of mechanical strength; thus, implants prematurely fail before their expected service life [77, 78]. Furthermore, hydrogen evolution, which is the main cathodic reaction and concurrent with the corrosion of Mg alloys, can significantly interfere with the bone-healing process [79]. Because Mg alloys can dissolve or corrode away in chloride-containing aqueous solutions (like the physiological environment), the need for a second surgery may be obviated. Nevertheless, before this application, it will be required to ensure the mechanical integrity of the Mg alloy in the physiological environment, at least until the implant has served its purpose [77].

10.3.3 Titanium (Ti) and Its Alloys

Since 1970, Ti and its alloys have been utilized widely as orthopedic and dental implants with known properties including low density, high mechanical strength, biocompatibility, and corrosion resistance. Ti makes a stable layer of oxide on the surface which passivates and protects it from corrosion [80]. Pure Ti (CP Ti) can be mixed with aluminum (Al) and vanadium (V) and can be categorized into four grades (I–IV) in which the amount of oxygen and each element in the four grades is different. CP-Ti grade is currently being applied as the tibial, acetabular, and femoral component in the case of knee and hip arthroplasty and in fracture fixation. These Ti implants reduce the risk of loosening, provide stable fixation, and good biocompatibility [81]. Stress-shielding effects due to high elastic moduli lead to implant failure and Ti implants have been used with 110 GPa as an elastic modulus lower than SS (at 210 GPa) and cobalt–chromium alloys (at 240 GPa). Several applications have been reported for Ti-based alloys in almost all disciplines of medicine such as stents, heart, and dental implants, as well as orthopedic implants including plates and joint replacements [82].

10.3.4 Cobalt–Chromium Alloys

Cobalt–chromium (Co–Cr) alloys have been developed for joint replacement prostheses like the femoral component and head in total knee and hip replacements. Additionally, these alloys have been introduced for use as screws in systems of trauma plating because of their lower osteointegration compared to Ti alloys. The weak integration of Co–Cr screws with bone tissues could facilitate easier removal of the screw after healing of the bone fracture [83]. Co–Cr alloys have been known to be corrosion resistant, biocompatible, hard and tough metals with high specific strength. Carbide precipitation and the multiphase structure of these alloys leads to good mechanical (tensile strength of 145–270 MPa) and an enormous increase in hardness (550–800 MPa). Although the reported allergic reactions in patients are very low, the movement of joints may result in the release of tiny metal ions into the body which sometimes can cause the allergic reactions mostly in patients with allergies to special metals like nickel [84].

10.4 COATING MATERIALS

10.4.1 Ceramics

Ceramic materials are some of the most significant biomaterials for applications in biomedical engineering. Increasing medical demands have drawn researchers' attention to create new materials. Bio-ceramics are a group of ceramics which are in contact with living tissues and have played a significant role as implants, specifically in dentistry and orthopedics [85]. Calcium phosphates, alumina, zirconia, silicon nitride, and certain glasses and glass-ceramics are all examples of bio-ceramics. Ceramic coatings can make high-performance oxide layers on the surface of alloys and metals to solve problems of corrosion, wear, heat, insulation and friction issues. Ceramic film thickness can range from 50 nanometers to several micrometers, depending on application and coating techniques [86]. Bio-ceramics are categorized into three groups as bioactive, bio-inert, and bio-resorbable based on their interaction with the biological environment [87]. For dental and orthopedic implants and the artificial synthesis of ceramics, the main goal is to implant a bioactive ceramic material that is able to create a bond between the damaged bone and surrounding tissues. Otherwise, if the ceramic is inert, the bone will be replaced by a material that the organism can tolerate, but which cannot substitute it by means of bone regeneration [85]. These materials are not suitable for heavy load applications, due to their low fracture toughness and tensile strength, so they can be used as coating materials on metals and for joint replacement applications [87].

10.4.1.1 Titanium Dioxide (TiO_2)

Tumor resection, fracture, and osteoarthritis lead to bone disorders and orthopedic or dental implants are usually used as an initial treatment. Currently, Ti (and its natural oxide TiO_2) is the most common material for such implants [88] and among the different Ti biomaterials, commercially pure Ti with a single-phase alpha microstructure, Ti6A14V and ASTM F67 with a two-phase alpha-beta microstructure have been extensively utilized in dental and orthopedic fields, respectively [89]. Nevertheless, the modification of Ti-based implants is required to make them suitable for osteogenesis, osteoconduction, and osteointegration, due to their difficulty in making a good chemical bond with surrounding bone and also a mismatch between Young's modulus of cortical bone and Ti [90].

10.4.1.2 Zinc Oxide (ZnO)

Among different nanoparticles, zinc oxide has low toxicity, significant antibacterial activities including adhesion and biofilm formation, good biocompatibility, and high resistance to corrosion [91]. Moreover, they also induce a positive effect on the attachment and growth of osteoblast-like cells [92]. The optical, thermal, and antibacterial properties of ZnO thin films can be improved by adding different elements such as Nd, Na, Mg, Fe, Co, Ag, B, Al, and Cd [93]. A zinc oxide coating has been used widely in the biomedical field to enhance the antibacterial and biocompatibility properties of biomaterials, because the zinc ions can improve the proliferation and adhesion of healthy tissue cells, and inhibit activities of the bacteria life cycle, such as glycolysis and transmembrane proton translocation [94]. The antibacterial properties of ZnO, with their deodorizing ability, make them applicable for different products as cotton fabrics, diaper rash preventing agents, antiseptic ointments, and anti-dandruff shampoos [52, 95].

10.4.1.3 Hydroxyapatite (HAp)

One of the main components of teeth and bone in vertebrates is calcium phosphate salts. Among calcium phosphate salts, hydroxyapatite (HAp) with the chemical formula of $[Ca_{10}(PO_4)_6(OH)_2]$ is the most stable crystalline phase in body fluids, and is similar to the mineral part of bone [96]. It has been well known that HAp has the ability to promote bone ingrowth without any inflammation or toxicity and can be used as a dental material, middle ear and bioactive coating on metallic osseous implants [97–99]. A HAp coating can be deposited on the surface of metal alloys to support the osseointegration of implants with the surrounding bone [100]. Some mechanical properties, like load-bearing ability, can be maintained by using a HAp coating and they also show good biocompatibility with bone [101]. Some disadvantages such as poor thickness uniformity, poor adhesion to an underlying substrate, high porosity, impurity in phases, and limited crystallinity are common in HAp coatings. Nevertheless, low coating adhesion has been known as the main disadvantage in using HAp as an implant commercially; thus, as a general requirement, the strength of bonding between a ceramic coating and metallic substrate should be improved, regardless of the method utilized [100].

10.4.1.4 Bio-Glass

It has been shown that bioactive glasses can stimulate the formation of bone both in *in vivo* and *in vitro*. In the 1960s, Hench discovered the first composition of 45S5 Bioglass® [102, 103]. Bio-glasses have been used as replacements for the middle ear, periodontal surgeries, and coatings for prosthetic metallic implants. Moreover, it has been incorporated into toothpastes which lead to a reduction in dental hypersensitivity [104]. The use of a bio-glass coating has two main purposes: improvement of implant ossteointegration and as a protectant against corrosion from body fluids. Bio-glasses help the prosthesis adapt to the cavity of bones, preventing fibrous tissue formation at the bone-prosthetic interface making a strong chemical bond between bone tissue and implants [105]. However, their low mechanical properties have limited their application in load-bearing bone defects. So, numerous attempts have been made to improve their mechanical strength and brittle behavior by modifying the chemical composition and addition of some metal oxides like magnesium oxide, zinc oxide, aluminum oxide, and boric oxide [106].

10.4.1.5 Zirconium Dioxide (ZrO_2)

Among the different ceramic materials used for biomedical applications, zirconium dioxide (ZrO_2) is considered a potential material for medical applications mostly focused on orthopedic implants and bone tissue engineering applications as well [107]. ZrO_2 has been utilized extensively, because of its excellent properties including good fracture strength and compression resistance, good cell culture and proliferation, biocompatibility, and non-cytotoxicity [108, 109]. A study completed by Degidi *et al.* showed the effect of zirconium dioxide in biomedical implants leading to the creation of an inflammatory substrate in comparison with that of metal alloys demonstrating the formation of soft tissue on implant surfaces [110]. Zirconium dioxide has also been documented as a surface coating to improve cell adhesion, proliferation, and culturing properties on the surface of a metal alloy, as the material shows no cytotoxicity [107]. Compared to titanium implants, they showed enhanced mechanical strength and esthetic appearance as well as a lower accumulation of plaque [111].

10.4.1.6 Carbon-Based Materials

In 2004, graphene, as a new carbon-based material, was first deposited via mechanical exfoliation, and has been developed as one of the most studied topics because of its excellent properties including high conductivity, transparency, chemical stability, mechanical strength, and good tribological properties [112]. In several studies, a graphene film was deposited on the substrates such as Cu and Al which improved the corrosion resistance [113, 114]. Graphene also plays an important role in tissue engineering, which helps scaffold materials improve their biological activity such as cell adhesion and proliferation [115].

Graphene oxide (GO) is known as one of the main graphene derivatives which is a famous material for biomedical applications because of its excellent properties such as high surface area, and conductivity, good biocompatibility, and strong mechanical strength [112, 116]. Moreover, GO shows significant antibacterial activity against Gram-positive and Gram-negative bacteria due to the sharp nano-edge of graphene which leads to a physical break in the microorganism cycle by cutting and/or penetrating the membrane of cells [117]. These properties make it a good candidate for coating materials for advanced implants and industry fields. Graphene-based materials have the ability to induce osteogenic and also chondrogenic differentiation of stem cells; thus, it can be used for the repair and regeneration of bones [118, 119].

10.4.2 Composites

The use of composite materials has led to rapid growth in the tissue engineering and biomedical fields due to their synergistically improved applications with a combination of corrosion resistance, stiffness, and light weight [120]. A composite is a formation of two or more component materials having remarkably different chemical or physical properties when combined together or creating a material that has unique characteristics different from component elements [121]. The toughness properties of individual ceramic biomaterials are lower than those of natural hard tissue, which limits their use in high load-bearing applications. The development of several composite coatings like ceramic–ceramic or polymer–ceramic coatings for metallic implants has led to better mechanical properties for enhanced biomedical applications [52].

10.4.3 Polymers

As mentioned before, metallic implants are the most utilized materials in the orthopedic and dental fields, and they may fail because of their inability to resist bacteria colonization or promote and maintain osseointegration. Polymeric coatings have been adopted as an effective technique to tailor the corrosion rate of implants without changing their bulk properties. Also, they can provide other functionalities to the substrate like enhanced biocompatibility and cellular responses, ability to load small molecules, self-healing, and antibacterial properties [122, 123].

Several synthetic polymers, such as polylactic acid (PLA), polyethylene glycol (PEG), poly(lactic-co-glycolic) acid (PLGA), polydopamine, and polycaprolactone (PCL), as well as natural polymers (like chitosan and silk fibroin), have been utilized as protective coatings on the surface of implants. While polymeric coatings can increase the corrosion resistance of implants, they are susceptible to damage in the human body, which leads to polymer degradation and deterioration of protective barrier properties [124]. Polyethylene glycol (PEG) is one of the most used polymers in medicine to obtain both an improvement in the elasticity and hydrophilicity of coatings. It is known that PEG addition improves implant biocompatibility by increasing hydrophilicity of a material and, therefore, enhancing cell adhesion and growth. Furthermore, PEG has reduced immunogenicity and little toxic effects [125, 126].

With introducing corrosion inhibitors (such as molybdate, cerium ions, vanadate, and benzotriazole) into polymeric coatings, their anticorrosion properties and durability have demonstrated improvement [127]. After damaging barrier coatings by corrosive species, these inhibitors are released into defected sites to reduce the progression of corrosion by creating insoluble metal complexes [122, 128]. Although being effective in reducing corrosion, most of these synthetic inhibitors are not appropriate for implants because of their potential cytotoxic and carcinogenic effects in the human body [129].

10.5 RELATIONSHIPS BETWEEN OSSEOINTEGRATION AND SURFACE MODIFICATION

Osseointegration has been described as the direct connection between human living bone and the implant [130] which has an important effect on implant stability ensuring implant success. Once the device is implanted into bone, the bone grows around implant's porous structure. Usually, the osseointegration process can range from a few weeks to a few months. The healing process can be divided into four steps: Hemostasis, Inflammation, Proliferation, and Remodeling [64, 131]. With surface modification of an implant, corrosion and wear resistance of the implant increases, and makes the implant more biocompatible which leads to better integration with bone tissue. Dental implants with rough surfaces have been shown to improve bone fixation and the bone-to-implant contact percentage over that of commercially available implants [132].

10.6 CORROSION OF BIOCOMPATIBLE METALS

Metallic materials are unique candidates for implants due to their excellent toughness, strength, and ductility, but, implant corrosion and metallic ions leach into surrounding tissues causing a significant disadvantage [133]. Corrosion is known as material degeneration into its constituent atoms because of chemical reactions happening between the materials and its surroundings [134]. Several research groups have shown that corrosion occurs slowly in the human body due to an electrochemical reaction after implantation of a metal [135]. A passive coating on the implant surface should control the corrosion process via the release of low levels of corrosion products for effective implantation [136]. The interstitial fluid around cells in tissue and the ionic response to biomaterials (because such fluid contains various ions, water, amino acids, and proteins) leads to implant corrosion [137]. The release of metal ions has been found to be highly toxic to human cells and to cause serious infection. Some elements, including arsenic, lead, beryllium, and mercury, have been found to be toxic and their use is to be avoided in clinical applications. They may result in proteins and enzymes denaturing, an extensive immunological response and even organ damage (mostly cardiac and renal tissues because of the ion accumulation) [136]. The wear and/or initial cracks created on the surface of an implant would initiate and accelerate corrosion rate [134].

Generally, there are four types of corrosion: galvanic, fretting, pitting, and crevice. In galvanic corrosion, direct contact happens between two dissimilar metals in an electrolytic solution. General corrosion is the most common and includes uniform regular metal removal from the surface of an implant, and localized corrosion, which occurs at the imperfections of a resistant surface known as pitting corrosion. Crevice corrosion occurs because of the geometry of the implant/prosthesis assembly when the local oxygen concentration is low, frequently in a crack or crevice (implant interconnections) [138].

Recently, numerous research and clinical diagnostic studies confirmed that these metallic materials are still susceptible to corrosion. Thus, before implementation, they should be subjected to a corrosion screening test to clarify their behavior in different environments [139, 140]. Table 10.1 presents several American Society for Testing and Materials (ASTM) standards under different conditions.

TABLE 10.1
Standards for Corrosion Resistance Testing of Biomaterials [134]

ASTM Standards	Specifications
ASTM F2129-01	Cyclic potentiodynamic polarization measurements
ASTM G71-81	Galvanic corrosion in electrolytes
ASTM F746-87	Pitting or Crevice Corrosion of metallic surgical implant materials
ASTM G 5-94 and ASTM G 61-86	Corrosion performance of metallic biomaterials

An ideal corrosion-resistant implant should be one that releases reduced metal ions in the harshest conditions of the human body, and keep the implant in a properly low stage of reactivity above a long service period, such as over 30 years in typical physiological situations [141].

10.7 CORROSION OF ORTHOPEDIC IMPLANTS

Orthopedic implants could be permanent or temporary. For the substitution of a toe, shoulder, knee, hip, section of the spine, and finger, permanent implants are being used. Temporary implants include plates, screws, pins, and rods for fracture fixation, trauma, or spinal curvature correction surgery. Wear is a major reason for malfunctioning of orthopedic implants that hasten corrosion; thus, materials with high wear resistance including Co–Cr alloys and ceramics are commonly used as orthopedic implants that are susceptible to constant shear and friction. For the fabrication of the femoral component, Ti and its alloys have been applied, while the ball is made of either a Co–Cr alloy or other hard ceramic materials for hip implants [142]. Nevertheless, the environment and allocation of corrosion leading to ion discharge within the body from this orthopedic implant are of significant concern [135].

10.7.1 Corrosion of Cardiovascular Implants

The artificial heart is known to be a solution for ischemic heart disease. Cardiovascular implants must have complete biocompatibility with blood to reduce the risk of adverse thrombogenic clotting. These implants can be made up of non-bioprosthetics like Ni–Co alloys, stainless steel, Co–Cr alloys, and polypropylene or natural materials. However, for the prevention of extra surgeries and patient fatality, suitable attention should be given to the entire design and selection of the implant [141]. The most frequently used material in stents is 316L stainless steel, which is susceptible to corrosion and its nickel release leads to an allergic reaction. In particular, the discharge of Mo, Ni, and Cr ions from stainless steel stents can activate limited immune and inflammatory reactions [143].

Nitinol is another material that has been used frequently as a cardiovascular medical implant because of its super-elastic properties that lead to higher reversible deformation in comparison with other metals, biocompatibility, thermal shape memory, fatigue resistance, and non-interference with magnetic resonance imaging diagnoses [144]. Nitinol corrosion in implants is vital for cardiovascular community. The corrosion is usually assessed by pitting corrosion testing (per ASTM F2129), nickel ion release rate testing (per ASTM F3306), and, in instances where the nitinol is in direct contact with a dissimilar metal, galvanic corrosion testing (per ASTM F3044). Several studies have evaluated nitinol biocompatibility, including the effect of various surface finishes and treatment techniques on the resultant corrosion behavior [145]. Recently, Nagaraja et al. [146] studied the effect of surface processing of nitinol on the in vivo release of nickel and its impact on the response of local biological reactions. They found that the concentration of nickel in urine and systemic serum were not different between processing groups 180 days after implantation. Although their findings demonstrated that the different surface processing of stents did not increase systemic nickel concentrations or negatively affect the liver, kidney, or hematological function.

10.7.2 Corrosion of Dental Implants

Dental implants are categorized as the endosseous (inside the bone) and subperiosteal (on the top of the bone) implants [141]. In dental implants, corrosion performance is assessed using artificial saliva [134]. Galvanic corrosion happens often in dental implants which occurs among a pair of biocompatible metal implants. Carcinogens discharge into the body and arise from the pitting corrosion of Co-based alloys [147]. On the other hand, Ti and its alloys are very resistant to pitting corrosion in dissimilar *in vivo* conditions but experience corrosion in solutions with elevated fluoride utilized for dental cleaning [148]. The corrosion leads to discoloration in adjacent soft tissue, rashes and allergic reactions in some cases. *In vitro* application of high corrosion-resistant materials may be unsuccessful in medical usages, mostly when they include acute situations such as changes in saliva composition, oral hygiene, diet, and stress.

High corrosion-resistant materials in the *in vitro* state may be unsuccessful for medical applications which feature acute situations like changes in the composition of saliva, oral hygiene, diet, stress, and differences in tooth brushing methods. Therefore, it is vital that a material is observed in favor of corrosion performance in a complete complex situation to reduce this failure risk in real clinical applications [141].

10.8 ANTIBACTERIAL PROPERTIES

Adhesion of planktonic bacteria to the surface of implants is the first step in biofilm formation and implant-associated infections (IAIs). For instance, 2–7% of spinal implants [149], and 2 and 2.4% for hip and knee [150] defects become infected. Extensive bacteria attachment and growth are the main reasons for infections of medical implants, although recently, fungus and mixed infections have gained a lot of attention as well [151]. Thus, prevention of initial bacterial attachment by using an antibacterial coating is one of the main concerns [152]. Generally, bacterial contamination and the formation of a biofilm at the implant surface are the main risks in the initial step of implantation that may lead to persistent infection and failure after surgery [153]. An exocellular polysaccharide matrix is produced by attached and growing bacterial colonies which play a protective role against antibiotics and the host body's innate defense system [154]. Improved sterilization procedures have remarkably reduced the frequency of early-stage implant infections, but delayed infections that take place several weeks or months after surgery have barely decreased and continue to result in a serious problem [155]. The usual technique of using local delivery of antibacterial agents can cause antibiotic resistance and has an effect on the growth of normal tissue; thus, it is ideal to create surface coatings with the antibacterial and biocompatibility properties simultaneously [156]. As shown in Figure 10.12, antibacterial coatings could be classified into four groups including repelling bacterial attachment or biofilm development; contact killing materials; releasing of antibacterial agents; and stimuli responsive release of bacteria presence.

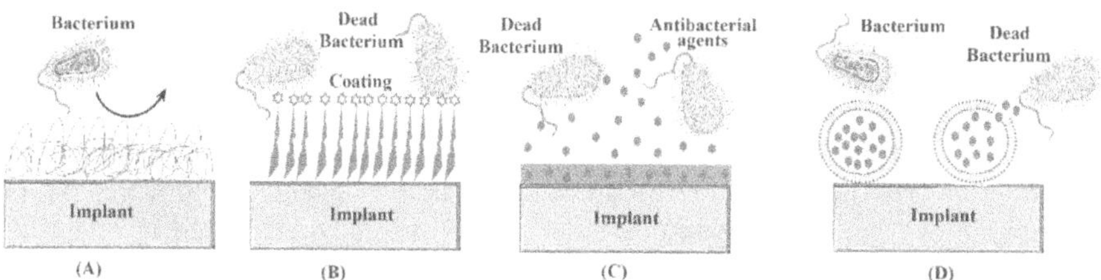

FIGURE 10.12 Antibacterial coating types: (A) repelling bacterial attachment or the development of a biofilm; (B) contact killing materials; (C) antibacterial agent releasing; and (D) stimuli responsive release in the presence of bacteria, redrawn from [152].

10.8.1 Repelling Bacterial Attachment

The first group of antibacterial coatings on medical devices includes those that bacteria do not like to attach to, such as polyethylene glycol, oxazolines, nitroxide radicals, or chlorinated plasma polymers. This type of coating provides an appealing option for the modification of any medical implant surfaces, like catheters, artificial heart valves, and pacemakers [157, 158].

10.8.2 Bacterial Killing Materials

Surfaces (such as bioactive polymers, hydrogels, peptides, and metal ions or nanoparticles) with an ability to kill bacteria upon contact have gained remarkable scientific attention. Polymers that have covalently bonded antimicrobial moieties like immobilized quaternary ammonium compounds, show a unique property of bacterial contact killing. If the density of a surface cationic charge is more than 10^{14} positive charges per cm^2 and is formed via alkylated ammonium groups, adhering bacteria will be killed upon contact because of membrane disruption through very strong electrostatic attractions [159, 160]. These materials have demonstrated enhanced bioavailability and biocompatibility without leaching of antibacterial agents. Nevertheless, this technique has inherent drawbacks, like limited long-term antibacterial activity, due to the accumulation of bacterial debris that play a shielding role and prevent the action of bactericidal agents. Thus, an ideal anti-biofilm medical device surface should be capable of inactivating bacteria and removing their dead debris [161, 162].

10.8.3 Releasing Surfaces

Surfaces that release antibacterial compounds are more frequently used because of their simplicity and efficiency to eliminate pathogens both on the device surface and in its vicinity.

A persistent and localized concentration of nitric oxide (NO) in the vicinity of an invasive medical implant may prove to be a unique approach for reducing implant-associated infection. By using coatings capable of releasing NO, the natural antimicrobial ability of the immune system may be augmented to prevent the survival of pathogenic bacteria at the implant surface [163, 164]. The NO molecule has the ability to inhibit the formation of a biofilm; nevertheless, it does not kill the bacteria and the antibacterial effect persists only until it is released. Moreover, materials incorporating silver (Ag) are categorized as releasing compounds. Silver (nanoparticles or metallic layers) oxidizes and dissolves when placed in a physiological environment, and then the released Ag ions kill bacteria (Gram-positive and Gram-negative). These ions can bind to DNA and block their replication or interfere with proteins and enzymes that influence with bacterial metabolism. Because of this multifaceted mechanism, bacterial resistance against Ag is not as easy in comparison with conventional antibiotics [152].

10.8.4 Responsive Coatings

Recently, nanotechnology has been used to develop and improve biomedical implants, mostly, for bioactive coating fabrication at the nanometer scale because their distinctive features make them suitable for several implantable materials. The essential requirements for bioactive coatings utilized on implants include biocompatibility with a protective ability and favorable mechanical properties (such as adhesion to the underlying implant) [165]. Responsive antibacterial coatings on implants that have two or more killing mechanisms, or that have the capability of triggering a cleaning mechanism are known as promising solutions for bacterial infection [166].

10.9 IMPLANT FAILURE

Several reasons lead to the failure of implants. Infection, improper design, lack of enough tests before implantation, and poor surgical technique during insertion and/or removal are just some of the causes of implant failure [167]. One of the main reasons for failure, amputation, or even death of a patient is implant-associated infection [168]. Infection known as a complex process related to the adhesion of bacterial onto material surfaces, aggregation or dispersion of their colonies and subsequent biofilm formation. Nevertheless, biofilm elimination becomes very challenging once it is created, due to its strong resistance to antimicrobial agents. Commonly, two types of bacteria, *Staphylococcus aureus* and *Staphylococcus epidermidis*, are responsible for most infections of implants in which both exist as natural pathogens of tissue and in the human skin, respectively [169]. Implant rejection and/or loosening happens because of bacterial colonization. Even by applying bioactive materials as coatings on implants, the dissolution rate of such coatings may decrease implant lifetimes through the formation of particle debris, which one must be very careful to avoid [170].

Third-body wear is another mechanism of implant failure, which leads to osteolysis (extensive bone degradation because of osteoclasts). Third-body wear, where free particles are trapped between articulating surfaces, can cause severe damage. Third-body particles may also damage the surfaces of metals [171]. Mechanical stress on a hip implant is one example of third-body wear [52].

10.10 CONCLUSION

Recently, biomaterials have undergone rapid development due to our improved understanding of their acceptance (and rejection) in the human body. Biomaterials have several prominent applications for improving human health including dental, orthopedic, and cardiovascular applications. They are usually composed of metals, polymers, and ceramics, in which metals are the most frequently utilized because of their significant mechanical strength along with biocompatibility. Thus, a comprehensive investigation is required to find the best metallic compound or alloy to obtain the best implant performance. However, corrosion, infection, and low biocompatibility provide important contributions to implant failure which may produce toxicity and adverse effects or even damage tissues.

Coating methods are among the most important techniques used to solve these problems. Selection and application of coatings should be carefully considered based on target tissues and the factors that govern the respective tissue regeneration process. Infections have developed as the main reason for the revision of implants; thus, finding a solution for this issue is urgently needed. An implant with an antibacterial surface which inhibits bacterial attachment and the formation of a biofilm could be an ideal solution. To make the implants more biocompatible, it is required to modify the surface of the implant to increase the osseointegration rate and corrosion resistance. The future of surface modification methods includes the development of nanoparticles and multifunctional coatings that combine the benefits of several coatings into one to enhance chances of implant success.

FUNDING INFORMATION

This research received no external funding.

ACKNOWLEDGMENTS

The authors would like to thank Northeastern University (NEU), Boston, MA, USA for funding some of the authors of this manuscript. The authors also would like to thank Fumie Yusa and Takafumi Komatsu for their help and support.

CONFLICTS OF INTEREST

The authors declare no conflicts of interest.

REFERENCES

1. Mylvaganam, K., et al., Hard thin films: Applications and challenges, in *Anti-Abrasive Nanocoatings*. 2015, Elsevier. p. 543–567.
2. Mozafari, M., et al., Thin films for tissue engineering applications, in *Thin Film Coatings for Biomaterials and Biomedical Applications*. 2016, Elsevier. p. 167–195.
3. Jilani, A., M.S. Abdel-Wahab, and A.H. Hammad, Advance deposition techniques for thin film and coating. *Modern Technologies for Creating the Thin-film Systems and Coatings*, 2017. **2**: p. 137–149.

4. Gould, R.D., S. Kasap, and A.K. Ray, Thin films, in *Springer Handbook of Electronic and Photonic Materials*. 2017, Springer. p. 1.
5. Brinker, C.J. and G.W. Scherer, *Sol-gel Science: The Physics and Chemistry of Sol-gel Processing*. 2013, Academic Press.
6. Wang, D. and G.P. Bierwagen, Sol–gel coatings on metals for corrosion protection. *Progress in Organic Coatings*, 2009. **64**(4): p. 327–338.
7. Messaddeq, S., et al., Microstructure and corrosion resistance of inorganic–organic (ZrO2–PMMA) hybrid coating on stainless steel. *Journal of Non-Crystalline Solids*, 1999. **247**(1–3): p. 164–170.
8. Yilbas, B.S., A. Al-Sharafi, and H. Ali, *Self-Cleaning of Surfaces and Water Droplet Mobility*. 2019: Elsevier.
9. Sahu, N., B. Parija, and S. Panigrahi, Fundamental understanding and modeling of spin coating process: A review. *Indian Journal of Physics*, 2009. **83**(4): p. 493–502.
10. Neacşu, I.A., et al., Inorganic micro-and nanostructured implants for tissue engineering, in *Nanobiomaterials in Hard Tissue Engineering*. 2016, Elsevier. p. 271–295.
11. Narayanan, T.S., I.-S. Park, and M.-H. Lee, Surface modification of magnesium and its alloys for biomedical applications: opportunities and challenges, in *Surface Modification of Magnesium and its Alloys for Biomedical Applications*. 2015, Elsevier. p. 29–87.
12. Fotovvati, B., N. Namdari, and A. Dehghanghadikolaei, On coating techniques for surface protection: a review. *Journal of Manufacturing and Materials Processing*, 2019. **3**(1): p. 28.
13. Kyeremateng, N.A., T. Brousse, and D. Pech, Microsupercapacitors as miniaturized energy-storage components for on-chip electronics. *Nature Nanotechnology*, 2017. **12**(1): p. 7–15.
14. Minh, N.Q., Solid oxide fuel cell technology—features and applications. *Solid State Ionics*, 2004. **174**(1–4): p. 271–277.
15. Avcu, E., et al., Electrophoretic deposition of chitosan-based composite coatings for biomedical applications: A review. *Progress in Materials Science*, 2019. **103**: p. 69–108.
16. Zhitomirsky, I., Electrophoretic deposition of organic–inorganic nanocomposites. *Journal of Materials Science*, 2006. **41**(24): p. 8186–8195.
17. Corni, I., M.P. Ryan, and A.R. Boccaccini, Electrophoretic deposition: From traditional ceramics to nanotechnology. *Journal of the European Ceramic Society*, 2008. **28**(7): p. 1353–1367.
18. Ammam, M., Electrophoretic deposition under modulated electric fields: A review. *RSC Advances*, 2012. **2**(20): p. 7633–7646.
19. Abegunde, O.O., et al., Overview of thin film deposition techniques. *AIMS Materials Science*, 2019. **6**(2): p. 174–199.
20. Ricciardi, S., *Surface Chemical Functionalization Based on Plasma Techniques*. 2012, LAP Lambert Academic Publishing.
21. Lee, S., et al., Investigation of carrier localization in InAs/AlSb type-II superlattice material system. *Applied Physics Letters*, 2019. **115**(21): p. 211601.
22. Martín-Palma, R.J. and A. Lakhtakia, Vapor-Deposition Techniques, in *Engineered Biomimicry*. 2013, Elsevier Inc. p. 383–398.
23. Gao, W., Z. Li, and N. Sammes, *An Introduction to Electronic Materials for Engineers*. 2011, World Scientific Publishing Company.
24. Kelly, P.J. and R.D. Arnell, Magnetron sputtering: A review of recent developments and applications. *Vacuum*, 2000. **56**(3): p. 159–172.
25. Faraji, G., H.S. Kim, and H.T. Kashi, *Severe Plastic Deformation: Methods, Processing and Properties*. 2018, Elsevier.
26. Valiev, R.Z. and T.G. Langdon, Principles of equal-channel angular pressing as a processing tool for grain refinement. *Progress in Materials Science*, 2006. **51**(7): p. 881–981.

27. Dumitru, V., et al., Optical and structural differences between RF and DC AlxNy magnetron sputtered films. *Thin Solid Films*, 2000. **359**(1): p. 17–20.
28. Sojoudi, H., et al., Micro-/nanoscale approach for studying scale formation and developing scale-resistant surfaces. *ACS Applied Materials & Interfaces*, 2019. **11**(7): p. 7330–7337.
29. Maruyama, T. and S. Arai, Electrochromic properties of niobium oxide thin films prepared by radio-frequency magnetron sputtering method. *Applied Physics Letters*, 1993. **63**(7): p. 869–870.
30. Vernardou, D., M. Pemble, and D. Sheel, Vanadium oxides prepared by liquid injection MOCVD using vanadyl acetylacetonate. *Surface and Coatings Technology*, 2004. **188**: p. 250–254.
31. Mori, M., et al., Development of long YBCO coated conductors by multiple-stage CVD. *Physica C: Superconductivity and Its Applications*, 2006. **445**: p. 515–520.
32. Nie, X., A. Leyland, and A. Matthews, Deposition of layered bioceramic hydroxyapatite/TiO2 coatings on titanium alloys using a hybrid technique of micro-arc oxidation and electrophoresis. *Surface and Coatings Technology*, 2000. **125**(1-3): p. 407–414.
33. Zhao, L., et al., Growth characteristics and corrosion resistance of micro-arc oxidation coating on pure magnesium for biomedical applications. *Corrosion Science*, 2010. **52**(7): p. 2228–2234.
34. Pan, Y., D. Wang, and C. Chen, Effect of negative voltage on the microstructure, degradability and in vitro bioactivity of microarc oxidized coatings on ZK60 magnesium alloy. *Materials Letters*, 2014. **119**: p. 127–130.
35. Pawlowski, L., *The Science and Engineering of Thermal Spray Coatings*. 2008, John Wiley & Sons.
36. Wang, M., Composite coatings for implants and tissue engineering scaffolds, in *Biomedical Composites*. 2010, Elsevier. p. 127–177.
37. Dorfman, M.R., Thermal spray coatings, in *Handbook of Environmental Degradation of Materials*. 2018, Elsevier. p. 469–488.
38. Huang, R. and H. Fukanuma, Future trends in cold spray techniques, in *Future Development of Thermal Spray Coatings*. 2015, Elsevier. p. 143–162.
39. Chrisey, D.B. and G.K. Hubler, Pulsed laser deposition of thin films. 1994.
40. Huang, J., et al., Ferroelectric thin films and nanostructures: Current and future, in *Nanostructures in Ferroelectric Films for Energy Applications*. 2019, Elsevier. p. 19–39.
41. Chang, J. and Y. Zhou, Surface modification of bioactive glasses, in *Bioactive Glasses*. 2018, Elsevier. p. 119–143.
42. Liu, Y. and E.B. Hunziker, Biomimetic coatings and their biological functionalization, in *Thin Calcium Phosphate Coatings for Medical Implants*. 2009, Springer. p. 301–314.
43. Liu, Y., G. Wu, and K. de Groot, Biomimetic coatings for bone tissue engineering of critical-sized defects. *Journal of the Royal Society Interface*, 2010. **7**(suppl_5): p. S631–S647.
44. Aramesh, N., et al., PtSn nanoalloy thin films as anode catalysts in methanol fuel cells. *Inorganic Chemistry*, 2020. **59**(15): p. 10688–10698.
45. Hoseini, S.J., M. Bahrami, and S.M. Nabavizadeh, Inorganic nanostructures especially obtained from liquid/liquid interface by reduction of organometallic precursors: A mini review. *Inorganic Chemistry Research*, 2020. **4**(2): p. 140–182.
46. Hoseini, S.J., M. Bahrami, and S.M. Nabavizadeh, ZIF-8 nanoparticles thin film at an oil–water interface as an electrocatalyst for the methanol oxidation reaction without the application of noble metals. *New Journal of Chemistry*, 2019. **43**(39): p. 15811–15822.
47. Rao, C. and K. Kalyanikutty, The liquid–liquid interface as a medium to generate nanocrystalline films of inorganic materials. *Accounts of Chemical Research*, 2008. **41**(4): p. 489–499.

48. Zarbin, A.J., Liquid–liquid interfaces: a unique and advantageous environment to prepare and process thin films of complex materials. *Materials Horizons*, 2021. **8**(5): p. 1409–1432.
49. Booth, S.G. and R.A. Dryfe, Assembly of nanoscale objects at the liquid/liquid interface. *The Journal of Physical Chemistry C*, 2015. **119**(41): p. 23295–23309.
50. Hoseini, S.J., M. Rashidi, and M. Bahrami, Platinum nanostructures at the liquid–liquid interface: Catalytic reduction of p-nitrophenol to p-aminophenol. *Journal of Materials Chemistry*, 2011. **21**(40): p. 16170–16176.
51. Hoseini, S.J., et al., Thin film formation of platinum nanoparticles at oil–water interface, using organoplatinum (II) complexes, suitable for electro-oxidation of methanol. *Dalton Transactions*, 2013. **42**(34): p. 12364–12369.
52. Priyadarshini, B., et al., Bioactive coating as a surface modification technique for biocompatible metallic implants: A review. *Journal of Asian Ceramic Societies*, 2019. **7**(4): p. 397–406.
53. Su, Y., et al., Biofunctionalization of metallic implants by calcium phosphate coatings. *Bioactive Materials*, 2019. **4**: p. 196–206.
54. Niinomi, M., M. Nakai, and J. Hieda, Development of new metallic alloys for biomedical applications. *Acta Biomaterialia*, 2012. **8**(11): p. 3888–3903.
55. Su, Y., et al., Zinc-based biomaterials for regeneration and therapy. *Trends in Biotechnology*, 2019. **37**(4): p. 428–441.
56. Lee, J.-W., et al., Long-term clinical study and multiscale analysis of in vivo biodegradation mechanism of Mg alloy. *Proceedings of the National Academy of Sciences*, 2016. **113**(3): p. 716–721.
57. Wang, X., et al., Topological design and additive manufacturing of porous metals for bone scaffolds and orthopaedic implants: A review. *Biomaterials*, 2016. **83**: p. 127–141.
58. Dehghanghadikolaei, A. and B. Fotovvati, Coating techniques for functional enhancement of metal implants for bone replacement: A review. *Materials*, 2019. **12**(11): p. 1795.
59. Singh, A., G. Singh, and V. Chawla, Influence of post coating heat treatment on microstructural, mechanical and electrochemical corrosion behaviour of vacuum plasma sprayed reinforced hydroxyapatite coatings. *Journal of the Mechanical Behavior of Biomedical Materials*, 2018. **85**: p. 20–36.
60. Bistolfi, A., et al., Does metal porosity affect metal ion release in blood and urine following total hip arthroplasty? A short term study. *HIP International*, 2018. **28**(5): p. 522–530.
61. Dehghanghadikolaei, A., et al., Abrasive machining techniques for biomedical device applications. *Journal Material Science*, 2018. **5**: p. 1–11.
62. Ghoreishi, R., A.H. Roohi, and A.D. Ghadikolaei, Analysis of the influence of cutting parameters on surface roughness and cutting forces in high speed face milling of Al/SiC MMC. *Materials Research Express*, 2018. **5**(8): p. 086521.
63. Hanawa, T., Transition of surface modification of titanium for medical and dental use, in *Titanium in Medical and Dental Applications*. 2018, Elsevier. p. 95–113.
64. Kurup, A., P. Dhatrak, and N. Khasnis, Surface modification techniques of titanium and titanium alloys for biomedical dental applications: A review. *Materials Today: Proceedings*, 2021. **39**: p. 84–90.
65. Bekmurzayeva, A., et al., Surface modification of stainless steel for biomedical applications: Revisiting a century-old material. *Materials Science and Engineering: C*, 2018. **93**: p. 1073–1089.
66. Virtanen, S., et al., Special modes of corrosion under physiological and simulated physiological conditions. *Acta Biomaterialia*, 2008. **4**(3): p. 468–476.
67. Okazaki, Y., Comparison of fatigue properties and fatigue crack growth rates of various implantable metals. *Materials*, 2012. **5**(12): p. 2981–3005.

68. Balamurugan, A., et al., Corrosion aspects of metallic implants—An overview. *Materials and Corrosion*, 2008. **59**(11): p. 855–869.
69. Hermawan, H., D. Ramdan, and J.R. Djuansjah, Metals for biomedical applications. *Biomedical Engineering-From Theory to Applications*, 2011. **1**: p. 411–430.
70. Yuan, S., et al., Enhancing antibacterial activity of surface-grafted chitosan with immobilized lysozyme on bioinspired stainless steel substrates. *Colloids and Surfaces B: Biointerfaces*, 2013. **106**: p. 11–21.
71. Medilanski, E., et al., Influence of the surface topography of stainless steel on bacterial adhesion. *Biofouling*, 2002. **18**(3): p. 193–203.
72. Benvenuto, P., et al., Adlayer-mediated antibody immobilization to stainless steel for potential application to endothelial progenitor cell capture. *Langmuir*, 2015. **31**(19): p. 5423–5431.
73. Rahman, M., Y. Li, and C. Wen, HA coating on Mg alloys for biomedical applications: A review. *Journal of Magnesium and Alloys*, 2020. **8**(3): p. 929–943.
74. Chakraborty Banerjee, P., et al., Magnesium implants: Prospects and challenges. *Materials*, 2019. **12**(1): p. 136.
75. Wolf, F.I. and A. Cittadini, Chemistry and biochemistry of magnesium. *Molecular Aspects of Medicine*, 2003. **24**(1–3): p. 3–9.
76. Jacobs, J.J., et al., Metal degradation products: a cause for concern in metal-metal bearings? *Clinical Orthopaedics and Related Research (1976-2007)*, 2003. **417**: p. 139–147.
77. Choudhary, L. and R.S. Raman, Magnesium alloys as body implants: Fracture mechanism under dynamic and static loadings in a physiological environment. *Acta Biomaterialia*, 2012. **8**(2): p. 916–923.
78. Bobby Kannan, M., et al., Influence of circumferential notch and fatigue crack on the mechanical integrity of biodegradable magnesium-based alloy in simulated body fluid. *Journal of Biomedical Materials Research Part B: Applied Biomaterials*, 2011. **96**(2): p. 303–309.
79. Zeng, R., et al., Progress and challenge for magnesium alloys as biomaterials. *Advanced Engineering Materials*, 2008. **10**(8): p. B3–B14.
80. Bruni, S., et al., Effects of surface treatment of Ti–6Al–4V titanium alloy on biocompatibility in cultured human umbilical vein endothelial cells. *Acta biomaterialia*, 2005. **1**(2): p. 223–234.
81. Kaur, M. and K. Singh, Review on titanium and titanium based alloys as biomaterials for orthopaedic applications. *Materials Science and Engineering: C*, 2019. **102**: p. 844–862.
82. Qiu, Y., et al., Corrosion of high entropy alloys. *npj Materials Degradation*, 2017. **1**(1): p. 1–18.
83. Goharian, A. and M. Abdullah, Bioinert metals (stainless steel, titanium, cobalt chromium). *Trauma Plating System*. 2017, Elsevier. p. 115–142.
84. Sahoo, P., S.K. Das, and J.P. Davim, Tribology of materials for biomedical applications, in *Mechanical Behaviour of Biomaterials*. 2019, Elsevier. p. 1–45.
85. Cabañas, M.V., Bioceramic coatings for medical implants. *Bio-ceramics With Clinical Applications*. 2014, Wiley. p. 249–289.
86. Asmatulu, R., Nanocoatings for corrosion protection of aerospace alloys, in *Corrosion Protection and Control Using Nanomaterials*. 2012, Elsevier. p. 357–374.
87. Piconi, C. and G. Maccauro, Zirconia as a ceramic biomaterial. *Biomaterials*, 1999. **20**(1): p. 1–25.
88. Adell, R., et al., A long-term follow-up study of osseointegrated implants in the treatment of totally edentulous jaws. *International Journal of Oral & Maxillofacial Implants*, 1990. **5**(4).
89. Ahn, T.-K., et al., Modification of titanium implant and titanium dioxide for bone tissue engineering. *Novel Biomaterials for Regenerative Medicine*, 2018. p. 355–368.

90. Thieme, M., et al., Titanium powder sintering for preparation of a porous functionally graded material destined for orthopaedic implants. *Journal of Materials Science: Materials in Medicine*, 2001. **12**(3): p. 225–231.
91. Trino, L.D., et al., Zinc oxide surface functionalization and related effects on corrosion resistance of titanium implants. *Ceramics International*, 2018. **44**(4): p. 4000–4008.
92. Guo, Y., et al., A multifunctional polypyrrole/zinc oxide composite coating on biodegradable magnesium alloys for orthopedic implants. *Colloids and Surfaces B: Biointerfaces*, 2020. **194**: p. 111186.
93. Voicu, G., et al., Co doped ZnO thin films deposited by spin coating as antibacterial coating for metallic implants. *Ceramics International*, 2020. **46**(3): p. 3904–3911.
94. Wang, M. and J. Gao, Atomic layer deposition of ZnO thin film on ZrO2 dental implant surface for enhanced antibacterial and bioactive performance. *Materials Letters*, 2021. **285**: p. 128854.
95. Kołodziejczak-Radzimska, A. and T. Jesionowski, Zinc oxide—from synthesis to application: A review. *Materials*, 2014. **7**(4): p. 2833–2881.
96. Sadat-Shojai, M., et al., Synthesis methods for nanosized hydroxyapatite with diverse structures. *Acta Biomaterialia*, 2013. **9**(8): p. 7591–7621.
97. O'Hare, P., et al., Biological responses to hydroxyapatite surfaces deposited via a co-incident microblasting technique. *Biomaterials*, 2010. **31**(3): p. 515–522.
98. Sadat-Shojai, M., M. Atai, and A. Nodehi, *Method for Production of Biocompatible Nanoparticles Containing Dental Adhesive*. 2013, Google Patents.
99. Blackwood, D. and K. Seah, Electrochemical cathodic deposition of hydroxyapatite: improvements in adhesion and crystallinity. *Materials Science and Engineering: C*, 2009. **29**(4): p. 1233–1238.
100. Mohseni, E., E. Zalnezhad, and A.R. Bushroa, Comparative investigation on the adhesion of hydroxyapatite coating on Ti–6Al–4V implant: A review paper. *International Journal of Adhesion and Adhesives*, 2014. **48**: p. 238–257.
101. Silva, P., et al., Adhesion and microstructural characterization of plasma-sprayed hydroxyapatite/glass ceramic coatings onto Ti-6A1-4V substrates. *Surface and Coatings Technology*, 1998. **102**(3): p. 191–196.
102. Fathi, M. and A. Doostmohammadi, Bioactive glass nanopowder and bioglass coating for biocompatibility improvement of metallic implant. *Journal of Materials Processing Technology*, 2009. **209**(3): p. 1385–1391.
103. Brunello, G., H. Elsayed, and L. Biasetto, Bioactive glass and silicate-based ceramic coatings on metallic implants: Open challenge or outdated topic? *Materials*, 2019. **12**(18): p. 2929.
104. Du Min, Q., et al., Clinical evaluation of a dentifrice containing calcium sodium phosphosilicate (novamin) for the treatment of dentin hypersensitivity. *American Journal of Dentistry*, 2008. **21**(4): p. 210–214.
105. Oliva, A., et al., Behaviour of human osteoblasts cultured on bioactive glass coatings. *Biomaterials*, 1998. **19**(11-12): p. 1019–1025.
106. Tripathi, H., et al., Structural, physico-mechanical and in-vitro bioactivity studies on SiO2–CaO–P2O5–SrO–Al2O3 bioactive glasses. *Materials Science and Engineering: C*, 2019. **94**: p. 279–290.
107. Patil, N.A. and B. Kandasubramanian, Biological and mechanical enhancement of zirconium dioxide for medical applications. *Ceramics International*, 2020. **46**(4): p. 4041–4057.
108. Bauer, S., et al., Size selective behavior of mesenchymal stem cells on ZrO2 and TiO2 nanotube arrays. *Integrative Biology*, 2009. **1**(8-9): p. 525–532.
109. Afzal, A., Implantable zirconia bioceramics for bone repair and replacement: A chronological review. *Materials Express*, 2014. **4**(1): p. 1–12.

110. Degidi, M., et al., Inflammatory infiltrate, microvessel density, nitric oxide synthase expression, vascular endothelial growth factor expression, and proliferative activity in peri-implant soft tissues around titanium and zirconium oxide healing caps. *Journal of Periodontology*, 2006. **77**(1): p. 73–80.
111. Özkurt, Z. and E. Kazazoğlu, Zirconia dental implants: A literature review. *Journal of Oral Implantology*, 2011. **37**(3): p. 367–376.
112. Zhao, C., et al., The promising application of graphene oxide as coating materials in orthopedic implants: Preparation, characterization and cell behavior. *Biomedical Materials*, 2015. **10**(1): p. 015019.
113. Dong, Y., Q. Liu, and Q. Zhou, Corrosion behavior of Cu during graphene growth by CVD. *Corrosion Science*, 2014. **89**: p. 214–219.
114. Liu, Y., et al., Fabrication of a superhydrophobic graphene surface with excellent mechanical abrasion and corrosion resistance on an aluminum alloy substrate. *RSC Advances*, 2014. **4**(85): p. 45389–45396.
115. Rikhari, B., S.P. Mani, and N. Rajendran, Polypyrrole/graphene oxide composite coating on Ti implants: A promising material for biomedical applications. *Journal of Materials Science*, 2020. **55**(12): p. 5211–5229.
116. Li, M., et al., An overview of graphene-based hydroxyapatite composites for orthopedic applications. *Bioactive Materials*, 2018. **3**(1): p. 1–18.
117. Li, H., et al., Lysozyme (Lys), tannic acid (TA), and graphene oxide (GO) thin coating for antibacterial and enhanced osteogenesis. *ACS Applied Bio Materials*, 2019. **3**(1): p. 673–684.
118. Yoon, H.H., et al., Dual roles of graphene oxide in chondrogenic differentiation of adult stem cells: Cell-adhesion substrate and growth factor-delivery carrier. *Advanced Functional Materials*, 2014. **24**(41): p. 6455–6464.
119. Crowder, S.W., et al., Three-dimensional graphene foams promote osteogenic differentiation of human mesenchymal stem cells. *Nanoscale*, 2013. **5**(10): p. 4171–4176.
120. Rajak, D.K., et al., Recent progress of reinforcement materials: A comprehensive overview of composite materials. *Journal of Materials Research and Technology*, 2019. **8**(6): p. 6354–6374.
121. Sereni, J.G., Reference module in materials science and materials engineering. 2016.
122. Asadi, H., et al., A multifunctional polymeric coating incorporating lawsone with corrosion resistance and antibacterial activity for biomedical Mg alloys. *Progress in Organic Coatings*, 2021. **153**: p. 106157.
123. Li, L.-Y., et al., Advances in functionalized polymer coatings on biodegradable magnesium alloys–A review. *Acta Biomaterialia*, 2018. **79**: p. 23–36.
124. Kunjukunju, S., et al., A layer-by-layer approach to natural polymer-derived bioactive coatings on magnesium alloys. *Acta Biomaterialia*, 2013. **9**(10): p. 8690–8703.
125. Catauro, M., et al., Silica–polyethylene glycol hybrids synthesized by sol–gel: Biocompatibility improvement of titanium implants by coating. *Materials Science and Engineering: C*, 2015. **55**: p. 118–125.
126. Kim, G., et al., The biocompatability of mesoporous inorganic–organic hybrid resin films with ionic and hydrophilic characteristics. *Biomaterials*, 2010. **31**(9): p. 2517–2525.
127. Wang, X., et al., Active corrosion protection of super-hydrophobic corrosion inhibitor intercalated Mg–Al layered double hydroxide coating on AZ31 magnesium alloy. *Journal of Magnesium and Alloys*, 2020. **8**(1): p. 291–300.
128. Abbaspoor, S., A. Ashrafi, and R. Abolfarsi, Development of self-healing coatings based on ethyl cellulose micro/nano-capsules. *Surface Engineering*, 2019. **35**(3): p. 273–280.
129. Umoren, S.A., et al., Exploration of natural polymers for use as green corrosion inhibitors for AZ31 magnesium alloy in saline environment. *Carbohydrate Polymers*, 2020. **230**: p. 115466.

130. Shah, F.A., P. Thomsen, and A. Palmquist, Osseointegration and current interpretations of the bone-implant interface. *Acta Biomaterialia*, 2019. **84**: p. 1–15.
131. Terheyden, H., et al., Osseointegration–communication of cells. *Clinical Oral Implants Research*, 2012. **23**(10): p. 1127–1135.
132. Grassi, S., et al., Histologic evaluation of early human bone response to different implant surfaces. *Journal of Periodontology*, 2006. **77**(10): p. 1736–1743.
133. Syrett, B.C. and S.S. Wing, Pitting resistance of new and conventional orthopedic implant materials—effect of metallurgical condition. *Corrosion*, 1978. **34**(4): p. 138–148.
134. Asri, R., et al., Corrosion and surface modification on biocompatible metals: A review. *Materials Science and Engineering: C*, 2017. **77**: p. 1261–1274.
135. Kamachimudali, U., T. Sridhar, and B. Raj, Corrosion of bio implants. *Sadhana*, 2003. **28**(3): p. 601–637.
136. Zhang, B.G., et al., Bioactive coatings for orthopaedic implants—recent trends in development of implant coatings. *International Journal of Molecular Sciences*, 2014. **15**(7): p. 11878–11921.
137. Abdulhameed, Z.N., *Corrosion Behavior of Some Implant Alloys in Simulated Human Body Environment*. 2011, Department of Materials engineering, University of Technology.
138. Adya, N., et al., Corrosion in titanium dental implants: Literature review. *The Journal of Indian Prosthodontic Society*, 2005. **5**(3): p. 126.
139. Calin, M., et al., Designing biocompatible Ti-based metallic glasses for implant applications. *Materials Science and Engineering: C*, 2013. **33**(2): p. 875–883.
140. Wang, S., et al., Study on torsional fretting wear behavior of a ball-on-socket contact configuration simulating an artificial cervical disk. *Materials Science and Engineering: C*, 2015. **55**: p. 22–33.
141. Manam, N., et al., Study of corrosion in biocompatible metals for implants: A review. *Journal of Alloys and Compounds*, 2017. **701**: p. 698–715.
142. Yu, J., Z. Zhao, and L. Li, Corrosion fatigue resistances of surgical implant stainless steels and titanium alloy. *Corrosion Science*, 1993. **35**(1-4): p. 587–597.
143. Holmes Jr, D.R., et al., The PARAGON stent study: A randomized trial of a new martensitic nitinol stent versus the Palmaz-Schatz stent for treatment of complex native coronary arterial lesions. *The American Journal of Cardiology*, 2000. **86**(10): p. 1073–1079.
144. Coon, M.E., et al., Nitinol thin films functionalized with CAR-T cells for the treatment of solid tumours. *Nature Biomedical Engineering*, 2020. **4**(2): p. 195–206.
145. Rosenbloom, S.N., P. Kumar, and C. Lasley, The role of surface oxide thickness and structure on the corrosion and nickel elution behavior of nitinol biomedical implants. *Journal of Biomedical Materials Research Part B: Applied Biomaterials*, 2021.
146. Nagaraja, S., et al., Impact of nitinol stent surface processing on in-vivo nickel release and biological response. *Acta Biomaterialia*, 2018. **72**: p. 424–433.
147. Clerc, C.O., et al., Assessment of wrought ASTM F1058 cobalt alloy properties for permanent surgical implants. *Journal of Biomedical Materials Research*, 1997. **38**(3): p. 229–234.
148. Pröbster, L., W. Lin, and H. Hüttemann, Effect of fluoride prophylactic agents on titanium surfaces. *International Journal of Oral & Maxillofacial Implants*, 1992. **7**(3).
149. Sampedro, M.F., et al., A biofilm approach to detect bacteria on removed spinal implants. *Spine*, 2010. **35**(12): p. 1218–1224.
150. Slane, J., et al., Mechanical, material, and antimicrobial properties of acrylic bone cement impregnated with silver nanoparticles. *Materials Science and Engineering: C*, 2015. **48**: p. 188–196.
151. Kurtz, S.M., et al., Economic burden of periprosthetic joint infection in the United States. *The Journal of Arthroplasty*, 2012. **27**(8): p. 61–65. e1.

152. Vasilev, K., Nanoengineered antibacterial coatings and materials: A perspective. *Coatings*, 2019. **9**(10): p. 654.
153. Kazemzadeh-Narbat, M., et al., Multilayered coating on titanium for controlled release of antimicrobial peptides for the prevention of implant-associated infections. *Biomaterials*, 2013. **34**(24): p. 5969–5977.
154. Darouiche, R.O., Treatment of infections associated with surgical implants. *New England Journal of Medicine*, 2004. **350**(14): p. 1422–1429.
155. Vasilev, K., J. Cook, and H.J. Griesser, Antibacterial surfaces for biomedical devices. *Expert Review of Medical Devices*, 2009. **6**(5): p. 553–567.
156. Zhao, L., et al., Antibacterial coatings on titanium implants. *Journal of Biomedical Materials Research Part B: Applied Biomaterials*, 2009. **91**(1): p. 470–480.
157. Michl, T.D., et al., Plasma polymerization of TEMPO yields coatings containing stable nitroxide radicals for controlling interactions with prokaryotic and eukaryotic cells. *ACS Applied Nano Materials*, 2018. **1**(12): p. 6587–6595.
158. Cavallaro, A.A., M.N. Macgregor-Ramiasa, and K. Vasilev, Antibiofouling properties of plasma-deposited oxazoline-based thin films. *ACS Applied Materials & Interfaces*, 2016. **8**(10): p. 6354–6362.
159. van de Lagemaat, M., et al., Comparison of methods to evaluate bacterial contact-killing materials. *Acta Biomaterialia*, 2017. **59**: p. 139–147.
160. Asri, L.A., et al., A shape-adaptive, antibacterial-coating of immobilized quaternary-ammonium compounds tethered on hyperbranched polyurea and its mechanism of action. *Advanced Functional Materials*, 2014. **24**(3): p. 346–355.
161. Gao, L., et al., Biomimetic biodegradable Ag@ Au nanoparticle-embedded ureteral stent with a constantly renewable contact-killing antimicrobial surface and antibiofilm and extraction-free properties. *Acta Biomaterialia*, 2020. **114**: p. 117–132.
162. Banerjee, I., R.C. Pangule, and R.S. Kane, Antifouling coatings: recent developments in the design of surfaces that prevent fouling by proteins, bacteria, and marine organisms. *Advanced Materials*, 2011. **23**(6): p. 690–718.
163. Nablo, B.J., A.R. Rothrock, and M.H. Schoenfisch, Nitric oxide-releasing sol–gels as antibacterial coatings for orthopedic implants. *Biomaterials*, 2005. **26**(8): p. 917–924.
164. Nablo, B.J., et al., Inhibition of implant-associated infections via nitric oxide release. *Biomaterials*, 2005. **26**(34): p. 6984–6990.
165. Kumar, A.M., et al., Biocompatible responsive polypyrrole/GO nanocomposite coatings for biomedical applications. *RSC Advances*, 2015. **5**(121): p. 99866–99874.
166. Wei, T., Q. Yu, and H. Chen, Responsive and synergistic antibacterial coatings: fighting against bacteria in a smart and effective way. *Advanced Healthcare Materials*, 2019. **8**(3): p. 1801381.
167. Bane, M., et al. Failure analysis of some retrieved orthopedic implants based on materials characterization, in *Solid State Phenomena*. Vol. 188 (2012) pp 114–117, Trans Tech Publications, Switzerland.
168. Yang, L., *Nanotechnology-enhanced Orthopedic Materials: Fabrications, Applications and Future Trends*. 2015, Woodhead Publishing.
169. Mah, T.-F.C. and G.A. O'Toole, Mechanisms of biofilm resistance to antimicrobial agents. *Trends in Microbiology*, 2001. **9**(1): p. 34–39.
170. Talik, N.A., et al. Review on recent developments on fabrication techniques of distributed feedback (DFB) based organic lasers. *Journal of Physics: Conference Series*. 2017. IOP Publishing.
171. Que, L. and L.T. Topoleski, Third-body wear of cobalt-chromium-molybdenum implant alloys initiated by bone and poly (methyl methacrylate) particles. *Journal of Biomedical Materials Research*, 2000. **50**(3): p. 322–330.

11 Flexible Sensors for Biomedical Applications Based on Elastic Polymers

Hui Li, Jing Chen, Lin Li, and Lei Wang
Shenzhen Institute of Advanced Technology, Chinese Academy of Sciences, Shenzhen, People's Republic of China

CONTENTS

DOI: 10.1201/9781003141358-11

11.1 INTRODUCTION

With the rapid development of modern society, new and disruptive technologies such as the artificial intelligence, wearable medical devices, and Internet of Things have emerged in recent years, improving the well-being and quality of life. Flexible electronics plays an important role in the aforementioned applications in the form of healthcare monitoring, bio-integrated devices, electronic skin, human–machine interface, etc. Compared with conventional microelectronic devices fabricated on rigid silicon substrates, flexible electronics not only show comparable performance but also have some unique features beyond rigid devices, such as ultrathin thickness, outstanding flexibility. This unique advantage has attracted tremendous research interests on flexible electronics, and various applications have been developed, such as stretchable transistors, light-emitting diodes, energy harvesters, energy storage devices, and flexible sensors.

Flexible pressure sensors that convert external pressure input into recordable electrical signals are an essential member of the flexible electronics family and have been developed into a multidisciplinary field that combines structural engineering, material science, system integration, and signal processing. After years of extensive investigation, the rapid developments in sensing materials, structural design, and fabrication techniques have contributed to significant progress of flexible pressure sensors.

In this chapter, we are dedicated to unfold the recent progress of the development of flexible pressure sensors. It covers basic concepts of flexible pressure sensors, sensing mechanisms, functional materials, manufacturing methods, and key performance parameters. Finally, we demonstrate the potential biomedical applications of flexible pressure sensors, such as physiological parameter detection, human activity monitoring, and human–machine interaction.

11.2 TRANSDUCTION MECHANISMS

The flexible pressure sensors can mimic the tactile sense of the human body by generating electrical signals to capture the external stimulus[1]. According to the transduction mechanisms, pressure sensors can be divided into three main categories including piezoresistivity, capacitance, and piezoelectricity. These principles have been widely studied for pressure sensing duo to easy fabrication and high performances. The details of three sensing mechanisms are discussed in the following subsections. The other working mechanisms, such as triboelectricity, are mentioned as well.

11.2.1 Piezoresistive Sensors

Piezoresistive sensors transduce mechanical stimulus into a resistance value change. Recently, the resistive-type sensing devices have been extensively researched due to their simple structure, read-out signal easy, and board responding range. As shown in Figure 11.1(a), these types of devices consist of a pair of flexible electrodes face-to-face with conductive material. The initial resistance (R) of the sensors can be given as

$$R = \frac{\rho L}{S} \tag{11.1}$$

where ρ, L, and S are the resistivity, original length, and contact area of the electrodes. As the applied pressure further increased, the contact area of compressed electrodes would be sharply increased and create lots of charge transport paths between two conductive layers, resulting in an obvious decrease in resistive signals. The relative change in resistance can be expressed as[2]

$$\frac{\Delta R}{R} = (1+2\nu)\varepsilon + \frac{\Delta \rho}{\rho} \tag{11.2}$$

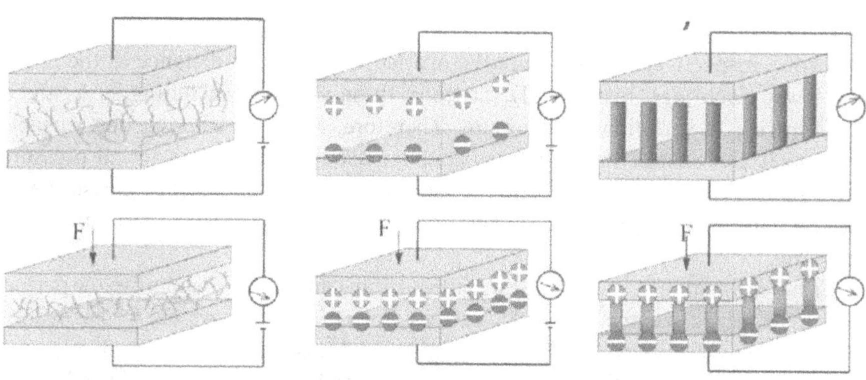

Advantages/Disadvantages	Advantages/Disadvantages	Advantages/Disadvantages
☑Wide measurement range	☑High sensitivity	☑High sensitivity
☑Simple structures and manufacturing methods	☑Fast response time	☑ Fast response time
☑Easy read-out signal	☑Low power consumption	☑ Low power consumption or self-power ability
☑Low cost	☑Low hysteresis	☒Static responses are unreliable
☒Low sensitivity	☒Complex circuitry	☒Easily affected by the temperature
☒High hysteresis	☒High susceptible to parasitic influence	☒Drift output
☒Lack of stability	☒Crosstalk between sensing unit	
☒Temperature dependent		

FIGURE 11.1 Schematic illustrations of three typical transduction mechanisms and summary of their performances.

where v and ε is the Poisson's ratio and compressive strain, respectively. From Equation 11.2 we know that the relative change in resistance is dependent on two parts. The first part is caused by geometrical changes of sensors due to applied pressure, and another is change in resistivity because of the phenomenon of piezoresistivity. Compared with resistivity change, small geometrical changes can often lead to larger resistance variations through increase in conductive paths between two poorly contacted electrodes. However, piezoresistive pressure sensors also have disadvantages that cannot be ignored, such as low sensitivity, high hysteresis, and lack of stability.

11.2.2 Piezocapacitive Sensors

Piezocapacitive sensors are another widely studied type of pressure devices based on parallel plate capacitor and exhibit the advantages of high sensitivity, fast response time, and low power consumption, etc. Similar to piezoresistive pressure sensors, typical piezocapacitive sensors are in the framework of two flexible electrodes that sandwich a dielectric (Figure 11.1(b)). The capacitance (C) is expressed by

$$C = \frac{\varepsilon_0 \varepsilon_r A}{d} \tag{11.3}$$

where ε_0 and ε_r is the permittivity of vacuum and the relative permittivity of the dielectric, respectively. A and d are the overlapping area and distance of the upper and lower electrodes, respectively. Three of these variables including ε_r, A, and d are sensitive to variations in pressure. Therefore, slightly changing one variable or another can change the output capacitance accordingly. The external pressure causes flexible pressure sensor compressed, which is usually used to measure normal pressure by changing d[3]. Change in A is typically utilized to measure shear pressure[4].

Easy deformation or specially designed structure can effectively improve the performances of piezocapacitive sensors[5]. Various dielectric layers have been designed for lower modulus of compression, such as air contained elastomer, porous foam, micropatterned surface, and multi-layered structure. It is worth mentioning that air-embedded elastomer may dramatically improve sensitivity. When the sensors are compressed, the effective dielectric constant is increased due to the volume of air removed. Besides, the formation of air gap structure will be farther reducing the viscoelastic behavior of materials and resulting in faster response and relaxation time. But the requirement of miniaturization endows this kind of sensors with small sensing capacitance, which causes high susceptible to parasitic influence and crosstalk between sensing units.

11.2.3 Piezoelectric Sensors

Piezoelectricity refers to the phenomenon where the polarization of the internal charges of dielectric material undergoes applied mechanical stress. Piezoelectric sensors make full use of this effect to produce voltage in response to stimulation.

As shown in Figure 11.1(c), once the pressure is applied, the positive and negative charges are spatially separated in the opposite direction of the materials and form electric dipoles. Then, the devices will return to the initial state when the external pressure unloads. Piezoelectric pressure sensors exhibited high sensitivity, fast response time, and low power consumption or self-power ability. Because of their variation-dependent transduction mechanism, this type of sensors especially suits measurement of dynamic pressure.

Many piezoelectric materials have been developed in recent years; whether natural or synthetic materials, organic or inorganic materials, they all can be used as active layer for piezoelectric sensors, such as ZnO, lead zirconate titanate, $BaTiO_3$, polyvinylidenefluoride (PVDF), and their copolymers. Piezoelectric coefficient d_{33} of piezoelectric material is used for evaluating energy conversion efficiency of mechanical energy into electrical energy or electrical energy into mechanical energy. Generally, inorganics possess a high d_{33} and high modulus; these materials lack sufficient flexibility. By contrast, piezoelectric polymers exhibit a high stretchability but low d_{33}.

Some piezoelectric materials can be hybridizing to create piezophotonic effect, which is effected by deformation-induced piezopotential that trapped charges in vacancy or surface states drop to the valence band, leading to photon emission. The piezophotonic effect based on mechanluminescent materials has wide applications in the flexible pressure sensor matrix for tactile imaging. The sensor matrix can not only detect dynamic pressure but also measure two-dimensional pressure distributions with high spatial resolution. It should be noted that the response of piezoelectric sensors is easily affected by the pyroelectric effect; thus, protection sensors from thermal interference is necessary. Moreover, piezoelectric sensors are unreliable for static pressure because the output voltage is a pulse signal.

11.2.4 Other Transduction Mechanisms

Besides the previous categories, there are several other transduction mechanisms that can sense pressure changes. One of them is based on the triboelectric effect; these kinds of sensors generate voltage after frictional contact with different materials. Triboelectric pressure sensors, similar to sensors based on piezoelectric, also have the ability of low power consumption or self-power. The response of triboelectric pressure sensors depends on the magnitude and frequency of pressure, so they are suitable for dynamic detecting as well. Pressure sensors based on magnetic transduction can convert mechanical stress into corresponding electrical signals by a change in magnetic property when they are pressed. However, this mechanism has limited further development in pressure sensors because of being restricted by materials and susceptibility to the environment. Another choice for pressure sensing is optical transduction. Light coupling mechanical information passes down through waveguides or optical fiber can be modulated and captured by photodetector. Optical sensors have high resolution and vision sensing, but they are not sensitive to small changes and need considerable power consumption.

11.3 MATERIALS

11.3.1 Substrate Materials

For traditional rigid sensors, ceramics, silicon, and glass have been utilized as substrates. However, these materials are brittle, limiting certain real-world applications of pressure sensors. Thus, the substrate materials play an important role in the sensor structure. Flexible pressure sensors are expected to provide reliable and accurate sensing without causing discomfort and compromising the natural movements of the users. To achieve these properties, various commercially available polymers, such as polycarbonate (PC), polyethylene terephthalate (PET), polyurethane (PU), and polyethylene (PEN), have been selected for flexible substrate owing to their superior deformability and easy fabrication process. Among polymer substrates, Polyimide (PI) exhibits excellent mechanical properties, thermal stability, and insulating property. It has been reported that PI has a glass-transition temperature (T_g) between 360°C and 410°C[6]. The optimal thermo-stability of PI allows for fabricating flexible electronics that can work at high temperatures. In addition, PI is compatible with the manufacturing processes of micro-electro-mechanical systems. These properties make PI a preferred substrate material for flexible sensing devices. However, PI cannot recover under large strains, such that it cannot be utilized as the substrate of highly stretchable electronic devices.

Another type of materials that has been employed for the fabrication of the soft substrates of flexible electronics is the class of silicone elastomers, for instance, polydimethylsiloxane (PDMS), and the trademarked silicone rubbers such as Dragonskin and EcoFlex. Recent research has demonstrated that PDMS is a promising and prevalent substrate material for flexible/stretchable devices on account of its high transparency, chemical inertness, large stretchability, simple process, and stable physicochemical properties[7]. Typically, PDMS film exhibits a high transmittance (>95%) in the visible wavelength region, making it an ideal substrate for optoelectronic devices. PDMS is also biocompatible and nontoxic, and thus can be directly attached to human skin or implanted into the human body without causing an infection. The PDMS film surface can be modified by various surface treatment technologies including chemical functionalization, ultraviolet irradiation, and oxygen plasma to make it more hydrophilic, and thus to improve the adhesion between the PDMS substrate and the attached components. Moreover, PDMS can mix with other conductive materials, such as silver[8], silver nanowire[9], CNT[10], and graphene[11], offering the opportunity to fabricate PDMS-based composites for flexible pressure sensors.

Apart from the above materials, some ordinary materials, such as papers and textiles, have also been developed as substrates for flexible electronics. As a renewable resource, paper possesses features including light weight, good flexibility, low price, and biodegradable, and has already been used in flexible pressure sensors. Recently, textile-based wearable sensors have also been intensively studied because of their high mechanical flexibility and good air permeability. Furthermore, besides the synthetic materials, some natural polymers such as silk fiber have attracted considerable attention due to their wide application potential in environmental friendly flexible electronic devices. Silk fiber exhibits distinguished features including biocompatibility, abundance in nature, solution-processability, biodegradability, as well as unique electrical, optical, and mechanical properties, thus promoting the development of flexible electronic devices toward next-generation environmental and organism benignity bio-integrated electronics.

11.3.2 Active Materials

11.3.2.1 Carbon-Based Nanomaterials

Over the past few years, the use of carbon-based nanomaterials including carbon blacks, carbon nanotube (CNT), and graphene in flexible electronics has been at the center of academic and industrial research due to their excellent physical and chemical properties. Carbon-based nanomaterials have been widely served as active materials for flexible pressure sensors.

Carbon black (CB) particles can be dispersed into flexible polymer matrixes to construct conductive networks. For example, by dispersing spherical carbon black into environmentally friendly thermoplastic polyurethane (TPU) to produce a TPU/CB slurry as the conductive active materials, Jiang et al. developed a piezoresistive flexible pressure sensor with high sensitivity and wide working range[12]. The flexible pressure sensor was fabricated by introducing a knoll-like microstructured surface into the printed TPU/CB film using a facile and efficient screen-printing technology (Figure 11.2(a)). Benefiting from the knoll-like microstructured sensing surface of the TPU/CB thin film, the sensor exhibited a high sensitivity of 5.205 kPa^{-1} (in the range of 0–100 kPa), an ultrawide pressure range of 0–1500 kPa, and excellent durability. In a separate study, Han et al. reported a flexible pressure sensor by simply drop-casting CB solution onto airlaid paper (AP) to construct conductive CB@AP composites as the sensing materials (Figure 11.2(b))[13]. By stacking multilayer CB@APs with an unique fiber-network structure, the sensor can provide more electric contact points between conductive CB@AP fibers under external pressure, which contributes to outstanding performance for the piezoresistive pressure sensors. The as-prepared pressure sensor presented an ultrahigh sensitivity of 51.23 kPa^{-1}, a fast response time, and a low detection limit of 1 Pa, and thus can accurately perceive subtle pressure such as weak gas blow, phonation, and wrist pulse.

CNT, as a carbon allotrope with one-dimensional cylindrical nanostructure, have been widely used as active materials for pressure sensors due to its high electrical conductivity (10^4 S cm^{-1}), remarkable intrinsic charge carrier mobility, superior chemical stability, and excellent mechanical performance (elastic modulus on the order of 1 TPa)[14]. To achieve large-area and low-cost production of CNT-based flexible electronic devices, facile and reliable synthesis techniques are of utmost importance. To this end, CNTs can be produced by various high-volume and low-purity techniques, such as arc discharge[15], laser ablation[16], and chemical vapor deposition (CVD)[17]. In addition, the alignment of CNTs could be used to further optimize the device performance. Therefore, a number of techniques including mechanical rolling[18], magnetic processing[19], Langmuir–Blodgett techniques[20], and mechanical shear techniques[21] have been developed to achieve large-scale alignment.

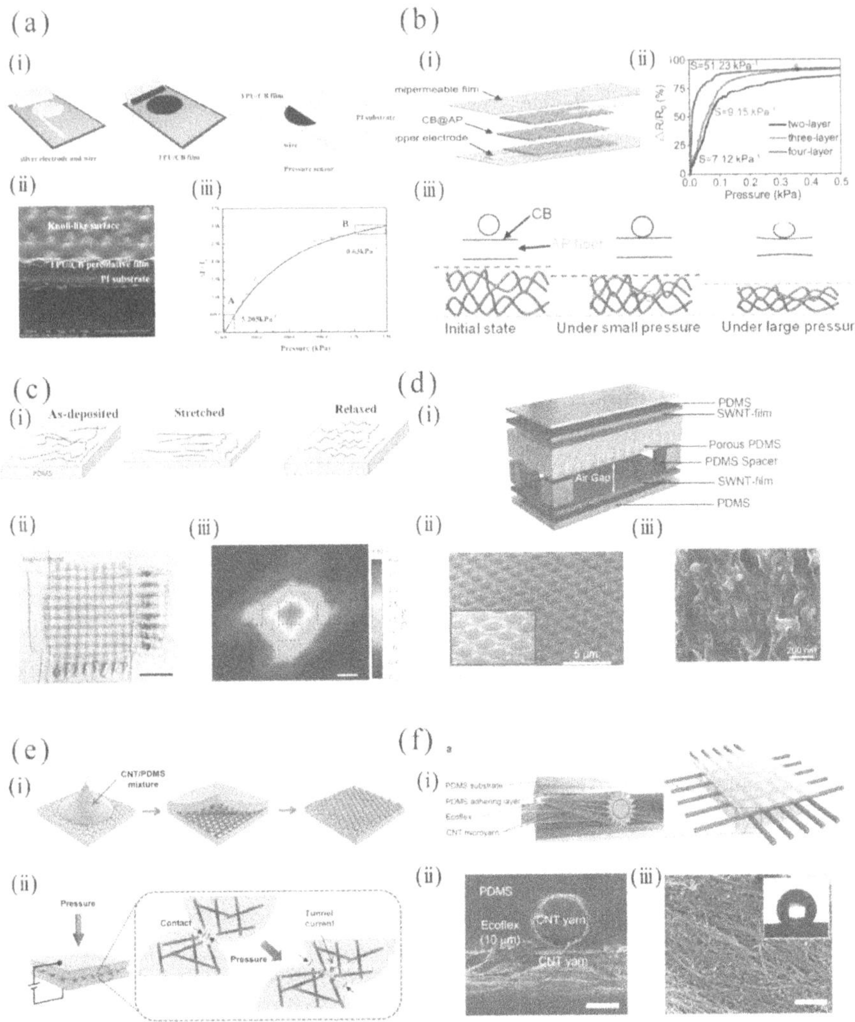

FIGURE 11.2 (a) (i) Schematics of the fabrication process of the TPU/CB pressure sensors. (ii) SEM images of the surface morphologies of TPU/CB films. (iii) Pressure sensitivity of the sensor[12]. (b) (i) Schematic illustration of the flexible pressure sensor based on CB@AP composites. (ii) Sensitivity of the flexible pressure sensors composed of two-layer, three-layer, and four-layer CB@AP composites. (iii) Structural change of two CB@AP layers under pressure[13]. (c) (i) Evolution of morphology of films of CNTs with stretching. (ii) Photograph of capacitive flexible sensor arrays. (iii) Corresponding capacitance change for pressure mapping[26]. (d) (i) Schematic structure of the stretchable e-skin. (ii) SEM image of the porous PDMS surface. (iii) SEM image buckled SWNTs on PDMS surface[27]. (e) (i) Schematic of the fabrication procedure of microstructured CNT-composite elastomers. (ii) Schematic showing the working principle of the piezoresistive pressure sensor[28]. (f) (i) Schematic illustration of the CNT microyarn-based sensor. (ii) Cross-section electron micrograph of the sensor. (iii) Field-emission scanning electron micrographs of CNT microyarn surface[29].

CNT networks can be directly deposited onto the surface of flexible substrates using commonly solution deposition methods such as spin-coating[22], spray-coating[23], vacuum filtration[24], and inkjet printing[25]. Bao et al. reported a flexible capacitive pressure sensor based on transparent and conducting single-walled carbon nanotubes films[26]. As shown in Figure 11.2(c)i, by applying strains on spray-deposited films of carbon nanotubes along each axis followed by releasing the strains, the morphology of CNTs could be rearranged upon this process, which generated a buckled and isotropic spring-like structure of CNT bundles. These processed CNT films can accommodate strains of up to 150% and demonstrate a high conductivity of 2,200 S cm^{-1} in the stretched state, and were used as electrodes in a 64-pixel array of pressure sensors which can map applied external pressure (Figure 11.2(c)ii–iii). In addition, the sensitivity of capacitive pressure sensors can be improved by changing the dielectric layer using different structures and materials. For example, Park et al. fabricated a stretchable e-skin that is capable of detecting multiple mechanical stimuli including pressure, bending, and strain[27]. Single-walled carbon nanotubes (SWNT) films were served as top and bottom electrodes, while porous PMDS and air gap were utilized as the dielectric layer (Figure 11.2(d)). Owing to the unique porous PDMS and air gap structure, the pressure sensor can achieve a high sensitivity of 1.5 kPa^{-1} in the pressure region <1 kPa.

CNTs can be dispersed into flexible polymer matrixes to construct conductive networks. A representative example was reported by Park et al. who developed a flexible piezoresistive pressure sensor based on interlocked microdome-patterned CNT/PDMS conductive composite films[28]. As shown in Figure 11.2(e)i, the composite elastomer films with microdome arrays were fabricated by micromolding a liquid mixture composed of PDMS and CNTs with a microstructure-patterned silicon mold. The unique structure of the pressure sensor leads to a giant tunneling resistance, thus enabling an extremely high sensitivity of 15 kPa^{-1} (Figure 11.2(e)ii).

CNTs assembled into microscopic or macroscopic yarns or fibers, in which the CNTs are hierarchically engineered and highly aligned, have various mechanical, electrical, and thermal properties different from individual CNT and CNT networks. For instance, Kim and coworkers developed a highly sensitive and multimodal all-carbon skin sensor based on hierarchically engineered elastic and highly conductive CNT microyarns[29]. The wearable capacitive sensor is capable of simultaneously sensing versatile external physical stimuli such as tactile, temperature, humidity, and biological variables in a single pixel. Architecturally, the sensor consists of a highly stretchable CNT microyarn circuitry incorporated with a flexible Ecoflex dielectric onto PDMS substrates (Figure 11.2(f)). The surface of the CNT microyarns exhibited a hierarchically structured fibrous network, which contributed to its hydrophobic characteristic, as well as outstanding fatigue resistance under external stress, thereby enabling the sensor with superior mechanical robustness under various mechanical deformations.

Since the first isolation by Geim and Novoselov in 2004, graphene has drawn tremendous attention as a promisng material in flexible electronic devices due to its extraordinary multiple properties. Benefited from the sp^2-hybridized carbon atoms arranged in a honeycomb lattice structure, graphene exhibits unique properties including ultrahigh intrinsic carrier mobility (2.0×10^5 $cm^2V^{-1}S^{-1}$)[30], large theoretical specific surface area (~2,630 m^2/g)[31], excellent mechanical properties (fracture strains of up to 25% and a Young's modulus of 1 TPa)[32], high thermal conductivity (5,300 W/mK)[33], chemical inertness, as well as exceptional optical transmittance (97.7%)[34]. Additionally, the perfect two-dimensional (2D) honeycomb structure of graphene allows it to provide abundant active sites on the basal plane to react with functional groups; thus, graphene can be assembled to various types of macrostructure to achieve application-adaptable electromechanical performance.[35] These properties make graphene an ideal active material for the development of flexible pressure sensors.

To fabricate graphene-based flexible sensors, the preparation method of pristine graphene is of great importance. So far, many approaches have been developed for the fabrication of pristine graphene, such as micromechanical exfoliation[36], epitaxial growth at high temperature[37], the reduction of graphite oxide[38], and chemical vapor deposition[39]. Among those methods, the CVD-grown graphene have attracted significant attention owing to the advantages of controllable size, high quality, and can be transferred to various flexible substrates. By using highly conductive aligned carbon nanotubes/graphene (ACNT/G) hybrid films that were prepared by CVD growth of graphene on aligned CNT-coated copper foil as the active membranes, and microstructured PDMS (m-PDMS) that molded from natural leaves as the flexible substrate, Jian and coworkers reported a high-performance piezoresistive pressure sensor (Figure 11.3(a))[40]. Owing to the unique hierarchical microstructures of ACNT/G films and m-PDMS substrate, the obtained sensor exhibited a high sensitivity of 19.8 kPa^{-1} (<0.3 kPa), and an ultralow detection limit of 0.6 Pa.

In recent years, graphene and its derivatives have been widely used as the conductive electrode materials for flexible pressure sensors. Yang et al. demonstrated a tunable flexible capacitive pressure sensor based on three-dimensional (3D) microconformal graphene electrodes[41]. The microstructured graphene electrode (MGrE) was controllably fabricated by the microconformal transfer method (Figure 11.3(b) i). By symmetric sandwiching the PDMS dielectric layer between the top and bottom MGrEs (Figure 11.3(b) ii), the obtained pressure sensor can achieve an ultrasensitivity of 7.68 kPa^{-1}, which is much higher than the smooth graphene electrodes (SGrEs)-based sensor (Figure 11.3(b) iii). Graphene oxide (GO) and reduced graphene oxide (rGO) have the advantages of chemically derived surface and large-scale production, making them attractive active materials for flexible sensors[42]. For instance, Zhu et al. reported a flexible resisitive pressure sensor consisting of the microstructured-reduced graphene oxide (rGO)/PDMS film and flat indium tin oxide (ITO)-coated PET film.[43] As shown in Figure 11.3(c), the graphene film was deposited on the microstructured PDMS substrate through a simple layer-by-layer (LBL) assembly process. By virtue of the anisotropic micropyramid structures of the graphene arrays, the sensor demonstrated a high sensitivity (–5.5 kPa^{-1}) at low pressure range (<100 Pa), and an ultra-fast response time (0.2 ms).

Graphene can be used as a good dielectric material for capacitive pressure sensors. For instance, Wan and coworkers fabricated a novel all-graphene capacitive pressure sensor, wherein rGO was used as the electrodes, and thin GO foam with excellent elastic property and highly relative dielectric permittivity was used as the dielectric layer[44]. Due to the porous microstructure of the GO foam, the distance

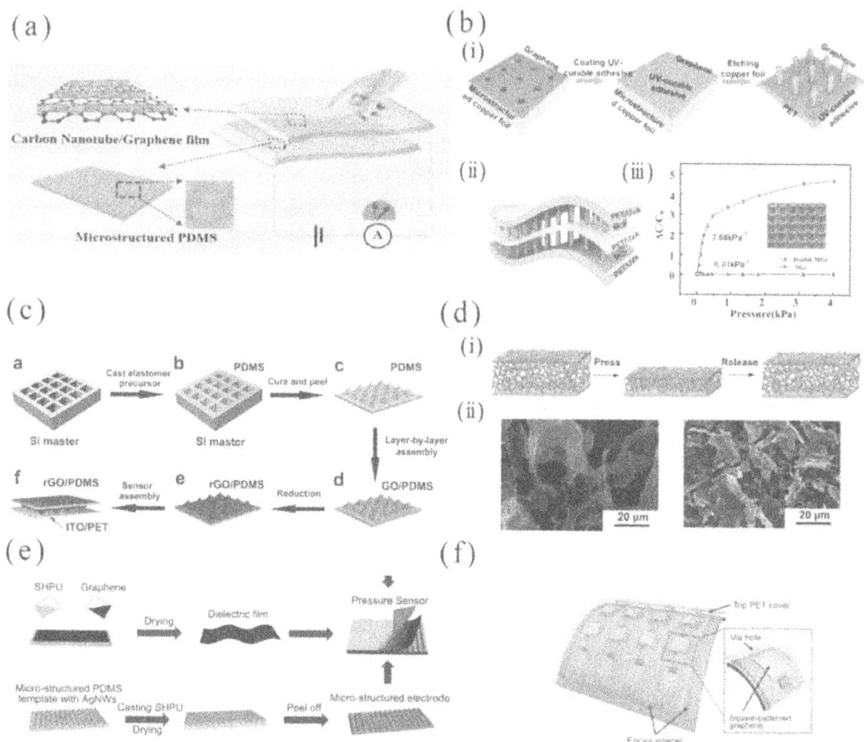

FIGURE 11.3 (a) Schematic structure of the ACNT/G pressure sensor[40]. (b) (i) Schematic diagram of the fabrication process for microstructured graphene electrodes. (ii) Schematic illustration of the sensor with double MGrEs. (iii) Sensitivity of pressure sensors with single MGrE and double MGrEs[41]. (c) Schematic illustration of the fabrication progress of the microstructured graphene-based pressure sensor[43]. (d) (i) Schematic of the loading–unloading cycle for the GO-based pressure sensor. (ii) SEM images of the original GO foam (left) and GO foam under pressure (right)[44]. (e) Schematic illustration of the micro-patterned pressure sensor fabrication[45]. (f) Schematic diagram of the pressure sensor matrix based on coplanar gate GFETs [46]

between the top and bottom electrode plates decreased and the effective dielectric constant increased under applied external pressure, thus leading to an increase in the capacitance of the sensor (Figure 11.3(d)). As a result, a flexible pressure sensor with high sensitivity (0.8 kPa^{-1}) and low detection limit (0.24 Pa) was achieved. By combining healable polyurethane (HPU) with graphene as a dielectric and Ag nanowires as an electrode, an omni-healable capacitive pressure sensor was obtained (Figure 11.3(e))[45]. Owing to the wave microstructured electrode and highly dielectric constant graphene composited dielectric layer, the sensor exhibited an outstanding pressure sensitivity (1.9 kPa^{-1}) in a low-pressure regime (<3 kPa).

In addition, graphene can also be utilized as the channel materials in the flexible field-effect transistor (FET). Sun et al. developed a flexible graphene field-effect transistor (GFET)-based pressure sensor matrix (4 × 4 pixels) for e-skin applications

(Figure 11.3(f))[46]. Due to the unique coplanar gate geometry and the use of graphene in the semiconducting channels and electrodes, the GFET pressure sensor exhibited a high pressure sensitivity of 0.12 kPa^{-1}, a low operating voltage of less than 2 V, and a high transparency of ~80%.

11.3.2.2 Conductive Polymers

Conductive polymers have drawn increasing attention due to the fact that they present many excellent properties, such as mechanical flexibility, charge transfer capability of conductive domains, and compatibility with flexible solid supports. An attractive feature of conductive polymers is their highly tunable physical and chemical properties, which can be controlled by altering the molecular structures and components. Additionally, conductive polymers can be synthesized by large-area solution-processing techniques including screen printing, spin coating, inkjet printing, and roll-to-roll fabrication, enabling them to offer low-cost, large-area electronic device arrays. These unique characteristics endow conductive polymers with the ability to fabricate various flexible sensing devices. Commonly used conductive polymers, such as polyaniline (PANI), poly(3,4-ethylenedioxythiophene):polystyrene sulfonate (PEDOT:PSS), polypyrrole (PPy), poly(3-hexylthiophene) (P3HT), and their derivatives, have been employed for constructing flexible tactile sensors.

Conductive polymer can be used as active sensing material sandwiched between two electrodes. For instance, Bao et al. presented an ultra-sensitive piezoresistive pressure sensor using PPy hydrogel with a morphology of hollow spheres as the active element (Figure 11.4(a) i), a copper foil as the top electrode, and an indium tin oxide (ITO)-coated conductive PET sheet as the bottom electrode[47]. The hollow-sphere microstructure enabled the PPy to elastically deform upon the application of external pressure (Figure 11.4(a) ii), thereby endowing the pressure sensor with reproducible and stable sensing performance. Due to the structure-derived elasticity and the polymer-electrode contact-resistance mechanism, the pressure sensor exhibited a high sensitivity of 133 kPa^{-1} (Figure 11.4(a) iii), and can detect an ultra-low pressure of 0.8 Pa within a short response time. Various hollow-structured nanomaterials have been used in bulk piezoresistivity pressure sensors. For example, by using polyaniline hollow nanospheres composite films (PANI-HNSCF) as the conductive active layer, Lou et al. demonstrated an ultrathin piezoresistive pressure sensor (Figure 11.4(b) i)[48]. The unique hollow structure endowed the flexible thin films with structure-derived elasticity as well as a low effective elastic modulus of 0.213 MPa. Due to the increased deformability, the sensor displayed a high sensitivity of about 31.6 kPa^{-1} in the low pressure range (<0.25 kPa) (Figure 11.4(b) ii). These excellent sensing properties enabled hollow polymer materials possess promising application in flexible sensing devices.

Conductive polymers can be applied directly to active electrode element of flexible pressure sensors. Yang and coworkers reported a highly sensitive piezoresisitive pressure sensor based on hierarchically patterned PPy electrode films[49]. Due to the stimulus-responsive characteristic and self-adaptive ability of three-scale nested surface topographical morphologies in PPy films, the contact area between the two microstructured PPy electrode layers will increase significantly under applied

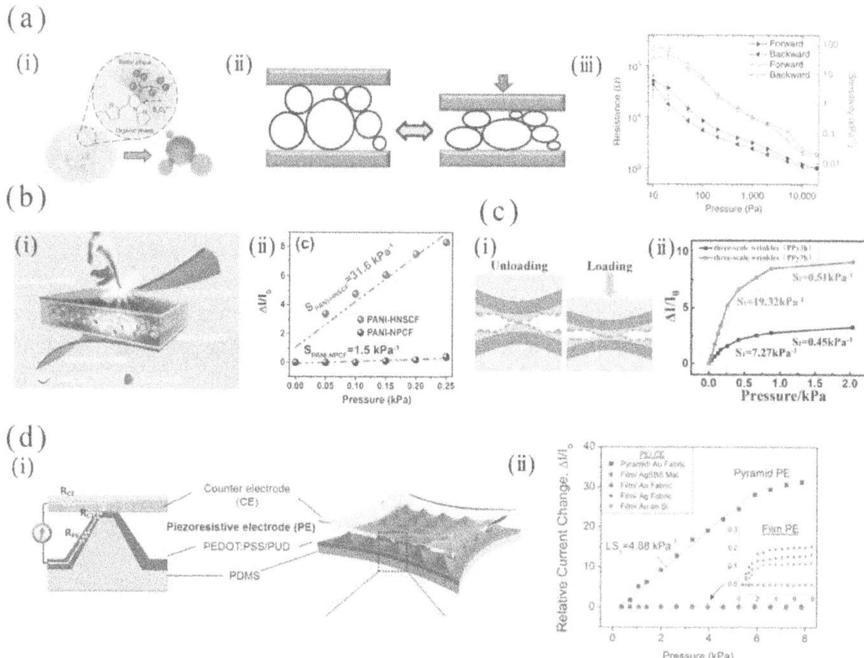

FIGURE 11.4 (a) (i) Schematic illustration of the synthesis mechanism of hollow-sphere microstructure PPy hydrogels. (ii) Pressure response process of the hollow-sphere-structured PPy. (iii) Resistance response and sensitivity of the pressure sensor[47]. (b) (i) Schematic illustration of the PANI-HNSCF-based pressure sensor. (ii) Sensitivity of the flexible pressure sensors[48]. (c) (i) Schematic illustration of the pressure sensor based on microstructured PPy films. (ii) Sensitivity of the pressure sensor based on the three-scale nested wrinkling PPy films[49]. (d) (i) Circuit model and schematic structure of the micropyramid-based resistive pressure sensor. (ii) Pressure sensitivity of the pyramid-structured sensor[50].

external minute pressure/force and can provide more conductive paths (Figure 11.4(c) i), thus leading to an enhancement of the device sensing performance. The fabricated sensor exhibited a high sensitivity (19.32 kPa^{-1}) (Figure 11.4(c) ii), an ultrafast response time (20 ms), and a low detection limit (1 Pa).

PEDOT:PSS, a classic conductive polymer consisting of conducting conjugated PEDOT and insulating PSS with negative charges has been widely used as active materials owing to its optical transparency, chemical stability, and biocompatibility. For example, Choong et al. developed a piezoresistive pressure sensor based on a stretchable microstructured substrate[50]. In this device, a reformulated PEDOT:PSS/polyurethane dispersion (PUD) composite polymer was coated on the surface of the compressible micro-pyramid-structured PDMS arrays to serve as a conductive electrode (Figure 11.4(d) i). With an external pressure applied, the contact area between the two electrode will increase significantly, thus increasing the current conduction. As shown in Figure 11.4(d) ii, the pressure sensor exhibited a sensitivity of 10.3 kPa^{-1} even stretched by 40%.

11.3.2.3 Metallic Nanomaterials

Metallic materials, such as metallic nanowires (NWs), nanoparticles (NPs), and thin films, are particularly attractive for the construction of flexible electronic devices due to their inherently high electrical conductivity. Metallic nanomaterials can be directly deposited on flexible substrates to form conductive thin films or serve as conductive fillers in conductive polymer composites.

Over the past decade, metallic nanowires (e.g., Ag, Cu, and Au) with high aspect ratios have been widely applied as active materials for flexible pressure sensors because they possess great electrical and optical properties. For example, Gong et al. used Au NWs-impregnated tissue paper sandwiched between a PDMS sheet patterned with interdigitated electrode arrays and a blank PDMS sheet to construct a highly sensitive pressure sensor (Figure 11.5(a))[51]. Upon pressure loading, the compressive deformation of tissue paper enabled more Au NWs in contact with the

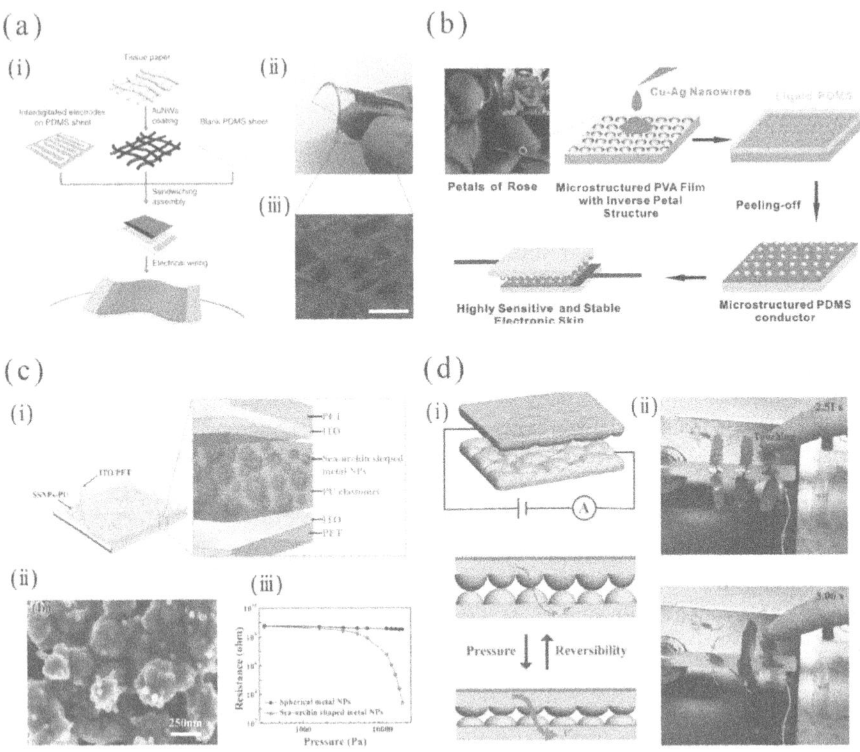

FIGURE 11.5 (a) Pressure sensor based on the AuNWs-coated tissue paper. (i) Schematic illustration of the fabrication progress of a flexible sensor. (ii) Optical image of the proposed sensor. (iii) SEM image of the morphology of gold nanowires-coated tissue fibers[51]. (b) Schematic of fabrication process of flexible micro-structured sensors[52]. (c) (i) Schematic illustration of the sensor composed of SSNPs and PU. (ii) SEM image of SSNPs in the PU elastomer. (iii) Piezoresistive characteristics of sensors using SSNPs and spherical NPs under applied pressure[53]. (d) (i) Schematic illustration of the structure and sensing mechanism of the Mimosa-inspired flexible pressure sensor. (ii) Optical photographs of the artificial mimosa leaves[54].

interdigitated electrodes, thus leading to more conductive pathways. The flexible sensor based on the above piezoresistive sensing mechanism exhibited a sensitivity of 1.14 kPa^{-1}, and a fast response time (<17 ms). Cu NWs were also developed for flexible pressure sensors owing to their low cost, high conductivity, and flexibility. By coating a thin silver layer on the surface of Cu NWs through a facile galvanic replacement reaction, Cu–Ag core-shell nanowires (Cu–Ag NWs) with excellent oxidation resistance were fabricated by Liu and coworkers[52]. As shown in Figure 11.5(b), two microstructured PDMS/Cu–Ag NWs composite layers were overlapped to construct a piezoresistive pressure sensor. The obtained sensor with petal-molded microstructures exhibited excellent sensitivity (1.35 kPa^{-1}), fast response/relaxation time, and low detection limit (<2 Pa).

Metallic nanoparticles-based conductive polymer composites can be used as active sensing elements as well. For instance, Lee et al. fabricated a bioinspired piezoresistive pressure sensor composed of insulating PU elastomer and sea-urchin-shaped metal nanoparticles (SSNPs) (Figure 11.5(c))[53]. When applied external pressure, the distance between adjacent nanoparticles became closer; thus, current transports from one particle to another through the quantum tunneling effect. Due to the sharp spike microstructures in the SSNPs, this composite-based pressure sensor showed much higher sensitivity than spherical metal NPs filled composite ones.

The metallic thin film is typically served as the electricity active layer to make interconnection. For instance, Su et al. developed a piezoresistive flexible pressure sensor by coating a thin layer of gold onto the microstructured PDMS films[54]. As shown in Figure 11.5(d) i, two layers of coated PDMS films with bio-inspired microstructures were placed face-to-face to construct the flexible pressure sensors. Due to the existence of the protuberant microdomains, the sensor exhibited a high sensitivity of 50.17 kPa^{-1} and a fast response time of 20 ms, which enabled the fabrication of "artificial mimosa leaves" that mimic the touch-sensitive capability of natural mimosa leaves (Figure 11.5(d) ii).

11.4 FABRICATION OF FLEXIBLE PRESSURE SENSORS

Conventional pressure sensors based on rigid mental and semiconductor materials have limited their application. By contrast, flexible sensors have many unique advantages, which attracted extensive research efforts on fabrication methods. Flexible sensors require novel approaches in structure design to realize high sensitivity, flexibility, quick response, and stability, etc. due to material science and microelectromechanical system (MEMS) technology progress; researchers have explored many feasible methods to fabricate flexible sensors. In this section, the typical methods are summarized as follows.

11.4.1 MICROMOLDING

Micromolding is commonly used to fabricate microstructure pressure sensors. Generally, the liquid polymer, such as PDMS or PI, is first spin-coated onto the prepared micromold, after high-temperature cured, peeled off to obtain the desired substrate. Next, the microstructure substrate is combined with conductive materials to form a hybrid electrode. Finally, pressure sensors are assembled by multi-conductive layer stacking.

Structured mold can be fabricated by MEMS technology. As shown in Figure 11.6(a), an etched pyramidal silicon mold was fabricated by using photolithography and etching. Then, the solution consisting of polystyrene (PS) beads poured into the silicon mold and was blade coated. Next, the pyramidal mold was placed on soft substrate and hot-pressed. Thereafter, pyramid-shaped substrate peeled off and combined with another ITO/PET film to complete the sensor[55]. These microstructured pressure sensors exhibited ultrahigh sensitivity for various e-skin applications. The micro-hemisphere is another classical microstructure used for pressure sensors. The procedures for the fabrication of pressure sensors with the novel hierarchical structure are illustrated in Figure 11.6(b)[56]. First, the acrylonitrile butadiene styrene (ABS) mold with micro-hemisphere array was marked via a laser-marking machine. Then the PDMS was drop-casted onto the mold. After curing and peeling, graphene oxide (GO) sprayed its surface. In the end, the flexible pressure sensor was completed after structured film and AgNWs substrate were assembled and encapsulated.

Microstructured silicon molds can create high-performance pressure sensors based on regular and precise patterns. However, such silicon molds typically need complicated equipment, resulting in relatively high fabrication costs. It is therefore

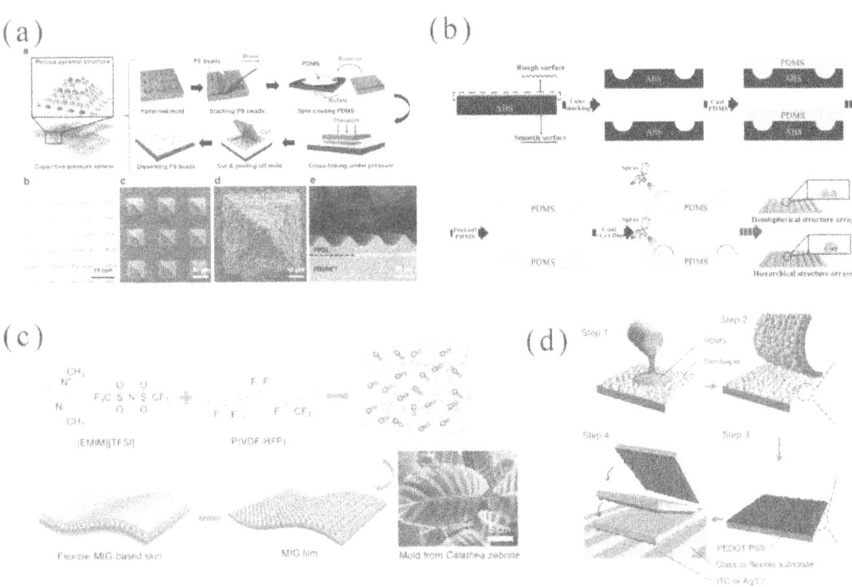

FIGURE 11.6 Flowcharts illustrate the micromolding fabrication procedures for flexible pressure sensors. (a) The capacitive pressure sensor was fabricated by silicon mold with etched pyramid micropatterns[55]. (b) The flexible (RGO/PDMS) sensing films were obtained by PDMS drop-casted onto the micro-hemispherical mold[56]. (c) The fabrication of pressure sensor with biomimetic dielectric layer templated form leaf[57]. (d) The irregular microhumps pressure sensor achieved by the pattern transfer of sandpaper molding [58].

necessary to find a cost-effective way to make micro-mold. Researchers have turned their attention toward the objects around daily life which can cast as microstructured mold. For example, Figure 11.6(c) shows the fabrication process of the biomimetic pressure sensor; the patterned film with array of micro-cones was patterned from fresh leaf[57]. In simple terms, the uncured PDMS was cast on the calathea zebrine leaf and peeled off as a second molding template after being cured. Then the molded pattern is replicated again by spinning liquid ionic gel onto the above template. In the end, the biomimetic ionic gel film as the active dielectric layer embedded into flexible electrodes to form a highly sensitive pressure sensor. Moreover, the silicon carbide waterproof sandpapers can also be used as molding templates to develop the microstructures pressure sensors[58]. As shown in Figure 11.6(d), uncured PDMS was deposited on the top of pretreated sandpaper and annealed at high temperature. The patterned PDMS then peeled off and coated with PEDOT:PSS; finally pressure sensor was built by putting microhump electrode on ITO glass.

In a word, micromolding does not involve toxic chemicals; it has the characteristics of rapid prototyping and repeated use, which is widely used for high sensitivity pressure sensors.

11.4.2 Sacrificial Template Method

Beyond micromolding, the sacrificial template method is also widely used to fabricate flexible pressure sensors. This method can produce porous conductive structures to form high-performance sensors. Porous structures are usually developed either by using a sacrificial template or by dissolving special materials such as PS beads in solvent once the elastomer has cured. The inner microspores sizes generally are 1–1000 μm; they increase the compressibility of layers due to the introduction of air voids, which course sensors are more sensitive to external stimulus.

Figure 11.7(a) shows a schematic illustration of the fabrication process of multilayered porous polydimethylsiloxane/silver nanoparticle (PDMS/AgNP) sensor[59]. First, AgNP, PDMS solution, and citric acid monohydrate (CAM) were mixed with three specific ratios, and stacked from bottom to top, respectively.

Then the mixture was cured and cut into cubes, one of the cube was immersed into organic solvent to dissolve the CAM completely. Finally, the porous structure was covered with the thermosensitive element to form a broad range pressure sensor. We can also control the morphology of porous pressure sensors by changing the pore sizes. As shown in Figure 11.7(b), a facile approach to fabricating sponge-like pressure sensors has been described[60]. PS beads were stacked in multiple layers by drop-casting on a Si substrate and then uncured PDMS were coated on the stacking microbeads. After the film was cured, the PS beads and the sacrificial layer were etched by immersing it in an organic solution. In the end, the porous PDMS layer was transferred onto an indium tin oxide(ITO)/polyethylene phthalate(PET) film. A porous microstructure pressure sensor was obtained.

The sacrificial template method is simple and inexpensive but very useful. The pressure sensors based on porous structure can not only solve the bottleneck problems of low sensitivity but also reduce the effective Young's modulus of substrate, improving the deformation behavior of the sensors.

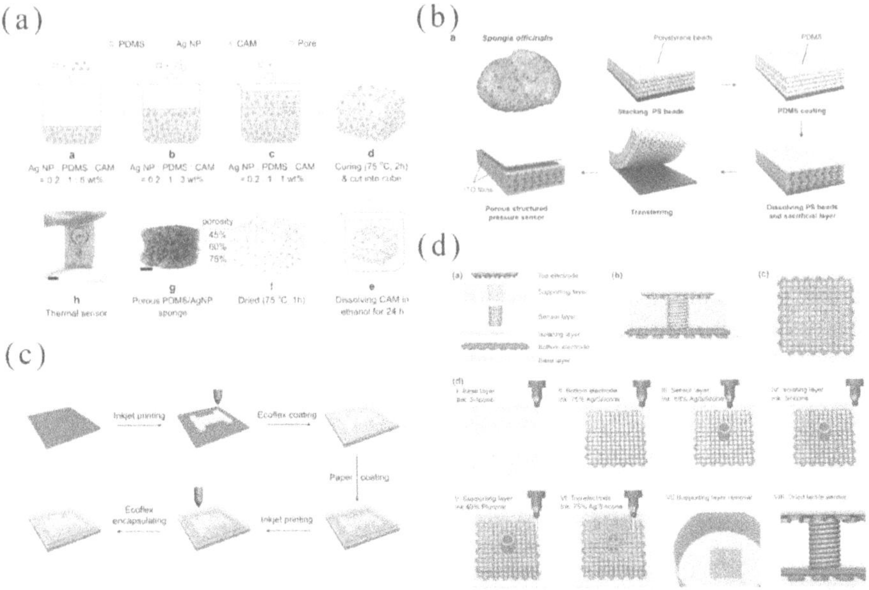

FIGURE 11.7 Schematic illustration of sacrificial template method and ink-jet printing fabrication process for flexible pressure sensors. (a) The high sensitivity pressure sensor using multilayered porous PDMS/AgNP sponge which was prepared mixing PDMS solution with citric acid monohydrate and dissolving in ethanol[59]. (b) Cured PDMS/PS mixture was etched by immersing in an organic solution to obtain porous structured pressure sensor[60]. (c) Fabrication of large-area bimodal pressure sensor using an all-inkjet-printing method[61]. (d) Tactile sensor was printed on polyimide film with ink-jet printing[62].

11.4.3 Inkjet Printing

Inkjet printing is a rapid direct patterning technique via liquid phase material deposits on substrate, which simplifies the processing steps as well as reduces the manufacturing cost. During the inkjet printing, the print head moves following the programmed trajectory and tiny droplets of ink are propelled through a micrometer-sized nozzle head. Many materials can be used as ink for printing; it greatly expands the scope of the use of this method. Thus, plenty of hybrid pressure sensors can be prepared by inkjet printing.

Figure 11.7(c) presents the fabrication procedure of bimodal pressure sensor by inkjet printing[61]. Silver nanoparticles (AgNPs) suspension as the conductive ink was directly printed on a cleaned polyethylene naphthalate (PEN) and cured. After particle network densification, the Ecoflex and paper were successively coated onto the film; finally another electrode was fabricated by inkjet printing on paper and the whole device was encapsulated by Ecoflex. The inkjet printing is also suitable for manufacturing complex multi-dimensional structures. As shown in Figure 11.7(d), to fabricate the multi-dimensional tactile sensors at one continuous process, sinter-free inks with adjustable viscosities and electrical conductivities were manipulated by different, independently addressable nozzles, all the printing paths were controlled by using commands. After the printing process, the sensor was cured at room temperature and bonded together with a thin, flexible cable[62].

Inkjet printing has many attractive features including cost-effective, simple manufacturing process, mask-free, and nontoxic. However, the fabrication difficulty is greatly increased when a microstructure with small diameters, e.g., <50 μm.

11.4.4 Vacuum Filtration Transfer

The vacuum filtration transfer technique makes wearable sensor fabrication more diversified, and it has been successfully used for flexible conductive films by nanometer solution vacuum filtration on the patterned filter paper. To put it simply, the liquid substrate is poured onto the prepared nanometer solution which is already patterned on filter paper, and then further impregnates into the conductive network after vacuum filtration, making excellent adhesion between the two parts. The pressure sensors fabricated by vacuum filtration transfer are highly reliable for long-term electrophysiological signal recording.

The vacuum filtration transfer was used to fabricate reduced graphene oxide fiber fabrics (rGOFF)/PDMS pressure sensor as determined in Figure 11.8(a), the rGOFF prepared by wetting-spinning and cut with scissors to obtain the short GO fiber. Then short GO fibers were assembled and collected to obtain GO fiber fabric by vacuum filtration. After drying, the uncured PDMS was poured into the surface of rGOFF until it was fully covered. Last, the obtain rGOFF/PDMS composites were dried and cut into proper sizes; the highly compressible pressure sensor was successfully prepared[63]. Vacuum filtration transfer can still be able to pattern the desired shape touch device[64]. As shown in Figure 11.8(b), the prepared inverse pattern was printed onto the membrane surface. The pattern area can clog the pores of the membrane to prevent suspension filtration. Next, the solution of GO was filtered. After the steps of drying and activating, the GO pattern was finally transferred on the PET substrate to form a touch-sensitive device.

This technique is usually considered a simple and safe method to get a uniform and ultrathin electrode, the whole process of transfer without the use of organic toxic compounds. Moreover, the thickness of film can be controlled by adjusting the concentration and amount of the nanometer solution during the vacuum filtration process.

11.4.5 Other Manufacturing Methods

Flexible pressure sensors fabricated by other methods have also been explored. 3D printing has provided a new strategy for pressure sensor. Figure 11.8(c) demonstrated a pressure sensor array fabricated with coaxial extrusion 3D-printing[65]. The inner and outer materials were loaded into different syringes, after applied pressure; the ink was extruded from the nozzle that was connected to the syringes with rubber hose. Then fibers were printed onto the sandpapers and cured at high temperature. After that, the peeled off fibers from the sandpaper were assembled as a pressure

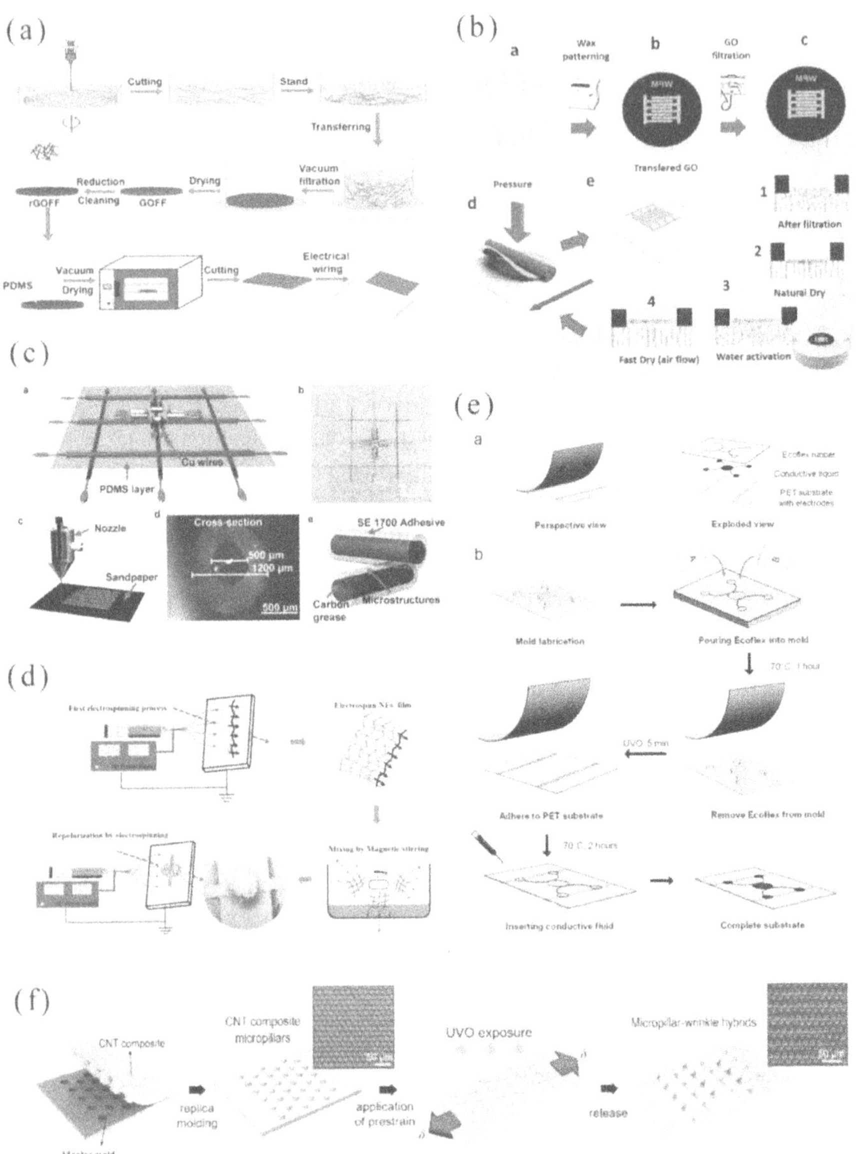

FIGURE 11.8 Vacuum filtration transfer and other methods were described for pressure sensors fabrications. (a) The simplified process for the preparation of rGOFF/PDMS sensor via a facile vacuum filtration method [63]. (b) A touch-sensitive device was fast patterned by graphene oxide printing technology using membranes filtration [64]. (c) A schematic of the wearable sensor array fabricated by coaxial extrusion 3D-printing technology[65]. (d) The diagrams of eletrospinning of the rGO modified P(VDF-TrFE) nanofibers to form self-powered sensor[66]. (e) The microfluidic pressure sensor was produced by introducing liquid metallic alloy into the microchannel with two needle syringes[67]. (f) Incorporating CNT into a hierarchical topography of wrinkle microstructure that mimic in human skin could produce the hybrid electronic skin[68].

sensor array. 3D printing is highly efficient, simple, and scalable; it's particularly suitable for the fabrication of complex microchannel structures.

Electro spinning is one of the other common manufacturing methods; it is a simple method that combines conductive solution casting with electric field polarization. As presented in Figure 11.8(d), after applying high voltage to the droplets at the nozzle, the homogeneous precursor solution was outflowed and formed nanofibers due to electric force larger than the surface tension force. Nanofibers were further mixed with rGO and repolarization to form film. Finally, the pressure sensor was fabricated by the composite film fixed on the copper electrodes and encapsulated with PDMS[66]. Furthermore, there are only limited materials available for the electrospinning because uniform dispersion of the filler is still a challenge.

Microfluidic technology can fabricate the flexible tactile sensor by introducing highly conductive liquid into soft substrate. The device schematic is illustrated in Figure 11.8(e); the soft silicone rubber was poured on the master mold and cured at high temperature. Then the pattern layer was assembled with PET film to form a closed internal structure. The next liquid metallic alloy was injected into microchannel using two syringes. One injected the conductive liquid from the inlet, while the other one evacuated the trapped air from the outlet. After the wires were connected to the ends of the ports, the working pressure sensor was finally obtained[67]. This method could avoid unwanted air bubbles with liquid metal in the inner structure and enable the filling process more easily.

Prestretching is another simple method to fabricate microstructure for pressure sensors. The approach utilizes ultraviolet–ozone (UVO) to fabricate periodic surface wrinkle structure because UVO can turn the surface of film into a dense layer, making mismatch elastic modulus between the relatively stiff surface layer and the more compliant elastomeric layer underneath. As shown in Figure 11.8(f), as-prepared nanocomposite elastomer was unidirectionally stretched, then followed by an exposure to UVO. After releasing the applied strain, a 3D wavy structure can be generated on the composite surface. Last, a highly sensitive pressure sensor was assembled by two hybrid films that were brought together with structured sides face to face[68].

11.5 FEATURES OF FLEXIBLE SENSORS

The sensing performance of flexible pressure sensors can be estimated through several key parameters, such as sensitivity, limits of detection, hysteresis, response time, and power consumption.

11.5.1 Sensitivity

Sensitivity, which is defined as the slope of the relative change in electrical signal (current, voltage, resistance, or capacitance) versus applied external pressure curve, is one of the most important parameters in determining the sensing performance of a flexible pressure sensor, especially for practical applications that require subtle force detection (e.g. hearing aids, wearable blood, and pulse monitor devices). With respect to the design optimization and fabrication of flexible pressure sensors with high sensitivity, various strategies have been reported. Herein, we summarized several typical optimization concepts in this section.

The first approach to realize high sensitivity relies on the employment of nanomaterials with intrinsic giant piezoresistive or piezoelectric effect, such as CNTs, Si nanowires, and ZnO nanowires[69]. However, due to the limitation of processing technology, building patterned macroscopic networks or arrays with identical performance using such nanomaterials still remains a big challenge. Another approach toward highly sensitive pressure sensors is based on the utilization of nano/microstructured active materials, which can contribute to significantly changed contact resistance under external tactile stimuli. This strategy is mainly applied to piezoresistive pressure sensors. For instance, Yao et al. reported a highly sensitive piezoresistive pressure sensor based on a fractured graphene-nanosheet-wrapped PU sponge[70]. The sensing mechanism of this sensor depends on the resistance changes caused by conductive rGO-wrapped PU microfibers during compressive (Figure 11.9(a) i). The sensitivity of the RGO–PU sponge sensor was greatly enhanced by creating a contact fractured microstructure (Figure 11.9(a) i–ii).

The sensitivity of flexible piezoresistive pressure sensors can also be improved by increasing the change in contact area or the number of contact points between two electrodes (or the electrode-active layer interface) to alter the contact resistance[71]. Accordingly, piezoresistive pressure sensors with high sensitivity are usually realized by using a microstructured active layer, which is achieved by the utilization of a mold (e.g., nature leave[72,73], abrasive paper[74], and etched silicon[75,76]) to shape the surface morphology. Park et al. compared the sensitivity of CNT/PDMS composite films with different geometries (i.e., planar, micropyramid, micropillar, and microdome) (Figure 11.9(b) i)[77]. As shown in Figure 11.9(b) ii–iii, the simulation results for the four different structures demonstrated that the microdome structure shows the highest sensitivity due to the largest contact area variation under external pressure. It was also found that forming an interlocked geometry can further increase the sensitivity.

As for capacitive-type pressure sensors, the sensitivity can be adjusted by changing the distance variation between the electrodes under pressure. Therefore, the dielectric and electrode layer can be microstructured to improve compressibility. For example, Zhang et al. utilized nature lotus leaves as templates to mold the unique surface structures on PDMS film through a simple replicating process (Figure 11.9(c) i)[78]. As shown in Figure 11.9(c) ii, the flexible capacitive pressure sensor was fabricated by sandwiching the polystyrene microsphere dielectric layer between two micropatterned PDMS/Au electrodes. When applied external pressure, the microcaves in the PDMS/Au electrodes will contribute to the reduced distance between electrodes as well as the rearrangement of PS microspheres in the dielectric layer, thus endowing the flexible sensor with high sensitivity (0.815 kPa^{-1}) and fast response time (38 ms) (Figure 11.9(c) iii). Based on a three-dimensional microporous dielectric elastomer, a flexible capacitive pressure sensor with high sensitivity was fabricated[79]. The microporous dielectric elastomer is highly deformable by even small pressure input levels, thus leading to an increase in its sensitivity (Figure 11.9(d)). Upon external pressure loading, the closure of micropores within the elastomeric dielectric layer can increase the effective dielectric constant, thereby further improving the sensitivity of the sensor.

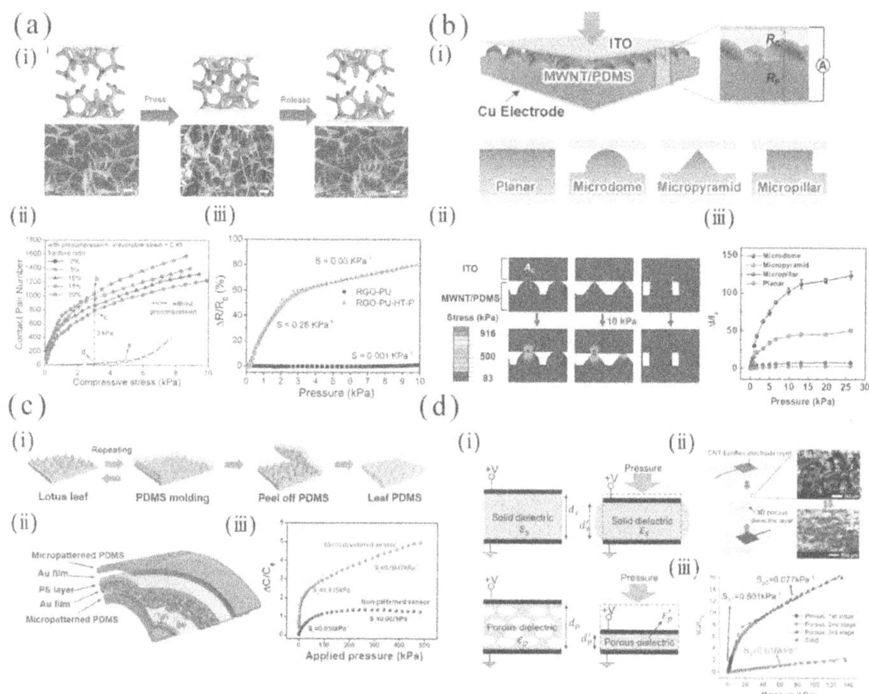

FIGURE 11.9 (a) (i) Graphene-polyurethane sponge with a contact fractured microstructure. (ii) The plots of contact pair number – compressive stress response curves for the fiber network with or without pre-compression under different fiber fracture ratios, respectively. (iii) Pressure-response curves for the sensors with hydrothermally treated and pressed RGO-PU (RGO-PU-HT-P) sponge and RGO-PU sponge[70]. (b) (i) Schematic illustration of the piezoresistive pressure sensor with four different surface morphologies. (ii) Finite-element analysis of localized stress distributions for different surface microstructures under external pressure. (iii) Relative current changes of sensors with different surface morphologies in response to normal pressure[77]. (c) (i) Schematic of the fabrication process for micropatterned PDMS film molded with lotus leaf. (ii) Schematic structure of the capacitive pressure sensor. (iii) Pressure–response curves of the micropatterned sensor (red) and nonpatterned sensor (black)[78]. (d) (i) Schematic illustration of the deformation process of capacitive pressure sensors with and without the air-embedded dielectric layer in response to external pressure. (ii) Schematic diagram of the sensor consisting of two CNT–Ecoflex electrode layers and a porous dielectric layer. (iii) Pressure–response curves of the pressure sensors using microporous and solid Ecoflex dielectric layers[79].

11.5.2 Limits of Detection

Limit of detection (LOD), another key performance parameter, is typically defined as the minimum pressure that produces a distinguishable signal change. Improvement on the LOD of pressure sensors is critical for sensitive and accurate detection of subtle pressure, which is important for some specific applications. In fact, the construction of pressure sensors that can detect ultra-low pressure, such as acoustic pressure and subtle touch stimuli, plays a significant part in promoting the development of hearing aids, microelectromechanical system (MEMS) microphones, touch screen devices, and other intelligent products.

Microengineering the active layer of flexible pressure sensors is one of the most common approaches to improve the limit of detection. For instance, Chen et al. reported a resistive pressure sensor based on PDMS/Ag microstructures and rough PI/Au interdigital electrodes (Figure 11.10(a))[80]. Compared with the sensors with flat electrodes, the proposed configuration exhibited a significantly enhanced sensing performance, and the detection limit of the resulting pressure sensors is only 0.36 Pa. This is because the increased compressibility of the microstructured active layer enabled it to respond to lighter loads.

11.5.3 Hysteresis

Hysteresis, an important feature in flexible pressure sensors, refers to the phenomenon of the inconsistent sensing performance of the pressure sensor between loading and unloading. Hysteresis is defined as the difference in the signal versus pressure curves under loading and unloading of pressure. Hysteresis behavior leads to different measurement under applied external pressure and should be minimized for practical usability.

Hysteresis phenomenon exists in all polymer-based flexible pressure sensors due to the viscoelastic nature of the polymers. The hysteresis of a polymer-based sensor is mainly affected by the properties of the polymers, the morphology of the active materials, the structure of the sensors, the sensing mechanism, and the interaction between the polymer substrate and the active sensing element. To reduce hysteresis,

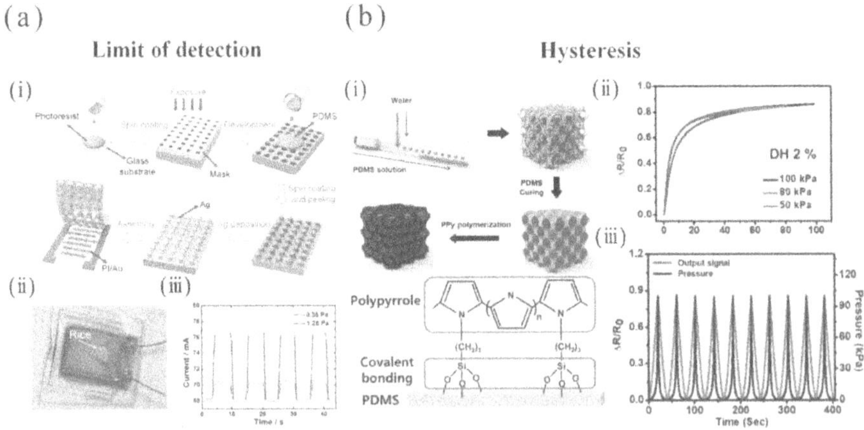

FIGURE 11.10 (a) (i) Schematic illustration of the fabrication process of the microstructured pressure sensor. (ii) Optical image of the proposed sensor pressed by a rice grain. (iii) A demonstration of the LOD by loading/unloading a grain of rice with different weights[80]. (b) (i) Schematic depiction of the sensor fabrication process. (ii) Hysteresis loops of sensor at different pressure levels. (iii) Relative resistance change under loading/unloading cycles with an applied pressure of 100 kPa[81].

one approach is to use a microstructured elastomer substrate to lower its viscoelastic properties. Furthermore, strengthening the bonding between the conductive materials and the elastomer matrix can prevent their relative sliding and displacement, thus resulting in low hysteresis. For instance, Park and coworkers reported a piezoresistive pressure sensor by chemically grafting conductive PPy polymer on the surface of porous PDMS elastomer with uniformly sized and arranged pores (Figure 11.10(b) i)[81]. Due to the microporous structure of the PDMS elastomer template and strong covalent bonding between the PPy and PDMS, the obtained pressure sensor exhibited extremely low degree of hysteresis (DH) of 2% (Figure 11.10(b) i) and stable output signals under loading and unloading of applied pressure (Figure 11.10(b) iii).

Despite its importance, there is inconsistency in the quantification criteria of hysteresis for the currently reported flexible pressure sensors. The hysteresis value is influenced by the rate of increase or decrease in pressure, the applied load of the sensor, and the number of cycles the sensor has previously undergone. Therefore, a global quantification standard of hysteresis should be utilized to properly compare different pressure sensors' hysteretic performance.

11.5.4 Response Time

The response time is defined as the time consumption from applying pressure to producing a stable signal output and reflects the speed at which a pressure sensor can respond to step input. Response time is a key parameter to estimate the flexible pressure sensors' performance upon dynamic loadings, which can be important for applications including real-time pulse waveform detection and instant-response displays.

The response time is related to the viscoelastic property of the elastomer as well as the contact conditions between the active material and the electrode. Bao et al. demonstrated that the response speed of the capacitive pressure sensor with micropyramid elastomeric films far surpassed that exhibited by unstructured rubber films of the same thickness[82]. Recently, various microstructures (pyramid, hemisphere, porous) have been utilized to reduce response time to tens of milliseconds. For instance, Qiu and coworkers developed a highly sensitive capacitive pressure sensor using porous conductive composites as the dielectric layer (Figure 11.11(a) i)[83]. The major advantage of the porous structure is the increased compressibility and elasticity of the dielectric layer due to the introduction of air voids, thus resulting in a rapid-response capacity (~45 ms) of the capacitive sensor (Figure 11.11(a) ii). Furthermore, the flexible sensor can be used for real-time and high-accuracy radial artery pulse wave monitoring (Figure 11.11(a) iii), demonstrating its potential applications in wearable healthcare monitoring devices.

11.5.5 Power Consumption

Fabricating flexible pressure with low-power consumption is of great importance for practical applications. Various strategies have been reported to reduce the power consumption of flexible sensors.

The first approach relies on reducing the operating current and operating voltage of the sensor. The power consumption of piezoresistive pressure sensors can

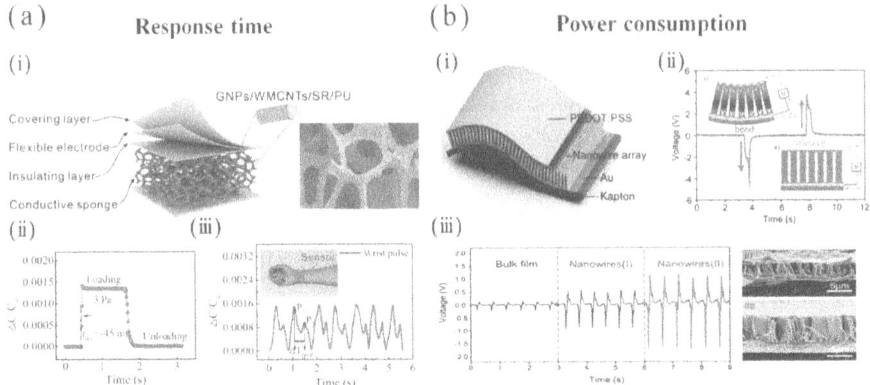

FIGURE 11.11 (a) (i) Schematic structure of the 3D porous dielectric layer-based pressure sensor. (ii) Response time of the pressure sensor under a loading pressure of 3 Pa. (iii) Real-time wrist pulse wave monitoring[83]. (b) (i) Schematic illustration of the flexible piezoelectric sensor. (ii) The sensing mechanism of vertically aligned P(VDF-TrFE) nanowire arrays under mechanical bending deformation. (iii) Output voltages for sensors based on nanowires with two aspect ratios and for a bulk P(VDF-TrFE) film[85].

be reduced by choosing appropriate active materials. A representative example is the piezoresistive pressure sensor based on vertically aligned gold nanowires[84]. Benefiting from the high conductivity of AuNWs, the operation voltage of the device was only 0.1 V, and the current was <2 mA, resulting in the energy consumption below 0.2 mW. For capacitive pressure sensors, utilizing dielectric materials with high dielectric constants can lead to enhanced capacitance, thereby decreased operation voltage and reduced power consumption.

Piezoelectric and triboelectric phenomena provide another approach to resolve the power consumption issues of flexible sensors. Therefore, self-powered pressure sensors have attracted increasing attention in recent years. One interesting example is the piezoelectric pressure sensor based on vertically aligned P(VDF-TrFE) nanowire arrays (Figure 11.11(b))[85]. The device can generate a peak voltage of ~4.8 V and a current density of ~0.11 μA/cm^2 under mechanical bending deformation, showing its strong power-generating performance. Furthermore, the vertically well-aligned nanowire arrays can be applied as self-powered pressure sensors to detect some human activities such as heartbeat pulse, and finger movement, demonstrating its potential applications in wearable electronic devices.

11.6 BIOMEDICAL APPLICATIONS

11.6.1 Physiological Parameter Detection

With the increasing demand for personal healthcare, remote and personalized health monitoring is of great importance for the timely and efficient management of health events and prevention of disease[86–89]. Flexible pressure sensors can serve as a low-cost and noninvasive tool to provide various physiological information including heart rate[90], pulse[91], blood pressure[92], and intraocular pressure[93], all of which are significant health indicators.

Heartbeat or pulse, which is defined as the frequency of the cardiac cycle, is one of the vital parameters to assess the physical and mental states of a person. Recently, flexible pressure sensors with high sensitivity have demonstrated the capability of accurate heart rate or pulse measurement. For instance, Pang et al. developed a wearable capacitive pressure sensor for real-time pulse signal detection[94]. The unique microhair structure of the device allowed exquisite sensing of various mechanical forces from the skin, and can maximize the effective contact area between the sensors and irregular epidermis to enhance the signal-to-noise ratio (Figure 11.12(a) i). Furthermore, the sensor was capable of continuous measuring radial artery and

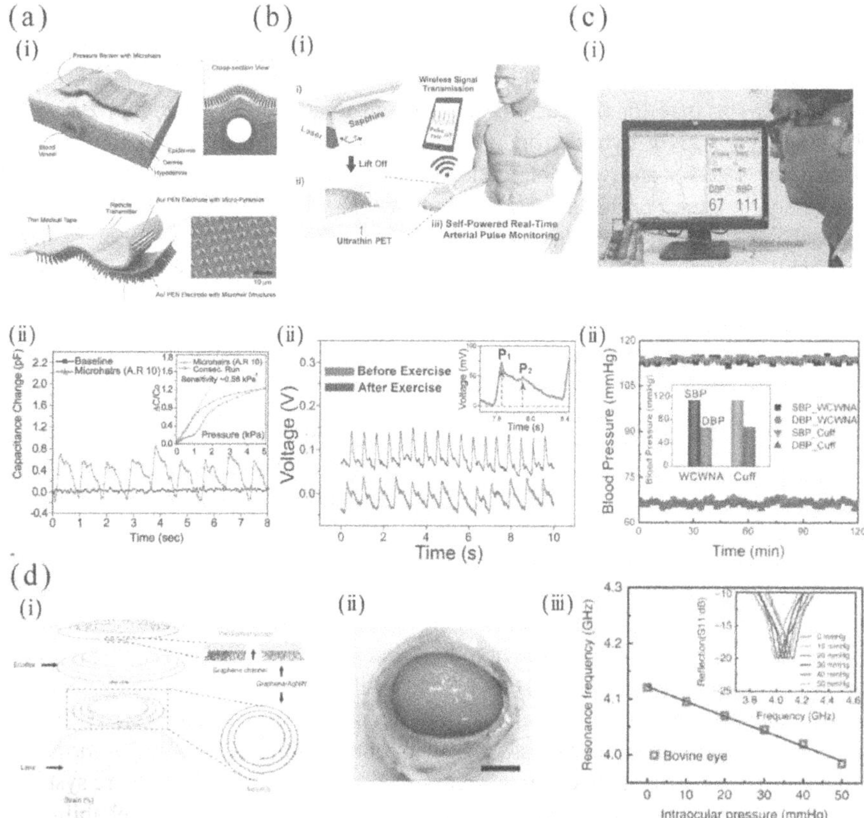

FIGURE 11.12 (a) (i) Schematic illustration of the pressure sensor with a microhair structure. (ii) Pulse waves of the radial artery detected by the pressure sensor[94]. (b) (i) Schematic illustration of the self-powered pressure sensor. (ii) Radial artery pulse signals measured by the sensor[95]. (c) (i) The sensor is worn against human finger and ear for real-time pulse and blood pressure measurement. (ii) Beat-to-beat systolic blood pressure (SBP) and diastolic blood pressure (DBP) calculated from the sensor[96]. (d) (i) Schematic structure of the wearable contact lens sensor. (ii) Photograph of the sensor integrated onto a bovine eyeball. (iii) Frequency response of the intraocular pressure sensor on the bovine eye[97].

jugular venous pulse signals (Figure 11.12(a) ii), demonstrating its potential applications toward human-health monitoring. Sensors with low-power consumption are essential for wearable applications. Park et al. fabricated a self-powered piezoelectric pressure sensor based on PZT film for real-time radial/carotid artery pulse signals monitoring (Figure 11.12(b))[95]. In addition, the detected pulse signals can be wirelessly transmitted to a smartphone through a Bluetooth transmitter.

Blood pressure is one of the most important markers for the evaluation of human cardiovascular function. Continuous and long-term monitoring of the blood pressure waveform can provide remarkable insights for the diagnosis and therapy of cardiovascular diseases, such as arteriosclerosis, hypertension, and arrhythmias[92]. Meng and coworkers developed a weaving constructed self-powered triboelectric pressure sensor that could provide real-time blood pressure monitoring by capturing the subtle mechanical change in the vessel (Figure 11.12(c))[96].

Elevated intraocular pressure is the main factor in the pathogenesis of glaucoma[93]. Therefore, accurate and continuous intraocular pressure monitoring is of great importance for the early diagnosis and treatment of glaucoma. Based on the graphene–AgNW hybrid structure, Kim et al. developed a transparent (>91%), stretchable (~25%), and wearable contact lens sensor that could continuously and wirelessly monitor intraocular pressure and glucose within tears (Figure 11.12(d))[97]. Furthermore, a bovine eyeball was utilized to test in-vitro to prove its reliable operation, which demonstrated the potential of the wearable contact lens sensors for next-generation ocular diagnostics.

11.6.2 Human Activity Monitoring

When flexible pressure sensors are attached to a person's muscles or joints, they are capable of measuring body motions via electrical or visual signals, and therefore enable various applications such as posture correction, sport training, and rehabilitation treatment[98]. As the most active parts in human bodies, movements of limbs and hands create most of our daily activities. Monitoring different motion patterns of limbs and hands requires the detection and differentiation of the types and direction of forces[99]. Park and coworkers reported a piezoresistive pressure sensor with interlocked microdome arrays for wearable wrist-motion detection (Figure 11.13(a) i)[100]. Owing to the unique interlocked geometry, the sensor could easily differentiate multiple tactile stimuli, such as pressure, shear force, lateral stretch, bending, and torsion (Figure 11.13(a) ii).

Monitoring of the plantar pressure distribution could provide useful information in the fields of sports biomechanics, gait analysis, and wearable healthcare systems. In particular, analysis of gait patterns enables the early diagnosis and rehabilitation assessment of several diseases such as Parkinson's disease and diabetic foot ulcers[101]. In addition, gait monitoring devices could be used in sports training such as running and golf to improve the performance of athletes[102]. Generally, the gait status could be monitored by measuring the force exerted by the periodic foot stepping. Because of the heavy bodyweight, flexible pressure sensors with high sensitivity, broad pressure range, and great durability are desired. Lee et al. developed a flexible ferroelectric pressure sensor with a multilayer interlocked microdome geometry[103]. Due to the multilayer structure of interlocked microdome arrays, the sensor can maintain high sensitivity and linear response over an ultrabroad pressure range. As shown, a smart

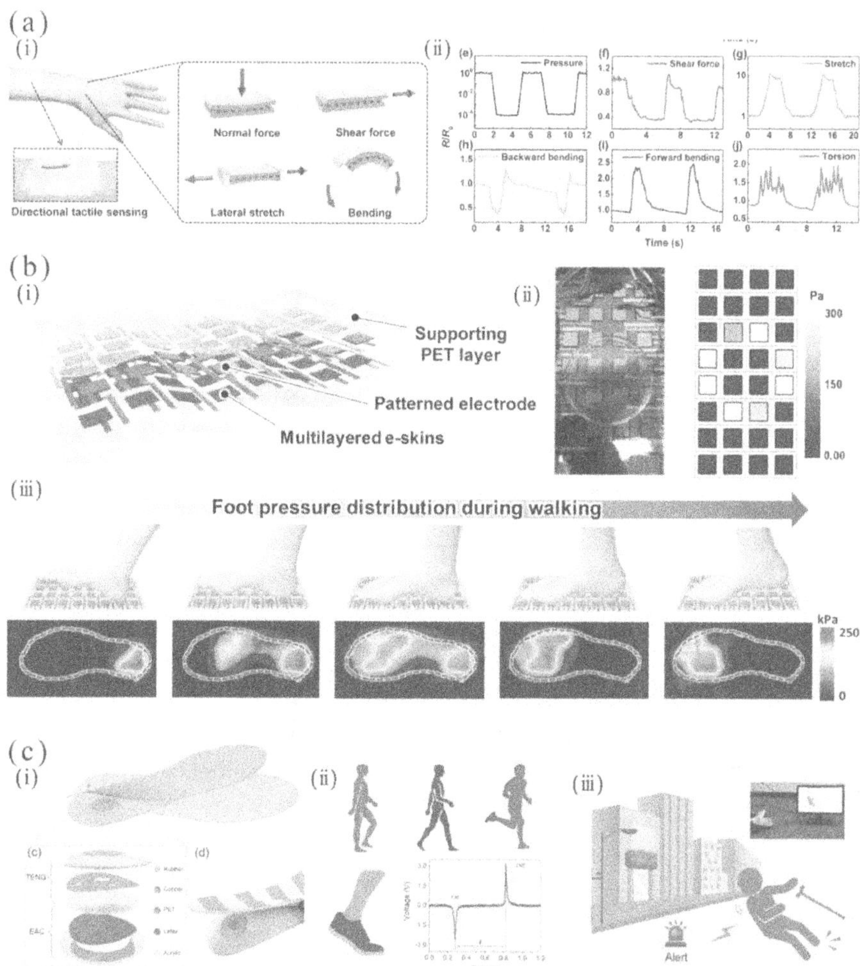

FIGURE 11.13 (a) (i) Differentiation of wrist motion using the pressure sensor with interlocked microdome structure. (ii) Relative electrical resistance changes of the sensor in response to different mechanical stimuli[100]. (b) (i) Schematic of the smart insole consisting of 4×8 pixel sensor arrays, (ii) photograph and the corresponding pressure map of the pressure sensor arrays under a Petri dish. (iii) Schematics of the foot pressure distribution of five different walking motions[103]. (c) (i) Schematic structure of the TENG-based sensors integrated in the insoles. (ii) The gait monitoring signals of the TENG-based insole. (iii) Schematic illustration showing the TENG-based insole for warning of fall down[104].

insole consisting of 4×8 pixel sensor arrays was designed to monitor the plantar pressure distribution experienced during walking motions (Figure 11.13(b)). Wang et al. reported a wearable triboelectric nanogenerator (TENG)-based smart insole for real-time human gait monitoring (Figure 11.13(c)). By analyzing the electrical signals generated from the smart insole, various gait patterns such as stepping, walking, and running could be accurately monitored and distinguished. More importantly, the TENG-based insole can serve as a fall-down alert system for patients or elder people.

11.6.3 Human Interaction

Recently, with the rapid development of new materials and fabrication processes, significant progress has been made in flexible pressure sensors for human interaction, related embodiments including touch screens, smart keyboards, and robotics control.

Inspired by the responsive color change in biological skins, visible detection of applied external tactile stimuli by endowing flexible pressure sensors with light-emitting property is in the spotlight. Due to the increasing demand for human-interactive devices, intensive efforts have been devoted to visualize e-skins by integrating flexible pressure sensing elements with active display components such as electro-/thermochromic devices[105], light-emitting diodes (LEDs)[106], and phosphor materials[107]. For example, Wang et al. reported a flexible pressure sensor matrix (PSM) for both 2D planar pressure mapping and single-point dynamic pressure recording through direct conversion of mechanical stress into visible light emission using the mechanoluminescence property of ZnS:Mn particles. (Figure 11.14(a) i)[108].

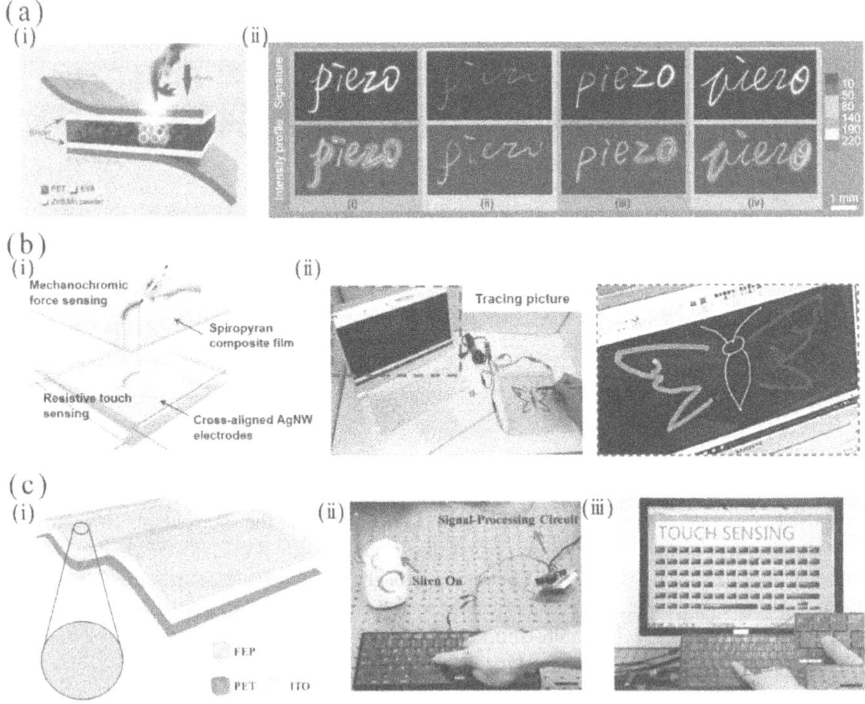

FIGURE 11.14 (a) (i) Schematic structure of the PSM device using ZnS:Mn particles. (ii) Demonstrations of signature recording of four signees by the PSM device[108]. (b) The flexible touch screen is capable of (i) monitoring dynamic writings, (ii) tracing a picture of a butterfly[109]. (c) (i) Schematic illustration of the triboelectric tactile sensor. (ii) Triggering a wireless alarm system by gentle finger tapping on the keyboard. (iii) Photograph of the intelligent keyboard for real-time keystroke tracing and recording[110].

Using an image acquisition and processing system, the PSM devices could record both the handwritten signatures and the signing habits of signees (Figure 11.14(a) ii), demonstrating its potential applications in human–machine interfaces and security systems.

Touch screens may require flexibility to fulfill the requirements of next-generation intelligent human–computer interactions. Cho et al. developed a flexible and transparent touch screen using a resistive touch sensor based on cross-aligned AgNW electrodes to sense the contact position[109]. The touch screen could monitor dynamic writings, tracing, and drawing of underneath pictures, and perception of handwriting patterns with locally different writing forces (Figure 11.14(b)). Enabled by the single-electrode triboelectric pressure sensor (Figure 11.14(c) i), Chen and coworkers reported a self-powered keyboard that can convert tactile stimuli applied to the keyboard into local electronic signals[110]. As shown, the intelligent keyboard can not only sensitively trigger a wireless alarm by gentle finger tapping (Figure 11.14(c) ii), but also trace and record the typing motions by detecting both the force used for each typing action and the dynamic time intervals between the inputting of letters (Figure 11.14(c) iii).

11.7 CONCLUSIONS

With the continuous development of intelligent technology and the Internet of Things, flexible pressure sensors, as an indispensable component of wearable electronic devices and electronic skin, have shown an increasingly broad market prospect. In this chapter, we highlighted the recent advances in flexible pressure sensors as well as the efforts carried out to develop flexible pressure sensors for potential applications in physiological parameter detection, human activity monitoring, and human–machine interaction. In terms of system-level planning of theoretical and experimental strategies, we have reported the common materials (conductive polymers, carbon-based and metallic nanomaterials etc.) used in the construction of flexible pressure sensors with different types of configurations (piezoresistive, piezocapacitive, and piezoelectric). In addition, we summarized the main manufacturing technology of flexible pressure sensors and their key performance parameters.

REFERENCES

1. Lee, Y.; Park, J.; Choe, A.; Cho, S.; Kim, J.; Ko, H., Mimicking human and biological skins for multifunctional skin electronics. *Adv Funct Mater* 2019, *30* (20), 1904523.
2. Wan, Y.; Wang, Y.; Guo, C. F., Recent progresses on flexible tactile sensors. *Mater Today Phys* 2017, *1*, 61–73.
3. Chun, S.; Choi, I. Y.; Son, W.; Bae, G. Y.; Lee, E. J.; Kwon, H., Park, W., A highly sensitive force sensor with fast response based on interlocked arrays of indium tin oxide nanosprings toward human tactile perception. *Adv Funct Mater* 2018, *28* (42), 1804132.
4. Boutry, C. M.; Negre, M.; Jorda, M.; Vardoulis, O.; Chortos, A.; Khatib, O.; Bao, Z., A hierarchically patterned, bioinspired e-skin able to detect the direction of applied pressure for robotics. *Sci Robot* 2018, *3* (24), eaau6914.
5. Luo, Z.; Chen, J.; Zhu, Z.; Li, L.; Su, Y.; Tang, W., Li, H., High-resolution and high-sensitivity flexible capacitive pressure sensors enhanced by a transferable electrode array and a micropillar-PVDF Film. *ACS Appl Mater Interfaces* 2021, *13* (6), 7635–7649.

6. Rim, Y. S.; Bae, S. H.; Chen, H.; De Marco, N.; Yang, Y., Recent progress in materials and devices toward printable and flexible sensors. *Adv Mater* 2016, *28* (22), 4415–4440.
7. Chen, J.; Zheng, J.; Gao, Q.; Zhang, J.; Zhang, J.; Omisore, O.; Li, H., Polydimethylsiloxane (PDMS)-based flexible resistive strain sensors for wearable applications. *Appl Sci* 2018, *8* (3), 345.
8. Li, H.; Zhang, J.; Chen, J.; Luo, Z.; Zhang, J.; Alhandarish, Y.; Wang, L., A supersensitive, multidimensional flexible strain gauge sensor based on Ag/PDMS for human activities monitoring. *Sci Rep* 2020, *10* (1), 4639.
9. Li, Y.; Han, D.; Jiang, C.; Xie, E.; Han, W., A facile realization scheme for tactile sensing with a structured silver nanowire-PDMS composite. *Adv Mat Technol* 2019, *4* (3), 1800504.
10. Yoon, S. G.; Chang, S. T., Microfluidic capacitive sensors with ionic liquid electrodes and CNT/PDMS nanocomposites for simultaneous sensing of pressure and temperature. *J Mater Chem C* 2017, *5* (8), 1910–1919.
11. Kou, H.; Zhang, L.; Tan, Q.; Liu, G.; Lv, W.; Lu, F.; Xiong, J., Wireless flexible pressure sensor based on micro-patterned Graphene/PDMS composite. *Sens Actuat A: Phys* 2018, *277*, 150–156.
12. Jiang, S.; Yu, J.; Xiao, Y.; Zhu, Y.; Zhang, W., Ultrawide sensing range and highly sensitive flexible pressure sensor based on a percolative thin film with a knoll-like microstructured surface. *ACS Appl Mater Interfaces* 2019, *11* (22), 20500–20508.
13. Han, Z.; Li, H.; Xiao, J.; Song, H.; Li, B.; Cai, S.; Chen, Y.; Ma Y.; Feng, X., Ultralow-cost, highly sensitive, and flexible pressure sensors based on carbon black and airlaid paper for wearable electronics. *ACS Appl Mater Interfaces* 2019, *11* (36), 33370–33379.
14. Zeng, W.; Shu, L.; Li, Q.; Chen, S.; Wang, F.; Tao, X. M., Fiber-based wearable electronics: a review of materials, fabrication, devices, and applications. *Adv Mater* 2014, *26* (31), 5310–5336.
15. Cai, X.; Cong, H.; Liu, C., Synthesis of vertically-aligned carbon nanotubes without a catalyst by hydrogen arc discharge. *Carbon* 2012, *50* (8), 2726–2730.
16. Ebbesen, T. W.; Ajayan, P. M. J. N., Large-scale synthesis of carbon nanotubes. *Nature*, 1992, *358* (6383), 220–222.
17. Hashempour, M.; Vicenzo, A.; Zhao, F.; Bestetti, M., Direct growth of MWCNTs on 316 stainless steel by chemical vapor deposition: Effect of surface nano-features on CNT growth and structure. *Carbon* 2013, *63*, 330–347.
18. Xu, W.; Chen, Y.; Zhan, H.; Wang, J. N., High-strength carbon nanotube film from improving alignment and densification. *Nano Lett* 2016, *16* (2), 946–952.
19. Kumar, N.; Curtis, W.; Hahm, J. I., Laterally aligned, multiwalled carbon nanotube growth using Magnetospirillium magnetotacticum. *Appl Phys Lett* 2005, 86 (17), 173101.
20. Li, X.; Zhang, L.; Wang, X.; Shimoyama, I.; Sun, X.; Seo, W. S.; Dai, Langmuir-blodgett assembly of densely aligned single-walled carbon nanotubes from bulk materials. *J Am Chem Soc* 2007, *129* (16), 4890–4891.
21. Goh, G. L.; Agarwala, S.; Yeong, W. Y., Directed and on-demand alignment of carbon nanotube: a review toward 3D printing of electronics. *Adv Mat Interfaces* 2019, *6* (4), 1801318.
22. Kim, S. H.; Song, W.; Jung, M. W.; Kang, M. A.; Kim, K.; Chang, S. J.; An, K. S., Carbon nanotube and graphene hybrid thin film for transparent electrodes and field effect transistors. *Adv Mater* 2014, *26* (25), 4247–4252.
23. Ahmed, A.; Jia, Y.; Huang, Y.; Khoso, N. A.; Deb, H.; Fan, Q.; Shao, J., Preparation of PVDF-TrFE based electrospun nanofibers decorated with PEDOT-CNT/rGO composites for piezo-electric pressure sensor. J Mater Sci Mater Electron 2019, *30* (37), 14007–14021.

24. Urper, O.; Çakmak, İ.; Karatepe, N., Fabrication of carbon nanotube transparent conductive films by vacuum filtration method. *Mater Lett* 2018, *223*, 210–214.
25. Kholghi Eshkalak, S.; Chinnappan, A.; Jayathilaka, W. A. D. M.; Khatibzadeh, M.; Kowsari, E.; Ramakrishna, S., A review on inkjet printing of CNT composites for smart applications. *Appl Mater Today* 2017, *9*, 372–386.
26. Lipomi, D. J.; Vosgueritchian, M.; Tee, B. C.; Hellstrom, S. L.; Lee, J. A.; Fox, C. H.; Bao, Z., Skin-like pressure and strain sensors based on transparent elastic films of carbon nanotubes. *Nat Nanotechnol* 2011, *6* (12), 788–792.
27. Park, S.; Kim, H.; Vosgueritchian, M.; Cheon, S.; Kim, H.; Koo, J. H., Bao, Z., Stretchable energy-harvesting tactile electronic skin capable of differentiating multiple mechanical stimuli modes. *Adv Mater* 2014, *26* (43), 7324–7332.
28. Park, J.; Lee, Y.; Hong, J.; Ha, M.; Jung, Y. D.; Lim, H., Kim, S.Y., Ko, H., Giant tunneling piezoresistance of composite elastomers with interlocked microdome arrays for ultrasensitive and multimodal electronic skins. *ACS Nano* 2014, *8* (5), 4689–4697.
29. Kim, S. Y.; Park, S.; Park, H. W.; Park, D. H.; Jeong, Y.; Kim, D. H., Highly sensitive and multimodal all-carbon skin sensors capable of simultaneously detecting tactile and biological stimuli. *Adv Mater* 2015, *27* (28), 4178–4185.
30. Morozov, S. V.; Novoselov, K. S.; Katsnelson, M. I.; Schedin, F.; Elias, D. C.; Jaszczak, J. A.; Geim, A. K., Giant intrinsic carrier mobilities in graphene and its bilayer. *Phys Rev Lett* 2008, *100* (1), 016602.
31. Stoller, M. D.; Park, S.; Zhu, Y.; An, J.; Ruoff, R. S. J. N. L., Graphene-based ultracapacitors. *Nano Lett* 2008, *8* (10), 3498–3502.
32. Lee, C.; Wei, X.; Kysar, J. W.; Hone, J., Measurement of the elastic properties and intrinsic strength of monolayer graphene. *Science* 2008, *321* (5887), 385–388.
33. Balandin, A. A.; Ghosh, S.; Bao, W.; Calizo, I.; Teweldebrhan, D.; Miao, F.; Lau, C. N. J. N. L., Superior thermal conductivity of single-layer graphene, *Nano Lett* 2008, *8* (3), 902.
34. Nair, R. R.; Blake, P.; Grigorenko, A. N.; Novoselov, K. S.; Booth, T. J.; Stauber, T., Geim, A. K., Fine structure constant defines visual transparency of graphene. *Science* 2008, *320* (5881), 1308–1308.
35. Allen, M. J.; Tung, V. C.; Kaner, R. B. J. C. R., Honeycomb carbon: a review of graphene. *Chem Rev*, 2009, *110* (1), 132–145.
36. Novoselov, K. S.; Geim, A. K.; Morozov, S. V.; Jiang, D.; Zhang, Y.; Dubonos, S. V.; Firsov, A. A., Electric field effect in atomically thin carbon films. *Science* 2004, *306* (5696), 666–669.
37. Yang, W.; Chen, G.; Shi, Z.; Liu, C. C.; Zhang, L.; Xie, G.; Zhang, G., Epitaxial growth of single-domain graphene on hexagonal boron nitride. *Nat Mat* 2013, *12* (9), 792–797.
38. Wei, Y.; Chen, S.; Dong, X.; Lin, Y.; Liu, L., Flexible piezoresistive sensors based on "dynamic bridging effect" of silver nanowires toward graphene. *Carbon* 2017, *113*, 395–403.
39. Khan, A.; Islam, S. M.; Ahmed, S.; Kumar, R. R.; Habib, M. R.; Huang, K.; Yang, D., Direct CVD growth of graphene on technologically important dielectric and semiconducting substrates. *Adv Sci (Weinh)* 2018, *5* (11), 1800050.
40. Jian, M.; Xia, K.; Wang, Q.; Yin, Z.; Wang, H.; Wang, C.; Zhang, Y., Flexible and highly sensitive pressure sensors based on bionic hierarchical structures. *Adv Funct Mater* 2017, 27 (9), 1606066.
41. Yang, J.; Luo, S.; Zhou, X.; Li, J.; Fu, J.; Yang, W.; Wei, D., Flexible, tunable, and ultrasensitive capacitive pressure sensor with microconformal graphene electrodes. *ACS Appl Mater Interfaces* 2019, *11* (16), 14997–15006.
42. Pang, Y.; Yang, Z.; Yang, Y.; Ren, T. L., Wearable electronics based on 2D materials for human physiological information detection. *Small* 2020, *16* (15), 1901124.

43. Zhu, B.; Niu, Z.; Wang, H.; Leow, W. R.; Wang, H.; Li, Y.; Chen, X., Microstructured graphene arrays for highly sensitive flexible tactile sensors. *Small* 2014, *10* (18), 3625–3631.
44. Wan, S.; Bi, H.; Zhou, Y.; Xie, X.; Su, S.; Yin, K.; Sun, L., Graphene oxide as high-performance dielectric materials for capacitive pressure sensors. *Carbon* 2017, *114*, 209–216.
45. Liu, F.; Han, F.; Ling, L.; Li, J.; Zhao, S.; Zhao, T.; Wong, C. P., An omni-healable and highly sensitive capacitive pressure sensor with microarray structure. *Chemistry* 2018, *24* (63), 16823–16832.
46. Sun, Q.; Kim, D. H.; Park, S. S.; Lee, N. Y.; Zhang, Y.; Lee, J. H.; Cho, J. H., Transparent, low-power pressure sensor matrix based on coplanar-gate graphene transistors. *Adv Mater* 2014, *26* (27), 4735–4740.
47. Pan, L.; Chortos, A.; Yu, G.; Wang, Y.; Isaacson, S.; Allen, R., Bao, Z., An ultra-sensitive resistive pressure sensor based on hollow-sphere microstructure induced elasticity in conducting polymer film. *Nat Commun* 2014, *5*, 3002.
48. Lou, Z.; Chen, S.; Wang, L.; Shi, R.; Li, L.; Jiang, K.; Shen, G., Ultrasensitive and ultra-flexible e-skins with dual functionalities for wearable electronics. *Nano Energy* 2017, *38*, 28–35.
49. Yang, C.; Li, L.; Zhao, J.; Wang, J.; Xie, J.; Cao, Y.; Lu, C., et al., Highly sensitive wearable pressure sensors based on three-scale nested wrinkling microstructures of polypyrrole films. *ACS Appl Mater Interfaces* 2018, *10* (30), 25811–25818.
50. Choong, C. L.; Shim, M. B.; Lee, B. S.; Jeon, S.; Ko, D. S.; Kang, T. H.; Chung, U. I., Highly stretchable resistive pressure sensors using a conductive elastomeric composite on a micropyramid array. *Adv Mater* 2014, *26* (21), 3451–3458.
51. Gong, S.; Schwalb, W.; Wang, Y.; Chen, Y.; Tang, Y.; Si, J.; Cheng, W., A wearable and highly sensitive pressure sensor with ultrathin gold nanowires. *Nat Commun* 2014, *5*, 3132.
52. Wei, Y.; Chen, S.; Lin, Y.; Yang, Z.; Liu, L., Cu–Ag core–shell nanowires for electronic skin with a petal molded microstructure. *J Mater Chem* 2015, *3* (37), 9594–9602.
53. Lee, D.; Lee, H.; Jeong, Y.; Ahn, Y.; Nam, G.; Lee, Y., Highly sensitive, transparent, and durable pressure sensors based on sea-urchin shaped metal nanoparticles. *Adv Mater* 2016, *28* (42), 9364–9369.
54. Su, B.; Gong, S.; Ma, Z.; Yap, L. W.; Cheng, W., Mimosa-inspired design of a flexible pressure sensor with touch sensitivity. *Small* 2015, *11* (16), 1886–1891.
55. Yang, J. C.; Kim, J. O.; Oh, J.; Kwon, S. Y.; Sim, J. Y.; Kim, D. W.; Park, S., Microstructured porous pyramid-based ultrahigh sensitive pressure sensor insensitive to strain and temperature. *ACS Appl Mater Interfaces* 2019, *11* (21), 19472–19480.
56. Li, Z.; Zhang, B.; Li, K.; Zhang, T.; Yang, X., A wide linearity range and high sensitivity flexible pressure sensor with hierarchical microstructures via laser marking. *J Mater Chem* 2020, *8* (9), 3088–3096.
57. Qiu, Z.; Wan, Y.; Zhou, W.; Yang, J.; Yang, J.; Huang, J.; Guo, C. F., Ionic skin with biomimetic dielectric layer templated from Calathea zebrine leaf. *Adv Funct Mater* 2018, *28* (37), 1802343.
58. Wang, Z.; Wang, S.; Zeng, J.; Ren, X.; Chee, A. J.; Yiu, B. Y.; Chan, P. K., High sensitivity, wearable, piezoresistive pressure sensors based on irregular micro-hump structures and its applications in body motion sensing. *Small* 2016, *12* (28), 3827–3836.
59. Zhao, S.; Zhu, R., High sensitivity and broad range flexible pressure sensor using multilayered porous PDMS/AgNP sponge. *Adv Mater Technol* 2019, *4* (9), 1900414.
60. Kang, S.; Lee, J.; Lee, S.; Kim, S.; Kim, J. K.; Algadi, H.; Lee, T., Highly sensitive pressure sensor based on bioinspired porous structure for real-time tactile sensing. *Adv Electron Mater* 2016, *2* (12), 1600356.

61. Fu, S.; Tao, J.; Wu, W.; Sun, J.; Li, F.; Li, J.; Pan, C., Fabrication of large-area bimodal sensors by all-inkjet-printing. *Adv Mater Technol* 2019, *4* (4), 1800703.
62. Guo, S. Z.; Qiu, K.; Meng, F.; Park, S. H.; McAlpine, M. C., 3D printed stretchable tactile sensors. *Adv Mater* 2017, *29* (27), 1701218.
63. Jiang, X.; Ren, Z.; Fu, Y.; Liu, Y.; Zou, R.; Ji, G.; Hu, N., Highly compressible and sensitive pressure sensor under large strain based on 3D porous reduced graphene oxide fiber fabrics in wide compression strains. *ACS Appl Mater Interfaces* 2019, *11* (40), 37051–37059.
64. Baptista-Pires, L.; Mayorga-Martinez, C. C.; Medina-Sanchez, M.; Monton, H.; Merkoci, A., Water activated graphene oxide transfer using wax printed membranes for fast patterning of a touch sensitive device. *ACS Nano* 2016, *10* (1), 853–860.
65. Gao, Y.; Yu, G.; Shu, T.; Chen, Y.; Yang, W.; Liu, Y.; Xuan, F., 3D-printed coaxial fibers for integrated wearable sensor skin. *Adv Mater Technol* 2019, *4* (10), 1900504.
66. Li, P.; Zhao, L.; Jiang, Z.; Yu, M.; Li, Z.; Li, X., Self-powered flexible sensor based on the graphene modified P(VDF-TrFE) electrospun fibers for pressure detection. *Macromol Mater Eng* 2019, *304* (12), 1900504.
67. Yeo, J. C.; Yu, J.; Koh, Z. M.; Wang, Z.; Lim, C. T., Wearable tactile sensor based on flexible microfluidics. *Lab on a Chip* 2016, *16* (17), 3244–3250.
68. Sun, K.; Ko, H.; Park, H. H.; Seong, M.; Lee, S. H.; Yi, H.; Jeong, H. E., Hybrid architectures of heterogeneous carbon nanotube composite microstructures enable multiaxial strain perception with high sensitivity and ultrabroad sensing range. *Small* 2018, *14* (52), 1803411.
69. Zang, Y.; Zhang, F.; Di, C. A.; Zhu, D., Advances of flexible pressure sensors toward artificial intelligence and health care applications. *Mater Horizons* 2015, *2* (2), 140–156.
70. Yao, H. B.; Ge, J.; Wang, C. F.; Wang, X.; Hu, W.; Zheng, Z. J.; Yu, S. H., A flexible and highly pressure-sensitive graphene-polyurethane sponge based on fractured microstructure design. *Adv Mater* 2013, *25* (46), 6692–6698.
71. Ruth, S. R. A.; Feig, V. R.; Tran, H.; Bao, Z., Microengineering pressure sensor active layers for improved performance. *Adv. Funct. Mater* 2020, *30* (39), 2003491.
72. Nie, P.; Wang, R.; Xu, X.; Cheng, Y.; Wang, X.; Shi, L.; Sun, J., High-performance piezoresistive electronic skin with bionic hierarchical microstructure and microcracks. *ACS Appl Mater Interfaces* 2017, *9* (17), 14911–14919.
73. Shi, J.; Wang, L.; Dai, Z.; Zhao, L.; Du, M.; Li, H.; Fang, Y., Multiscale hierarchical design of a flexible piezoresistive pressure sensor with high sensitivity and wide linearity range. *Small* 2018, *14* (27), 1800819.
74. Pang, Y.; Zhang, K.; Yang, Z.; Jiang, S.; Ju, Z.; Li, Y.; Ren, T. L., Epidermis microstructure inspired graphene pressure sensor with random distributed spinosum for high sensitivity and large linearity. *ACS Nano* 2018, *12* (3), 2346–2354.
75. Li, H.; Wu, K.; Xu, Z.; Wang, Z.; Meng, Y.; Li, L., Ultrahigh-sensitivity piezoresistive pressure sensors for detection of tiny pressure. *ACS Appl Mater Interfaces* 2018, *10* (24), 20826–20834.
76. Yuan, L.; Wang, Z.; Li, H.; Huang, Y.; Wang, S.; Gong, X.; Hu, W., Synergistic resistance modulation toward ultrahighly sensitive piezoresistive pressure sensors. *Adv Mater Technol* 2020, *5* (4), 1901084.
77. Park, J.; Kim, J.; Hong, J.; Lee, H.; Lee, Y.; Cho, S.; Ko, H., Tailoring force sensitivity and selectivity by microstructure engineering of multidirectional electronic skins. *NPG Asia Mat* 2018, *10* (4), 163–176.
78. Li, T.; Luo, H.; Qin, L.; Wang, X.; Xiong, Z.; Ding, H., Zhang, T., Flexible capacitive tactile sensor based on micropatterned dielectric layer. *Small* 2016, *12* (36), 5042–5048.
79. Kwon, D.; Lee, T. I.; Shim, J.; Ryu, S.; Kim, M. S.; Kim, S.; Park, I., Highly sensitive, flexible, and wearable pressure sensor based on a giant piezocapacitive effect of three-dimensional microporous elastomeric dielectric layer. *ACS Appl Mater Interfaces* 2016, *8* (26), 16922–16931.

80. Chen, M.; Li, K.; Cheng, G.; He, K.; Li, W.; Zhang, D.; Yang, C., Touchpoint-tailored ultrasensitive piezoresistive pressure sensors with a broad dynamic response range and low detection limit. *ACS Appl Mater Interfaces* 2019, *11* (2), 2551–2558.
81. Oh, J.; Kim, J. O.; Kim, Y.; Choi, H. B.; Yang, J. C.; Lee, S.; Park, S., Highly uniform and low hysteresis piezoresistive pressure sensors based on chemical grafting of polypyrrole on elastomer template with uniform pore size. *Small* 2019, *15* (33), 1901744.
82. Mannsfeld, S. C.; Tee, B. C.; Stoltenberg, R. M.; Chen, C. V.; Barman, S.; Muir, B. V.; Bao, Z., Highly sensitive flexible pressure sensors with microstructured rubber dielectric layers. *Nat Mater* 2010, *9* (10), 859–864.
83. Qiu, J.; Guo, X.; Chu, R.; Wang, S.; Zeng, W.; Qu, L.; Xing, G., Rapid-response, low detection limit, and high-sensitivity capacitive flexible tactile sensor based on three-dimensional porous dielectric layer for wearable electronic skin. *ACS Appl Mater Interfaces* 2019, *11* (43), 40716–40725.
84. Zhu, B.; Ling, Y.; Yap, L. W.; Yang, M.; Lin, F.; Gong, S.; Cheng, W., Hierarchically structured vertical gold nanowire array-based wearable pressure sensors for wireless health monitoring. *ACS Appl Mater Interfaces* 2019, *11* (32), 29014–29021.
85. Chen, X.; Shao, J.; An, N.; Li, X.; Tian, H.; Xu, C.; Ding, Y., Self-powered flexible pressure sensors with vertically well-aligned piezoelectric nanowire arrays for monitoring vital signs. *J Mater Chem* 2015, *3* (45), 11806–11814.
86. Kim, J.; Campbell, A. S.; de Avila, B. E.; Wang, J., Wearable biosensors for healthcare monitoring. *Nat Biotechnol* 2019, *37* (4), 389–406.
87. Trung, T. Q.; Lee, N. E., Flexible and stretchable physical sensor integrated platforms for wearable human-activity monitoringand personal healthcare. *Adv Mater* 2016, *28* (22), 4338–4372.
88. Gao, Q.; Zhang, J.; Xie, Z.; Omisore, O.; Zhang, J.; Wang, L.; Li, H., Highly stretchable sensors for wearable biomedical applications. *J Mater Sci* 2018, *54* (7), 5187–5223.
89. Li, L.; Zheng, J.; Chen, J.; Luo, Z.; Su, Y.; Tang, W.; Li, H., Flexible pressure sensors for biomedical applications: from ex vivo to in vivo. *Adv Mater Interfaces* 2020, *7* (17), 2000743.
90. Fan, X.; Huang, Y.; Ding, X.; Luo, N.; Li, C.; Zhao, N.; Chen, S. C., Alignment-free liquid-capsule pressure sensor for cardiovascular monitoring. *Adv Funct Mater* 2018, *28* (44), 1805045.
91. Dagdeviren, C.; Su, Y.; Joe, P.; Yona, R.; Liu, Y.; Kim, Y. S.; Rogers, J. A., Conformable amplified lead zirconate titanate sensors with enhanced piezoelectric response for cutaneous pressure monitoring. *Nat Commun* 2014, *5*, 4496.
92. Wang, C.; Li, X.; Hu, H.; Zhang, L.; Huang, Z.; Lin, M.; Xu, S., Monitoring of the central blood pressure waveform via a conformal ultrasonic device. *Nat Biomed Eng* 2018, *2* (9), 687–695.
93. Araci, I. E.; Su, B.; Quake, S. R.; Mandel, Y., An implantable microfluidic device for self-monitoring of intraocular pressure. *Nat Med* 2014, *20* (9), 1074–1078.
94. Pang, C.; Koo, J. H.; Nguyen, A.; Caves, J. M.; Kim, M. G.; Chortos, A.; Bao, Z., Highly skin-conformal microhairy sensor for pulse signal amplification. *Adv Mater* 2015, *27* (4), 634–640.
95. Park, D. Y.; Joe, D. J.; Kim, D. H.; Park, H.; Han, J. H.; Jeong, C. K.; Lee, K. J., Self-powered real-time arterial pulse monitoring using ultrathin epidermal piezoelectric sensors. *Adv Mater* 2017, *29* (37), 1702308.
96. Meng, K.; Chen, J.; Li, X.; Wu, Y.; Fan, W.; Zhou, Z.; Wang, Z. L., Flexible weaving constructed self-powered pressure sensor enabling continuous diagnosis of cardiovascular disease and measurement of cuffless blood pressure. *Adv Funct Mater* 2019, *29* (5), 1806388.
97. Kim, J.; Kim, M.; Lee, M. S.; Kim, K.; Ji, S.; Kim, Y. T.; Park, J. U., Wearable smart sensor systems integrated on soft contact lenses for wireless ocular diagnostics. *Nat Commun* 2017, *8*, 14997.

98. Lee, Y.; Park, J.; Choe, A.; Cho, S.; Kim, J.; Ko, H., Mimicking human and biological skins for multifunctional skin electronics. *Adv Funct Mater* 2019, *30* (20), 1904523.
99. Chen, J.; Zhang, J.; Luo, Z.; Zhang, J.; Li, L.; Su, Y.; Li, H., Superelastic, sensitive, and low hysteresis flexible strain sensor based on wave-patterned liquid metal for human activity monitoring. *ACS Appl Mater Interfaces* 2020, *12* (19), 22200–22211.
100. Park, J.; Lee, Y.; Hong, J.; Lee, Y.; Ha, M.; Jung, Y.; Ko, H., Tactile-direction-sensitive and stretchable electronic skins based on human-skin-inspired interlocked microstructures. *ACS Nano* 2014, *8* (12), 12020–12029.
101. Park, J. H.; Wu, C.; Sung, S.; Kim, T. W., Ingenious use of natural triboelectrification on the human body for versatile applications in walking energy harvesting and body action monitoring. *Nano Energy* 2019, *57*, 872–878.
102. Kim, S.; Amjadi, M.; Lee, T. I.; Jeong, Y.; Kwon, D.; Kim, M. S.; Park, I., Wearable, ultrawide-range, and bending-insensitive pressure sensor based on carbon nanotube network-coated porous elastomer sponges for human interface and healthcare devices. *ACS Appl Mater Interfaces* 2019, *11* (26), 23639–23648.
103. Lee, Y.; Park, J.; Cho, S.; Shin, Y. E.; Lee, H.; Kim, J., Ko, H., Flexible Ferroelectric Sensors with Ultrahigh Pressure Sensitivity and Linear Response over Exceptionally Broad Pressure Range. *ACS Nano* 2018, *12* (4), 4045–4054.
104. Lin, Z.; Wu, Z.; Zhang, B.; Wang, Y. C.; Guo, H.; Liu, G.; Wang, Z. L., A Triboelectric nanogenerator-based smart insole for multifunctional gait monitoring. *Adv Mater Technol* 2019, *4* (2), 1800360.
105. Chou, H. H.; Nguyen, A.; Chortos, A.; To, J. W.; Lu, C.; Mei, J.; Kurosawa, T.; Bae, W. G.; Tok, J. B. H.; Bao, Z., A chameleon-inspired stretchable electronic skin with interactive colour changing controlled by tactile sensing. *Nat. Commun* 2015, *6* (1), 8011.
106. Bao, R.; Wang, C.; Dong, L.; Yu, R.; Zhao, K.; Wang, Z. L.; Pan, C., Flexible and controllable piezo-phototronic pressure mapping sensor matrix by ZnO NW/p-polymer LED array. *Adv Funct Mater* 2015, *25* (19), 2884–2891.
107. Larson, C.; Peele, B.; Li, S.; Robinson, S.; Totaro, M.; Beccai, L.; Shepherd, R., Highly stretchable electroluminescent skin for optical signaling and tactile sensing. *Science* 2016, *351* (6277), 1071–1074.
108. Wang, X.; Zhang, H.; Yu, R.; Dong, L.; Peng, D.; Zhang, A.; Wang, Z. L., Dynamic pressure mapping of personalized handwriting by a flexible sensor matrix based on the mechanoluminescence process. *Adv Mater* 2015, *27* (14), 2324–2331.
109. Cho, S.; Kang, S.; Pandya, A.; Shanker, R.; Khan, Z.; Lee, Y.; Ko, H., Large-area cross-aligned silver nanowire electrodes for flexible, transparent, and force-sensitive mechanochromic touch screens. *ACS Nano* 2017, *11* (4), 4346–4357.
110. Chen, J.; Zhu, G.; Yang, J.; Jing, Q.; Bai, P.; Yang, W.; Wang, Z. L., Personalized keystroke dynamics for self-powered human–machine interfacing. *ACS Nano* 2015, *9* (1), 105–116.

Index

Q

R

S

T

V

W

Z

For Product Safety Concerns and Information please contact our EU representative GPSR@taylorandfrancis.com
Taylor & Francis Verlag GmbH, Kaufingerstraße 24, 80331 München, Germany

www.ingramcontent.com/pod-product-compliance
Lightning Source LLC
Chambersburg PA
CBHW060804220326
41598CB00022B/2532

* 9 7 8 0 3 6 7 6 9 3 2 3 7 *